Lecture Notes on Data Engineering and Communications Technologies

181

Series Editor

Fatos Xhafa, *Technical University of Catalonia, Barcelona, Spain*

The aim of the book series is to present cutting edge engineering approaches to data technologies and communications. It will publish latest advances on the engineering task of building and deploying distributed, scalable and reliable data infrastructures and communication systems.

The series will have a prominent applied focus on data technologies and communications with aim to promote the bridging from fundamental research on data science and networking to data engineering and communications that lead to industry products, business knowledge and standardisation.

Indexed by SCOPUS, INSPEC, EI Compendex.

All books published in the series are submitted for consideration in Web of Science.

Zhengbing Hu · Ivan Dychka · Matthew He
Editors

Advances in Computer Science for Engineering and Education VI

Set 1

 Springer

Editors
Zhengbing Hu
Faculty of Applied Mathematics
National Technical University of Ukraine
"Igor Sikorsky Kiev Polytechnic Institute",
Ukraine
Kyiv, Ukraine

Ivan Dychka
Faculty of Applied Mathematics
National Technical University of Ukraine
"Igor Sikorsky Kiev Polytechnic Institute",
Ukraine
Kyiv, Ukraine

Matthew He
Halmos College of Arts and Sciences
Nova Southeastern University
Fort Lauderdale, FL, USA

ISSN 2367-4512   ISSN 2367-4520 (electronic)
Lecture Notes on Data Engineering and Communications Technologies
ISBN 978-3-031-36117-3   ISBN 978-3-031-36118-0 (eBook)
https://doi.org/10.1007/978-3-031-36118-0

This Springer imprint is published by the registered company Springer Nature Switzerland AG
The registered company address is: Gewerbestrasse 11, 6330 Cham, Switzerland

Preface

Modern engineering and educational technologies have opened up new opportunities in computer science, such as conducting computer experiments, visualizing objects in detail, remote learning, quickly retrieving information from large databases, and developing artificial intelligence systems. Governments and science and technology communities are increasingly recognizing the significance of computer science and its applications in engineering and education. However, preparing the next generation of professionals to properly use and advance computer science and its applications has become a challenging task for higher education institutions due to the rapid pace of technological advancements and the need for interdisciplinary skills. Therefore, higher education institutions need to focus on developing innovative approaches to teaching and providing opportunities for hands-on learning. Additionally, international cooperation is crucial in facilitating and accelerating the development of solutions to this critical subject.

As a result of these factors, the 6th International Conference on Computer Science, Engineering, and Education Applications (ICCSEEA2023) was jointly organized by the National Technical University of Ukraine "Igor Sikorsky Kyiv Polytechnic Institute," National Aviation University, Lviv Polytechnic National University, Polish Operational and Systems Society, Warsaw University of Technology, and the International Research Association of Modern Education and Computer Science on March 17–19, 2023, in Warsaw, Poland. The ICCSEEA2023 brings together leading scholars from all around the world to share their findings and discuss outstanding challenges in computer science, engineering, and education applications.

Out of all the submissions, the best contributions to the conference were selected by the program committee for inclusion in this book.

March 2023

Zhengbing Hu
Ivan Dychka
Matthew He

Organization

General Chairs

Felix Yanovsky — Delft University of Technology, Delft, Netherlands

Ivan Dychka — National Technical University of Ukraine "Igor Sikorsky Kyiv Polytechnic Institute", Ukraine

Online conference Organizing Chairs

Z. B. Hu — National Technical University of Ukraine "Igor Sikorsky Kyiv Polytechnic Institute", Ukraine

Solomiia Fedushko — Lviv Polytechnic National University, Lviv, Ukraine

Oleksandra Yeremenko — Kharkiv National University of Radio Electronics, Kharkiv, Ukraine

Vadym Mukhin — National Technical University of Ukraine "Igor Sikorsky Kyiv Polytechnic Institute", Ukraine

Yuriy Ushenko — Chernivtsi National University, Chernivtsi, Ukraine

Program Chairs

Matthew He — Nova Southeastern University, Florida, USA

Roman Kochan — University of Bielsko-Biała, Bielsko-Biala, Poland

Q. Y. Zhang — Wuhan University of Technology, China

Publication Chairs

Z. B. Hu — National Technical University of Ukraine "Igor Sikorsky Kyiv Polytechnic Institute", Ukraine

Matthew He — Nova Southeastern University, USA

Ivan Dychka — National Technical University of Ukraine "Igor Sikorsky Kyiv Polytechnic Institute", Ukraine

Publicity Chairs

Sergiy Gnatyuk	National Aviation University, Kyiv, Ukraine
Rabah Shboul	Al-al-Bayt University, Jordan
O. K. Tyshchenko	University of Ostrava, Czech Republic

Program Committee Members

Artem Volokyta	National Technical University of Ukraine "Igor Sikorsky Kyiv - Polytechnic Institute", Kyiv, Ukraine
Rabah Shboul	Al-al-Bayt University, Jordan
Krzysztof Kulpa	Warsaw University of Technology, Warsaw, Poland
Bo Wang	Wuhan University, China
Anatoliy Sachenko	Kazimierz Pułaski University of Technology and Humanities in Radom, Poland
Ihor Tereikovskyi	National Technical University of Ukraine "Igor Sikorsky Kyiv Polytechnic Institute", Kyiv, Ukraine
E. Fimmel	Mannheim University of Applied Sciences, Germany
G. Darvas	Institute Symmetron, Hungary
A. U. Igamberdiev	Memorial University of Newfoundland, Canada
Jacek Misiurewicz	Warsaw University of Technology, Warsaw, Poland
X. J. Ma	Huazhong University of Science and Technology, China
S. C. Qu	Central China Normal University, China
Ivan Izonin	Lviv Polytechnic National University, Lviv, Ukraine
Y. Shi	Bloomsburg University of Pennsylvania, USA
Feng Liu	Huazhong Agricultural University, China
Sergey Petoukhov	MERI RAS, Moscow, Russia
Z. W. Ye	Hubei University of Technology, China
Oleksii K. Tyshchenko	University of Ostrava, Czech Republic
C. C. Zhang	Feng Chia University, Taiwan
Andriy Gizun	National Aviation University, Kyiv, Ukraine
Yevhen Yashchyshyn	Warsaw University of Technology, Warsaw, Poland
Vitaly Deibuk	Chernivtsi National University, Chernivtsi, Ukraine

| J. Su | Hubei University of Technology, China |
| G. K. Tolokonnikova | FNAT VIM of RAS, Moscow, Russia |

Conference Organizers and Supporters

National Technical University of Ukraine "Igor Sikorsky Kyiv Polytechnic Institute", Ukraine
National Aviation University, Ukraine
Lviv Polytechnic National University, Ukraine
Polish Operational and Systems Society, Poland
Warsaw University of Technology, Poland
International Research Association of Modern Education and Computer Science, Hong Kong

Contents

Perfection of Computer Algorithms and Methods

Advances in Technological and Educational Approaches

Computer Science for Manage
of Natural and Engineering Processes

Construction and Operation Analysis of Logistics Sorting Simulation Experimental Platform

Yuexia Tang[✉] and Guangping Li

School of Intelligent Manufacturing, Nanning University, Guangxi 530200, China
Fortangyuexia@163.com

Abstract. The logistics sorting system, which undertakes the function of automatic sorting, is an important equipment for realizing logistics automation. In order to improve the efficiency of the logistics sorting system debugging and reduce the risk of debugging, this paper designs a three-dimensional simulation experimental platform for the logistics sorting system, establishes a three-dimensional virtual simulation model, sets its multiple attributes and compiles the behavior code during operation, takes the programmable controller PLC as the core controller, designs and makes a virtual-reality conversion module, which converts the virtual sensor signal of the model into a real electrical signal, Convert the real PLC electrical signal into the digital information of the virtual model, and realize the real-time interaction between the virtual simulation model and the PLC. The test results show that the designed logistics sorting simulation experimental platform works stably and reliably with good real-time interaction, and effectively solves practical problems in learning and training in related fields.

Keywords: Logistics sorting system · Simulation experiment platform · 3D model · Virtual reality conversion · PLC

1 Introduction

As an important equipment of modern logistics, logistics sorting system plays an important role in the process of goods warehousing, distribution and transportation. Its increasingly popular application also puts forward higher requirements for professional and technical personnel [1–3]. However, in the study and training in related fields, it was found that the large area of the logistics sorting system, the high cost of training consumables and maintenance, limited the number of sets of training equipment, resulting in short operation time and poor learning effect [4]. At the same time, it is also very risky for beginners to debug directly on the equipment [5–8]. Therefore, the establishment of logistics sorting simulation experimental platform has important practical significance for solving the above problems.

At present, most logistics sorting simulation experimental platforms are realized by configuration software. However, most of the practical training carried out on the simulation interface of configuration software development is limited to software design

Z. Hu et al. (Eds.): ICCSEEA 2023, LNDECT 181, pp. 3–12, 2023.
https://doi.org/10.1007/978-3-031-36118-0_1

and cannot be carried out on hardware design [9, 10]. At the same time, the debugging results cannot be directly applied to the actual equipment. This problem can be effectively solved by using the simulation experiment platform [11–14].

This design takes the logistics sorting system at five crossings as an example, uses UG software to build its three-dimensional model, takes Mitsubishi PLC as the core controller, designs and makes the virtual-reality conversion module, and constructs the logistics sorting simulation experimental platform. The simulation experimental platform can visually verify the circuit and program design, and effectively improve the efficiency and safety of equipment operation and debugging.

2   Design and Three-Dimensional Model Construction of Logistics Sorting Experimental Platform

2.1   Overall Design of Experimental Platform

The design of the overall framework is the first step in the design of the experimental platform. As shown in Fig. 1, the logistics sorting simulation experimental platform in this design consists of three parts: the three-dimensional model of the logistics sorting system, the PLC controller and the virtual-reality conversion module.

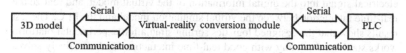
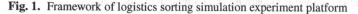

Fig. 1. Framework of logistics sorting simulation experiment platform

Three-dimensional model is to measure the size of the logistics sorting system according to its entity, use UG software to first establish the three-dimensional model of each component, then assemble the model of each component into a whole according to their motion position, and set various model parameters. The virtual-reality conversion module is to realize the communication between the entity PLC and the virtual model. The module reads the virtual sensor signal of the model and converts it into a real electrical signal to drive the PLC, and converts the electrical signal output by the PLC into a virtual execution signal to drive the actuator action of the model. The experimental platform uses Mitsubishi FX series PLC as the core controller, reads the switch value and sensor signal, and outputs the signal to control the model action. The three parts use serial port mode for two-way communication.

2.2   Construction of Three-Dimensional Model of Logistics Sorting System

The function of the logistics sorting platform is to distribute parcels from different regions to different collection bins. The parcels are sent to the identification unit through the main conveyor. If the sorting conditions are met, the baffle in front of the sorting conveyor extends and the parcels are transferred to the corresponding collection box. The platform is mainly composed of one main conveyor belt and four sorting conveyor belts. Use UG software to build the three-dimensional model of each part, and assemble it into a whole according to its relative position to get the overall model of the logistics sorting platform, as shown in Fig. 2.

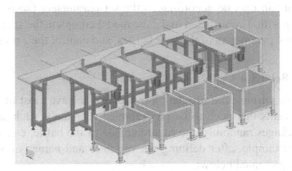

Fig. 2. Logistics sorting platform model

2.2.1 Model Physical Attribute Setting

Physical attribute is to add physical properties to each part based on the geometric model. First of all, the model parts that need to move, such as the baffle of the parcel and sorting conveyor, are rigid bodies. Any geometric object can only accept the influence of external forces if it has added rigid body attributes [15–17]. Secondly, model parts with contact action such as package, main conveyor, sorting conveyor, baffle of sub-conveyor and collecting box are set as collision body. Models without collision bodies will cross each other during simulation, and cannot simulate the real motion of objects. Finally, set the package model as the object source. The setting of the object source allows the model to copy and generate the original model in the set way, so that the package can be automatically generated continuously in the simulation.

2.2.2 Model Motion Attribute Setting

The motion attribute is to add motion pairs to the geometric model [18]. Kinematic pair refers to the movable connection between two parts in the model that can contact directly and make relative motion, mainly including hinge pair, sliding pair, cylindrical pair, spiral pair, plane pair, etc. Set the baffle on the sorting conveyor as a hinge pair to realize the rotation function.

2.2.3 Model Electromechanical Object Setting

Electromechanical object setting is to add sensor and actuator attributes on the model. The sensor is a kind of detection device, which can convert the information of the object to be measured into electrical signal output according to certain rules [19–21]. Set a camera in front of the main conveyor model to enable it to identify the barcode of the package during the simulation process; A position sensor is set in front of each of the four sorting conveyors to detect the position of the parcel model on the main conveyor [22, 23]. The actuator setting can make the model move, including transmission surface, position control, etc. The transmission surface can make the rigid body on its surface move. Set the main conveyor belt and the four sorting conveyor belts as the transmission surface, so that it can operate according to the set parameters (such as transmission speed, etc.). The position control can make the model complete the specified action, set the position control parameters of the four baffles, and control their rotation speed.

2.2.4 Runtime Behavior Code

Through the above settings, the simulation model can achieve most of the actions, but there are some details that need to be programmed separately, such as the automatic generation of packages randomly, the rotation angle of the baffle, etc. Taking package generation as an example, after defining the variables and names of all modules, the operation code is presented below:

```
if (dtsum > 0.05)
        {    dtsum = 0;
             if(f.Pin[6]==true || f.Pin[26] == true)
                 time++;
                 if(time>100)
             {   int a1 = r1.Next(99)/20;
                 SB[a1].Active = true;
                 time = 0;
             }
        }
```

3   Design of Virtual-Reality Conversion Module

When using ForceControl, MCGS and other configuration software to build a virtual model, the control signal sent by the PLC can control the action of the simulation model, but the virtual sensor signal triggered after the simulation model moves in place cannot return to the PLC, and the interaction between virtual and reality is one-way. Therefore, a virtual-reality conversion module is designed to realize the two-way transmission of information between virtual and reality.

3.1   Hardware Design

The core of virtual-reality conversion module hardware mainly includes communication circuit, virtual signal virtualization module and actual signal virtualization module.

Considering the cost and real-time of communication, the communication circuit adopts CH340 chip as the communication chip of IO interface circuit board, which can realize the direct conversion between serial port and USB communication. In the virtual signal realization module, the optocoupler is used for electrical isolation. When the virtual sensor is connected, the sensor port in the module will pull the IO port level down after receiving the signal, and the optocoupler output end will be connected to achieve the purpose of converting data signals into electrical signals. There are two kinds of control signals output by PLC: switching value and pulse value. In the actual signal virtualization module, the switch signal is directly read after optoelectronic coupling; The pulse quantity signal is interrupted at the special port, the pulse quantity is read, and the corresponding addition and subtraction calculation is performed to realize the digitalization of the electrical signal.

3.2 Data Conversion Software Design

The OPC server can be directly used to communicate between the virtual simulation model and the PLC, but the PLC manufacturers of each major brand have their own OPC server. If the PLC is replaced, the OPC server of the same brand needs to be reinstalled. To solve this problem, a data conversion software is designed for the virtual-reality conversion module.

First, in UG software, create runtime behavior code, establish the interface of data conversion software, and map it to the sensor and actuator in the virtual simulation model. Secondly, in order to ensure the continuity of animation and the communication margin, the international standard communication rate of 115200bps is selected as the communication baud rate. On the premise of communication guarantee, using C # language as a development tool, we can control the corresponding virtual sensors and actuators by writing programs and controlling the runtime behavior code. Finally, the data conversion software is embedded into the STEP function of UG software, and the UG software calls the STEP function regularly. During the execution of the STEP function, the data conversion software continuously collects information, reads and writes data to the hardware module, and realizes real-time data exchange without passing through the OPC server, and is not affected by the PLC replacement.

4 Operation of Material Sorting Simulation Experimental Platform

After the construction of the logistics sorting simulation experiment platform, the simulation experiment can be carried out.

The control requirements of the logistics sorting control system are as follows: after the system is started, packages are transmitted on the main drive belt, their barcode information is identified by the camera, and they are sorted to the corresponding collection box according to the identification results.

According to the above control requirements, the I/O address allocation table of the logistics sorting control system is shown in Table 1. The entire logistics sorting equipment control system needs five input ports for camera identification results, representing sorting addresses 1–5 respectively; 4 position detection sensors to detect whether the package has arrived at the sorting conveyor; 8 baffle position detection sensors to detect whether the baffle rotates in place and returns to its original position; With the start and stop buttons, a total of 19 digital input points are required. The logistics sorting control system needs to control the action of four baffles and five drive belts, and the number of digital output points is 9. Considering the 10%–15% I/O point margin, the PLC model is Mitsubishi FX$_{3U}$-48MT.

Some words in Table 1 are abbreviated, for example, Baffle is abbreviated as "B", Solenoid Valve is expressed as "SV" and Conveyor Belt is indicated as "CB".

Table 1. Input and output port assignment

Input signal				Output signal	
Address	Name	Address	Name	Address	Name
X000	Sort Address 1	X011	B-1 in place detect	Y000	B-1 SV
X001	Sort Address 2	X012	B-1 reset detect	Y001	B-2 SV
X002	Sort Address 3	X013	B-2 in place detect	Y002	B-3 SV
X003	Sort Address 4	X014	B-2 reset detect	Y003	B-4 SV
X004	Sort Address 5	X015	B-3 in place detect	Y004	Main CB
X005	Position 1 detection	X016	B-3 reset detect	Y005	Sort CB 1
X006	Position 2 detection	X017	B-4 in place detect	Y006	Sort CB 2
X007	Position 3 detection	X020	B-4 reset detect	Y007	Sort CB 3
X010	Position 4 detection	X021	Start	Y010	Sort CB 4
		X022	Stop		

According to the I/O address allocation table, the hardware circuit of the logistics sorting control system is designed, as shown in Fig. 3.

The workflow of the logistics sorting control system is shown in Fig. 4. After the system is started, all conveyor belts start to run. Packages are transported on the main conveyor belt. After being identified by the camera one by one, their address information will be stored in five data stacks in the same time period. Before the package runs to the sorting drive belt, it is detected by the position sensor, and the control system compares the current position information of the package with the barcode information stored in the stack. When the comparison results are consistent, the baffle of the current sorting conveyor is opened, the package is received, and the address information of the package in all stacks is deleted. If the comparison results are inconsistent, the address information stored in the current position data stack of the package will be deleted, the package will continue to be transmitted forward, and the information comparison will be performed again before the next sorting conveyor.

The control program is completed using the sequential function diagram (SFC) of GX Developer software, which divides a complete control process into several stages. Each stage is responsible for completing different actions. The conversion between different stages is determined by the set conversion conditions. When the conditions are met, the stage transfer is realized. The action of the previous stage ends and the action of the next stage begins.

After the control program is written, virtual debugging is conducted on the logistics sorting simulation platform to simulate the process of parcel sorting by region. By observing the working state of the logistics sorting system on the simulation experiment platform, the correctness of the PLC hardware circuit and control program can be verified, and the effectiveness of the logistics sorting simulation experiment can also be verified.

Due to the verification of the designed hardware circuit and control program in the simulation experiment, it can be quickly completed in the actual operation of the

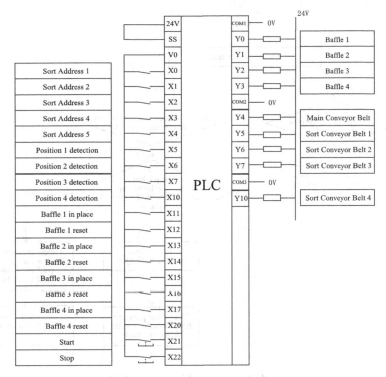

Fig. 3. Connection between PLC hardware

logistics sorting system. The conveyor belt runs smoothly and the parcel sorting meets the requirements. The actual operation results reflect the application value of the simulation experiment platform, improve the efficiency of programming and debugging, break the limitation of insufficient equipment in training and learning, solve most of the problems that may occur in the actual operation process in advance, and ensure the safety of equipment and personnel.

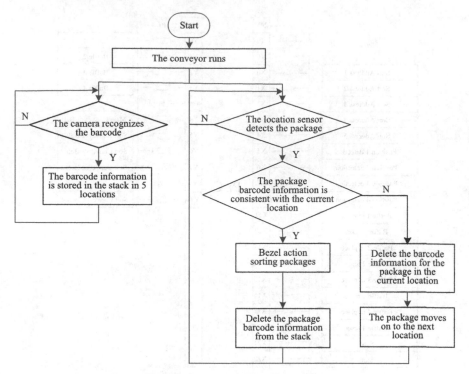

Fig. 4. Flow chart of logistics sorting

5 Conclusion

Sorting is an important work in the logistics system, which affects and even determines the function and efficiency of the system to a certain extent. With the progress of science and technology, the technicians in charge of sorting work need to constantly learn and improve their skills and quality. It is an effective way to use the simulation experimental platform of logistics sorting for professional training.

The logistics sorting simulation experimental platform realizes the three-dimensional dynamic simulation of virtual and real interaction. The logistics sorting system is modeled, and its geometric model is given physical and motion attributes, and electromechanical objects are added and runtime behavior code is written to obtain a realistic three-dimensional virtual simulation model. The virtual-reality conversion module enables PLC entities to seamlessly interface with virtual simulation models and real-time data interaction. The simulation experiment can carry out hardware design and software design at the same time, find and solve problems such as PLC hardware wiring and control program errors in the process of logistics sorting system debugging in advance, optimize the sorting scheme, and save debugging time. The security of the experiment is improved after the simulation experiment is completed. The logistics sorting simulation platform has certain practical significance for the study of professional courses in colleges and universities and the induction and further training of professional technicians.

References

1. Xiang, H., Wang, Y., Gao, Y., et al.: Research on key technology of logistics sorting robot. Lect. Notes Bioinform. (12595), 121–132 (2020)
2. Grambo, P., Mullick, T., Furukawa, T., et al.: Automatic sorting-and-holding for stacking heterogeneous packages in logistic hubs. IFAC-Papers OnLine 52(10), 109–114 (2019)
3. Jouglet, A., Nace, D., Outteryck, C.: Timetabling of sorting slots in a logistic warehouse. Ann. Oper. Res. 239(1), 295–316 (2016)
4. Benotmane, Z., Belalem, G., Neki, A.: A cost measurement system of logistics process. Int. J. Inform. Eng. Electron. Bus. 10(5), 23–29 (2018)
5. Zheng, K., Dai, F., Lian, L.: Virtual debugging of robot grinding system based on NX MCD. Mod. Mach. Tool Autom. Mach. Technol. (12), 57–60+64 (2019). (in Chinese)
6. Qu, Y., Ruan, X., Wang, S.: Construction and design of computer hardware virtual simulation experimental platform. Lab. Res. Explor. 37(12), 116–119 (2018). (in Chinese)
7. Li, X., Lv, L., Sheng, R., et al.: An effective construction method of modular manipulator 3d virtual simulation platform. 3D Res. 9(2), 24 (2018)
8. He, Z., Huang, T.: Exploration of virtual simulation experiment teaching project construction. Experim. Technol. Manage. 35(02), 108–111+116 (2018). (in Chinese)
9. Gjorgjevikj, F., Iakimoski, K.: System monitoring addon analysis in system load simulation. Int. J. Comput. Netw. Inform. Secur. 14(1), 40–51 (2022)
10. Li, Q., Mi, C., Wang, C., Shenyang.: Semi-in-the-loop simulation system of immersive PLC control program based on virtual reality technology. China J. Construct. Mach. 11(01), 41–45 (2013). (in Chinese)
11. Lu, D.: Design of modular automatic production line based on PLC. Labor. Res. Explor. 35(10), 56–59 (2016). (in Chinese)
12. Wang, D., Chen, X., Gao, H.: Application of virtual reality technology in PLC gas-fired boiler training system. J. Syst. Simul. 29(12), 3010–3015 (2017). (in Chinese)
13. Wang, H., Li, H., Cheng, Z., Chen, X.: Design of greenhouse automatic monitoring system based on PLC. Labor. Res. Explor. 36(05), 21–23+71 (2017). (in Chinese)
14. Xiao, Z., Liu, H., Li, B., Rao, X.: Application and research of UG NX in conceptual design of electromechanical products. Mod. Mach. Tools Autom. Process. Technol. 07, 27–30 (2014). (in Chinese)
15. Tang, Y., Guo, X., Zhang, Z., et al.: Parameter optimization method of ADC model based on real-time simulation. Power Autom. Equip. 4(3), 214–218 (2020). (in Chinese)
16. Klucik, S., Taraba, J., Orgon, M., Adamko, D.: The use of PLC technology in broadband services offered to households. Int. J. Inform. Technol. Comput. Sci. 4(4), 1–8 (2012)
17. Afram, R.M., Marie, M.J.: Design and implementation of optimal PID controller using PLC for Al-Tahady ESP. Int. J. Image Graph. Signal Process. 12(5), 1–12 (2020). https://doi.org/10.5815/ijigsp.2020.05.01
18. Wang, X., Tang, X., Dong, Z., et al.: Research on rapid development platform of PLC control system. High-tech Commun. (English Version) 27(2), 210–217 (2021)
19. Wu, Y., Wang, Y., Zheng, G., Zhang, J., Li, Y.: Conceptual design and control simulation of electrical integration of CNC lathe loading and unloading machinery based on MCD platform. Mach. Tool Hydraul. 46(15), 99–104 (2018). (in Chinese)
20. Zhang, J., Tan, R., Zhang, Z., Dai, J., Yang, X.: Research on the integration of product conceptual design and detailed design driven by computer-aided innovation technology. J. Mech. Eng. 52(05), 47–57 (2016). (in Chinese)
21. Tang, P., Tian, H.: Research on the layout of national economic mobilization logistics centers. Int. J. Intell. Syst. Appl. 2(1), 44–50 (2010)

22. Mohamed, M.: Smart Warehouse management using hybrid architecture of neural network with barcode reader 1D/2D vision technology. Int. J. Intell. Syst. Appl. **11**(11), 16–24 (2019)
23. Shieh, J., Zhang, J., Liao, Y., Lin, K.: Applications of barcode images by enhancing the two-dimensional recognition rate. Int. J. Image Graph. Signal Process. **4**(7), 26–32 (2012)

Analysis on Ventilation Design of the Negative Pressure Isolation Rooms in Wuhan Leishenshan Hospital

Xiang Lei[1], Yilei Liu[2], Min Xu[1], Zheng Yang[1], Qiong Dan[1], and Xueqin Yan[1(✉)]

[1] Wuhan Railway Vocational College of Technology, Wuhan 430225, China
458798669@qq.com
[2] The University of Hong Kong, Hong Kong, China

Abstract. In response to the outbreak of the novel corona virus pneumonia (COVID-19 pneumonia) in late 2019, many temporary emergency hospitals have been built or rebuilt in some key areas in China. The environment of hospital negative pressure isolation room is closely related to medical workers' health and safety, while ventilation system has been the major means of environmental control. The design points of Wuhan Leishenshan Hospital (also named as the Thunder Mountain Hospital, which was specifically built to address the COVID-19 pneumonia in February 2020) was briefly introduced and analyzed in the article, as well as some essential elements of the ventilation design to build negative pressure isolation rooms, such as ventilation volume design standard, differential pressure control standard and airflow direction strategy. The measures of differential pressure control are also illustrated in the article. The points in the assay could be used as reference for those engaged in the future designing and construction of infectious disease hospitals.

Keywords: COVID-19 pneumonia · negative pressure isolation room · pressure difference · air distribution

1 Introduction

In the spring of 2020, corona virus pneumonia flared up widely in the world. According to the statistics, 1,500,830 cases of the virus have been confirmed globally by 9th April 2020. Heavy pressure was imposed to current medical system due to the numerous infected patients. In order to cope with the public health emergency, a large number of temporary medical facilities have been constructed or rebuilt in China. Such construction program is of great social significance with a heavy work burden in limited time. Instructed by the Wuhan government, the author's team has took over the working mission of HVAC designing as well as epidemic prevention research of the Wuhan Leishenshan Temporary Hospital [1].

The architectural designing core of such infectious disease hospitals is to cut off the infection source and avoid cross-infection among medical staff and patients. Two important factors have to be taken into consideration: firstly, the functional zones and flow lines of the hospital should be reasonably arranged to achieve "three zones, two channels"; secondly, the environment of each zone should be well controlled, the crux of which is to control the pressure of each zone to achieve an orderly pressure gradient [2]. A reasonable and effective ventilation system is one of the important means to control the ward's environment [3].

As the firstly built temporary emergency hospital, Wuhan Leishenshan Hospital is positioned to treat severely ill patients. It could accommodate up to 1500 beds with all the isolation rooms of negative pressure. The box-type board room structure has been adopted, and the overall dimensions (length × width × height) of each unit is 3 × 6 × 2.6 m. 2000 patients of the COVID-19 pneumonia have been adopted and treated up to 19th March 2020. Figure 1 is the Wuhan Leishenshan Hospital.

Fig. 1. The constructing Leishenshan hospital

2 Brief Introduction of Ventilation System

Wuhan, where the Leishenshan Hospital is situated, was suffering from a low temperature during the widespread of COVID-19 pneumonia. Heating facilities are of great need in hospitals. Due to the heavy workload as well as limited time, suitable AHU for fresh air system is in shortage. Therefore, the configuration measure, "split air conditioner + centrifugal supply fan + electric heating", has been adopted in isolation rooms in order to balance indoor ambient temperature. Every air supply and exhaust system serve 4–6 isolation rooms, and each room's air supply/exhaust branch pipe is equipped with constant air volume valve, ensuring that the wards' pressure be controlled through the curbing of air supply/exhaust volume differential, thus the miniaturization and simplification of the system could be realized. Figure 2 is the schematic diagram of air supply/exhaust system in Leishenshan's isolation room zones. The ward is equipped with independent air supply/exhaust system, while the air exhaust system in the bathroom is integrated into the ward's, where HEPA outlet is installed. A common air supply system is shared between doctor corridor and anteroom, and the doctor corridor enjoys an independent air exhaust system. The ward's exhaust outlet is 5.4 m above the ground and 20 m horizontal from the supply inlet.

Due to the net height (2.4 m) of the ward, no air pipe is installed indoor. Air supply and return inlet are on one side, when supply inlet is on the top and return inlet is near the floor. Return air terminal was equipped with HEPA inlet. Figure 3 is the negative pressure isolation room of Leishenshan Hospital. Figure 4 is the key factors of ventilation system.

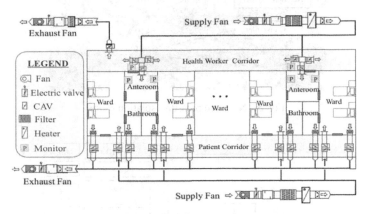

Fig. 2. Schematic diagram of ventilation system

3   Method of Ventilation Design of the Negative Pressure Isolation Rooms

3.1   Isolation Room's Ventilation Designing Standard and Differential Pressure Control

3.1.1   Air Changes Rate

It was emphasized by China National Health Commission in Chinese Clinical Guidance for COVID-19 Pneumonia Diagnosis and Treatment (6th edition) that COVID-19 is transmitted through respiratory droplets and close contact; aerosol transmission is plausible when patients are exposed to high concentration virus-containing aerosols for a long period of time and in a relatively closed environment. Studied by current domestic and foreign researches, short-distance airborne transmission possibly exists in relatively closed environment; airborne transmission is also acknowledged by WHO. Hence, lower the virus concentration of contaminated air in negative pressure isolation room is a crucial method to lower the infectious rate. In the fight against SARS in 2003, it was put forward by domestic experts that the virus-containing air was no longer infectious when diluted above 10000 times [4], showing that virus activity and toxicity decreased when the virus concentration reduced in diluted air. Therefore, the suitable air changes rate has become the essential parameter during isolation room's design and construction. Many countries have regulated the minimized air changes rate (see Table 1).

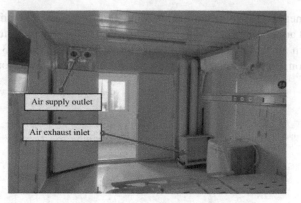

Fig. 3. Negative pressure isolation room

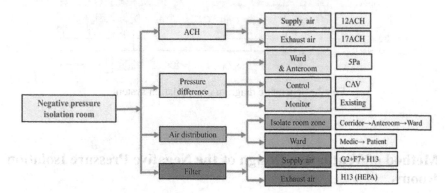

Fig. 4. Key factors of ventilation system

Table 1. Minimized air changes per hour (ACH)

Object	Air changes (ACH)	Source
Airborne infectious isolation room	12	ASHRAE Standard 17012
Airborne infectious isolation room	6–12★	The U.S. CDC11
Airborne infectious isolation room	12 and ≥145L/(s·P)	Australia17
Negative pressure isolation room	12	China14

★12 air changes per hour (ACH) new or renovation, 6 ACH existing.

Wells Riley has put forward the noted formula is:

$$P = C/S = 1 - \exp(-\frac{Ipqt}{Q})$$ (1)

In the formula, P(C/S) is the infection probability of the most susceptible group where I is each infector's quanta; p is the expiratory volume of susceptible group (m³/h); t is exposure time (h), Q is indoor ventilation volume (m^3/h). The mode is used to predict the probability of airborne respiratory infectious diseases. In a relatively closed environment, airborne transmission possibly exists in a short distance. The basis of isolation room air change rate value could be illustrated by Wells-Riley mode. The infectious rate of medical workers who enter isolation room without protective equipment could be calculated through formula 1. According to the number of SARS cases, supposed that quanta takes the same value of 4680 [5], p equals to 0.3 m^3/h, t equals to 0.5 h; all the patients of COVID-19 virus are requested to wear surgical masks, the infiltrating rate of which could reach above 95%. Hence the quanta of such patients spreading in isolation room could take the value of 234. Figure 5 shows the result. It could be found from Fig. 5 that when ventilation volume is between 0–500 m^3/h, the increase of ventilation volume evidently reduces the infectious risk of susceptible people in isolation room. However, the constant increase contributes little to the decline of infectious rate [6]. If patients did not wear surgical masks, the infectious rate could enjoy a drastic growth.

In order to save investment and air-conditioning system operating costs, excessive air changes should not be adopted. When the fresh air volume reaches 500 m^3/h, 12ACH will be correspondent, which is suitable for Leishenshan Hospital. So far, the hospital has been operating for 2 months, the air of which is of high quality due to the well control of the differential pressure with none medical worker infected. Actual testing data of CO^2 and PM2.5 in hospital's isolation room is demonstrated in Figs. 6 and 7.

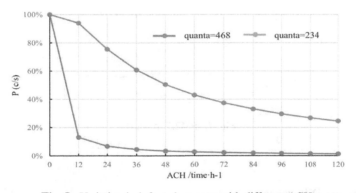

Fig. 5. Variation in infected persons with different ACHs

During the outbreak of COVID-19 pneumonia, a large amount of temporary negative pressure isolation rooms (TNPIs) have been rebuilt. In order to cope with the volume shortage of fresh air, these isolation, including 19 Fangcang shelter hospitals, have been equipped with portable HEPA. Those devices could provide more than 13ACH air changes rate and remote more than 90% particles larger than 0.3 mm in ten minutes. 24 fan filter units (FFUs) providing 15ACH air changes rate have been used in another hospital. HEAP has been used in FFU which could arrest 99.97% of 0.3 mmm particles, which makes particle concentration 12% less than before. In such TNPI, fresh air volume is around 3–6ACH, and none of the medical workers has got infected. Therefore, it is advised that when air supply unit equipped with HEPA is adopted to dilute contaminated

air, the fresh air volume in TNPI should be reduced to 3ACH and total air changes rate
should be more than 13ACH.

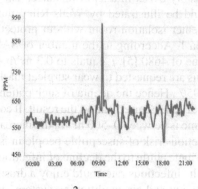

Fig. 6. CO^2 concentration in ward

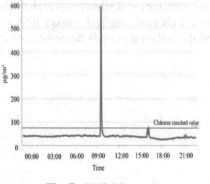

Fig. 7. PM2.5 in ward

3.1.2 Pressure Difference

The pressure difference can be analyzed from four aspects.

1) Purpose of differential pressure control

The main non-medical prevention treatment to respiratory infectious disease is to
control airborne route and cut off airborne chain, physically separating and isolating
infected patients. Reasonable partition is one of the physical separating method. Rooms
shall be divided into clean zone, potentially contaminated zone and contaminated zone;
independent patient passage and doctor passage could be used to separate patients and
medical workers. Although such measures are largely helpful to reduce airborne and
contact transmission, virus transmission within the hospital cannot be severed. Cross-
infection might be caused without effective control of disorderly air flow, while indoor
temperature, outdoor wind, opening or closing the door/window, and medical workers'

passing by might all produce air flow. Air isolation is to maintain an appropriate differential pressure between clean and contaminated zone so as to guarantee the orderly air flow, preventing the spread of pathogenic micro-organisms from contaminated zone. In conclusion, air flow control is essential to cross-infection prevention, and the purpose of differential pressure control is to thoroughly cut off the airborne transmission chain.

2) Standard of differential pressure control

Differential pressure control is the crux of isolation room's design. It might be hard to prevent spread of contaminated air when differential pressure is too low, while energy consumption and system investment might increase if it is too high. Table 2 demonstrates differential pressure setting standard in three countries. It is specifically regulated in China's standard that no less than 5 pa differential pressure should be maintained between negative pressure isolation room and its adjacent or connected anteroom (or corridor). Differential pressure schematic in various districts is included in the standard. In the U.S, recommendable differential pressure has been 0.25 Pa since 2000 [7]. Scholar Xu Z and his team has studied the relationship between differential pressure and pollutant leakage, discovering that concentration ratio of pollutant leakage only reduce 0.03 when differential pressure changes to −6 Pa from 0, while the ratio remains scarcely unchanged when it turns to −30 Pa from −6 Pa [8]. Scholar Li Y compares ventilation performance of 9 hospitals in Hong Kong. They found that among the 38 tested rooms, differential pressure in 97% rooms complies with the standard −2.5 Pa (all tested rooms have the same average pressure of −7.7 Pa) [9].

Table 2. Smallest differential pressure between adjacent rooms

Object	Pressure difference (Pa)	Source
Airborne infectious isolation room	2.5	ASHRAE Standard 17012
Airborne infectious isolation room	2.5	The U.S. CDC11
Airborne infectious isolation room	15	Australia17
Negative pressure isolation room	5	China14

The formality of differential pressure is relevant to building's air sealing performance and supply/exhaust volume. Given that gap remains unchanged, the greater the differential pressure is, the more the air exhaust volume will be. The relation between open flow and differential pressure could be illustrate in formula below:

$$L = 3600\mu A\left(^{\Delta p}/\rho\right)^{1/2} \tag{2}$$

L is air leakage volume (m³/h); μ is flow coefficient; A is gap width; Δp is differential pressure between two sides of the gap (Pa); ρ is air density, select the value of 1.2 kg/m³.

In engineering project, differential pressure method is used to calculate air volume infiltrated from door and window gap, the simplified formula is [10]:

$$L = 1.05 \, A\Delta P^{1/2} \tag{3}$$

The infiltrated air volume in ward could be calculated by formula (2). When the air supply volume in negative pressure isolation room is fixed, air exhaust volume will be directly influenced by different negative pressure value. Since air infiltration exists among ward and its adjacent anteroom and bathroom, the relation between negative pressure and demanded air supply volume in isolation room is not the functional formula (2) and (3) above. Supposed that the air supply volume is 500 m³/h (12ACH) in one isolation room in Leishenshan Hospital, the relation between negative pressure and demanded air supply volume under particular condition is shown in Fig. 8. The demanded air supply volume in Fig. 8 is determined by Leishenshan Hospital's box-type board structure and designed differential pressure between adjacent rooms. Suppose the door gap were 0.0025 m.

It could be seen from Fig. 8 that approximate linear relationship exists between ward negative pressure and demanded air exhaust volume. Greater exhaust volume could lead to more investment cost and energy consumption. Apart from that, huge amount of engineering work will extend much more time, which might be impractical to the emergent medical facilities. Therefore, excessive differential pressure is not suitable for temporary hospital. In conclusion, the differential pressure between isolation room, anteroom and corridor shall be no less than −5 Pa in Leishenshan Hospital. According to the actual operating statistics, differential pressure between isolation rooms and anterooms have been maintained between the range of −6 Pa to −10 Pa. Figure 9 shows the differential pressure value on measuring gauge.

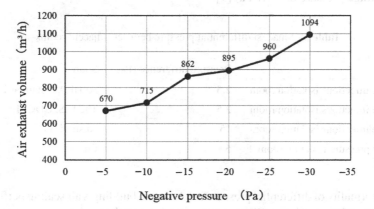

Fig. 8. Relations between air exhaust volume and differential pressure

It has been instructed in American CDC Standard [11] and ASHRAE Standard 170 [12] that differential pressure between adjacent rooms shall be no less than 2.5 Pa. The air sealing performance are sometimes not so good in temporary hospitals; according to national current standard, as well as ASHRAE Standard 170, the smallest differential pressure in different zones varies: 5 Pa between isolation room and anteroom; 5 Pa between anteroom and corridor; 2.5 Pa between medical corridor and outside; 2.5 Pa between patient corridor and outside.

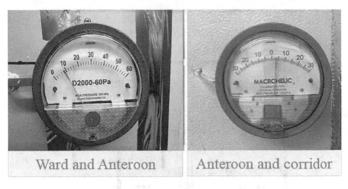

Fig. 9. Differential pressure measuring gauge

3) Measures to control differential pressure

After ward's air density has been confirmed, the main measure controlling differential pressure is to control air supply and exhaust volume. The primary task for ventilation system is to build a pressure gradient, making air flow from clean zone to contaminated zone. There are two ways:

Variable-frequency control: Supposed that air supply volume remains unchanged, variable-frequency control means to adjust air exhaust volume of the ward according to differential pressure among ward, anteroom and corridor. Constant air volume valve shall be installed on air supply branch pipes [13].

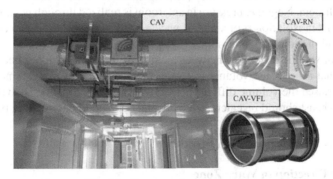

Fig. 10. Constant air volume valve

Constant air volume control: High precision constant air volume valve shall be stalled on every air supply/exhaust branch pipe, which shall be pressure-independent and air volume control deviation shall not be greater than 5%. During the tight construction time, BA system control might not be put into use in temporary hospital. Constant air volume valve could effectively compensate system air volume change produced by high and medium efficiency particulate filter resistance change [14]; hence it has concise procedures and useful function. Figure 10 shows a CAV installed on air supply branch pipe.

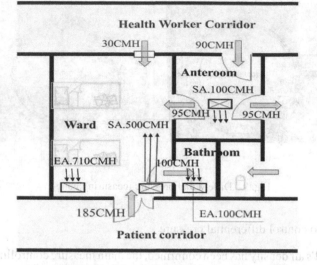

Fig. 11. Air volume balance in ward

4) Air volume balance in isolation rooms

Air volume balance plays a crucial part in differential pressure control. Figure 11 demonstrates an air volume balance schematic in one Leishenshan isolation room. SA represents air supply and EA represents air exhaust. Air supply inlet is installed in isolation room, anteroom and corridor [15]; air exhaust outlet is installed in isolation room and bathroom. Negative pressure in anteroom is realized through negative pressure suction exhaust in isolation room, and, if necessary, 10–20 mm door gap could be left between isolation room and anteroom.

In terms of volume difference between air supply and exhaust, it has been instructed in standard that air exhaust volume is 150 m^3/h greater than supply. American CDC standard has raised the volume difference to 210 m^3/h from 85 m^3/h. The volume of air supply and exhaust should be calculated by actual demand in engineering practice.

4   Analysis on Airflow Direction

4.1   Airflow Direction in Ward Zone

The directed and organized air flow is one of the main ways to prevent cross-infection in relevant hospital. Airflow direction shall be strictly controlled to make clean air flow from clean zone to potentially contaminated zone and eventually to contaminated zone under the pressure gradient. It could efficiently make pathogen and other contaminated substantiate spread in the smallest range.

4.2   Airflow Distribution Interior to Ward

It is required interior the isolation room that the clean airflow shall firstly go through the zone where medical staff pass or work, then the contaminated sources and finally the exhaust outlet. Figures 12 and 13 show two airflow direction in negative pressure isolation room recommended by American CDC standard as well as ASHRAE Standard 170. Figure 12 is aimed at infectious patients with normal immunity, while Fig. 13 is for susceptible infectious patients. In terms of COVID-19 pneumonia, the airflow direction shown in Fig. 12 is suitable for relevant negative pressure isolation rooms, which is complied with targeted airflow direction principle and adopted by Leishenshan Hospital [16].

As for the location of air supply/exhaust exit, it has been instructed in relative standard that air supply outlet shall be installed on the top of the ward; exhaust air grilles of the ward and clinical room shall be installed on the bottom of the room; the distance between exhaust outlet and the ground shall be no less than 100 mm. The layout of air supply/exhaust exit shall be corresponding to targeted airflow direction principle; air supply outlet shall be installed on the top of the ward, while exhaust air grilles shall be installed near the bed facilitating the emission of contaminated air. Air supply outlet shall be installed above where medical works normally stand, and exhaust air grilles shall be installed below the bed opposite to the inlet. All the standard above aim at protecting medical workers' safety by practicing targeted airflow principle, making sure that clean air flow past medical zone at first and finally flow outside from patient zone. The location of air supply outlet has not been specifically regulated in ASHRAE Standard 170, as for exhaust outlet, it has been instructed in the standard that the exhaust air grilles in the patient room shall be located directly above the patient bed [17], on the ceiling or on the wall near the head of bed, unless it can be demonstrated that such a location is not practical. Scholar Qian Hua has also argued about the importance of upper air exhaust [18]. During winter, the fine particles exhaled by patient might rise to the ceiling through the hot wind. In order to emit the turbid air on the top of the isolation, exhaust air grilles shall be installed at the lower part of the room and the ceiling above the patient's head. Electric valve could be installed on the air exhaust branch pipe, which could stop exhausting when medical workers come in and avoid disturbing the targeted airflow. The combination achieves the ventilation type of "partial exhaust, entirely ventilation".

The optimum airflow distribution could not be achieved due to the isolation's limited storey height, but actual effect of airflow direction in Leishenshan has made isolation's wind speed and temperature stay at reasonable level. CFD simulation has shown that fresh air orderly flow past medical workers first and then patients.

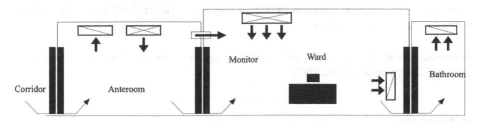

Fig. 12. Example 1 of ward control for airborne infection isolation

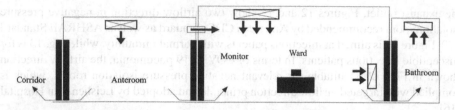

Fig. 13. Example 2 of ward control for airborne infection isolation

5 Conclusion

The wide spread of COVID-19 pneumonia has been threatening global people's health. The fight against corona virus pneumonia highlights the importance of infectious disease prevention and, particularly, the significance of medical facility construction for infectious disease. Summarizing relevant experience largely contributes to future's construction of infectious disease hospitals and protection of medical workers, patients as well as the environment. Key factors in ventilation system design have been analyzed in the article, as well as design and construction difference between temporary and permanent hospitals. Main advises and conclusion are listed below:

According to the statistics, in terms of indoor ambient isolation supervision, 12ACH air volume could provide a suitable indoor environment for isolation. As is calculated, 10–15ACH air changes rate could serve as economic and proficient ventilation volume.

Differential pressure control is the key factor to controlling isolation. The differential pressure between isolation and adjacent room shall be not less than 5 Pa. Given that temporary building's air sealing performance might not be so good, and considerable windows are installed along contaminated corridor, negative pressure differential value could be lower to no less than −2.5 Pa. Auxiliary measures shall also be adopted in anteroom to strengthen air isolation.

With the fixed building's configuration and air sealing performance, the differential value is determined by air exhaust volume. Excessive differential pressure needs more air exhaust volume, which will lead to some disadvantages such as much more investment cost, energy consuming and construction burden.

The collective effect of anteroom isolation and fresh air diluting makes medical corridor remain relatively clean where virus concentration is approximately 0.0031% of isolation. The air flow direction shall be scientifically determined from both medical workers and patient protection.

Acknowledgment. This project is supported by Projects of Hubei Institute for Vocational and Technical Education (ZJGB2022095).

References

1. Odji, E.: Graphic design principles and theories application in rendering aesthetic and functional installations for improved environmental sustainability and development. Int. J. Eng. Manufac. **1**(9), 21–37 (2019)

2. Gaomin, D.: Positive crankcase ventilation system. Int. J. Eng. Manufac. **1**(1), 13–19 (2011)
3. Dinggen, L., Xiaozhong, L.: Study on gasoline intake model modification based on PCV. Vehicle Eng. **1**(1), 31–35 (2009)
4. Jiang, Y., Zhao, B., Li, X., et al.: Investigating a safe ventilation rate for the prevention of indoor SARS transmission: an attempt based on a simulation approach. Build. Simul. **2**(4), 281–289 (2009). (in Chinese)
5. Qian, H., Zheng, X.: Prediction of risk of airborne transmitted diseases. J. Univ. **42**(3), 468–472 (2012)
6. Memarzadeh, F., Xu, W.: Role of air changes per hour (ACH) in possible transmission of airborne infections. Build. Simul. **5**(1), 15–28 (2011). https://doi.org/10.1007/s12273-011-0053-4
7. Marko, H., Anna, R., Pertti, P., et al.: Airborne infection isolation rooms-a review of experimental studies. Indoor Built Environ. **20**(2), 584–594 (2011)
8. Xu, Z., Zhang, Y., Wang, Q., et al.: Study on isolation effects of isolation wards. J. HV&AC **36**(4), 1–4 (2006)
9. Li, Y., Ching, W.H., Qian, H., et al.: An evaluation of the performance of new SARS isolation wards in nine hospitals in Hong Kong. Indoor Built Environ. **16**(1), 400–410 (2007)
10. Technical standard for smoke management systems in buildings: GB51251-2017. Ministry of public security of the People's Republic of China, Beijing (2015). (in Chinese)
11. ASHRAE. Ventilation of health care facilities: ANSI/ASHRAE/ASHE standard 107–2017. Beijing, Victorian advisory committee on infection control (2017). (in Chinese)
12. Xu, Z., Zhang, Y., Wang, Q., et al.: Isolation principle of isolation wards. J. HV&AC **36**(1), 1–7 (2006)
13. Code for design of infectious diseases hospital: GB/T 50849-2014. National health commission of the Peoples Republic of China, Beijing (2015). (in Chinese)
14. SAC/TC319. Requirements of environmental control for hospital negative pressure isolation ward: GB/T 35428-2017[S]. Beijing, National health commission of the Peoples Republic of China (2017). (in Chinese)
15. The design standard of infectious disease emergency medical facilities for novel corona virus (2019-nCoV) infected pneumonia. T/CECS 661-2020. Beijing, China Ippr international Co. Ltd. (2020). (in Chinese)
16. Victorian advisor committee on infection control: guidelines for the classification and design of isolation rooms in health care facilities. Victoria, Victorian advisory committee on infection control (1999)
17. Aganovic, A., Cao, G.: Evaluation of airborne contaminant exposure in a single bed isolation ward equipped with a protected occupied zone ventilation system. Indoor Built Environ. **28**(8), 1092–1103 (2019)
18. Qian, H., Li, Y.: Removal of exhaled particles by ventilation and deposition in a multiple airborne infection isolation room. Indoor Air **20**(3), 284–297 (2010). (in Chinese)

Multi Object Infrared Image Segmentation Based on Multi-level Feature Fusion

Chengquan Liang[✉] and Ying Zhang

School of Intelligent Manufacturing, Nanning University, Nanning 530200, China
759137827@qq.com

Abstract. Aiming at the problem that the segmentation accuracy may be too low when using a single feature to segment infrared image multi objects, a multi object segmentation method based on multi-level feature fusion is proposed in this paper. The entropy feature, contrast feature and gradient feature of infrared image are extracted respectively, and the multi-level features of the extracted infrared image are fused using the parallel weighted feature fusion method to build the multi-level feature fusion of the infrared image. The multi-level feature fusion space of the infrared image is set as the ergodic space of Mean shift algorithm, and all feature points in the multi-level feature fusion space are subject to mean shift processing, Obtain infrared image multi object segmentation results. This method can use the multi-level features of the extracted infrared image to segment the multi objects of the infrared image, and the multi object segmentation accuracy of the infrared image is high.

Keywords: Multi-level feature fusion · Infrared image · Multi object segmentation · Contrast characteristics · Gradient characteristics · Mean shift algorithm

1 Introduction

Image segmentation is an important step in image processing [1, 2]. Infrared image object segmentation is often used in object tracking, fault detection and other applications [3, 4]. The infrared image obtained by the infrared imaging technology has the defects of low signal-to-noise ratio and low definition [5]. When the infrared image is affected by light and ambient temperature, the object area has a high similarity to the ambient temperature. When the infrared image background area has a high similarity to the object area, the object segmentation accuracy is affected [6]. Extract the multi-level features contained in the infrared image, improve the defect that the object cannot be accurately presented due to the extraction of a single feature [7–9], and achieve the multi object segmentation of the infrared image through multi-level feature fusion. Multi object segmentation of infrared image is to divide the infrared image into different regions, so that the same region that has been segmented in the infrared image has certain feature similarity [10–12], and different regions have high differences. Infrared image segmentation is an important basis for infrared image analysis and infrared image pattern recognition, and

Z. Hu et al. (Eds.): ICCSEEA 2023, LNDECT 181, pp. 26–36, 2023.
https://doi.org/10.1007/978-3-031-36118-0_3

an important basis for computer vision fields such as infrared image object tracking and object recognition [13]. When infrared image is segmented by multiple objects, contrast features and gradient features are widely used [14]. The infrared image contrast features and gradient features are applied to infrared image multi object segmentation to improve the performance of infrared image multi object segmentation.

At present, many researchers have studied image segmentation. Zhang Yishu and others have applied watershed algorithm to laser image segmentation [15]. Experiments have verified that this method can accurately segment inductive laser thermal imaging images, but there is a defect of too much background information in the segmentation results; Zeng Yanyang et al. realized weak light image segmentation by using intercept histogram [16]. This method can effectively segment weak light images without being affected by background interference, but it has the defect of poor real-time segmentation; Ding Yongqian and others carried out adaptive segmentation for narrowband images of spectral images [17]. This method has a high adaptive performance, but it has the defect of small application range. Aiming at the defects of the above research methods when they are applied to image segmentation, the infrared image multi object segmentation method based on multi-level feature fusion is studied. The experimental results show that this method can achieve infrared image multi object segmentation, make full use of the multi-level features of the extracted infrared image, and improve the infrared image object segmentation performance.

2 Multilevel Feature Extraction and Fusion of Infrared Image

2.1 Extraction of Multi-level Features of Infrared Image

2.1.1 Infrared Image Entropy Feature Extraction

Entropy feature of infrared image is an important feature to measure texture details of infrared image. When the texture evenness of the infrared image is low, the absolute value of the infrared image entropy feature is large, otherwise the absolute value of the infrared image entropy is small [18]. The gray level co-occurrence matrix method is selected to extract the entropy features of infrared images. If the original infrared image size is $M \times N$, and the infrared image pixel exists in the S_{xy} area, $P(i,j)$ means that the infrared image sliding pixel block area is. When the pixel gray level is i, the probability of the pixel moving from a fixed position to the target gray j level is $P(i,j)$. See (1) for the calculation formula.

$$P(i,j) = \frac{g_{ij}}{(1+L)^2} \tag{1}$$

In the formula, g_{ij} and L respectively signify the number of pixels separated from the infrared image and the total number of gray levels.

The entropy characteristic expressions of infrared image pixels in different directions are shown in (2).

$$ent(x,y) = -\sum_{i}^{L}\sum_{j}^{L} P(i,j) \log_c P(i,j) \tag{2}$$

The mean value expression of infrared image entropy features is shown in (3).

$$\overline{ent}(x, y) = -\sum_\theta ent(x, y)/4 \tag{3}$$

Traverse all pixels of the infrared image [19], and obtain the entropy characteristic image formula of the infrared image as shown in (4).

$$e_1(x, y) = \overline{ent}(x, y) \tag{4}$$

In the formula, $ent(x, y)$ and θ separately represent the entropy characteristics of the infrared image and the calculation direction of the infrared image pixels.

2.1.2 IR Image Contrast Feature Extraction

The gray level co-occurrence matrix is used to extract the contrast features of infrared images. Linear weighting of different components in the infrared image, establish new components of the infrared image, and use the new components to convert the infrared image into a grayscale image [20]. The gray image expression of the infrared image is shown in (5).

$$I_1(x, y) = 0.288 \times R_{f0} + 0.576 \times G_{f0} \tag{5}$$

Where, R_{f0} and G_{f0} respectively mean the R component and G component of the original infrared image.

The contrast value expression of the infrared image pixel in the moving area is visible from (6).

$$con(x, y) = \sum_i^L \sum_j^L (i - j)^2 P(i, j) \tag{6}$$

Calculate the average contrast of pixels moving in different directions, and set the obtained average contrast $\overline{con}(x, y)$ to the gray level co-occurrence matrix of pixels (x, y) in the infrared image area S_{xy}. The matrix $c_1(x, y)$ is established by using the average contrast value of pixels in the infrared image, the contrast feature extraction threshold K_0 is set, and the infrared image contrast feature map $c_2(x, y)$ is obtained by $c_1(x, y)$ binary processing. The obtained contrast feature map is the contrast feature extraction result of the infrared image [21]. The contrast feature can reflect the contour of the object area to be segmented in the infrared image.

2.1.3 Infrared Image Gradient Feature Extraction

The infrared image has a small imaging range, poor gray level distribution uniformity of the infrared image, and the gray level distribution of the infrared image is prone to extreme distribution [22]. When extracting gradient features of infrared images, it is necessary to reduce the impact of infrared image brightness on infrared images [23, 24].

Select coverage operator to reduce the influence of infrared image brightness on infrared image. The calculation formula of coverage operator is shown in (7).

$$f_2(x, y) = (1 - \eta)f_0(x, y) + \eta f_1(x, y) \tag{7}$$

Where, $f_0(x, y)$ and $f_1(x, y)$ separately represent the original infrared image and the all black image after cross fusion of the original infrared image; η and $f_2(x, y)$ represent the attenuation factor and the infrared image with reduced brightness respectively. The value interval of the attenuation factor μ is [0,1]. The change of the attenuation factor is used to reduce the influence of brightness on the original infrared image. Cross fusion of all black image and infrared image is adopted to reduce the influence of brightness on infrared image, and the infrared image after brightness reduction is obtained and represented by $f_2(x, y)$ [25, 26].

Convert the infrared image $f_2(x, y)$ after the brightness reduction processing from RGB mode to HSI mode, and then complete the brightness reduction processing. The expression of I component of $f_2(x, y)$ is shown in (8).

$$I_2(x, y) = 0.288 \times R_{f_2} + 0.576G_{f_2} + 0.103B_{f_2} \tag{8}$$

Where, R_{f_2}, G_{f_2}, and B_{f_2} represent the R, G, and B components of the infrared image $f_2(x, y)$.

Use the I component of the infrared image to obtain the gradient feature image expression of the infrared image. See (9).

$$g(x, y) = |I_2(x, y) - I_2(x + 1, y)| + |I_2(x, y) - I_2(x, y + 1)| \tag{9}$$

The gradient feature of the infrared image is extracted by using the difference processing of the horizontal and vertical components of the infrared image through Eq. (9).

2.2 Multi Level Feature Fusion of Infrared Image

A parallel weighted feature fusion method is used to fuse the multi-level features of the extracted infrared image. A multi-level feature fusion space is designed for the infrared image, which includes the entropy features, contrast features and gradient features of the extracted infrared image [27–29]. When using parallel feature fusion method to fuse multi-level feature vectors of infrared images, use to represent imaginary units. When the dimensions of the two level feature vectors are different, use the imaginary units to fill zeros behind the feature vectors with lower dimensions, so that the extracted infrared image features have the same dimensions. The feature dimension of infrared image fused by parallel feature fusion method is expressed by $\max(m, n)$.

The parallel weighted feature fusion method fully considers the influence of the relationship between multi-level features of infrared images on multi object segmentation, provides a basis for multi object segmentation of infrared images, and makes full use of the feature dimension and distinguishability of the extracted multi-level features of infrared images. According to the multi-level features of the extracted infrared image, the feature vector space expression of the fused infrared image is obtained as (10).

$$F = f_1 + if_2 + jf_3 \tag{10}$$

Where, i and j are imaginary numbers, f_1, f_2 and f_3 respectively refer to the entropy feature, contrast feature and gradient feature of the infrared image.

Due to the difference in the contribution rate of the extracted infrared image multi-level features to the infrared image multi object segmentation, it is necessary to set different weights w for the extracted infrared image multi-level features. See (11) for the histogram vector expression of multi-level feature fusion of infrared image multi object segmentation.

$$F = w_1 f_1 + w_2 i f_2 + w_3 j f_3 \tag{11}$$

When the units of feature vectors in the same sample space of infrared images are different, the multi-level features of infrared images need to be normalized. See (12) for infrared image multi-level feature normalization processing method.

$$f_n = (f_o - \lambda)/(1 + \delta) \tag{12}$$

In the formula, δ and λ separately imply the variance and mean value of the infrared image feature vector; f_n and f_o respectively represent the infrared image feature vector that has completed the normalization process and the original infrared image feature vector.

The inner product of multi-level feature vectors of infrared images is defined as $(X, Y) = X^T Y$, both the feature vector X and the feature vector Y are $\in F$, F representing the complex vector of multi-level features of infrared images, T representing the conjugate transpose symbol. In the multi-level feature fusion space of infrared image, the distance expression of two feature vectors is shown in (13).

$$\|F_1 - F_2\| = \sqrt{(F_1 - F_2)^T (F_1 - F_2)} \tag{13}$$

It can be seen from Formula (13) that the distance between the two complex vectors in the multi-level feature fusion space is only related to the value of the real part and the value of the imaginary part in the infrared image feature space. The selection order of the real part and the imaginary part of the infrared image does not affect the distance between the complex vectors, which verifies that the infrared image multi-level feature fusion has high stability.

3 Multi Object Segmentation of Infrared Image Based on Mean Shift Algorithm

Mean shift algorithm is selected as the algorithm for multi object segmentation of infrared images. This algorithm is a nonparametric kernel density estimation algorithm. Weight coefficients and kernel functions are introduced to achieve multi object segmentation of infrared images using mean shift vectors in the multi-level feature fusion space of infrared images. Mean shift algorithm sets the sample mean calculation formula in the multi-level feature fusion space of infrared image, as shown in formula (14).

$$m_z(x) = \frac{\sum_{x_i \in S_z} K\left(\frac{x_i - x}{z}\right) z w(x_i) x_i}{\sum_{x_i \in S_z} K\left(\frac{x_i - x}{z}\right) w(x_i)} \tag{14}$$

Where, K and z purport the kernel function and mean shift bandwidth; $w(x_i)$ represents the weight coefficient.

Set the multi-level feature fusion space of infrared image as the traversal space of Mean shift algorithm, and perform mean shift processing on all feature points in the multi-level feature fusion space. All feature points of the infrared image have two-dimensional space coordinate feature information. In the multi-level feature fusion space, the mean shift bandwidth is used to implement the mean shift infrared image clustering segmentation, and the multi object segmentation results are converted from the feature space to the infrared image space to obtain the infrared image after the mean shift multi object segmentation. The expression of mean shift vector of infrared image is (15).

$$M_z(x) = m_z(x)z - x \tag{15}$$

When the mean shift algorithm is used to segment infrared image multi objects, it is necessary to mean shift all feature points in the multi-level feature fusion space to obtain pixels in the multi-dimensional sphere window that meet the feature bandwidth conditions. $m_z(x)$ is used to represent the sample mean of the pixel points in the multidimensional sphere obtained through screening, and the mean shift vector $M_z(x)$ is calculated using the sample mean. When the mean shift vector obtained is less than the set threshold, it means that the feature point has drifted to the point with the maximum probability density. The sample mean value obtained at this time is used to replace the value of the completed drift, and the feature point will complete the drift, and the next feature point will drift. When the mean shift vector of the multi-level feature fusion space of the infrared image is not less than the set threshold, assign the sample mean value to the feature point, continue the drift of the feature point until the point with the maximum probability density is obtained, and complete the drift of the feature point.

When mean shift is carried out by using multi-level feature fusion results, the feature points in the multi-level feature fusion space of infrared image are regarded as multidimensional vectors. The expression of adding multidimensional kernel function to mean shift vector is shown in (16).

$$K_{z_s,z_r}(x) = \frac{c}{z_f z_s^2 z_r^p} K\left(\left\|\frac{x^s z_r^p}{z_s z_f}\right\|^2\right) K\left(\left\|\frac{x^r}{z_r z_f}\right\|^2\right) \tag{16}$$

In the formula, x^s and x^r individually mean the distance between the points in the multi-dimensional sphere and the feature dimension s and r in the multi-level feature fusion space; z_s and z_r separately represent the spatial bandwidth and spectral bandwidth of mean shift; z_f indicates the characteristic bandwidth of the infrared image.

Multidimensional Gaussian kernel function is used as the kernel function of infrared image multi object segmentation, and formula (16) is transformed into the expression of formula (17).

$$K_{z_s,z_r}(x) = \frac{1}{\sqrt{2\pi}} e^{-\left\|\frac{x^s}{z_s} z_f^p\right\|} e^{-\left\|\frac{x^r}{z_r} z_f^p\right\|} \tag{17}$$

The expression of the multidimensional mean shift sample mean value of the infrared image multi object segmentation obtained according to formula (17) is as follows (18).

$$m_{z_s,z_r}(x) = \frac{\sum \frac{1}{\sqrt{2\pi}} e^{-\left\| \frac{x^s}{z_s} z_r^p \right\|^2} e^{-\left\| \frac{x^r}{z_r} z_r^p \right\|^2} w(x_i) x_i}{\sum \frac{1}{\sqrt{2\pi}} e^{-\left\| \frac{x^s}{z_s} \right\|^2} e^{-\left\| \frac{x^r}{z_r} \right\|^2} w(x_i)} \tag{18}$$

In the formula, $\| \ \|^2$ represents the similarity between the points in the high-dimensional sphere obtained through screening and the points in the multi-level feature fusion space in the feature dimensions.

The steps for multi object segmentation of infrared image using multi feature mean shift algorithm are as follows.

(1) The multi-level features of the extracted infrared image are combined into a multi-level feature fusion space;
(2) Set the spatial bandwidth and spectral bandwidth of mean shift, as well as the characteristic bandwidth of infrared image. Mean shift all feature points in the infrared image. Use the current feature point coordinates to preliminarily screen the index of spectral bandwidth points, calculate the spectral characteristics between the points in the multi-level feature fusion space of infrared images and the current feature points, as well as the distance from the third type of features, and obtain the feature vector screening results. The cubic window filtering method is used to obtain the feature points that can be applied to mean shift in the multi-dimensional sphere;
(3) Calculate the mean value of infrared image feature samples and the drift amount of infrared image feature samples in the mean shift. Set the error threshold. When the drift between the sample point and the sample point is less than the error threshold, set a new center point to replace the current point, set the center point feature value obtained through calculation in the corresponding position feature space, and obtain the multi-dimensional feature space of the infrared image after the mean shift. In the multi-dimensional feature space of the infrared image with mean shift, the regional growth method is selected to calculate the multi-level feature similarity and determine the growth point. The region growing points of the infrared image are merged, and the multi object segmentation results of the infrared image are obtained after the multi-level feature point traversal of the infrared image is completed.

4 Application of Infrared Image Multi Object Segmentation Method in Fault Diagnosis System

In order to verify the effectiveness of the multi object segmentation method of infrared image based on multi-level feature fusion, the method is applied to the fault diagnosis system of an electric power enterprise. The power enterprise uses infrared imaging system to collect many power equipment running in the power system, and the performance of multi object segmentation of infrared image has a great impact on the performance of power equipment fault diagnosis. As an important technology of power equipment fault diagnosis, the multi object segmentation result of infrared image determines the performance of power equipment fault diagnosis, which also has an important impact on the operation of power enterprises.

The infrared image of the current contactor of the power system is selected as the experimental object. The infrared image of the current contactor collected by the infrared imaging system has low definition, which improves the difficulty of multi object segmentation of the infrared image. The multi-level features of the infrared image of the current contactor are extracted by this method, and the multi-level features of the extracted infrared image are fused. The multi object segmentation of the infrared image of the current contactor is better by using the multi-level feature fusion result of the infrared image.

The infrared image of the current contactor of the power system on November 20, 2020 is chosen randomly as the experimental object. The original infrared image of the current contactor collected by the infrared imaging system is shown in Fig. 1. It can be seen from the infrared image of the original current contactor in Fig. 1 that the infrared image of the original current contactor has low definition, which improves the difficulty of multi object segmentation of the infrared image.

Fig. 1. Infrared image of original current contactor

The multi-level features of the infrared image of the current contactor are extracted by the method in this paper, and the multi-level features of the infrared image extracted are fused. The infrared image of the current contactor is segmented by using the multi-level feature fusion result of the infrared image. The result of multi object segmentation is shown in Fig. 2. It can be seen from the experimental results in Fig. 2 that this method can effectively segment multiple objects in the infrared image of the current contactor, and this method can effectively segment the four objects contained in the infrared image of the current contactor. The result of multi object segmentation of infrared image of current contactor in this method can be used as an important basis for fault diagnosis of power equipment and provide a good image basis for fault diagnosis of power system.

In order to further verify the multi object segmentation performance of this method for infrared images, this method is used to segment the infrared images of different power equipment in the power system. Statistics of the performance of infrared image multi object segmentation using this method are shown in Table 1. It can be seen from the experimental results in Table 1 that this method can effectively segment multiple objects in the infrared images of different power equipment. Using this method to segment infrared image multi objects, there are only a few wrong segmentation object pixels, and the multi object segmentation accuracy is higher than 99.5%. The experimental results show that the proposed method can effectively use the multi-level features contained in

Fig. 2. Multi object segmentation result of infrared image

the infrared image to achieve accurate segmentation of multiple objects in the infrared image.

Table 1. Infrared Image Multi object Segmentation Results

Name of electric equipment	Number of divided objects/piece	Pixels of segmentation object/piece	Actual object pixels/piece	Number of pixels wrongly segmented object/piece
Oil immersed transformer	2	1852	1857	5
Current transformer	3	1678	1680	2
Voltage transformer	2	1984	1988	4
Alternator	3	2052	2055	3
Arrester	4	948	952	4
Motor	3	1085	1090	5
Insulator	2	1385	1381	4
Current limiting reactor	2	1258	1261	3
Circuit breaker	3	1648	1655	7
Isolating switch	7	2354	2361	7
Fuse	5	1852	1858	6

Dice similarity coefficient is selected as the evaluation index to measure the performance of infrared image multi object segmentation. This index is an important index used in image segmentation evaluation and an important index for measuring the similarity between infrared image multi object segmentation results and original infrared image objects. See (19) for the calculation formula of Dice similarity coefficient.

$$Dice = \frac{2|p \cap q|}{|p| + |q|} \tag{19}$$

In Formula (19), p and q separately represent the pixel value of the infrared image segmentation object and the pixel value of the original infrared image object. The value of Dice similarity coefficient is in the [0,1] range. The higher the Dice similarity coefficient, the better the infrared image multi object segmentation performance. The infrared image multi object obtained by segmentation can completely retain the edge feature details of the object, and solve the defect that the infrared image cannot accurately segment the object due to its low definition.

5 Conclusion

Multi object infrared image segmentation is an important content of infrared image analysis and infrared image pattern recognition, and also an important research topic in the field of computer vision. The infrared image contrast feature and gradient feature are applied to infrared image multi object segmentation, which can effectively improve the performance of infrared image multi object segmentation.

The multi object segmentation method of infrared image based on multi-level feature fusion is studied. Multilevel features are extracted from infrared image, and the result of multi-level feature fusion is taken as the basis of infrared image multi object segmentation. The experimental results show that the multi object segmentation of infrared image using multi-level feature fusion results has high segmentation accuracy and efficiency. The proposed method can accurately extract multiple objects from the background area of infrared image, and the object information of the infrared image multi object segmentation result is high. The method studied has the advantage of simple calculation, and can be applied to infrared image segmentation occasions that require high segmentation efficiency and high segmentation accuracy.

References

1. Tereikovskyi, I., Hu, Z., Chernyshev, D., et al.: The method of semantic image segmentation using neural networks. Int. J. Image Graph. Signal Proc. **14**(6), 1–14 (2022)
2. Kurama, V., Alla, S., Vishnu, K.R.: Image semantic segmentation using deep learning. Int. J. Image Graph. Signal Process. **10**(12), 1–10 (2018)
3. Mei, F., Chen, S., Li, Y., et al.: Investigation of infrared image prediction for subsonic exhaust plume. Int. J. Eng. Manufac. **2**(3), 46–52 (2012)
4. Mei, F., Chen, S., Jiang, Y., Cai, J.: A preliminary model of infrared image generation for exhaust plume. Int. J. Image Graph. Signal Process. **3**(4), 46–52 (2011)
5. Zeng, Y., Xie, G., Zhang, J.: Weak light image segmentation based on intercept histogram and Otsu fusion. Adv. Laser Optoelectron. **58**(20), 219–227 (2021). (in Chinese)
6. Chegeni, M.H., Sharbatdar, M.K., Mahjoub, R., Raftari, M.: New supervised learning classifiers for structural damage diagnosis using time series features from a new feature extraction technique. Earthquake Eng. Eng. Vibrat. **21**(1), 169–191 (2022)
7. Ding, Y., Xing, Z., Jiang, Y., et al.: Adaptive light intensity segmentation method for narrowband image in spectral index measurement. J. Nanjing Agric. Univ. **43**(03), 574–581 (2020). (in Chinese)
8. Anh, N.T.N., Thang, T.N., Solanki, V.K.: Product feature extraction from the descriptions. SpringerBriefs in Applied Sciences and Technology pp. 43–51 (2022)

9. Long, Y.-H., Chen, Y.-C., Chen, X.-P., Shi, X.-H., Zhou, F.: Test-driven feature extraction of web components. J. Comput. Sci. Technol. (English) **37**(2), 389–404 (2022)
10. Zhang, L., Xu, X., Cao, C., et al.: Robot pose estimation method based on image and point cloud fusion of dynamic feature elimination. China Laser **49**(06), 58–69 (2022). (in Chinese)
11. Li, C., Lan, H., Wei, X.: Attention based millimeter wave lidar fusion target detection. Comput. Appl. **41**(07), 2137–2144 (2021). (in Chinese)
12. Jiang, B., Ma, Y., Wan, J., et al.: Local feature extraction of lidar point cloud based on hemispherical neighborhood. Signal Process. **38**(02), 329–339 (2022). (in Chinese)
13. Li, Y., Li, F., Qu, H., et al.: Infrared image segmentation method based on improved artificial bee colony sine cosine optimization. Laser Infrared **51**(08), 1076–1080 (2021). (in Chinese)
14. Yao, Y., Cheng, G., Xie, X., et al.: Target detection of optical remote sensing image based on multi-resolution feature fusion. J. Remote Sens. **25**(05), 1124–1137 (2021). (in Chinese)
15. Yang, X., Gao, L.: Hyperspectral image classification based on reduced dimension Gabor feature and decision fusion. Comput. Appl. Res. **37**(03), 928–931 (2020). (in Chinese)
16. Shen, H., Meng, Q., Liu, Y.: Facial expression recognition based on lightweight convolution network multilayer feature fusion. Adv. Laser Optoelectron. **58**(06), 148–155 (2021). (in Chinese)
17. Xi, Z., Yuan, K.: Image super-resolution reconstruction based on residual channel attention and multi-level feature fusion. Progress Laser Optoelectron. **57**(04), 262–270 (2020). (in Chinese)
18. Wu, J., Lu, Z., Guan, Y., et al.: SAR target recognition method based on feature fusion of 2D compressed sensing multi projection matrix. Infrared Laser Eng. **50**(06), 314–320 (2021). (in Chinese)
19. Wang, F., Sun, H., Wang, Y., et al.: Bilateral filtering enhancement and FCM segmentation of femtosecond laser ablation spot image. Laser Infrared **50**(05), 630–633 (2020). (in Chinese)
20. Zhang, Y., Zhang, B., Zhao, Y., et al.: Remote sensing image classification based on dual channel depth dense feature fusion. Laser Technol. **45**(01), 73–79 (2021). (in Chinese)
21. Yu, X., Shan, D., Yu, Y., et al.: Hyperspectral image hybrid convolution classification based on multi feature fusion. Infrared Technol. **44**(01), 9–20 (2022). (in Chinese)
22. Ma, L., Gou, Y., Lei, T., et al.: Small target detection in remote sensing image based on multi-scale feature fusion. Optoelectron. Eng. **49**(04), 49–65 (2022). (in Chinese)
23. Liu, J., Zheng, C., Zhang, L., et al.: Hyperspectral image classification method based on image reconstruction feature fusion. China Laser **48**(09), 203–212 (2021). (in Chinese)
24. Zhang, Y., Wang, X., Hou, D., et al.: Image segmentation of inductive laser thermal imaging based on watershed algorithm. Infrared Technol. **43**(04), 367–371 (2021). (in Chinese)
25. Xu, X., Xue, D., Qi, G.: Laser thermal imaging image segmentation based on regional differences. Laser J. **43**(10), 83–86 (2022). (in Chinese)
26. Xu, W., Jin, G., Miao, Z., Yang, J.: Analysis of on-line transformer fault detection technology based on infrared image. Bonding **49**(09), 193–196 (2022). (in Chinese)
27. Lu, Y.: Thermal defect detection method for buildings based on infrared image features. Dalian University of Technology, Dalian (2022). (in Chinese)
28. Islam, S.M.M., Debnath, R.: A Comparative Evaluation of feature extraction and similarity measurement methods for content-based image retrieval. Int. J. Image Graph. Signal Process. **12**(6), 19–32
29. Yasaswini, V., Baskaran, S.: An optimization of feature selection for classification using modified bat algorithm. Int. J. Inform. Technol. Comput. Sci. **13**(4), 38–46 (2021)

Spear-Man Correlation Data Analysis of Scale of Intelligent Logistics and Investment and Financing of Logistics Technology

Ni Cheng[1]($\boxtimes$) and Meng Lei[2]

[1] Wuhan Railway Vocational College of Technology, Wuhan 430205, Hubei, China
342603058@qq.com
[2] Hubei Changjiang Media and Press Group, Wuhan 430079, Hubei, China

Abstract. In recent years, with the development of technologies such as the Internet and artificial intelligence, as well as higher requirements of new retail and intelligent manufacturing, the market scale of intelligent logistics in our country continues to expand. At the same time, the development prospect of intelligent logistics has been favored by social capital of all parties, and the amount of invest ment and financing in China's logistics technology is on the rise. Whether there is a correlation between the scale of intelligent logistics and the investment and financing of logistics technology is very important for promoting logistics to enter the smart age and seeking the opportunity of listing in the logistics industry in the future. To solve this problem, This paper firstly collects and collates the data of the scale of intelligent logistics and the amount of investment and financing in logistics technology, and then uses spear man data analysis to explore the correlation between them, finally proposes to promote the diversified development of various subdivided fields of logistics technology, closely grasp the three mainstream fields of investment and financing, and combine the Internet, big data and artificial intelligence to promote the digitized process of logistics.

Keywords: Spear-man · Data Analysis · Intelligent logistic · Investment and Financing of Logistics Technology · Artificial Intelligence

1 Introduction

Intelligent logistics was first proposed by IBM. In December 2009, the Information Center of China Logistics Technology Association, Hua Xia Internet of Things and editorial Department of Logistics Technology and Application jointly proposed the concept. It refers to the modern logistics mode that realizes the refined, dynamic, and visual management of each link of logistics, improves the intelligent analysis, decision making and automatic operation and execution ability of logistics system, and improves the efficiency of logistics operation through intelligent technology means such as intelligent hardware and software, Internet of Things, big data and so on [1].

According to the data of China Federation of Logistics and Purchasing, logistics data, logistics cloud and logistics equipment are the three major areas that logistics enterprises

Z. Hu et al. (Eds.): ICCSEEA 2023, LNDECT 181, pp. 37–46, 2023.
https://doi.org/10.1007/978-3-031-36118-0_4

demand for intelligent logistics at present. From 2015 to 2019, the scale of intelligent logistics in China maintained a double-digit growth rate. In 2019, the market size has reached 500 billion yuan, with a year-on-year growth of 23.1%. Even under the impact of the global epidemic, the scale of intelligent logistics in China reached 584 billion yuan in 2020 and 647.7 billion yuan in 2021, and the market size is expected to exceed one trillion yuan by 2025. In recent years, with the development of technologies such as the Internet of Things and artificial intelligence, as well as higher requirements for logistics in fields such as new retail and intelligent manufacturing, rapidly growing demand is forcing the traditional logistics industry to transform to smart logistics [2]. As a result, the market size of intelligent logistics will continue to expand. On the other hand, the development prospect of intelligent logistics has been favored by social capital of all parties, and the amount of investment and financing in China's logistics technology is on the rise. In 2018, due to the development of logistics technology towards the direction of refinement, the industry is becoming more mature, and the investment and financing competition of various circuits is becoming increasingly fierce. In this context, although the amount of investment and financing steadily rises, the number of investment and financing cases has decreased significantly. On the other hand, whether the increase of the amount of investment and financing of logistics technology can affect the scale of domestic intelligent logistics is also a question worth discussing. Therefore, it is of great practical significance to clarify the relationship between the scale of intelligent logistics and the investment and financing of logistics technology.

2 Literature Review

The research on intelligent logistics at home and abroad mainly focuses on: Professor Wang Zhitai believes that the new generation of information technology and modern management theory are deeply integrated with the Internet and applied to the logistics industry, and finally realize an innovative form of logistics, namely intelligent logistics [3]. The research on the definition and development of intelligent logistics is relatively mature, which is related to enterprise management and talent training. For example, Qi Jianxiu pointed out that the use of automation technology to replace manual operation can improve the comprehensive efficiency of logistics enterprises, and realize the reasonable scheduling and distribution of resources through the effective integration of social resources [4]. Yang Hongyue analyzed the influence on the training of logistics financial talents under the background of intelligent logistics, and proposed the training ideas of logistics financial talents under the "1 + X" certificate system [5]. With the development of informatization, networking and intelligence of smart logistics technology, research on smart logistics mainly focuses on big data analysis, AI technology, blockchain technology, Internet of technology and specific case studies. Dong Haifang uses big data technology to analyze logistics data, thereby improving the efficiency of logistics management [6]. 5G can meet the massive data exchange between management equipment, products, vehicles, and workers, and substantially support the development of smart logistics [7]. Bi Ying proposed the future development trend of logistics industry from the perspective of "intelligent logistics + AI technology" [8]. Based on the characteristics of blockchain technology, Zhou Xinming et al. analyzed its role in the development of intelligent logistics platform [9]. Feng Yayun pointed out that the application of Internet of Things technology changes the traditional management mode and promotes the development of intelligent logistics field to the direction of intelligence and information [10]; Hao Shuchi took the green smart logistics in the Guangdong-Hong Kong Greater Bay Area as the research object to build the evaluation index system of green smart logistics distribution capability [11].

To sum up, the research on intelligent logistics has attracted the full attention and attention of scholars at home and abroad. For example, measuring systems are used to evaluate the costs associated with each logistics process. There are positive implications for the sustainable development of logistics [12]. Most of the research focuses on specific logistics technology fields and case studies, but lacks the overall grasp of logistics technology investment and financing.

3 Research Methods

In this paper, the Spearman Correlation Coefficient is used to measure the dependence correlation research between the two variables of Scale of Intelligent Logistics (Y)and Investment and Financing of Logistics Technology (X).

The Spearman's Rank Coefficient of Correlation, or Spearman Correlation Coefficient, is a nonparametric measure of rank correlation. Named after the British statistician Charles Spearman, it is often written as the Greek letter 'ρ' (rho).

Full version of the calculation formula:

$$\rho = \frac{\frac{1}{n} \sum_{i=1}^{n} \left(\left(R(x_i) - \overline{R(x)}\right) \cdot \left(R(y_i) - \overline{R(y)}\right) \right)}{\sqrt{\left(\frac{1}{n} \sum_{i=1}^{n} \left(R(x_i) - \overline{R(x)}\right)^2\right) \cdot \left(\left(\frac{1}{n} \sum_{i=1}^{n} \left(R(y_i) - \overline{R(y)}\right)^2\right)\right)}} \tag{1}$$

Thereinto:

- R(x) and R(y) are the positions of x and y, respectively
- $\overline{R(x)}$ and $\overline{R(y)}$ indicate average ranks, respectively

A simpler and easier calculation formula is as follows:

$$\rho = 1 - \frac{6 \sum_{i=1}^{N} d_i^2}{N(N^2 - 1)} \tag{2}$$

Where:

d_i Represents the difference in the bit value of the i-th data pair.

n Total number of observed samples.

If Y tends to increase when X increases, the Spearman Correlation Coefficient is positive. If Y tends to decrease as X increases, the Spearman Correlation Coefficient is negative. The Spearman Correlation Coefficient is zero, indicating that there is no tendency for Y to increase as X increases.

3.1 Data Collection

Based on the Spearman Correlation Coefficient, this paper selects indicators of the Scale of China's Intelligent Logistics (Scale of CIL), the Total Annual Revenue of Logistics Industry (TAR of LI), the Amount of Investment and Financing in China's Logistics Technology (Amount of I&F), and the Investment and Financing Events in China's logistics technology as indicators (I&F Events). In order to ensure the scientific, authoritative, and objective evaluation results, the statistical data in this paper are from 2015–2021 China Federation of Logistics and Purchasing, National Bureau of Statistics, Statistical bulletins of National Economic and Social development over the years and related websites of China Logistics and Purchasing Website. Some index data are obtained through calculation.

Table 1. CIL from 2015 to 2021 (RMB 100 million), TAR of LI (trillion), Amount of I&F (RMB 100 million), I&F Events (number)

Year	CIL (RMB 100 million)	TAR of LI (one trillion yuan)	Amount of I&F (RMB 100 million)	I&F Events (number)
2015	2205	7.6	355.1	16
2016	2790	7.9	760.9	143
2017	3375	8.8	1770.3	117
2018	4060	10.1	1345.8	106
2019	4885	10.3	606.4	137
2020	5840	10.5	413.2	90
2021	6477	11.9	1815.7	113

Table 1 shows that from 2015 to 2021, the Scale of China's Intelligent Logistics (Scale of CIL) and the Total Annual Revenue of Logistics Industry (TAR of LI) had an increasing trend year by year. In 2021, the Scale of China's Intelligent Logistics (Scale of CIL) exceeded 647.7 billion yuan, hitting a record high. From 2015 to 2019, the Amount of Investment and Financing in China's Logistics Technology (Amount of I&F) has been on the rise.

As shown in Fig. 1, in 2018, due to the development of Logistics Technology in the direction of refinement and the increasingly mature industry, the number of major financing events in the logistics subdivision industry, intelligent logistics continued to stay popular, freight O2O and warehouse logistics gradually received capital attention, and intelligent express cabinets, cold chain logistics, cross-border logistics and logistics drones were favored.

Whether there is any correlation between the Scale of China's Intelligent Logistics (Scale of CIL) and Investment and Financing in China's Logistics Technology (I&F in CLT) is very important to promote logistics into the age of smart, shape the new pattern of Logistics Technology, and seek the opportunity of future listing in the Logistics Industry.

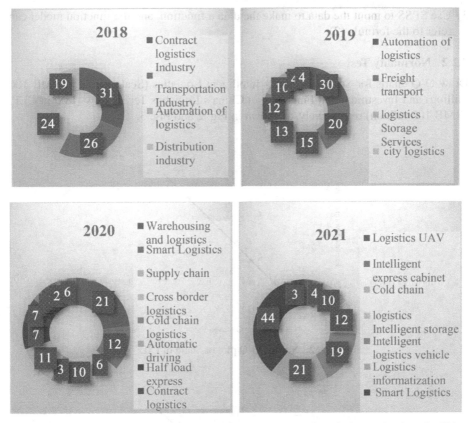

Fig. 1. Investment and Financing situation of each subsector of Logistics Technology in China from 2015 to 2021

3.2 SPSS Determination of Applicable Conditions

3.2.1 Solving Model

References [13–15] use Pearson's Correlation to make statistical measures of the strength of the linear relationship between two random variables, giving only the correlation of X and Y described by the linear equation. Spearman Correlation analysis is used when examining the strength of the monotonic relationship between two variables.

References [16–22] Through Spearman Correlation Analysis, we explore the extent to which the two variables tend to become larger or smaller in step, to accurately obtain the correlation between the sampling probability distributions of X and Y.

Main steps are listed:

1) Normality test: The variable contains a hierarchical variable, or the variable does not follow a normal distribution or the distribution type is unknown. This is the first condition that satisfies Spearman Correlation Analysis.

2) Monotonicity judgment: There is a monotonic relationship between two variables. This is the second condition that satisfies Spearman Correlation Analysis.

3) Use SPSS to input the data to make the data a function, and the function model can refer to the formula (2)

3.2.2 Normality Test

Draw Q-Q charts for Scale of China's Intelligent Logistics (Scale of CIL) (RMB 100 million) and Investment and Financing in China's Logistics Technology (I&F in CLT) (RMB 100 million) respectively, as shown in Fig. 2 and Fig. 3.

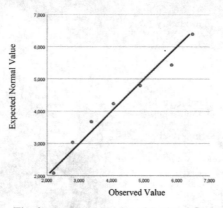

Fig. 2. Normal Q-Q plot of Scale of CIL

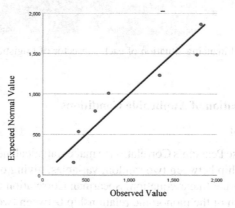

Fig. 3. Normal Q-Q plot of I & F in CLT

The comparison shows that the scattered points of variables deviate from the diagonal distribution more, indicating that the two variables did not obey the normal distribution. Apply spearman conditions.

3.2.3 Monotonically Judgment

Monotonic judgment is a meaningful analysis.

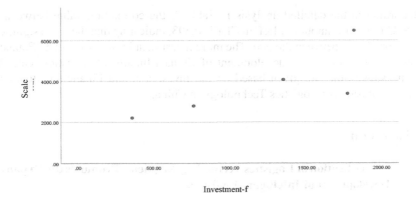

Fig. 4. Scatter plot of Scale by Investment-f

Draw a scatter plot for Scale of China's Intelligent Logistics (Scale of CIL) (RMB 100 million) and t Investment and Financing in China's Logistics Technology (I&F in CLT) (RMB 100 million) (due to the impact of the epidemic, the data in 2019 and 2020 are especially excluded), and the results are shown in Fig. 4.

As shown in the Fig. 4, the Scale of CIL (RMB 100 million) and the I & F in CLT (RMB 100 million) increase with the increase, and the two variables show a monotonous relationship. And the relationship between them is positive. In conclusion, the data in this case can be analyzed by Spearman Correlation.

3.2.4   SPSS Calculation of Spearman Correlation

Set the Scale of CIL (RMB 100 million) as the analysis item Y, and set the I&F in CLT (RMB 100 million) as X. Use SPSS27.0 statistical software to calculate:

Table 2. Correlations Coefficient

			Scale	Investment_f
Spearman's rho	Scale	Correlation Coefficient	1.000	.393
		Sig. (2-tailed)	–	.383
		N	7	7
	Investment_f	Correlation Coefficient	.393	1.000
		Sig. (2-tailed)	.383	–
		N	7	7

Correlation Analysis was used to study the correlation between the Scale of China's Intelligent Logistics (Scale of CIL) and Investment and Financing in China's Logistics Technology (I&F in CLT), and Spearman's Correlation Coefficient was used to indicate the strength of the correlation.

According to the detailed analysis in Table 2, the correlation value between the Scale of CIL and the amount of I&F in CLT is 0.393, indicating that there is a significant positive correlation between the two. The more investment in Investment and Financing will promote the continuous development of China's Intelligent Logistics scale. The following suggestions are given based on the Investment and Financing situation of various subdivisions of Logistics Technology in China.

4   Discussion

4.1   The Diversification of Logistics Technology Segments Promotes the Expansion and Development of Intelligent Logistics

According to Fig. 1, the investment and financing of logistics technology in 2018 are mainly concentrated in the contract logistics industry, transportation industry, logistics automation and distribution. With the continuous deepening and expansion of China's intelligent logistics, the subdivided fields of logistics technology have increased to 8 fields in 2019 and 13 fields in 2020. The specific contents include supply chain, cross-border logistics, cold chain logistics, automatic driving, air trunk, intra-city distribution, etc. Diversified development will become an important factor affecting the scale of intelligent logistics.

4.2   Investment and Financing in the Three Major Areas have Become the Mainstream, Promoting the Application and Development of the Intelligent Logistics Industry

In 2019, 2020 and 2021, warehousing, transportation and distribution have become the major financing and investment areas of Logistics Technology. The main reasons are as follows: first, the traditional labor force cannot fully meet the warehouse side's demand for efficiency improvement and refined management; Secondly, in recent years, transportation costs account for more than 50% of the total logistics costs. Reducing transportation costs is an important means to reduce costs in the Logistics Industry. However, logistics costs, which are mainly labor costs and fuel costs, can be compressed. Third, in the terminal distribution, distribution needs are diverse, distribution time conflict, low efficiency and high cost and other problems are prominent. Therefore, investment and financing events in the three areas are highly concerned by the capital.

4.3   Internet of Things, Big Data, and Artificial Intelligence to Promote the Intelligent Logistics

The Internet of Things technology connects the virtual network world with real objects, providing part of the data source for Big Data, and Artificial Intelligence technology is the deep learning technology under Big Data. Transportation management platforms, Fintech companies, instant logistics companies, and e-commerce platforms have applied digital technologies in their daily operations and management. With the continuous progress of science and technology and the support of macro policies and economy,

Artificial Intelligence will replace human labor and brain power in the warehousing link. In the transportation sector, the application of unmanned driving technology can reduce labor costs and achieve complete autonomous driving. In the delivery sector, drones will become a part of smart delivery in vast rural areas and urban communities.

5 Conclusion

With the development of logistics technology in the direction of refinement, the number of major financing events in the logistics segmentation industry and Intelligent Logistics continue to stay hot. Warehousing, transportation, and distribution have become the main areas of Financing and Investment in Logistics Technology. For example, the truck formation technology in the transportation link, automatic sorting technology in the storage link, AGV handling technology, intelligent express cabinet in the distribution link, unmanned distribution technology and so on have gradually attracted the attention of the capital. Logistics Technology helps logistics operations to realize intelligent and efficient operation, and realize the efficient flow of goods from the production place to the consumption place. Therefore, the Investment and Financing of Logistics Technology will shape the new pattern of Logistics Technology and promote the expansion of Intelligent Logistics.

References

1. Qian, H.-M., He, J., Guan, J.: Research on the coupling effect evaluation of Intelligent + Sharing Logistics. China Bus. Market (11), 3–16 (2019). (in Chinese)
2. Wang, T.: Internet of things and big data in smart logistics. In: Proceedings of the 2022 2nd International Conference on Education, Information Management and Service Science (EIMSS 2022), pp. 1265–1271 (2022)
3. Wang, Z.: Smart Logistics Is needed by urbanization. China Bus. Market 3, 4–8 (2014). (in Chinese)
4. Qi, J.: Research on the development status and improvement countermeasures of intelligent logistics integration. Transport. Manager World 20, 50–52 (2022). (in Chinese)
5. Yang, H., Ding, C.: Analysis of problems and countermeasures faced by logistics financial talent training in the context of smart logistics. Logist. Technol. 45(11), 164–166 (2022). (in Chinese)
6. Dong, H.: Discussion on the application mode of smart logistics based on big data. China Logist. Procurem. 20, 72–73 (2022). (in Chinese)
7. Alexandra, L., Chiara, C., Roberto, P., Sergio, C.: 5G in Logistics 4.0: potential applications and challenges. Procedia Comput. Sci. (217), 650–659 (2023)
8. Bi, Y.: Smart logistics + AI technology drives the transformation of the logistics industry. China's Storage Transport. 08, 142–143 (2020). (in Chinese)
9. Zhou, X., Hu, Y.: Research on the development of intelligent logistics platform based on blockchain technology. Logistics Eng. Manage. 24–26 (2022)
10. Ya, Y.: Research on the application of internet of things in smart logistics management. China Logist. Procurem. (06), 93–94 (2022). (in Chinese)
11. Hao, S.: Research on the development of green and smart logistics distribution in the Guangdong-Hong Kong-Macao greater bay area. Logist. Technol. 45(15), 37–41 (2022). (in Chinese)

12. Benotmane, Z., Belalem, G., Neki, A.: A cost measurement system of logistics process. Int. J. Inform. Eng. Electron. Bus. **10**(5), 23–29 (2018)
13. Strelnytskyi, O.O., Svyd, I.V., Obod, I.I., Maltsev, O.S., Zavolodko, G.E.: Optimization of secondary surveillance radar data processing. Int. J. Intell. Syst. Appl. **11**(5), 1–8 (2019)
14. Jagadesh, B.N., Srinivasa Rao, K., Satyanarayana, C.: A robust skin colour segmentation using bivariate pearson type ii$\alpha\alpha$ (bivariate beta) mixture model. Int. J. Image Graph. Signal Process. **4**(11), 1–8 (2012)
15. Lemmer, H.H.: The allocation of weights in the calculation of batting and bowling performance measures. South African J. Res. Sport Phys. Educ. Recreat. **29**(2), 75–85 (2007)
16. Dey, P.K., Ghosh, D.N., Mondal, A.C.: PL team performance analysis: a multi-criteria group decision approach in fuzzy environment. Int. J. Inform. Technol. Comput. Sci. **7**(8), 8–15 (2015)
17. Joanna, R., Andrzej, N., Piotr, H., Ewa, D.: Application of Spearman's method for the analysis of the causes of long-term post-failure downtime of city buses. Appl. Sci. **12**(6), 2921 (2022)
18. Jie, T.M.R., He, K.K.: Ultrasound measurements of rectus femoris and locomotor outcomes in patients with spinal cord injury. Life **12**(7), 1073–1073 (2022)
19. Irina, K., Asta, P., Ginas, C., Aldona, A., Berita, S., Ramunas, K.: The Impact of achievements in Mathematics on cognitive ability in primary school. Brain Sci. **12**(6), 736–736 (2022)
20. Nurul, F.A., Immanuel, M., Nur, H.: The correlation between sedentary lifestyle and physical fitness level in adolescents. Enfermería Clínica **31**(5), S668–S671 (2021)
21. Du, H., Sun, L., Chen, S., et al.: Noninvasive prenatal prediction of fetal haplotype with Spearman rank correlation analysis model. Mol. Genet. Genomic Med. **10**(8), 1988 (2022)
22. Andréas, H., Alfonso, V.: The Kendall and Spearman rank correlations of the bivariate skew normal distribution. Scand. J. Stat. **49**(4), 1669–1698 (2022)

Neural Network Model for Laboratory Stand Control System Controller with Parallel Mechanisms

Peter Kravets[1], Anatolii Novatskyi[1], Volodymyr Shymkovych[1](✉),
Antonina Rudakova[1], Yurii Lebedenko[2], and Hanna Rudakova[2]

[1] National Technical University of Ukraine «Igor Sikorsky Kyiv Polytechnic Institute», 37
Prosp. Peremohy, Kyiv, Ukraine
shymkovych.volodymyr@gmail.com
[2] Kherson National Technical University, 24 Beryslav Highway, Kherson, Kherson Region,
Ukraine

Abstract. The paper considers a frame installation with mechanisms of parallel
structure. In it the movement of the tool is carried out in the vertical plane. The
movement of the working body is set by the upper control level, each motor
is programmed through the local control system, and the sensors monitor their
position. To improve the control system, the integration of an automated system
using neural networks is proposed. A mathematical model of a frame installation
with mechanisms of parallel structure is constructed. The neural network model
of this object is constructed with the help of specialized software. This model has
the optimal structure for the error of reproduction of the object and the minimum
structure. Shows that the smallest error, equal to 0.001236, has a structure with 11
neural networks in the hidden layer, with 1 delayed input and 2 delayed outputs
and is the best structure for the selected system. The implementation of neural
network control system on hardware the Field-Programmable Gate Array (FPGA)
is proposed. This implementation will allow the neural network controller to adapt
to undefined system parameters in real time. That made it possible to increase the
efficiency of the control of this object. Due to the consideration by the neural
network of those parameters that cannot be described mathematically and taken
into account at the system design stage.

Keywords: Frame installation · manipulator · neural network · parallel
structure · control system · FPGA

1 Introduction

There are basically two types of manipulators: serial and parallel [1–4]. Serial manipulators are open structures consisting of several links connected in series [1, 2]. Such a
manipulator can be efficiently operated throughout the working space. The actuator must
basically work and shift the entire manipulator with its links and actuators. Implementing
fast and accurate movements with such manipulators is a difficult task. When solving

this problem, there are problems of poor rigidity and reduced accuracy. In contrast to serial manipulators, parallel manipulators consist of many circuits with a closed loop [3, 4]. These chains control the end effector together in a parallel structure. They can take a variety of forms. The most common form of parallel manipulators are platform manipulators. Platform manipulators have an architecture similar to the architecture of flight simulators. In such a system, we can distinguish two special links – the base and the moving platform [5, 6]. Parallel manipulators have better positioning accuracy, higher rigidity and higher load capacity. Because in parallel manipulators the total load on the system is distributed between the actuators.

The most important advantage of parallel manipulators is the ability to keep all their actuators fixed on the basis of [7, 8]. The moving mass in them can be much higher. This type of manipulator can perform fast movements. But their working spaces are much smaller, which limits the full operation of these preferred features.

Perfect control of drives is necessary for fast and exact movements of parallel manipulators. To minimize tracking errors, dynamic forces must be compensated by the controller. For accurate compensation it is necessary to know exactly the parameters of the dynamic model of the manipulator [8, 9]. Closed mechanical circuits make the dynamics of parallel manipulators extremely complex, and their dynamic models nonlinear. When constructing a control system for a parallel manipulator, informational uncertainty in the parameters of the object itself should be taken into account [10, 11]. There are parameters that can be determined, such as mass. And there are a number of parameters that cannot be accurately determined such as strength factors. Because of this, many control methods are ineffective. Therefore, it is necessary to determine the unknown parameters of the system during operation and adapt the control system to them.

The application of artificial neural network technology in control systems allows to take into account the uncertainties in the system [12–14, 26, 27]. Neural network control systems are a high-tech direction of control theory and belong to the class of nonlinear dynamic systems. High speed due to parallelization of input information combined with the ability to learn artificial neural networks makes this technology very attractive for creating control devices in automatic systems. Regulators built on the basis of artificial neural networks can be learned in the process and take into account all the uncertainties in the design of the system. Parallel frame installation is a multidimensional control object. Neural networks are a universal tool for modeling multidimensional nonlinear objects and finding solutions to incorrect problems. Using the reinforced learning paradigm, you can teach the neural network to deviate a given position of the platform from the current one.

By implementing the entire system on FPGA [15–19], can get an accurate and high-speed control system for the platform of the parallel manipulator. Therefore, the development of a control system for manipulators based on artificial neural networks will increase the efficiency of the parallel manipulator.

The aim of the work is to develop neural network model and controller of the control system of the frame installation with mechanisms of parallel structure, which allows to install the platform in a given position.

By implementing a neural network controller capable of learning in real time into the control system of this object, we want to achieve an increase in control accuracy compared to traditional methods. Due to additional training of the neural network during the operation of the stand, which will allow to take into account those parameters of the object that cannot be taken into account during its mathematical modeling.

2 Design of Frame Installation with Mechanisms of Parallel Structure

An example of such a frame installation of a parallel structure is a research stand with two guide rods, developed and constructed at the Kherson National Technical University, which is shown in Fig. 1.

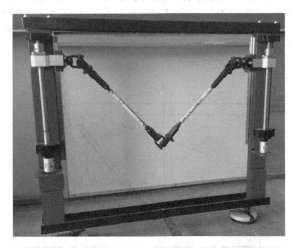

Fig. 1. General view of the installation of the frame structure

The installation consists of a metal frame, which is equipped with two stepper motors, drives, hinges and a working body. A spindle, manipulator and extruder are used as a working body.

The movement of the working body is set by the upper level of control, which may consist of a personal computer. Each motor is programmed separately via the local control system. The sensors located on the drives monitor their position and send data to the upper control level.

The frame layout equipment uses stepper motors that convert electrical pulses of control signals into angular movements of the rotor (discrete mechanical movements) with its fixation in a given position. Each stepper motor of the installation has the ability to perform precise positioning and speed control. This is well suited for a system that operates at low acceleration and with a relatively constant load.

All these systems do not allow detecting information uncertainties in the operation of the equipment. This can lead to accidents. It is advisable to improve the control system of a multi-drive frame layout by integrating an automated control system and functional diagnostics based on FPGA hardware using neural network technology [20–24].

3 Mathematical Model of a Laboratory Stand

The movement of the platform with the working tool is caused by movement of separate knots along directing rods. The vertical movement along the axes S1 and S2 is provided by the operation of individual motors. The kinematic system is considered when the movement of the tool is carried out in the vertical plane. The kinematic scheme of a symmetrical frame construction with two guide rods is shown in Fig. 2.

The geometric dimensions of the installation elements are set at the stage of design and manufacture. l_{c1}, $l_{c2} = l$ – length of beam elements. P – the platform on which the working tool is installed. The distance between the supports is known $d/2 = -x_{01} = x_{02}$, as well as the initial provisions $S_{01} = 0$ and $S_{02} = 0$ carriage k_1 i k_2.

The position of the work site, ie the coordinates of its center (x_c, z_c) and the angle of deviation of the normal from the vertical axis ϕ can be determined from the analytical description of the connection of the coordinates of the structure, which is based on the use of the dependences presented below [4, 5].

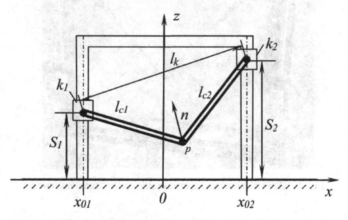

Fig. 2. Kinematic scheme of frame installation

1. Coordinates (xk_i, zk_i) carriage locations k_i can be determined from the ratios:

$$xk_i = S_i \cdot \cos \alpha + x_{0i}, \quad zk_i = S_i \cdot \sin \alpha + z_{0i}, \quad i = 1, 2 \tag{1}$$

2. Coordinates (xs_i, zs_i) hinge locations sh_i can be found from the equations of their possible movement in a circle:

$$(xs_i - xk_i)^2 + (zs_i - zk_i)^2 = l_{ci}^2, \quad i = 1, 2 \tag{2}$$

3. Equation of rigid connection of hinges due to the presence of a work platform p:

$$(xs_1 - xs_2)^2 + (zs_1 - zs_2)^2 = l_p^2$$

4. To form a complete system of equations relative to the unknown, you can use the relationship between the angles formed by the sides of the quadrilateral.

The maximum possibilities of moving the working body can be estimated on the basis of the solution of the direct problem of kinematics. To find the ranges of change of operating parameters of individual drives, limiting speeds and accelerations, it is necessary to solve the inverse problem of kinematics.

A stepper motor is an electric motor, the main feature of which is that its shaft rotates by performing steps moving a fixed number of degrees. This characteristic is obtained due to the internal structure of the engine. It allows you to know the exact angular position of the shaft by counting the steps performed without the need to use a sensor. This feature makes it suitable for a wide range of applications. By switching on one or more phases of the stator, under the action of current flowing in the coil, a magnetic field is generated, and the rotor is aligned with this field. By feeding the different phases in series, the rotor can be rotated by a certain amount to achieve the desired end position. This is the basic principle of operation of a stepper motor.

A two-phase hybrid stepper motor is a low-speed hydromagnetic synchronous motor. Due to its stable operation, it can achieve high-precision closed-loop positioning control and input pulse sequences to achieve digital control. Therefore, it is widely used in engineering fields such as robot motion control, antenna scanning and aircraft location control. In high-precision control systems, two-phase hybrid stepper motors are often used as actuators. For example, when establishing satellite-to-satellite communications, custom satellite relay antennas must track targets in real time, and two-phase hybrid stepper motors are often used as antenna drives. The drive of the tracking control system causes the antenna load to indicate the target of the tracking.

The two-phase hybrid stepped motor system has a high degree of nonlinearity, which does not contribute to the analysis and design of the engineering control system, so it simplifies the task of obtaining an approximate model of the system and its transfer function.

The block diagram of the servo control system with a closed cycle of the two-phase hybrid stepper motor is shown in Fig. 3 [25]. The inner circuit of a closed circuit is a current circuit. The purpose of this circuit is to implement tracking of the winding current of the hybrid stepper motor to a given current, so that the hybrid stepper motor can smoothly output torque under the microstep drive. The outer loop is a position loop, the goal is to track the output load at a given position.

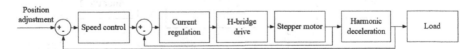

Fig. 3. Block diagram of the closed-loop control system of a hybrid stepper motor

The mathematical model of a two-phase hybrid stepper motor is the basis for the analysis and study of the control system of a two-phase hybrid stepper motor. Because a hybrid stepper motor is a type of highly nonlinear electromechanical device, there are many difficulties in accurately describing and accurately defining nonlinear parameters.

The flux relationship created by the approximate flux of the permanent magnet in the phase winding varies with the position of the rotor according to the sinusoidal law. The influence of the stator current on it is not taken into account, the influence of hysteresis and eddy currents is not taken into account, but only the average permeability of the air gap and the main component, ignoring the mutual inductance between two-phase windings.

The transfer function of each stepper motor component is described in the block diagram of the transfer function of the system and is shown in Fig. 4.

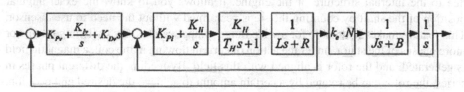

Fig. 4. Block diagram of the transfer function of the stepper motor system

4 Block Diagram of a Neural Network Controller

To solve the problem of designing a neural network controller, a block diagram based on FPGA is developed, the prototypes of this controller are the tools developed in [15, 16, 22, 28, 29]. In Fig. 5 present a generalized block diagram of a neurocontroller, where as a computing core is used a neurocomputer based on FPGA, which implements an artificial neural network with an algorithm for its training.

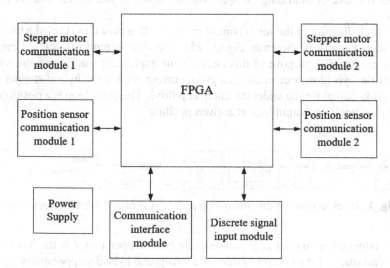

Fig. 5. Block diagram of a neurocontroller for controlling a frame installation with mechanisms of parallel structure

The neurocontroller for controlling a frame installation with mechanisms of parallel structure includes a computing core, means for input of analog and digital information, means for output of analog and digital information, means of data exchange with other computing devices, elements of visualization and operational control. The neurocontroller includes: FPGA – programmable logic integrated circuit; discrete signal input module; discrete signal output module; communication module with control interface; communication modules with position sensors 1 and 2; communication modules with stepper motors 1 and 2; generator.

FPGA is a central element of this structure, which implements neural network control algorithms, including the network learning algorithm in the "online" mode.

The computing core represented by the FPGA is assigned the role of a "fast" fragment calculator of the general algorithm of the controller associated with neurocomputations; complement the FPGA program on an equal footing, or be a purely technical element of the neural network controller that provides FPGA modes. However, in all cases, the FPGA is designed to implement the neural network elements of the control system, so in the future the main attention is paid to the computer on the FPGA [15, 16].

To model the neural network in MATLAB, system matrices in the form of state spaces are defined. To do this, the transfer function is converted

$$G(s) = \frac{16.47s^3 + 1648s^2 + 69.81s + 1}{0.00005845s^5 + 0.6525s^4 + 80.94s^3 + 1648s^2 + 69.81s + 1}$$

to the control object and the necessary coefficients are obtained using Workspace in MATLAB.

5 Construction of a Neural Network System

To synthesize the structure of the neural network model of a frame installation with a parallel structure, it is necessary to determine the "external" and "internal" structure of the neural network. The "external" structure of the neural network model is completely determined by the number of inputs and outputs of the object. When choosing the "internal" structure of the neural network model, it is necessary to determine:

– the number of hidden layers;
– the number of neurons in each hidden layer;
– type of activation function for each layer.

To solve the problem of choosing a neural network of the minimum structure from the set of possible, [24] describes the technology and software package "MIMO-Plant" for research and evaluation of neural network models of multidimensional control objects for further implementation on FPGA. Their essence is to model all possible variants of "internal" structures within the "external" structure and determine the structure that provides the least standard error, or a neural network of minimal structure whose standard error satisfies the condition of the problem (Fig. 6).

Random values in the range from 0 to 1 are fed to the input of the system. The block "RMS" is a requirements management system that calculates the total standard error,

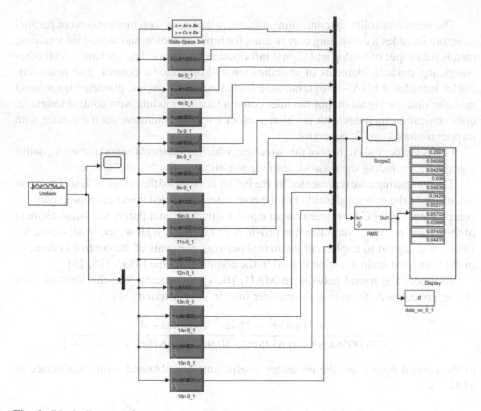

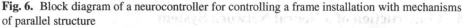

Fig. 6. Block diagram of a neurocontroller for controlling a frame installation with mechanisms of parallel structure

which is mathematically described by the formula:

$$\varepsilon = \sqrt{\frac{\sum_{i=1}^{z}(y'-y)^2}{z}},$$

where z is the number of error values (choose equal to the number of elements of the vector tout from Matlab workspace); y – output model of the object or system; y'is the output of the neural network model of the object or system.

The "Display" displays the error values of each of the neural networks. All values of root mean square errors are given in Table 1.

The table shows that the smallest error, equal to 0.001236, has a structure with 11 neural networks in the hidden layer, with 1 delayed input and 2 delayed outputs and is the best structure for the selected system. The block diagram of the proposed controller is shown in Fig. 7.

The control system works on the basis of a neuro-controller, which generates control signals according to the developed neural network. The movement of the working body is set by the upper level of control, which may consist of a personal computer or controller. The control drivers generate control signals according to the software. The software

Table 1. The results of the study of neural network models

i_j n	0_1	0_2	0_3	1_1	1_2	1_3	2_1	2_2	2_3
5	0,251	0,247	0,251	0,013	0,002	0,112	1,764	**0,259**	3,96
6	0,041	0,082	0,089	0,245	0,0024	0,085	0,31	0,29	0,485
7	0,043	0,079	0,095	0,2475	0,002	0,09	3,167	1,031	0,486
8	**0,039**	0,088	0,094	0,147	0,0025	0,084	0,0015	1,728	0,484
9	0,045	0,079	0,098	0,302	0,0016	0,016	1,053	1,39	0,852
10	0,043	0,083	**0,086**	**0,0051**	0,0019	0,089	**0,0013**	2,208	1,579
11	0,053	0,074	0,091	0,272	**0,0012**	0,014	0,152	2,211	0,486
12	0,057	**0,073**	0,1075	0,105	0,004	0,01	0,0014	2,21	**0,45**
13	0,04	0,082	0,107	0,015	0,0036	0,021	0,6546	0,944	0,487
14	0,075	0,087	0,097	0,06	0,0023	**0,005**	0,0027	2,203	0,662
15	0,044	0,104	0,1075	0,12	0,0029	0,015	0,085	2,21	0,485
16	0,043	0,085	0,096	0,125	0,0024	0,024	0,483	1,593	0,574

allows you to enter the trajectory of the working body and generates control signals for the two-coordinate machine, which come to the control controllers for the formation and control of actuators - stepper motors. Each motor is programmed separately through the local control system (using separate drivers), and the sensors located on the drives monitor their position and send data to the upper control level [30–32].

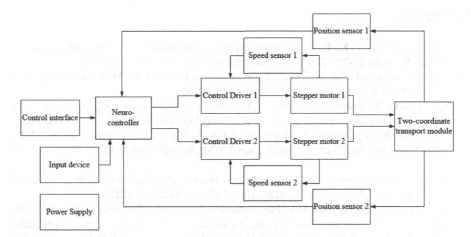

Fig. 7. Block diagram of a neurocontroller for controlling a frame installation with mechanisms of parallel structure

The neural network control system of the frame installation with parallel manipulators developed in this work differs from the existing works [6–8, 12] by the introduction of the optimal in complexity and error neural network. This neural network will accurately display a frame installation with parallel manipulators. And the choice of the minimum neural network architecture and its implementation on FPGA will allow you to control and adapt the controller in real time with high speed.

6   Conclusions

The structural scheme of the control system of the frame installation with the mechanisms of the parallel structure based on the neural network controller is developed. A neural network model was built using the MATLAB "MIMO-Plant" package. The root mean square error of neural networks is calculated and the parameters for the most accurate operation of the system are selected. The structure with 11 neural networks in the hidden layer, with 1 delayed input and 2 delayed outputs has the smallest error 0.001236 and is the best structure for the selected system. This model is used as a neural network regulator. The structure of the neural network controller using the FPGA chip has been developed, which will allow managing and adapting to various types of uncertainties in the system in real time with high speed. The controller is responsible for adjusting the control signal by stepper motors depending on the position sensor indicators.

In this paper, a control system with two parallel manipulators is modeled. In the following works, control systems with a larger number of manipulators, namely 4, 6, 8, 10 and 12, will be investigated. Neural network models will be selected for them. The prototype of the controller is realized and the experimental data on the workbench are received.

References

1. Wang, M., Ehmann, K.: Error model and accuracy analysis of a six-dof stewart platform. J. Manuf. Sci. Eng. **124**(2), 286–295 (2002). https://doi.org/10.1115/1.1445148
2. Patel, Y., George, P.: Parallel manipulators applications—a survey. Mod. Mech. Eng. **2**(3), 57–64 (2012). https://doi.org/10.4236/mme.2012.23008
3. Polyvoda, O., Rudakova, H., Kondratieva, I., Rozov, Y., Lebedenko, Y.: Digital acoustic signal processing methods for diagnosing electromechanical systems. In: Lytvynenko, V., Babichev, S., Wójcik, W., Vynokurova, O., Vyshemyrskaya, S., Radetskaya, S. (eds.) ISDMCI 2019. AISC, vol. 1020, pp. 97–109. Springer, Cham (2020). https://doi.org/10.1007/978-3-030-264 74-1_7
4. Kuznetsov, Y.N., Dmitriev, D.A.: Realization of frame-configurations of machine tools with mechanisms parallel structure. J. Techn. Univ. Gabrovo **54**, 27–31 (2017)
5. Kondratieva, I.U., Rudakova, H.V., Polyvoda, O.V., Lebedenko, Y.O., Polyvoda, V.V.: Using entropy estimation to detect moving objects. In: 2019 IEEE 5th International Conference Actual Problems of Unmanned Aerial Vehicles Developments (APUAVD), pp. 270–273 (2019). https://doi.org/10.1109/APUAVD47061.2019.8943839
6. Yang, C., Huang, Q., Jiang, H.: PD control with gravity compensation for hydraulic 6-DOF parallel manipulator. Mech. Mach. Theory **45**(4), 666–677 (2010). https://doi.org/10.1016/j. mechmachtheory.2009.12.001
7. Rudakova, H., Polyvoda, O., Omelchuk, A.: using recurrent procedures in adaptive control system for identify the model parameters of the moving vessel on the cross slipway. Data **3**(4), 1–60 (2018). https://doi.org/10.3390/data3040060

8. Le, T.D., Kang, H.-J., Suh, Y.-S., Ro, Y.-S.: An online self-gain tuning method using neural networks for nonlinear PD computed torque controller of a 2-dof parallel manipulator. Neurocomputing **116**, 53–61 (2013). https://doi.org/10.1016/j.neucom.2012.01.047

9. Rudakova, H., Polvvoda, O., Omelchuk, A.: Using recurrent procedures to identify the parameters of the large-sized object moving process model in real time. In: 2018 IEEE Second International Conference on Data Stream Mining & Processing (DSMP), pp. 247–250 (2018). https://doi.org/10.1109/DSMP.2018.8478506

10. Korniyenko, B., Ladieva, L.: Method of static optimization of the process of granulation of mineral fertilizers in the fluidized bed. In: Hu, Z., Petoukhov, S., Dychka, I., He, M. (eds.) ICCSEEA 2021. LNDECT, vol. 83, pp. 196–207. Springer, Cham (2021). https://doi.org/10. 1007/978-3-030-80472-5_17

11. Korniyenko, B., Ladieva, L.: Mathematical modeling dynamics of the process dehydration and granulation in the fluidized bed. In: Hu, Z., Petoukhov, S., Dychka, I., He, M. (eds.) ICCSEEA 2020. AISC, vol. 1247, pp. 18–30. Springer, Cham (2021). https://doi.org/10. 1007/978-3-030-55506-1_2

12. Doan, Q.V., Le, T.D., Le, Q.D., Kang, H.-J.: A neural network–based synchronized computed torque controller for three degree-of-freedom planar parallel manipulators with uncertainties compensation. Int. J. Adv. Rob. Syst. **15**(2), 1–13 (2018). https://doi.org/10.1177/172988141 8767307

13. Mugweni, D.T., Harb, H.: Neural networks-based process model and its integration with conventional drum level PID control in a steam boiler plant. Int. J. Eng. Manufac. **11**(5), 1–13 (2021). https://doi.org/10.5815/ijem.2021.05.01

14. Rababah, B., Eskicioglu, R.: Distributed intelligence model for iot applications based on neural networks. Int. J. Comput. Netw. Inform. Secur. **13**(3), 1–14 (2021). https://doi.org/10. 5815/ijcnis.2021.03.01

15. Shymkovych, V., Telenyk, S., Kravets, P.: Hardware implementation of radial-basis neural networks with Gaussian activation functions on FPGA. Neural Comput. Appl. **33**(15), 9467–9479 (2021). https://doi.org/10.1007/s00521-021-05706-3

16. Kravets, P., Shymkovych, V.: Hardware implementation neural network controller on FPGA for stability ball on the platform. In: Hu, Z., Petoukhov, S., Dychka, I., He, M. (eds.) ICCSEEA 2019. AISC, vol. 938, pp. 247–256. Springer, Cham (2020). https://doi.org/10.1007/978-3-030-16621-2_23

17. Artem, V., Volodymyr, S., Ivan, V., Vladyslav, V.: Research and development of a stereo encoder of a FM-transmitter based on FPGA. In: Hu, Z., Petoukhov, S., Dychka, I., He, M. (eds.) ICCSEEA 2018. AISC, vol. 754, pp. 92–101. Springer, Cham (2019). https://doi.org/ 10.1007/978-3-319-91008-6_10

18. Opanasenko, V.N., Kryvyi, S.L.: Synthesis of neural-like networks on the basis of conversion of cyclic hamming codes. Cybern. Syst. Anal. **53**(4), 627–635 (2017). https://doi.org/10.1007/ s10559-017-9965-z

19. Palagin, A., Opanasenko, V.: The implementation of extended arithmetics on FPGA-based structures. In: 2017 9th IEEE International Conference on Intelligent Data Acquisition and Advanced Computing Systems: Technology and Applications (IDAACS), pp. 1014–1019 (2017). https://doi.org/10.1109/IDAACS.2017.8095239

20. Rayhan Ahmed, Md., Robin, T.I., Shafin, A.A.: Automatic Environmental Sound Recognition (AESR) using convolutional neural network. Int. J. Mod. Educ. Comput. Sci. **12**(5), 41–54 (2020). https://doi.org/10.5815/ijmecs.2020.05.04

21. Abdisa, W.T., Harb, H.: A neural network based motor bearing fault diagnosis algorithm and its implementation on programmable logic controller. Int. J. Intell. Syst. Appl. **11**(10), 1–14 (2019). https://doi.org/10.5815/ijisa.2019.10.01

22. Shymkovych, V., Samotyy, V., Telenyk, S., Kravets, P., Posvistak, T.: A real time control system for balancing a ball on a platform with FPGA parallel implementation. Techn. Trans. **5**, 109–117 (2018). https://doi.org/10.4467/2353737XCT.18.077.8559

23. Awad, M., Zaid-Alkelani, M.: Prediction of water demand using artificial neural networks models and statistical model. Int. J. Intell. Syst. Appl. **11**(9), 40–55 (2019). https://doi.org/10.5815/ijisa.2019.09.05

24. Kravets, P.I., Lukina, T.I., Shymkovych, V.M., Tkach, I.I.: Development and research the technology of evaluation neural network models MIMO-objects of control. Visnyk NTUU "KPI" Informatics operation and computer systems **57**, 144–149 (2012)

25. Xu, W., Jianhong, Y.: Derivation of transmission function model of two-phase hybrid stepping motor. Space Electron. Technol. **3**, 50–53 (2011)

26. Vorotyntsev, P., Gordienko, Y., Alienin, O., Rokovyi, O., Stirenko, S.: Satellite image segmentation using deep learning for deforestation detection. In: 2021 IEEE 3rd Ukraine Conference on Electrical and Computer Engineering (UKRCON), pp. 226–231 (2021). https://doi.org/10.1109/UKRCON53503.2021.9575783

27. Shibli, M.A., Marques, P.: Artificial intelligent nonlinear auto-regressive external input neural network modeling, design and control of a sea wave electro-mechanical power generating system. Int. J. Intell. Syst. Appl. **11**(6), 1–12 (2019). https://doi.org/10.5815/ijisa.2019.06.01

28. Kravets, P., Nevolko, V., Shymkovych, V., Shymkovych, L.: Synthesis of high-speed neuro-fuzzy-controllers based on FPGA. In: 2020 IEEE 2nd International Conference on Advanced Trends in Information Theory (ATIT), 291–295 (2020). https://doi.org/10.1109/ATIT50783.2020.9349299

29. Shymkovych, V., Niechkina, V.: The criterion for determining the buffering time of the measuring channel for smoothing the variable changes of the sensor signal. In: 2020 IEEE 7th International Conference on Energy Smart Systems (ESS), pp. 343–346 (2020). https://doi.org/10.1109/ESS50319.2020.9160084

30. Loutskii, H., et al.: Topology synthesis method based on excess de bruijn and dragonfly. In: Hu, Z., Petoukhov, S., Dychka, I., He, M. (eds.) ICCSEEA 2021. LNDECT, vol. 83, pp. 315–325. Springer, Cham (2021). https://doi.org/10.1007/978-3-030-80472-5_27

31. Loutskii, H., Volokyta, A., Rehida, P., Oleksandr Honcharenko, V., Thinh, D.: Method for synthesis scalable fault-tolerant multi-level topological organizations based on excess code. In: Zhengbing, H., Petoukhov, S., Dychka, I., He, M. (eds.) ICCSEEA 2020. AISC, vol. 1247, pp. 350–362. Springer, Cham (2021). https://doi.org/10.1007/978-3-030-55506-1_32

32. Shakhovska, N., Montenegro, S., Kryvenchuk, Y., Zakharchuk, M.: The neurocontroller for satellite rotation. Int. J. Intell. Syst. Appl. **11**(3), 1 (2019). https://doi.org/10.5815/ijisa.2019.03.01

Web Application State Management Performance Optimization Methods

Liubov Oleshchenko[✉] and Pavlo Burchak

National Technical University of Ukraine "Igor Sikorsky Kyiv Polytechnic Institute",
Kyiv 03056, Ukraine
oleshchenkoliubov@gmail.com

Abstract. The use of various libraries often leads to a decrease in the speed of the web application and the complexity of the execution of the program code. The object of this research is the process of storing and managing the data of the client part of the web application, the subject of the research is the software methods of managing the local state of the data of the web application. The research objective is the data processing time reduction of web applications relative to existing software methods. The main idea of the proposed method is to use an atomic approach to the state of the web application data. Having an arbitrary entity, in the general state of the web application, a state fragment is created that is responsible only for this entity. Such a fragment is independent of other state fragments and can only work with the encapsulated entity. Using encapsulation, the configuration of an entity is passed to the React Context API as an object containing data and functions that modify it. The developed software method was compared with popular state management libraries Redux, MobXState-Tree and Recoil. Comparing each of the test scenarios in a percentage ratio, an average decrease in program execution time by 17% was obtained. The analysis of methods was performed using the SonarQube utility. To evaluate the results of the software methods, the Google Chrome browser utility DevTools was used. Based on decreased program execution time it can be stated that proposed optimized software method allows reduce the data processing time and optimize the state management of web applications. The research value is that such a software method can replace existing popular libraries with better proposed solution.

Keywords: Performance · web application · optimization · local state of web application data · Redux · MobXState-Tree · Recoil · React

1 Introduction

A feature of building web applications to support standard browser functions is that the functions must be performed independently of the client's operating system. Instead of writing software code for Microsoft Windows, Mac OS X, GNU/Linux and other operating systems, a web application is created once for an arbitrarily chosen platform and deployed on it. Different implementations of HTML, CSS, DOM and other specifications in browsers can cause problems when developing web applications and their

Z. Hu et al. (Eds.): ICCSEA 2023, LNDECT 181, pp. 59–74, 2023.
https://doi.org/10.1007/978-3-031-36118-0_6

further support. The ability of the user to configure several browser parameters at the same time (for example, font size, colors, disabling scripting support) can prevent the correct operation of the application. The shorter the waiting time, the more efficiently the web application works. A web application is a data store, which can store any type of data in any structure and provides tools to change this data. This is the main task is managing the local state of web application data. The speed of data storage and processing of web applications is also taken into account when ranking web pages, e-commerce web sites also affect the increase in profits along with the decrease in the loading time of the web application page. Therefore, the research of the optimization of the management of the local data state of web applications is an urgent task. A major challenge that arises when developing a client-side web application is managing the internal state of the web application's data. Currently, there are plenty of solutions exists that help manage the state of application and resolve an issue with storing a lot of application data. Here are some libraries, which are the most popular ones and usually deal with the problem: Redux, MobXState-Tree and Recoil. However, the use of such libraries often leads to a decrease in the speed of the application and the complexity of the code. It is necessary to conduct a study of existing methods to develop software in which the analysis of selected software methods will be performed and it will be done conclusion about their advantages and disadvantages. The major research objective is to find a better solution for state management of modern web application. First of all, it is required to make a research and analysis of popular libraries and based on their performance, propose a new optimized software method of state management. Then proposed method needs to be compared with existing solution, and made a conclusion whether implemented optimization was beneficial. It is necessary to determine the best solutions and, taking into account the advantages of each of the method.

1.1 Analysis of Recent Research and Related Work

Research [1] goal is to analyze the most popular state management library Redux and research an alternative to it. In this thesis React Hooks Method was introduced, which showed some significant performance improvements. Given results of the research was measured in a tool to measure performance – Google Chrome Dev Tools.

In research [2] Redux library performance was measured and compared to React Context API method. Although, Redux library is easy to use, it shows significant performance issues when working with big data objects. Also, hooks approach was analyzed in detail, proving its scalability. Research [3–14] contains libraries and frameworks for the modern Web Application software development. There was selected some base metrics for libraires to be compared with each other – time of state management method execution for unit of work and suitability for developer to use. Tools that provides that most accurate results, not depending on side-effects was presents, such as Google Chrome Lighthouse and Sonarqube.

The existing articles do not describe the methods of optimizing access time to web application data and their changes, so, research of the web application local state data management optimizing is an urgent task.

1.2 Analysis of Existing Software Solution

Consider the main principles of the web application local state data and what tasks it performs. First of all, such a repository must be a part web application, i.e. cannot be a remote database, the application must be fast and reliable access to data. The web application is not intended for storage large amount of data, it should be only local data that used by a specific user in a specific session. After closing the application, the data is automatically deleted, if not saved on the server. In this case, the use of a database in the system architecture will slow down the application.

The state of a web application must span the entire application. That is, each element, or component of the application, has access to the state and possibility to change it. The state in this case should be the same in one time slot for all web application components. It is important that system components be able to exchange state and its updates with each other without directly passing parameters to each other. State should not be passed to components that do not need it. That is, it would be inappropriate to pass state up the tree from the parent component to the lowest-level component, as this creates a lot of extra code. Components that use state are called subscribers, that is, those that subscribe to state updates.

The third requirement for the state of a web application is its reactivity. This means that when the state or one of its fields is updated, it is assumed that the subscribing components of the state receive these updates immediately and automatically, without the need to request them. Vuex is a state management library for the Vue.js JavaScript framework. The Angular framework includes its own RxJS library using Observables, while Redux is a common state management library that can be used with any of these frameworks or other libraries, but is very often used with the React library.

Redux library designed to manage the local data state of JavaScript programs. Most often used together with React or Angular for building user interfaces. Redux stores the data state of everything application in the tree of objects in the same repository. One state tree makes it easier debugging or testing the application, it allows to save the state of the application data in progress to speed up the development cycle.

Store object connects events that record the fact that something happened and reducers that update the state according to these events together. The storage containing the state of the application (application state) provides access to the state using the getState() function; can issue status updates and handles unregistering listeners using subscribe (listener) function. The way to change the state is single out the action, the describing object what happened. This ensures that neither views nor network callbacks will never change state. Instead, they only express intent to do it. All changes are centralized and occur in a clear manner sequence. Since they are simple objects in action, they can be registered, serialized, stored and subsequently reproduced for debugging or testing. Advantages of this method are Typescript language support, using the Redux Toolkit Query extension, availability of additional Redux DevTools for development. The main disadvantages of the method are a large amount of repeated code and lack of solutions that allow third party interception effects.

MobX-State-Tree (MST) is a container system state, built on the MobX functional reactive state library. MobX is a state management engine, the MobX-State-Tree provides it structure and general tools needed for specific web application. MST is useful for fast scaling code into programs. Compared to Redux, MST offers better performance and much less boilerplate code. MST combines the best features of immutability (transaction, traceability and composition) and approach to management state, based on changeability (openness, joint location, encapsulation). Advantages of this method are the use of several isolated states simplifies scaling, use of a tree structure, does not require writing a large amount of identical code. The main disadvantages of the method are complexity during software debugging, using their own types to describe data that can conflict with Typescript, the need to use the observer function in the necessary system components.

Recoil is a state management library for React that allows create a graph of the data flow that flows from atoms (shared state) via selectors to React components. Atoms are units of state to which components can subscribe. Selectors convert this state synchronously or asynchronously. Atoms are units of state that are updated and updated subscriptable. When an atom is updated, each component subscribed to it is re-rendered with the new value. Atoms can be used instead of the state of a local React component. If an atom is used of several components, all of these components have the same state. A selector is a function that takes atoms as input. When these atoms are updated, the function the selector is recreated. Components can subscribe to selectors just like atoms and will then be re-rendered when selectors change. Selectors are used to calculate the received data on based on the condition. This avoids redundant state, as it is minimal the set of states is stored in atoms and everything else is effectively computed as function of this state. Selectors track, what components they need and what state they depend on. Advantages of this method are Typescript language support and availability of tools that support asynchrony. The main disadvantages of the method are compatibility only with the React framework and tendency to duplicate code during web application development.

1.3 Software Requirements

The main requirements for the developed software are possibility building a web application to run it in a web browser; the presence of a graphical user interface with the ability to operate with the status data of the web application; the possibility of connecting selected libraries to the software; connecting instruments for measuring time program execution and the ability to connect tools for code quality assessment programs. Software for testing and evaluating the uptime of libraries should be compatible with libraries to manage the local state of web application data.

For the development of the software client part, the JavaScript language was chosen for creating web page scripts, which provides an opportunity on the client side to interact with the user, control the browser, and asynchronously exchange data with the server, change the structure and appearance of the web page. To simplify the process of setting up and developing a web application, the React library was chosen due to its compatibility with the selected libraries and the presence of React Dev Tools for measuring the time of program methods and evaluating their work quality. To evaluate the speed of the methods, the Google Chrome web browser for measurement the web application's operating time for a specified period was used and allows view a diagram of the application's operating time distribution for specific actions, such as rendering, system operations and program code execution. Taking into account the described requirements, in the software ensuring that each of the methods selected for research should be used in libraries and their work should be demonstrated on the basis of certain data. In a conclusion there are main Software Requirements to implement such software: Google Chrome 95+, React 17+, Node 16+. For the accuracy comparison, the methods should be used in the same conditions, work with the same data and according to their number, the same actions must also be performed on the data: adding data segments, editing, deleting, etc. The data, which will be operated by various methods, should be used for output to the user graphical interface. To estimate the methods operation time operating on the web application data local state, we use a large data volume. The data conform to an array structure with a large number of elements. In quality entities, to simplify the interface, the task entity Todo consists of a field name, a unique identifier and a status. In order to ensure the accuracy of testing methods, an additional, more complex ExtendedTodo data entity is also introduced (Table 1).

Table 1. Fields of the ExtendedTodo entity

Field name	Data type
id	UNIQUE IDENTIFIER
name	String
status	Boolean
tags	String []
author	{String, Number}
image	Buffer
description	String
createdAt	Date

ExtendedTodo entity fields include more complex data types. The software interface consists of identical segments, each of which is operated by different state management methods. The segment consists of the method name, the elements number in the data array, and the display of the list of tasks. For testing it is possible to manipulate data using the adding elements operations to the array, deleting elements, and editing. In the Todo task list, each task is available to change status and delete. The visual appearance of the software interface is shown in Fig. 1. By clicking on "+" and "×", respectively, the user can change the task status. Such a function provides an opportunity to test the finding of an element in an array with a large number of elements and its editing. The "Remove" button on each of the elements allows test the removal of the element from the array. The "Add" button to add items to the list for selecting different numbers of items and manipulating data is used. Based on the software requirements, there is a need to connect additional tools to evaluate the quality and methods effectiveness – Google Chrome browser utility DevTools. Chrome DevTools helps the developer edit web pages "on the fly" and quickly diagnose problems.

For this research the Performance panel of this utility and tools that allow recording and analyzing the program execution time during the program execution were used.

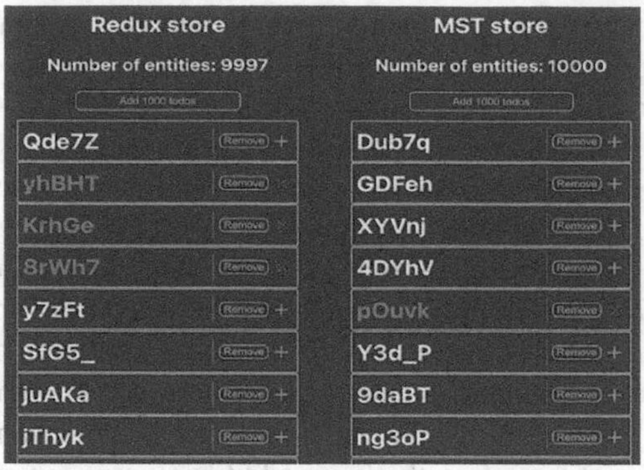

Fig. 1. Graphical software interface

After the analysis is completed, the tool provides a report in the time chart form, which allows drawing conclusions about the program operation. All execution time indicators are given in milliseconds. Figure 2 shows the Performance panel of the Google Chrome DevTools utility, which displays the application runtime, the frames per second scale, and the operation distribution diagram. To evaluate the program performance both for the entire time interval and for the allocated time, a diagram of the distribution of operations by time is used. Scripting is JavaScript code execution time, Rendering is time for the browser to display the interface components, Painting is time for the browser to build graphic elements of the program, System are system operations, Idle is idle execution time, Total is total measurement time.

Fig. 2. Performance panel of the Chrome DevTools utility

The Scripting time counter shows how much time was spent executing the JavaScript code. This time will be different when using different methods, comparing it, we can make a conclusion about the greater effectiveness of one or more methods when performing various types of tasks.

SonarQube is a code quality testing platform for performing automated reviews with static code analysis for bugs. SonarQube offers review reports on duplicate code, coding standards, unit tests, test code coverage, comments, bugs and security recommendations. With the help of SonarQube, it is possible to analyze the program code using the state management method on a file-by-file basis.

The developed software for comparing performance allows to measure the operating time and perform the following operations on the data: adding elements to the array, removing an element from the array, editing an element of the array.

The combination of such operations on the data allows exploring the advantages and disadvantages of the methods in different situations. Simple operations on the array are performed in a very short time slot with a small number of elements.

Judging by this, it will be appropriate to perform operations on arrays with a large number of elements, for example, thousands and more units. The local data state of the investigated web application before the start of the scripts will be equal to an empty array.

In order to correctly test the speed of operation of methods according to the general recommendations for testing the speed of software, it is necessary to carry out work scenarios several times. To calculate the performance index, we will take the arithmetic average of three tests. This approach will avoid possible errors and side effects during the operation of the software.

The following speed test scenarios for research are used:

- adding 1000, 5000, 10000, and 100000 tasks;
- deleting tasks from an array of 10,000 elements long;
- changing the status of several list tasks, with a length of 10,000;
- successive addition, deletion and changes of several elements;
- use of an extended task data structure for operations with the addition of various elements;
- a combination of adding, removing and editing elements in any order with a large number of elements (more than 1000).

Having analyzed the scenarios for evaluating the speed of execution of software methods, we can say about the readiness of testing using the developed software and Google DevTools tools.

After that, a conclusion will be made about the feasibility of using each of the methods in different situations. The results of this research will be used in the development of a combined optimized method for managing the local state of web application data.

1.4 Objective

The object of the research in this article is the reduction in the data processing time of web applications relative to existing software methods.

2 Research Method

2.1 Evaluation Criteria for Software Method

To ensure the accuracy of the estimation, the program code for the research was extended. This approach allows testing the complexity of the code when using a larger number of entities and data. High scalability can indicate that the method with an increase in the amount of software code and expansion of data entities will not lose in the speed of the code.

The software methods of the Redux, MobXState-Tree and Recoil libraries were chosen for the performance study. The best result was shown by the Redux library method (Table 2). The MobX library method showed the longest time, as the number of elements to add increases, the difference between MobX and the other methods increases.

Table 2. Results of speed analysis when adding elements to the array

Number of elements	The time of the operation of adding elements, using the method, ms		
	Redux	MobX-State-Tree	Recoil
1000 elements	239	336	274
5000 elements	1101	1483	1121
10000 elements	2101	3211	2294

To evaluate and compare the performance of state management libraries, different data and scenarios should be used in order to achieve the accuracy of the results and reach an unerring conclusion about the effectiveness of the method.

The developed software for comparing the speed of operation allows to perform and measure the operation time of the following data operations: adding elements to the array, removing an element from the array, editing an element of the array.

The combination of such operations on the data will allow to accurately investigate the advantages and disadvantages of the methods in different situations.

It will be important to say that simple operations, such as those given above, on the array are performed in a very short period of time, with a small number of elements. Judging by this, it will be appropriate to perform operations, by default, on arrays with a large number of elements, for example a thousand and more elements. Let's single out the following speed test scenarios for research:

- adding 1000, 5000, 10000, and 100000 tasks;
- deleting tasks from an array of 10,000 elements long;
- changing the status of several list tasks, with a length of 10,000;
- successive addition, deletion and changes of several elements;
- the use of an extended data structure of the task, for carrying out operations with the addition of a different number of elements;

- a combination of adding, deleting and editing elements in any order, with a large number of elements (more than 1000).

To ensure reliability and accuracy of our results, it is important to mention conditions in which performance tests will be conducted. This includes both hardware and software setup. Hardware: MacBook Pro, Processor 2 GHz Quad-Core Intel Core i5, Intel Iris Plus Graphics 1536 MB, 16 GB of memory 3733 MHz LPDDR4X. Operation system — macOS Ventura 13.0.1. Software: Google Chrome 108, React v18.0.1, Node v16.13.1.

When increasing the number of elements to 100,000, the result of the MobX library was obtained in almost 250 s, which is 3 times slower than analogues. The Redux and Recoil libraries show roughly the same results at ~80 s, which is still quite slow and needs improvement.

Analyzing the previous results, we can conclude that on the example of the Redux library, the increase in execution time increases disproportionately with the increase in the number of elements. As can be seen from Fig. 3, the curve of the ratio of the number of added elements to the time for which they were added is similar to the graph of an exponential function.

Analyzing the data obtained during the testing of the speed of adding elements, we can conclude that the Redux and Recoil libraries show better results when adding a large number of elements.

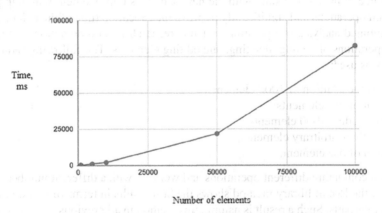

Fig. 3. Ratio of the number of added elements to time

Tables 3, 4 show the results of the speed analysis when editing and deleting elements from the array.

After analyzing the results of the element editing operation, the methods of the MobX library show better results than when adding a large number of elements. It should also be noted that the Redux library shows much worse results in the operation of editing elements than its counterparts. During the edit operation, it was observed that when trying to pass a new value for the status field that is equal to the current value, the methods change the current value to the new value. But in this case, such editing does not make sense, since the value of the field has not changed and the state has not changed either. In this case, for optimization, it makes sense to add an additional check for the

Table 3. Results of speed analysis when removing elements from an array of length 10000

Number of elements	The time of the operation of adding elements using the method, ms		
	Redux	MobX-State-Tree	Recoil
1 element	743	686	723
10 elements	5260	5245	5177

Table 4. Results of speed analysis during editing three elements from an array of different lengths

Number of elements	The time of the operation of adding elements using the method, ms		
	Redux	MobX-State-Tree	Recoil
1000 elements	457	239	289
10000 elements	2151	3211	2011

coincidence of the current state with the new one. This improvement will help prevent redundant operations and significantly improve the execution time of the edit operation. The combined analysis of operations for this research is a combination in any order of the operations of adding, deleting, and editing elements. The following sequence of actions was used:

1. 3 times the addition of 1000 elements;
2. removal of two elements;
3. 5 times adding 1000 elements;
4. editing three arbitrary elements;
5. deletion of one element.

When combining different operations and working with a different number of array elements, the Recoil library method shows the best results in terms of speed compared to its counterparts. Such a result is natural, this method in all previous examples showed effective work in comparison with analogues.

Various data structures were used to test the operation of methods for managing the local data state of the web application, and 2 data models, Todo and ExtendedTodo were used. To test the difference, each of the mentioned operations was used in a certain order. For example, the following sequence of actions was used:

1. adding 10,000 elements;
2. removal of two elements;
3. editing two arbitrary elements.

In this case, the difference in performance of methods with different data structures fluctuates within the margin of error and does not have a significant impact on the speed of the method. After analyzing the methods of each of the selected libraries for analysis and conducting all the planned test scenarios, the performance results were obtained. Each of the methods has shown its effectiveness in different situations. An important conclusion is the exponential increase in the algorithm's operating time when adding elements relative to their number. The results showed that the data structure used as an element of the array during the operation of the method is not fundamental. Analyzing the results of the application of combined operations, which included a different number of operations, a conclusion was made about the most efficient operation of the Recoil library method. Bearing in mind that this method performed well in all other performance test scenarios when measuring time, we can conclude that the Recoil library's method for managing the local state of web application data is the most efficient compared to other libraries. When testing the addition of one hundred thousand elements to the array, each of the methods demonstrated a rather low performance index.

2.2 Proposed Software Method

Fundamental the technology of the method, which will allow the developer to have the opportunity to create and configure a system for managing the local data state of a web application, is to use the React Context API technology framework. In the web application using React web application data are passed from top to bottom through parameters, but such usage can be cumbersome for certain data types, such as customizing the user interface theme. The Context API provides a way to share similar values between components without the need for explicit transmission parameter through each level of the tree. To ensure use method throughout the web application in the root file that contains itself calls to all other parts of the web application must be called React Context. The main idea of the method is to use an atomic approach to state. Having an arbitrary entity in the general state of the web application creates a state fragment that is only responsible for that entity. Such a fragment is independent of other state fragments and can only operate with an encapsulated essence.

Using encapsulation, in React the Context API is passed the configuration of the entity in the form of a containing object data and functions that change them. Based on the configuration of general state, a fragment is created that will be responsible for such essence. To interact with such a fragment, the method interface provides the hook function. When called, such a function provides access to the data interface and the ability to change it. The program interface can be arbitrary, since at the creation of a state fragment directly indicates its configuration.

Consider the features of the proposed software method.

1. Context API is used to provide operation in all files of the web application.
2. Atomic fragment of the state is configured.
3. Using the hook function it is possible to access the local state of the data and change the atomic fragment of the state.

Once such a feature has been used in a web application component, that component automatically subscribes to state updates. This means that with any change to the state fragment to which this component is subscribed, the interface component itself will be recalculated and changed using the updated data.

For a better understanding of how the method works, let's consider its application for a state fragment example.

$$\text{interface create} = (\text{storeConfig}) \Rightarrow \text{useStoreHook};$$

Syntax (storeConfig) => useStoreHook means that function takes a parameter of type storeConfig and returns a value useStoreHook.

Consider the interface of state creation functions:

```
interface create = (storeConfig) => useStoreHook;
interface storeConfig = (
set: (any) => void,
get: (void) => any)
=> (stateConfig:Record <string, any>);
interface useStoreHook =
(selector: (any) => object) => (stateConfig: Record <string, any>);
```

Type any means any type, Record <string, any> is object with key-value pairs, string and any type, respectively. The set and get methods can be called when configuration for further use of field values and for counting certain values in functions. The create method does two things functions: assignment of the initial value of the state and a description of the functions that perform operations on state data. The hook function generated by the useStore method is used as a way to get state data directly for use in the user interface. The main advantage of this approach is the possibility of using the technology of closing and transfer state handler functions. Such approach was inherited from the Recoil library. It is intuitive similar to the standard functions-hooks of the React framework, for which it was this method was developed. After analyzing and comparing popular libraries for managing the state of a web application, the following tasks for optimization were determined: prevention of unnecessary operations, if when editing the state, the current state is equal to the new one; improving the speed of work when adding a large number of elements; possibility of repeated use of similar parts of the code.

For better understanding of method flow algorithm diagram is provided (Fig. 4).

During the optimized method development, the mentioned disadvantages of analogues were taken into account. The problem of redundant operations is often the reason for the low method efficiency, because after each state change, even if the field values themselves have not changed, all the components that have signed up for the state change will be automatically recalculated. An operation is considered redundant, when attempting to change the state, the initial state and the new state do not differ from each other. To prevent such a problem, before changing the state, we need to check whether the current state is equal to the new state. For this purpose, the possibility of specifying an additional parameter of the useStore hook function for comparing state objects has been added to the implementation of the optimized method.

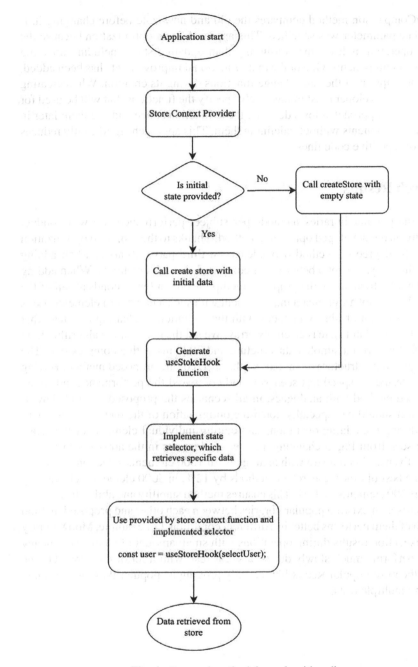

Fig. 4. Poposed methodology algorithm diagram

ShallowComparison method compares the old and new state before changing it, if the appropriate parameter was specified. This approach is an optimization because the comparison operation is less time-consuming than updating state, including updating state subscriber components. One of the main functional improvements has been added, namely the description of the state change functions during its creation. When creating a state object, the developer must immediately specify the functions that will be used for its operation. This approach allows describe these functions once and use them later in all subscriber components without redefining them. This approach significantly reduces the number of repetitive code lines.

3 Research Results

Comparing the popular libraries methods per subject performance, we will conduct similar testing on the developed optimized method. Thanks to the proposed optimization and low by the degree of method dependence on third-party libraries, when adding elements to the array, we get a better result compared to others methods. When adding too many elements to an array the proposed method is only a few seconds ahead of the analogues, which is not a significant improvement with such a number of elements. Let's compare the operation of library methods with the operation of clearing the state, that is, returning to the default state is an empty array. We get the result of the algorithm time execution 3844 ms with a simple data structure and 3979 ms with a complex one. The use of a complex data structure did not affect the speed of the proposed method. Having analyzed and tested all speed test scenarios and compared the performance indicators of the proposed method with analogues, in all scenarios the proposed method showed better results. It should be especially noted the optimization of the method, which was manifested during the adding operations and editing individual elements of the array. As could be seen from Fig. 5 changing number of elements in the array doesn't affect method effectiveness in compare with analogues. In 1000 elements selection method is faster than the best of other researched methods by 12%, in 5000 elements selection by 16% and in 10000 elements by 18%. This ensures method stability and ability to scale the store. Comparison of existing popular libraries between each other and proposed method shows which of them performs better in a different condition. For example, MobX library shows fast execution results during operations with small amount of elements or heavy objects, but performs much slowly during a stress test, with a lot of elements. On the other hand, the most popular Redux library, only proving its popularity, showing stable results during multiple tests.

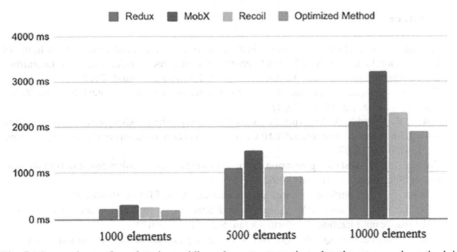

Fig. 5. Comparison of performing adding elements operations by the proposed method in comparison with analogues

4 Conclusion

The article analyzes the specialized software libraries Redux, MobXState-Tree and Recoil for managing the local state of web application data, according to with the help of which the time of access to data and their changes is optimized, their advantages and disadvantages are considered. A modified method of managing the local state of web application data is proposed for creating applications in the React framework ecosystem, which allows reduce program execution time by 17% on average comparing each of the method test scenarios. As proposed methodology, based on the conducted analysis, performed better than popular existing solutions, in a sense of execution time, it can be said that the research objective was achieved. Analysis of software methods and evaluation of the quality of the scanned code was performed using the SonarQube utility. To analyze and evaluate the results of the considered methods, the Google Chrome browser utility DevTools was used. The research makes some conclusions about use cases of several popular solutions. For instance, lately developed Recoil library, in most situation shows performance even better than other libraries. Also, further development and improvement of optimized method, may become mature enough for commercial usage and outclass current solutions. Further areas of research are the software methods complexity analysis using the Cyclomatic Complexity and Cognitive Complexity metrics, the analysis of each file associated with the use of library tools, testing of the metrics data using the proposed software method. Also, further examination in this area also includes in deeper analysis of other popular solutions and test cases for them. In future on base of received results, if it's effectiveness will be proved for most cases, the method could be published as library for public usage.

References

1. Pronina, D., Kyrychenko, I.: Comparison of redux and react hooks methods in terms of performance. In: COLINS-2022: 6th International Conference on Computational Linguistics and Intelligent Systems, May 12–13, 2022, pp. 3–7. Gliwice, Poland (2022)
2. Thanh Le, Comparison of State Management Solutions between Context API and Redux Hook in ReactJS, pp. 14–21 (2021)
3. Atterer, R., Schmidt, A.: Tracking the interaction of users with AJAX applications for usability testing. In: Proceedings of the SIGCHI Conference on Human Factors in Computing Systems, pp. 1347–1350 (2007)
4. Zhongsheng, Q.: Test case generation and optimization for user session-based web application testing. J. Comput. **5**(11), 1655–1662 (2010)
5. Souders, S.: High-performance web sites. Commun. ACM **51**(12), 36–41 (2008)
6. Shivakumar, S., Suresh, P.V.: A survey and analysis of techniques and tools for web performance optimization. J. Inform. Organ. **8**(2), 31–57 (2018)
7. Chi, X., Liu, B., Niu, Q., Wu, Q.: Web load balance and cache optimization design based nginx under high-concurrency environment. In: Third International Conference on Digital Manufacturing & Automation, pp. 1029–1032 (2012)
8. Oleshchenko, L., Burchak, P.: Analysis and optimization of methods for storing and processing data of web applications. Applied mathematics and computing. PMK - 2021. Kyiv, November 17–19, pp. 59–63 (2021)
9. Dychka, I., Legeza, V., Oleshchenko, L., Bohutskyi, D.: Method simultaneous using GAN and RNN for generating web page program code from input image. In: Hu, Z., Petoukhov, S., Dychka, I., He, M. (eds.) ICCSEEA 2020. AISC, vol. 1247, pp. 338–349. Springer, Cham (2021). https://doi.org/10.1007/978-3-030-55506-1_31
10. Wang, Y., Huarui, W., Huang, F.: High-performance concurrent Web application system analysis and research. Comput. Eng. Design Optim. **08**, 2976–2981 (2014)
11. Yao, Y., Xia, J.: Analysis and research on the performance optimization of Web application system in high concurrency environment. In: 2016 IEEE Information Technology, Networking, Electronic and Automation Control Conference, pp. 321–326 (2016). https://doi.org/10.1109/ITNEC.2016.7560374
12. Zagane, M.., Abdi, M..K..: Evaluating and comparing size, complexity and coupling metrics as web applications vulnerabilities predictors. Int. J. Inform. Technol. Comput. Sci. **11**(7), 35–42 (2019). https://doi.org/10.5815/ijitcs.2019.07.05
13. Bibi, R., Jannisar, M., Inayet, M.: Quality implication for prognoses success in web applications. IJMECS **8**(3), 37–44 (2016). https://doi.org/10.5815/ijmecs.2016.03.05
14. Khanna, M., Chauhan, N., Sharma, D., Toofani, A.: A novel approach for regression testing of web applications. Int. J. Intell. Syst. Appl. **10**(2), 55–71 (2018). https://doi.org/10.5815/ijisa.2018.02.06

Design and Implementation of Signal Acquisition System Based on FPGA and Gigabit Ethernet

Haiyou Wang[1(✉)], Xiaozhe Yang[2], and Qianxian Bao[3]

[1] Beihai Confidential Technology Service Center (Beihai Special Communication Technology Service Center), Beihai 536000, China
970423706@qq.com
[2] Faculty of Intelligent Manufacturing, Nanning University, Nanning 530299, China
[3] Nanning Branch of China Mobile Guangxi Co., Ltd., Nanning 530022, China

Abstract. In order to meet the demand of high-speed data transmission between PC and FPGA, a signal acquisition system based on FPGA and Gigabit Ethernet is designed and implemented. FPGA is used as the main control platform, SDRAM is used as the external data storage device, Gigabit Ethernet technology is used as the data transmission protocol, and the FPGA is programmed in the Quartus II design software provided by Altera using the hardware description language Verilog HDL, Through socket programming under Linux system and the implementation of UDP_IP_ARP protocol stack on FPGA, and (SOPC Builder) tool, the logic circuits such as microprocessor Nios II, data buffer, program memory, Ethernet controller are integrated in FPGA, and the MAC inside FPGA and the PHY outside the physical layer are sent to PC through RGMII interface for subsequent demodulation, analysis and other work, A programmable system on chip (SOPC) is formed. This design realizes the wireless broadband signal acquisition function with less hardware resources, and has the characteristics of simple circuit and high transmission rate.

Keywords: FPGA · Gigabit Ethernet · Programmable system-on-chip

1 Introduction

With the development of electronic technology, the system equipment is developing towards miniaturization, integration and networking. FPGA has the characteristics of rich logic and abundant pins, and is widely used in embedded systems of high-speed data processing and communication. The biggest advantage of FPGA is its programmability, that is, users can program the FPGA according to their own needs to achieve user logic, and FPGA can be repeatedly programmed, modified and upgraded without the need to modify the circuit board, only through algorithmic modification and update of the program, which greatly reduces the design work of developers and effectively shortens the development cycle [1–3].

© The Author(s), under exclusive license to Springer Nature Switzerland AG 2023
Z. Hu et al. (Eds.): ICCSEEA 2023, LNDECT 181, pp. 75–84, 2023.
https://doi.org/10.1007/978-3-031-36118-0_7

Combining the advantages of FPGA and Ethernet, this paper designs an embedded gigabit Ethernet device based on FPGA. In this design, we use Altera's high-end FPGA: EP3C10E144C8N chip to build and implement Nios II 32-bit embedded soft-core processor inside, with a working frequency of 185 MHz, and its rich RAM resources can provide high-speed data and code storage for the processor system. A wireless broadband acquisition system is designed around the Nios II processor and the IP core with MAC function, which is in line with the current technological development trend and has certain practical significance.

2 Hardware Design of the System

In this paper, the MTV818 of RAON company is selected as the tuner for receiving signals, the EP3C10E144 of Altera company's Cyclone III series is used as the main control chip, the 88E1111 of Marvell company is used as the physical layer chip of gigabit Ethernet, and the SDRAM chip of ISSI company's IS42S16160G is used as the external data storage chip [4, 5].

The system inputs the UHF band (frequency range: 470–806 MHz) wireless broadband signal received by the antenna to the tuner MTV818, which performs low-noise filtering, I/Q mixing and low-pass filtering on the input signal, and then performs ADC analog-to-digital conversion and Viterbi demodulation, and then transmits the 8-bit digital signal to the FPGA for further digital signal processing. In FPGA, first build a SOPC (programmable system on chip) with Nios II processor as the core and configured with SDRAM controller, Gigabit Ethernet MAC and other peripherals. SOPC will first receive data and store it in SDRAM, and then read the data stored in SDRAM through FIFO and transmit it to Gigabit Ethernet MAC after packaging the data packet of UDP protocol, and then transmit it to the physical layer chip through RGMII interface. The physical layer chip outputs UDP protocol data to the PC through the Ethernet interface RJ45 interface, and the received data can be further analyzed and processed on the PC. In this paper, a design scheme of wireless broadband signal acquisition system based on Ethernet is proposed by comprehensively using FPGA, SDRAM, Gigabit Ethernet transmission and other technologies. The overall block diagram of the system is shown in Fig. 1.

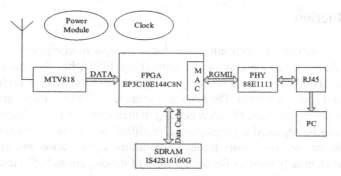

Fig. 1. Overall block diagram of the system

2.1 Tuner MTV818 Design

MTV818 is a high-integration system-on-chip receiving system applied to T-DMB, DAB, FM and ISDB-T standards. Tuner MTV818 is a portable mobile TV and baseband single chip solution; Internal integration of functional modules such as DCDC converter and AD analog-to-digital conversion module greatly simplifies peripheral devices [6]. The interface between MTV818 and external communication supports data stream interfaces of SPI, HPI, I2C, EBI2 and MPEG2.The connection diagram of MTV818 and FPGA through SPI interface is shown in Fig. 2.

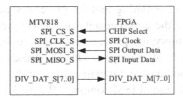

Fig. 2 MTV818 and FPGA connection diagram

2.2 FPGA Design

The system selects the Cyclone III series chip EP3C10E144C8N of Altera Company as the main control platform of the system, which mainly realizes data processing and control of various modules.EP3C10E144C8N support high-speed external memory interface, including DDR, DDR2, DDR3, SDRAM and SRAM.Build an embedded soft-core processor Nios II in FPGA with a three-speed Ethernet MAC core, SDRAM controller, EPCS Flash controller, JTAG debugging interface and self-designed UDP_IP_SOPC of peripheral devices such as ARP protocol module [7–9].

FPGA download circuit and configuration mode are divided into active configuration mode, passive configuration mode and JTAG configuration mode, which are three commonly used configuration modes for FPGA devices. The design selects AS mode and JTAG mode. JTAG and AS pin circuits on FPGA chip are shown in Fig. 3.

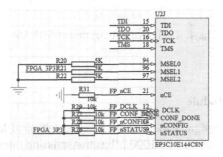

Fig. 3. JTAG and AS pin circuits on FPGA chip

2.3 SDRAM Module

Due to the limited internal storage resources of FPGA, real-time signal processing and transmission are required. The total RAM capacity of the FPGA device selected in this design is only 424 kbits, and the internal storage capacity is small, which cannot meet this requirement. Therefore, the system's requirements for data cache can be met by adding memory devices. In this design, large capacity data needs to be cached, so the system can meet the requirements of data caching by adding memory devices. Commonly used data memory devices mainly include solid state memory (FLASH), static random access memory (SRAM), synchronous dynamic random access memory (SDRAM) and other devices [10, 11]. However, considering that the price of solid state memory and static random access memory with the same capacity is much higher than that of synchronous dynamic random access memory, this paper selects synchronous dynamic random access memory chip as the external data storage device. At present, there are many SDRAM manufacturers. After comprehensive comparison in various aspects, this paper selects the more common SDRAM chip of ISSI company, the model is IS42S16160G, the total capacity is 256 Mbit, and the working voltage is 3.3 V. The connection diagram between chip and FPGA pin is shown in Fig. 4.

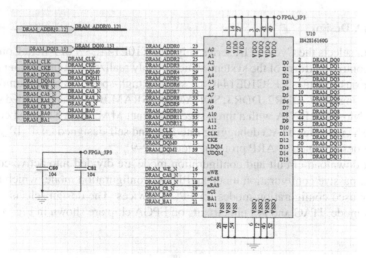

Fig. 4. Design circuit diagram of SDRAM

2.4 Physical Layer Module

The PHY chip in this design is the Gigabit Ethernet physical layer chip 88E1111 of Marvell. 88E1111 supports 10/100/1000 Mbps transmission rates. The communication rate of two network interfaces can be automatically negotiated under IEEE 802.3u standard. It supports GMII, RGMII, SGMII, TBI, RTBI and other interfaces. The connection diagram of MAC and PHY is shown in Fig. 5.

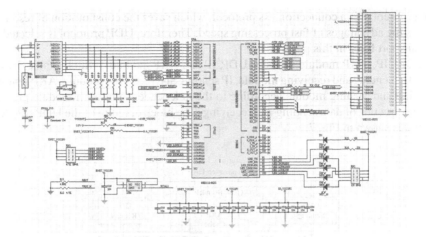

Fig. 5. Design circuit diagram of PHY chip 88E1111

3 Software Design of the System

3.1 Implementation of UDP_IP_ARP Protocol Stack

This paper builds a SOPC system based on Nios II, in which the processor Nios II, Ethernet MAC controller, SDRAM controller, EPCS Flash controller, JTAG debugging interface and other peripherals can be designed using the IP core provided by Altera, but UDP_ IP_ ARP protocol stack must be developed and designed by yourself. The software development flow chart of Nios II is shown in Fig. 6.

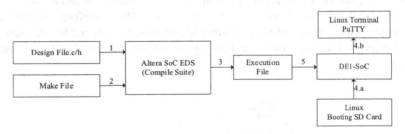

Fig. 6. Software development flow chart

This paper uses FPGA hardware to implement UDP_IP_ARP protocol can improve the data transmission rate of the system. The hierarchical model of TCP/IP network is divided into application layer, transport layer, network layer and link layer from top to bottom. Among them, the transport layer and network layer are the focus of this paper. The transport layer is responsible for packaging the data sent from the application layer into packets suitable for sending, and then transmitting them to the network layer for processing. The transport layer mainly includes TCP protocol and UDP protocol [12, 13]. UDP protocol does not have the handshake, confirmation and other mechanisms of

TCP protocol. It is a connectionless protocol, which saves the consumption of resources and has the advantages of fast processing speed. Therefore, UDP protocol is selected for the transport layer in this paper.

UDP_ IP_ ARP module includes UDP, IP and ARP modules, which can be divided into UDP sending and receiving module, IP sending and receiving module, ARP sending, receiving and caching module according to the independence of each module protocol and the direction of data sending and receiving. The block diagram of the UDP_IP_ARP module is shown in Fig. 7.

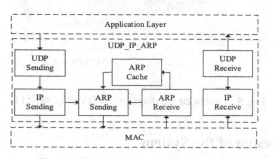

Fig. 7. Block diagram of UDP_IP_ARP

3.2 Main Functions of Socket Programming

Socket communication can be used for communication between different processes on the same device or between different devices. The principle of socket communication is similar to the principle of "everything is a file" of Unix/Linux. You can use the mode of "open – > read/write write/read – > close" to operate [14]. Therefore, you can regard the socket as a special file and open, read/write and close this special file through some socket functions. The flow chart of socket programming design based on Linux is shown in Fig. 8.

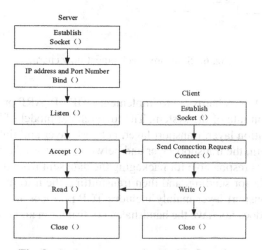

Fig. 8. Socket programming design flow chart

4 Simulation Case and Data Analysis

4.1 Implementation and Simulation of UDP_IP_ARP Protocol Stack

In the simulation file, we first design the protocol stack to receive an ARP query request, and then the module responds. Then input the length, destination IP address, destination port number, source port number, user data and other information, and observe the MAC frame output of the module. After that, input a MAC frame that is not ARP request to the module, and observe whether the extraction and output of each part of the frame are correct [15, 16]. The above three types of input and output constitute the entire test file. In order to comprehensively test the performance of the module, we have carried out a number of three types of input tests. It is observed that the output of the module meets the expectation. This shows that the module designed in this paper can receive ARP query requests from other devices and give correct responses. At the same time, it can add the data from the application layer first and generate MAC frames to send to the MAC layer. It can also unpack the data from the MAC layer by layer, and finally send it to the application layer [17]. Simulation is such an effective tool that can play an important role in system design [18, 19]. UDP_IP_ARP protocol stack simulation test results are shown in Fig. 9.

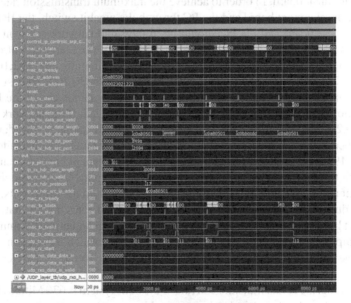

Fig. 9. UDP_IP_ARP protocol stack simulation test

4.2 Overall Test of the System

The DE1-SoC development board is used as the lower computer. The DE1-SoC development board and the upper computer form a Gigabit Ethernet communication platform through the Gigabit switch. As a server, the PC is used to receive the data transmitted from the development board. As a client, the development board is used to generate data and send data. The communication process of the whole communication system is as follows: first, a 16bit DDS signal is generated on the development board as the source signal of the client, and then a data connection request is sent to the upper computer through socket programming. When the server monitors the connection request, it will make a corresponding response message and send a reply agreeing to the connection request of the client, after receiving the reply from the server, the client can start sending data to the server. Finally, we can observe the transmission rate of the entire gigabit network in real time through third-party software tools. The overall system test schematic diagram is shown in Fig. 10.

Fig. 10. Schematic diagram of overall system test

After connecting the development board, gigabit switch and PC, set the IP address of the development board. In order to achieve the maximum transmission rate after connecting the development board and the PC through the gigabit switch, the IP address of the development board and the PC should be in the same network segment, so the IP address of the development board can be set to 192.168.80.200. Start the control terminal of the development board on the PC, execute the command ifconfig eth0 192.168.80.200, and set the IP address of the development board to 192.168.80.200. To ensure that the IP address of the development board is set successfully, you can check the IP address of the development board by executing the ifconfig command. If the network address is 192.168.80.200 in the execution result, it indicates that the IP address of the development board has been set correctly. Then open the DOS terminal on the PC, execute the command ping 192.168.80.169, and check whether the development board has successfully pinged with the PC through the gigabit switch. After the ping test, the number of data packets sent and received in the network test can be seen from the test results. When the number of data packets sent and received is the same, it indicates that the data transmission has not been lost, indicating that the network of the gigabit network communication system composed of the PC and the development board through the gigabit switch has been successfully connected. The transmission rate of the overall system test is shown in Table 1.

After testing, the actual transmission rate of the built system is 630 Mbit/s on average, which is less than the theoretical value of gigabit network, but it can meet the actual application transmission requirements. Table 1 shows the transmission rate of the overall system test.

Table 1. Transmission rate of overall system test

Sequential	Rate (Mbits / sec)	Time delay (ms)	Datagrams (bit)
1	615	0.007	53419
2	615	0.007	53389
3	617	0.015	53539
4	621	0.006	53949
5	623	0.013	54042
6	624	0.006	54129
7	623	0.013	54117
8	626	0.010	54308
9	625	0.013	54219
10	620	0.014	53829
11	622	0.014	53979

5 Conclusion

In this design, firstly, a SOPC (programmable system on chip) with Nios II processor as the core and including peripheral devices such as Gigabit Ethernet MAC is built in FPGA, and the received U-band wireless broadband signal is sent to the tuner chip MTV818. MTV818 performs low-noise filtering on the wireless broadband signal, followed by I/Q mixing and low-pass filtering, and then performs ADC analog-to-digital conversion Viterbi demodulation operation. Finally, the signal is sent to the PC through the RGMII interface between the MAC inside the FPGA and the PHY outside the physical layer for subsequent demodulation and analysis. The design realizes wireless broadband signal acquisition with less hardware resources, and has the characteristics of miniaturization and networking of equipment. Later, it can be upgraded and maintained according to the actual needs, which has a broad application prospect.

Acknowledgment. This project is supported by the project of scientific research and basic ability improvement for young and middle-aged teachers of Guangxi universities (2022KY1781).

References

1. Colak, A., Manabe, T., Shibata, Y., et al.: Peak detection implementation for real-time signal analysis based on FPGA. Circ. Syst. **9**(10), 148–167 (2018). (in Japan)
2. Dulnik, G., Grzelka, A., Luczak, A.: A Gigabit Ethernet interface with an embedded lossless data encoder on FPGA. Pomiary Automatyka Kontrola **61**(7), 364–366 (2015)
3. Windisch, D., Knodel, O., Juckeland, G., et al.: FPGA-based real-time data acquisition for ultrafast x-ray computed tomography. IEEE Trans. Nucl. Sci. **68**(12), 2779–2786 (2021)
4. Runqiu, X., Yang, L.: Design and implementation of PCI bus infrared image acquisition system based on FPGA. LCD Disp. **33**(09), 772–777 (2018). (in Chinese)

5. Jianyu, Z., Jiesi, N., Fuchun, G., Yuchen, J.: Design and analysis of multi-channel signal acquisition circuit based on FPGA. J. Phys: Conf. Ser. **2206**(1), 012034 (2022). (in Chinese)
6. Yihe, H., Dongdong, Z., Lei, D.: Design of Gigabit Ethernet data transmission system based on FPGA+MAC+PHY. Sci. Technol. Proj. **19**, 275–279 (2014). (in Chinese)
7. Qi, W., Jiangtao, Y., Xihong, M.: Design and implementation of high-speed signal acquisition circuit based on FPGA. Res. Explor. Lab. **34**(04), 124–128 (2015). (in Chinese)
8. Mingqing, D., Xiaohu, D., Bao, D.: Design and implementation of a multi-channel data acquisition system. Aeronaut. Comput. Tech. **48**(02), 109–112 (2018). (in Chinese)
9. Hao, Z., Haoquan, W., Shilei, R.: Design of portable signal acquisition system based on FPGA and NAND Flash. Appl. Elec. Tech. **44**(09), 82–86 (2018). (in Chinese)
10. Wei, W., Qiuyun, Z., Hong, J., et al.: Design and implementation of multiplex acquisition and switching system based on FPGA and TCP/IP. Appl. Elec. Tech. **45**(06), 125–129 (2019). (in Chinese)
11. Jinjing, Z., Tao, C., Dong, P., et al.: Design and implementation of 4-DOF manipulator control system based on FPGA. J. Mach. Des. **34**(07), 62–66 (2017). (in Chinese)
12. Bin, L., Chunwu, L., Zhiping, H., et al.: Design and implementation of SFI-5 signal source based on FPGA. Opt. Commun. Technol. **38**(01), 1–4 (2014). (in Chinese)
13. Wang Xuying, L., Yinghua, Z.L.: Design and implementation of high-speed real-time data acquisition system based on FPGA. J. China Univ. Posts Telecommun. **13**(4), 61–66 (2006). (in Chinese)
14. LiLinwei, Jiaoxu, M., Lei, W., et al.: Design and implementation of a data acquisition system based on FPGA and Ethernet. Elec. Des. Eng. **22**(07), 1–4 (2014). (in Chinese)
15. Jin, L.: Design and Implementation of FPGA High Speed Data Acquisition and Transmission System Based on Gigabit Ethernet. Beijing University of Posts and Telecommunications, Beijing (2019). (in Chinese)
16. Yuhang, T.: Design and Implementation of Multifunctional Data Acquisition System Based on DSP and Gigabit Ethernet Technology. Huazhong University of Science and Technology, Wuhan (2020). (in Chinese)
17. Xiaoyu, L., Xinyang, H., Peiyan, S.: Design and implementation of rotational speed signal acquisition system based on FPGA. China New Commun. **24**(07), 48–50 (2022). (in Chinese)
18. Kalpachka, G.: Computer modeling and simulations of logic circuits. Int. J. Mod. Educ. Comput. Sci. (IJMECS) **8**(12), 31–37 (2016)
19. Gjorgjevikj, F., Jakimoski, K.: System monitoring add on analysis in system load simulation. Int. J. Comput. Netw. Inf. Secur. (IJCNIS) **14**(1), 40–51 (2022)

Design of a Dynamic Probability-Based Automatic Test Paper Generation System

Fang Huang, Mingqi Wei, Lianju Su[(⊠)], and Dichen Guan

Global School of Software, Nanning University, Nanning 530022, Guangxi, China
23255294@qq.com

Abstract. Our work and life is closely related to computer technology, and computer applications are especially prevalent in the education sector. The traditional manual approach requires teachers to select questions from workbooks and books, compile and edit them, and finally layout the test papers. This process is tedious and time consuming, and it is difficult to ensure the level of knowledge coverage and the repetition rate of questions. This paper takes a dynamic probability-based test paper generation system as the background, follows the standard development model of software engineering, adopts an object-oriented approach to design the system, specifies the functional requirements, Design the system architecture and divide functional modules according to the principles of modularity, independence and abstraction, and realize the detailed design of the main functional modules such as user management, course management, and examination paper management builds a dynamic probability model to realize the automatic formation of papers, and makes the task of assigning questions more scientific.

Keywords: Test papers · Automatic generation · Dynamic probability

1 Introduction

As an important part of the assessment of the quality of teaching and learning, examinations have traditionally been highly valued by schools. It is a way for students to test their mastery of the knowledge they have learn [1]. For teachers, it is a way of giving feedback on the effectiveness of teaching and learning, and teachers use the quantitative analysis of examination results to make subsequent changes to the curriculum.

Traditionally, question papers rely heavily on manual questioning, which can take a lot of time and effort for teachers to consult a large number of books and materials. Once teachers have found the right questions, it takes a lot of time and effort to edit and layout the papers and write the marking scheme. In practice, this can lead to repetition of questions, incomplete coverage of assessment points, and errors and omissions due to manual processing. To this end, the software industry at home and abroad has carried out practical research on the software development of the test paper generation system earlier [2]. Moreover, with the continuous advancement of software and hardware technology and the continuous maturity of development concepts, the software of the test paper generation system is also improving day by day, which is conducive to realizing

Z. Hu et al. (Eds.): ICCSEEA 2023, LNDECT 181, pp. 85–94, 2023.
https://doi.org/10.1007/978-3-031-36118-0_8

the "separation of examination and teaching", improving the quality of teaching, and evaluating students' learning more fairly, which is one of the possible ways for schools to realize the automation, digitalization and information management of examinations [3, 4]. The design of a dynamic probability-based automatic test paper generation system can therefore not only improve the efficiency of teachers' work in proposing questions, but also ensure the objectivity and scientificity of test papers, which has important significance for improving the quality of teaching.

2 Analysis of Requirements

2.1 Analysis of Functional Requirements

Requirements analysis is the first stage of software development, the main task of this stage is to accurately determine "what the target system needs to do in order to solve this problem", this stage plays a crucial role in the success or failure of software development. The requirements analysis phase typically includes obtaining requirements, analyzing requirements, defining requirements, and validating requirements [5, 6]. Analysis and study of acquisition needs requires consideration of completeness, correctness, reasonableness, feasibility and sufficiency [7].

Designed and developed to meet the urgent needs of current school teachers for test papers, in view of the various problems existing in the management of test questions by many teachers, the test paper generation module is taken as the core research object, and a feasible system is built by modifying and improving the initial system.In order to solve the current problems in the area of question setting and paper assembly, we carried out an analysis of the system's operational requirements. We started from the actual situation of the system users, took the business process in teachers' question appointment work as the main reference basis, and carried out requirement analysis by means of offline visits to master the workflow of question paper appointment. The resulting dynamic probability-based test paper generation system provides the following functions:

(1) Test bank creation and maintenance. Entry, modification, deletion, query and statistics of test questions, setting of question types, realization of querying test questions based on course and question type.
(2) Course management. It includes the addition of new courses, modification of course-related information and editing relevant knowledge points covered by the courses.
(3) Course analysis. Analysis of each test paper by total number of questions, average difficulty, average mark, and by number of questions, difficulty, mark, and knowledge points, with graphical display of analysis results.
(4) Paper generation. A variety of ways to assemble papers, either manually by filtering and setting the key examination points, or automatically by setting the course question type and amount of questions or creating a paper template and modifying or adding questions on the basis of the paper template to meet the difficulty requirements, the paper can be saved as a word document for preview and download after generation.
(5) User management. The users are mainly administrators and teachers, and the different roles have different permissions. Administrators have all the administrative

rights of the system, and teachers have only some of the rights of the system. Figure 1 shows the diagram of system use cases.

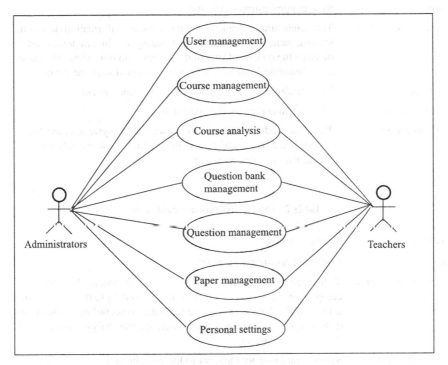

Fig. 1. Diagram of the system use cases

The system use case diagram describes the management correlation of the administrator and the teacher. The main use cases are analyzed and described below.

(1) Course management is to manage the information and knowledge points contained in the course. The table of course management use cases is shown in Table 1.
(2) Course analysis is to analyze the number of question banks, test papers, and questions included in the course. The table of use cases for course analysis is shown in Table 2.
(3) Question bank management is to manage the basic information of the question bank. The use case table of the question bank management is shown in Table 3.
(4) Examination paper management is to manage the examination paper composition and basic information of the questions contained in the system. The table of management use cases is shown in Table 4.

Table 1. Table of course management use cases

Use case name	Course management
Actor	System administrator / teacher
Use case description	The administrator manages the basic course information data of the system, including the functions of adding, editing, inquiring and deleting the course information; teachers can only view the course information and edit the knowledge information of the course
Prefix condition	System administrator / teacher to login to the system
Post-condition	The system completes the operation
Basic event flow	The system administrator / teacher logs in the system, enters the system home page, select the course management module for functional operation and save

Table 2. Course analysis use case table

Use case name	Course analysis
Actor	System administrator / teacher
Use case description	To manage the course information data of the system, through the comprehensive analysis of each course according to the total amount, average difficulty and average score, but also according to the quantity, difficulty, score and knowledge points, and the analysis results will be displayed graphically
Prefix condition	System administrator / teacher to login to the system
Post-condition	The system completes the operation
Basic event flow	The system administrator / teacher logs in the system, enters the system home page, select the course analysis module for functional operation and save

Table 3. Table of question bank management use cases

Use case name	Question bank management
Actor	System administrator / teacher
Use case description	Manage the basic information and data of the question bank of the system, including the addition, editing, query and deletion of the question bank
Prefix condition	System administrator / teacher to login to the system
Post-condition	The system completes the operation
Basic event flow	The system administrator / teacher logs in to the system, enters the main page of the system, and chooses the question bank management module for functional operation and saving

Table 4. Table of course management use cases

Use case name	Examination paper management
Actor	System administrator / teacher
Use case description	To manage the paper generation of the system, you can manually screen and set up the key knowledge points, or automatically set up the number of course questions, and the paper can be saved as a word document and downloaded
Prefix condition	System administrator / teacher to login to the system
Post-condition	The system completes the operation
Basic event flow	The system administrator / teacher logs in the system, enters the system home page, select the personal setting module for functional operation and save

3 System Design

3.1 System Architecture Design

Using a layer-by-layer architecture approach to architecture design, the system takes the front-end UI, view layer, business layer, data layer, database, and operating environment into account in the system architecture, clearly divides the architecture technology and functions of each layer, realizes the interaction of modules at each layer, and builds a complete architecture system for automatic test paper generation, as shown in Fig. 2.

3.2 Global Design

It is necessary to use a computer to centralize the management of examination papers and to be able to complete the work of question entry, examination paper enquiry, examination paper generation and examination paper management. The system can improve the low efficiency of traditional manual paper production, use the computer to build a test bank, achieve automatic computer selection of questions to form papers, and give full play to the advantages of computers in information processing. A test paper generation system with a high level of automation can realize the systematization, standardization and automation of test papers [9].

The design of the system is based on a mature software design and development model. A complete set of business functional requirements is analyzed and refined, and specific functional modules and physical realization schemes are planned on the basis of the "high cohesion and low coupling" system design principle. The functional module design of the system is shown in Fig. 3.

3.3 Database Design

As a key aspect of database design, a good database will enable the system to run efficiently and safely. Database design will have a direct impact on the efficiency of the system and the effectiveness of the implementation.

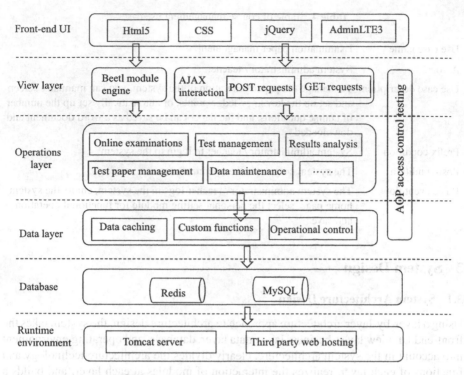

Front-end UI
Html5 | CSS | jQuery | AdminLTE3

View layer
Beetl module engine | AJAX interaction | POST requests | GET requests

Operations layer
Online examinations | Test management | Results analysis
Test paper management | Data maintenance | ......

Data layer
Data caching | Custom functions | Operational control

Database
Redis | MySQL

Runtime environment
Tomcat server | Third party web hosting

AOP access control testing

Fig. 2. Diagram of the system architecture

The system database is designed using the New Orleans method, which organizes, stores and manages data according to a data structure, abstracting the user requirements obtained during the requirements analysis phase into a conceptual model, which is represented by an E-R diagram. The E-R diagram, as a conceptual model, is the basis for both database design by designers and information modelling [10, 11]. In a dynamic probability-based test paper automatic generation system, the entity attributes are more complex. The E-R diagram of the system in general is shown in Fig. 4.

(1) A single course can contain multiple knowledge points, and a single knowledge point can exist in multiple courses, so there is a many-to-many relationship between the course and the knowledge points.

(2) A course can have multiple papers, but one paper belongs to only one course, so the course and the paper have a one-to-many relationship.

(3) One teacher can teach multiple courses, and one course can also be taught by multiple teachers. There is a many-to-many relationship between a teacher and a course.

(4) A teacher can create and manage multiple question banks, but a question bank can only be managed by the creator, so there is a one-to-many relationship between the teacher and the question bank.

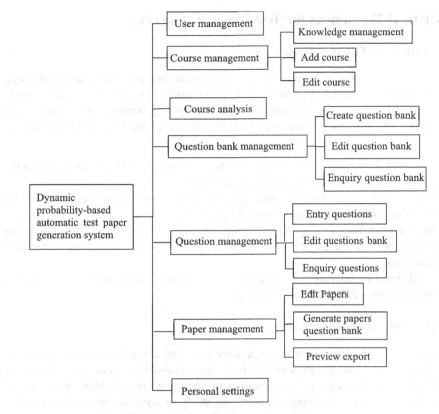

Fig. 3. Diagram of the system architecture

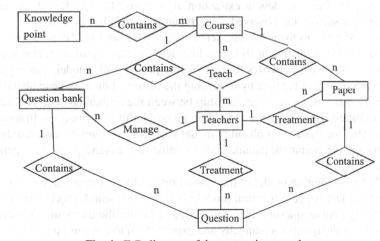

Fig. 4. E-R diagram of the system in general

4   Critical Techniques for Test Paper Extraction

4.1   Multi-layer Filtering Model

Test questions are extracted by filtering all test questions in layers until a paper satisfying the requirements is filtered. A multi-layer filtering model is used to filter all test questions in layers until a test paper satisfying the requirements is generated. The system uses a multi-layer filtering model based on which test questions are extracted, as described below.

(1) First level of filtering. Course filtering, which filters out the list of questions for the examination course from the list of all questions;
(2) Second level of filtering. Question type filtering, which filters out all questions belonging to different question types from all questions in the current course;
(3) Third level of filtering. Grouping filtering, which randomly groups all the questions for that question type and can ensure that different questions are selected in the same conditions;
(4) Fourth level of filtering. Question quantity filtering, which filters the number of questions from the current question type.

4.2   Dynamic Probability Models

The multi-layer filtering model involves more than one extraction process. The issue of how to extract individuals with relatively high probability from a set containing individuals with different probabilities of being extracted is a problem that needs to be studied, and this is the main focus of the study of dynamic probability models. As a result, the question of how to dynamically adjust the probability of a test being drawn different times from the set of results to be drawn becomes the core of the study of dynamic probability models in extraction algorithms [12, 13]. Based on the time-series characteristics of the observed data, dynamic probability models are able to use probability distributions to model time-series data, tap the statistical correlation of serial data and use it for predictive analysis. Unlike deterministic dynamic neural networks, such as recurrent neural networks RNN, dynamic probabilistic models can explore the potential uncertainty of the data by modelling the observed data, and can also describe the mutually non-deterministic relationship between the sequential observed data, and the random time-series hidden variables through probability distributions. In solving for the parameters, the dynamic probability model takes into account the uncertainty of the parameters and can obtain the parameter distribution rather than a point in the parameter space [14, 15].

The dynamic filtering model of the system relates to the extraction process several times, and the test paper generation module is the core research object; the questions are queried by course, question type and difficulty factor; the user can randomly draw questions or set the questions manually and extract them to compose the test paper; the total number of questions in the test paper can be adjusted at will, and the questions can even be added or deleted on the basis of the already generated test paper, so that the test paper can be previewed and exported after achieving the best results [16, 17].

Suppose the question pool of a course is drawn as a set {Q1, Q2,... Qm}, the corresponding decimation probability is denoted as {P1, P2,... Pn}. When drawing for the first time, because the probability of each topic being selected is the same, the system can automatically generate a random number for topic extraction, and after each draw of the topic, the number of draws is counted through the accumulation counter, and the scope of topic extraction is expanded according to the multiple of the number of draws, and the relationship between the number of draws and the expansion multiple is shown in Table 5.

Table 5. The relationship between the number of draws and the expansion multiplier

Number of draws	0	1	2	...	m−1
Enlarge the multiplier	nm	nm−1	nm−2	...	n

5 Conclusion

With the wide application of computer technology, digital campus construction has become the current development trend of education construction, and colleges and universities have begun to use computer technology to assist teaching. For schools, computer operation has become quite widespread. The traditional manual roll-out method has the problems of manpower, time, cumbersome process and low efficiency. How to use computer technology to reduce the light labor of the test taker and realize the scientific and standardized management of the test paper is particularly important.

This paper introduces the design of an automatic test paper generation system based on dynamic probability, and takes the actual operation process of question setting as the main line throughout the whole system design process. The system's functional requirements are outlined by first analyzing the workflow of question assignment, then dividing the functional modules according to the principle of "high cohesion and low coupling", and designing the system database. In view of the large number of question papers, system used Redis as the cache database to improve the speed of data retrieval, and analyzed the key techniques for extracting question papers. The implementation of the system design can, to a certain extent, reduce the workload of teachers and contribute to the improvement of teaching quality.

Acknowledgment. This project is supported by: (1) Nanning University's first-class undergraduate major cultivation project- software engineering (2020YLZYPY06); (2) The educational training program for 2020's teaching achievement award in Nanning University-The exploration and practice for application talents training mode based on the combination of curriculum system and practice system (2020JXCGPY09) and (3) The third batch of pilot project for professional certification in Nanning University -Educational certificate of software engineering (ZYRZ06).

References

1. Xingrong, L.I.U.: Design and implementation of automatic generation system of test paper based on dynamic probability. University of Electronic Science and Technology of China, Chengdu (2014). (in Chinese)
2. Changli, W.: Design and implementation of the exam paper generation management system. Des. Dev. **3**, 120–121 (2011). (in Chinese)
3. Mingqiu, Z., Hongyan, L.: Design and implementation of NET"s online paper-generating system based on ASP. China's Manag. Informatization **18**(22), 149 (2015) (in Chinese)
4. Haoqiang, T.: C++ Programming, (2nd edition). Tsinghua University Press, Beijing, pp. 16–20 (2011) (in Chinese)
5. Rajesh, K., Rakesh, K.: Optimizing requirement analysis by the use of meta-heuristic in search based software engineering. Int. J. Elec. Comput. Eng. (IJECE) **9**(5), 4336–4343 (2019)
6. Wang, Z., et al.: A novel data-driven graph-based requirement elicitation framework in the smart product-service system context. Adv. Eng. Inform. **42**(C), 100983 (2019)
7. Wang, Z., et al.: A novel requirement analysis approach for periodic control systems. Front. Comput. Sci. **7**(2), 214–235 (2013)
8. Xiange, L.: Design and realization of a test paper generation system. Softw. Guide **11**(06), 55–56 (2012). (in Chinese)
9. Rui, Z.: Design and realization of a dynamic probability-based automatic test paper generation system. Autom. Instrum. **11**, 210–211 (2016). (in Chinese)
10. Tao, Y.: Implementation of test paper library system and web online examination system under WEB. Sci. Technol. Perspect. (03), 145+173 (2016)
11. Bril, R.J., Altmeyer, S., van den Heuvel, M.M.H.P., Davis, R.I., Behnam, M.: Fixed priority scheduling with pre-emption thresholds and cache-related pre-emption delays: integrated analysis and evaluation. Real-Time Syst. **53**(4), 403–466 (2017). https://doi.org/10.1007/s11241-016-9266-z
12. Mian, F.: Study on test question extraction algorithm based on multi-layer filtering and dynamic probability model. J. Xichang College (Natural Science Edition) **33**(02), 59–62+124 (2019) (in Chinese)
13. Liancheng, G.: Design of the automatic generative system of examination papers based on ARM. Int. J. Adv. Netw. Monit. Controls **2**(4), 110–113 (2017)
14. Guan, L.: Design of the automatic generative system of examination papers based on ARM. Int. J. Adv. Netw. Monit. Controls **2**(4), 110–113 (2018)
15. Liu, Z., Liang, R., Pan, X.: Customized online test database systems and automatic generation system for testing papers. Int. J. Educ. Manag. Eng. (IJEME) **2**(8), 45–52 (2012)
16. Fu, Y.: Design and implementation of an automatic test paper generation system based on ant colony optimization. In: Proceedings of the 2015 International Conference on Intelligent Systems Research and Mechatronics Engineering, Zhengzhou, China (2015)
17. Jia, Z., Zhang, C., Fang, H.: Department of computer science and engineering North China institute of astronautic engineering, Langfang, China. The research and application of general item bank automatic test paper generation based on improved genetic algorithms. In: Proceedings of 2011 IEEE 2nd International Conference on Computing, Control and Industrial Engineering, pp. 30–34 (2011)

The Extended Fredkin Gates
with Reconfiguration in NCT Basis

Vitaly Deibuk[1]([✉]), Oleksii Dovhaniuk[2], and Taras Kyryliuk[1]

[1] Yuriy Fedkovych Chernivtsi National University, Chernivtsi 58012, Ukraine
v.deibuk@chnu.edu.ua
[2] School of Computer Science, University College Cork, Cork T12 R229, Ireland

Abstract. The Fredkin gate is a universal reversible logic gate widely used in designing low-power and quantum devices of various complexities. Recently, the gate's extension was proposed based on its elementary components. It provided variety in functionality and, at the same time, preserved the main features of the gate. This paper offers additional variations of the Fredkin gate, referring to the recent findings. The versatility of logic circuits is particularly beneficial in programmable logic. Therefore, a reconfiguration for the proposed gates has been designed. The device includes 24 newly obtained gates and 8 known extensions. The reconfigurable circuit has a significant advantage in flexibility over the classic gate of 32 extended 3-bit Fredkin gates with a relatively small increase in hardware complexity and delay time. The implementation of the circuit has been tested and verified in Active-HDL, confirming the design's correctness. Reconfigurable devices synthesized in the work make it possible to create programmable fault-tolerant reversible logic circuits effectively.

Keywords: Reversible logic · Fredkin gate · NCT gate library · Reconfiguration

1 Introduction

The rapid development of computer technologies caused a growing interest in reversible logic for low-power devices, quantum computing, signal processing, computer graphics, bioinformatics, and DNA computing [1, 2]. The processor overheating is a primary reason for researching alternatives to classic computing. According to Landauer's principle [3], each bit of erased information releases $kT \ln 2$ J of energy, and recent experiments fully confirm it [4]. In contrast to classic computations, reversible computations do not lead to loss of information, which reduces processor overheating and power dissipation [5].

The main goal of the synthesis of binary and quantum reversible devices is to obtain an optimal reversible circuit for a given bijective function. For the functions of several variables, analytical and diagram synthesis methods can produce optimal results [6]. Nevertheless, heuristic evolutionary algorithms, such as genetic, ant colony, and annealing algorithms [7–11], work more efficiently for functions with numerous variables. These approaches utilize the functionally complete reversible NCT gate library of NOT, controlled-NOT, and Toffoli (double-control NOT) gates. The reversible gates can have positive (black dot), negative (white dot), or mixed control lines that expand their functionality (Fig. 1a–i). The positive control line is activated at a signal of 1, and the negative control line is activated at 0. For multiple controls, activation occurs when all conditions are satisfied. These terms also apply to the Toffoli gates with disjunctive control or OR-Toffoli gates [12].

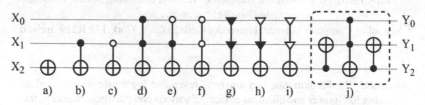

Fig. 1. Symbolic representation of the NCT basic gates with mixed polarity: (a) NOT, (b) CNOT (Feynman gate), (c) C'NOT, (d) CCNOT (Toffoli gate), (e) C'CNOT, (f) C'C'NOT, (g) OR-CCNOT, (h) OR-C'CNOT, (i) OR-C'C'NOT, (j) Fredkin gate on a NCT basis

The functionally complete reversible Fredkin gate (FRG) is of great interest to researchers and engineers since any reversible device can be designed using just this one gate [13]. Besides being functionally complete, the Fredkin gate preserves parity which is critical in quantum computing for simplifying error detection and ensuring fault tolerance. In literature, FRG is also notated as a control SWAP gate for "swapping" the target signals on the output when the control signal equals 1. The classic Fredkin gate contains one control input (X_0) and two target inputs (X_1, X_2) satisfying the following conditions:

$$Y_0 = X_0;$$
$$Y_1 = \overline{X_0}X_1 \oplus X_0X_2; \tag{1}$$
$$Y_2 = \overline{X_0}X_2 \oplus X_0X_1.$$

A few approaches were proposed to obtain new functionality and, at the same time, maintain the "swap" operation of the gate. First, adding extra control lines creates the generalized Fredkin gate (GFRG) with N controls and two targets [14]. This generalized gate interchanges data on the target lines when all control signals are 1. Second, changing activation conditions for gates with N controls in the way that the "swap" occurs when there is at least one activated control line [12]. Third, similarly to the NCT set, the polarity of controls can be positive (activation on 1), negative (activation on 0), or mixed for multiple controls [19]. A particularly compelling strategy was proposed in [15], where authors consider NCT implementation of the FRG (Fig. 1j), trying different polarity settings on the elementary gates and changing the activation condition for one of the gates. This approach led to the introduction of 8 unique FRG extensions. In the present paper, we analyzed the results of [15] and applied them to introduce 24 new extended Fredkin gates.

This paper aims to increase the number of extended Fredkin elements compared to those proposed in [15]. For this purpose, reconfigurable devices have been synthesized, allowing new reversible fault-tolerant circuit solutions to be created effectively. Such reconfigurable devices for advanced Fredkin gates are an innovation and are offered in the paper for the first time. Combining the FRG extensions into a reconfigurable circuit gives an advantage in flexibility. It can be utilized for assembling reversible encoders, decoders, or more complex devices such as field-programmable gate arrays [18], saving time and resources during production.

The rest of this paper is organized as follows. Section 2 explains the main features of the extended Fredkin gates. The reconfigurations for extended Fredkin gates are proposed in Sect. 3 with an analysis of the obtained circuits. Section 4 provides details of the reconfigurable devices described in hardware descriptive language, including simulation and verification of the models. The conclusion of the work is presented in Sect. 5.

2   The Model of Extended Fredkin Gate

Reversible logic circuits implement the bijective function $f : \{0, 1\}^n \rightarrow \{0, 1\}^n$, where n is the number of input and output binary signals. Function f can be defined as a permutation of n quantities in 2^n ways. Consequently, the number of input and output signals (lines) must be the same. A reversible circuit consists of reversible gates and implements the corresponding logical function, where branching at the output (fan-out) and fan-in at the input are prohibited. Reversible logic circuits do not lose information compared to classic logic circuits saving energy and allowing direct and reversible data flow, which are the main advantages.

The Fredkin gate is a three-input reversible logic gate that implements the logical function specified by (1). The conditions show that input signal X_0 (control signal) transmits to output Y_0 unchanged. If the control bit is not activated ($X_0 = 0$), two other input signals, X_1 and X_2 (target signals), transmit to the output unchanged ($Y_1 = X_1$, $Y_2 = X_2$). However, if $X_0 = 1$, the target signals exchange information at the output ($Y_1 = X_2$, $Y_2 = X_1$). The specified logical function of the Fredkin gate can be implemented on an NCT basis, as illustrated in Fig. 1j.

The gates of the NCT library invert a target signal with (controlled-NOT, Toffoli gate) or without (NOT gate) control. The input signals of the control lines transmit to the output unchanged for CNOT and Toffoli gate, in the same way as for Fredkin gate. The NCT set is also universal. However, unlike FRG, the NCT gates do not preserve parity. Thus circuits built with these gates require additional parity checks to ensure fault tolerance. The FRG modifications proposed in [15] are listed in Table 1. The table contains 16 modified Fredkin gates that implement 8 unique reversible functions. All modifications preserve features of "swapping" target gates on the output and transmitting the control signals without changes. It can be easily confirmed by comparing expressions in column "Function" with conditions (1). Nonetheless, target signals are inverted on the output for some configurations, indicating differences among the extended FRG functions. Additionally, circuits in the four last rows of Table 1 include an OR-Toffoli gate [12] instead of the classic Toffoli gate. This modification of the Toffoli gate has disjunctive control lines schematically represented as upside-down triangles (Fig. 1g–i). The notation is derived from analogy to sign '∨' of the logical OR operation since activation on one control line is enough for the OR-Toffoli gate to invert the target signal. An extended Fredkin gate that contains the OR-Toffoli gate is usually called the OR-Fredkin gate.

Table 1. Extended Fredkin gates (I) with activation keys [15]

$K_0\, K_1\, K_2$	Function ($Y_0 = X_0$)	Circuit	Equivalent Circuit
0 0 0	$Y_1 = \overline{X_0}\,X_1 \oplus X_0 X_2$ $Y_2 = \overline{X_0}\,X_2 \oplus X_0 X_1$		
0 0 1	$Y_1 = \overline{X_0}\,X_1 \oplus X_0\,\overline{X_2}$ $Y_2 = \overline{X_0}\,X_2 \oplus X_0\,\overline{X_1}$		
0 1 0	$Y_1 = \overline{X_0}\,\overline{X_1} \oplus X_0 X_2$ $Y_2 = \overline{X_0}\,X_2 \oplus X_0\,\overline{X_1}$		
0 1 1	$Y_1 = \overline{X_0}\,\overline{X_1} \oplus X_0 X_2$ $Y_2 = \overline{X_0}\,X_2 \oplus X_0 X_1$		
1 1 1	$Y_1 = \overline{X_0}\,X_2 \oplus X_0\,\overline{X_1}$ $Y_2 = \overline{X_0}\,X_1 \oplus X_0\,\overline{X_2}$		
1 1 0	$Y_1 = \overline{X_0}\,\overline{X_2} \oplus X_0\,\overline{X_1}$ $Y_2 = \overline{X_0}\,X_1 \oplus X_0 X_2$		
1 0 1	$Y_1 = \overline{X_0}\,X_2 \oplus X_0 X_1$ $Y_2 = \overline{X_0}\,X_1 \oplus X_0 X_2$		
1 0 0	$Y_1 = \overline{X_0}\,\overline{X_2} \oplus X_0 X_1$ $Y_2 = \overline{X_0}\,X_1 \oplus X_0\,\overline{X_2}$		

3 Reconfiguration of Extended Fredkin Gate

Reconfiguration is the ability of digital devices to change a functional circuit depending on the key signals applied to the corresponding input bits. A widely used example of a complex reconfigurable device is a field-programmable gate array (FPGA), which allows the programming of various circuits on one chip. FPGAs greatly facilitate and accelerate the development of computing systems [16–18, 21].

The circuit proposed in Fig. 2a activates one of the eight extended Fredkin gates presented in Table 1 by inputting key signals (K_0, K_1, K_2). The first three CNOT gates change the polarity of the corresponding input signals. The following two Toffoli gates invert the polarity of the target lines concerning the control signal X_0 if $K_2 = 1$ and do nothing if $K_2 = 0$. The three gates, outlined with a dashed rectangle, represent the classic Fredkin gate that performs a "swap" on activation. Finally, the last CNOT gate restores the control bit signal $(Y_0 = X_0)$.

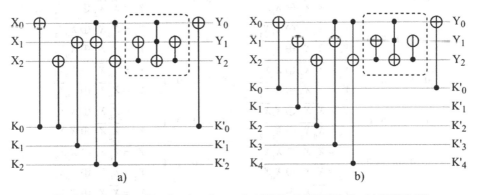

Fig. 2. Reconfigurable circuits of extended FRGs: (a) RFRG (I), (b) RFRG (II)

The eight extended gates proposed in [15] implement only a part of the possible 32 reversible functions based on the 3-bit Fredkin gate (1). Adding two extra key lines (K_3, K_4) and arranging the NCT gates makes it possible to obtain an extensive reconfigurable circuit RFRG (II) (Fig. 2b) with a similar configuration. It has enough key lines to represent 32 unique variations of extended Fredkin gates (Table 2). RFRG (II) fully incorporates RFRG (I). This attribute is proved by Table 2, which includes all the logic functions of Table 1, offering 24 additional unique logic functions and, thus, new reversible circuits of extended Fredkin gates. Proposed representations of the reconfigurable circuits can be easily adjusted to a specific problem by removing key lines and rearranging the NCT gates on these lines. For instance, a reconfigurable circuit with only the first 16 functions from Table 2 is an RFGR (II) without K_4 and the gate connected to this key line.

The key signals are transmitted to the outputs without changes since only the control bits of the reversible gates are connected to them. The hardware complexity of RFRG (I) is the same as that of RFRG (II) and equals three times more than the hardware complexities of the classic FRG, which is insignificant considering the 32-to-1 gates advantage.

Table 2. Activation keys of RFRG (II)

$K_0 K_1 K_2 K_3 K_4$	Function ($Y_0 = X_0$)	$K_0 K_1 K_2 K_3 K_4$	Function ($Y_0 = X_0$)
0 0 0 0 0	$Y_1 = \overline{X_0} X_1 \oplus X_0 X_2$ $Y_2 = \overline{X_0} X_2 \oplus X_0 X_1$	0 0 0 0 1	$Y_1 = \overline{X_0} X_1 \oplus X_0 \overline{X_2}$ $Y_2 = \overline{X_0} X_2 \oplus X_0 \overline{X_1}$
1 0 0 0 0	$Y_1 = \overline{X_0} X_2 \oplus X_0 X_1$ $Y_2 = \overline{X_0} X_1 \oplus X_0 X_2$	1 0 0 0 1	$Y_1 = \overline{X_0}\,\overline{X_2} \oplus X_0 X_1$ $Y_2 = \overline{X_0} X_1 \oplus X_0 X_2$
0 1 0 0 0	$Y_1 = \overline{X_0}\,\overline{X_1} \oplus X_0 X_2$ $Y_2 = \overline{X_0} X_2 \oplus X_0 \overline{X_1}$	0 1 0 0 1	$Y_1 = \overline{X_0}\,\overline{X_1} \oplus X_0 \overline{X_2}$ $Y_2 = \overline{X_0} X_2 \oplus X_0 \overline{X_1}$
1 1 0 0 0	$Y_1 = \overline{X_0} X_2 \oplus X_0 \overline{X_1}$ $Y_2 = \overline{X_0} X_1 \oplus X_0 X_2$	1 1 0 0 1	$Y_1 = \overline{X_0} X_2 \oplus X_0 \overline{X_1}$ $Y_2 = \overline{X_0} X_1 \oplus X_0 \overline{X_2}$
0 0 1 0 0	$Y_1 = \overline{X_0} X_1 \oplus X_0 \overline{X_2}$ $Y_2 = \overline{X_0}\,\overline{X_2} \oplus X_0 X_1$	0 0 1 0 1	$Y_1 = \overline{X_0} X_1 \oplus X_0 X_2$ $Y_2 = \overline{X_0}\,\overline{X_2} \oplus X_0 X_1$
1 0 1 0 0	$Y_1 = \overline{X_0} X_2 \oplus X_0 X_1$ $Y_2 = \overline{X_0} X_1 \oplus X_0 \overline{X_2}$	1 0 1 0 1	$Y_1 = \overline{X_0} X_2 \oplus X_0 X_1$ $Y_2 = \overline{X_0} X_1 \oplus X_0 \overline{X_2}$
0 1 1 0 0	$Y_1 = \overline{X_0}\,\overline{X_1} \oplus X_0 \overline{X_2}$ $Y_2 = \overline{X_0}\,\overline{X_2} \oplus X_0 \overline{X_1}$	0 1 1 0 1	$Y_1 = \overline{X_0}\,\overline{X_1} \oplus X_0 X_2$ $Y_2 = \overline{X_0}\,\overline{X_2} \oplus X_0 \overline{X_1}$
1 1 1 0 0	$Y_1 = \overline{X_0}\,\overline{X_2} \oplus X_0 \overline{X_1}$ $Y_2 = \overline{X_0} X_1 \oplus X_0 \overline{X_2}$	1 1 1 0 1	$Y_1 = \overline{X_0}\,\overline{X_2} \oplus X_0 \overline{X_1}$ $Y_2 = \overline{X_0} X_1 \oplus X_0 \overline{X_2}$
0 0 0 1 0	$Y_1 = \overline{X_0} X_1 \oplus X_0 X_2$ $Y_2 = \overline{X_0} X_2 \oplus X_0 \overline{X_1}$	0 0 0 1 1	$Y_1 = \overline{X_0} X_1 \oplus X_0 \overline{X_2}$ $Y_2 = \overline{X_0} X_2 \oplus X_0 \overline{X_1}$
1 0 0 1 0	$Y_1 = \overline{X_0} X_2 \oplus X_0 X_1$ $Y_2 = \overline{X_0} X_1 \oplus X_0 X_2$	1 0 0 1 1	$Y_1 = \overline{X_0}\,\overline{X_2} \oplus X_0 X_1$ $Y_2 = \overline{X_0} X_1 \oplus X_0 X_2$
0 1 0 1 0	$Y_1 = \overline{X_0}\,\overline{X_1} \oplus X_0 X_2$ $Y_2 = \overline{X_0} X_2 \oplus X_0 \overline{X_1}$	0 1 0 1 1	$Y_1 = \overline{X_0}\,\overline{X_1} \oplus X_0 \overline{X_2}$ $Y_2 = \overline{X_0} X_2 \oplus X_0 \overline{X_1}$
1 1 0 1 0	$Y_1 = \overline{X_0} X_2 \oplus X_0 \overline{X_1}$ $Y_2 = \overline{X_0} X_1 \oplus X_0 X_2$	1 1 0 1 1	$Y_1 = \overline{X_0} X_2 \oplus X_0 \overline{X_1}$ $Y_2 = \overline{X_0} X_1 \oplus X_0 \overline{X_2}$
0 0 1 1 0	$Y_1 = \overline{X_0} X_1 \oplus X_0 \overline{X_2}$ $Y_2 = \overline{X_0} X_2 \oplus X_0 \overline{X_1}$	0 0 1 1 1	$Y_1 = \overline{X_0} X_1 \oplus X_0 X_2$ $Y_2 = \overline{X_0}\,\overline{X_2} \oplus X_0 \overline{X_1}$
1 0 1 1 0	$Y_1 = \overline{X_0} X_2 \oplus X_0 X_1$ $Y_2 = \overline{X_0} X_1 \oplus X_0 \overline{X_2}$	1 0 1 1 1	$Y_1 = \overline{X_0} X_2 \oplus X_0 X_1$ $Y_2 = \overline{X_0} X_1 \oplus X_0 \overline{X_2}$
0 1 1 1 0	$Y_1 = \overline{X_0}\,\overline{X_1} \oplus X_0 X_2$ $Y_2 = \overline{X_0}\,\overline{X_2} \oplus X_0 X_1$	0 1 1 1 1	$Y_1 = \overline{X_0}\,\overline{X_1} \oplus X_0 X_2$ $Y_2 = \overline{X_0}\,\overline{X_2} \oplus X_0 X_1$
1 1 1 1 0	$Y_1 = \overline{X_0}\,\overline{X_2} \oplus X_0 \overline{X_1}$ $Y_2 = \overline{X_0} X_1 \oplus X_0 \overline{X_2}$	1 1 1 1 1	$Y_1 = \overline{X_0}\,\overline{X_2} \oplus X_0 \overline{X_1}$ $Y_2 = \overline{X_0} X_1 \oplus X_0 \overline{X_2}$

Table 3. Results of the RFRG (II) testing

Hex	Gate outputs	Hex	Gate outputs	Hex	Gate outputs	Hex	Gate outputs
00	$+X_1+X_2+X_2+X_1$	02	$+X_1+X_2+X_2-X_1$	01	$+X_1+X_2-X_2+X_1$	03	$+X_1+X_2-X_2-X_1$
10	$+X_2+X_1+X_1+X_2$	12	$+X_2-X_1+X_1+X_2$	11	$-X_2+X_1+X_1+X_2$	13	$-X_2-X_1+X_1+X_2$
08	$-X_1+X_2+X_2-X_1$	0A	$-X_1+X_2+X_2+X_1$	09	$-X_1+X_2-X_2-X_1$	0B	$-X_1+X_2-X_2+X_1$
18	$+X_2-X_1-X_1+X_2$	1A	$+X_2+X_1-X_1+X_2$	19	$-X_2-X_1-X_1+X_2$	1B	$-X_2+X_1-X_1+X_2$
04	$+X_1-X_2-X_2+X_1$	06	$+X_1-X_2-X_2-X_1$	05	$+X_1-X_2+X_2+X_1$	07	$+X_1-X_2+X_2-X_1$
14	$-X_2+X_1+X_1-X_2$	16	$-X_2-X_1+X_1-X_2$	15	$+X_2+X_1+X_1-X_2$	17	$+X_2-X_1+X_1-X_2$
0C	$-X_1-X_2-X_2-X_1$	0E	$-X_1-X_2-X_2+X_1$	0D	$-X_1-X_2+X_2-X_1$	0F	$-X_1-X_2+X_2+X_1$
1C	$-X_2-X_1-X_1-X_2$	1E	$-X_2+X_1-X_1-X_2$	1D	$+X_2-X_1-X_1-X_2$	1F	$+X_2+X_1-X_1-X_2$

The RFRG (I) delay time can be represented as:

$$\tau_I = 2\tau_{CNOT} + 2\tau_{TG} + \tau_{FRG} = 4\tau_{CNOT} + 3\tau_{TG}, \qquad (2)$$

where τ_{CNOT}, τ_{TG}, τ_{FRG} - delay time of CNOT, Toffoli gate, and classic Fredkin gate, respectively. Since second and third CNOTs operate in parallel, their combined delay time equals the delay of a single CNOT gate. The last two CNOTs also conduct signals in time of one CNOT. Similarly to RFRG (I), the delay time of RFRG (II) becomes:

$$\tau_{II} = \tau_{CNOT} + 2\tau_{TG} + \tau_{FRG} = 3\tau_{CNOT} + 3\tau_{TG}. \qquad (3)$$

RFRG (II) operates one CNOT gate faster than RFRG (I). At the same time, RFRG (II) allows the implementation of 32 logical functions (Table 2), including 8 logical functions (Table 1), for which RFRG (I) is assigned. Nevertheless, the advantage of RFRG (I) is that it contains only 3 permanent key lines, in contrast to 5 key lines in RFRG (II). Depending on the design task, both reconfigurable circuits can be utilized while considering their advantages and disadvantages.

4 FPGA Implementation

A circuit is coded in hardware descriptive language (HDL) to implement it in an FPGA. One of the HDLs is the VHSIC Hardware Description Language (VHDL), which can model the behavior and structure of digital designs at various conceptual levels, from logic primitives to complex independent systems for describing, documenting, and verifying digital circuits [20].

The proposed RFRG (I) and RFRG (II) have been described and tested using VHDL in the Active-HDL 13.0 engineering software. For RFGR (II), the input and output signals have been initialized as binary vectors $X[0..2]$, $Ki[0..4]$ and $Y[0..2]$, $Ko[0..4]$ correspondingly. The architecture of the reconfigurable circuit is presented in Fig. 3. First of all, NCT gates were synthesized in advance and then imported into the design as components (lines 2–5), similar to the previous work [14]. "CNOT" represents the controlled-NOT gate, and component "TG" stands for the Toffoli gate. This approach allows safe modifications in the architecture of the NCT gates without changing anything directly in the RFRG's design. Next, a "port map" operation, which manually links signals to components' input and output, has been used to connect gates among each other. This way, the links between 9 gates from Fig. 2b have been established (*G0 - G8*). Gates *G5*, *G6*, and *G7* represent the classic Fredkin gate when the rest of the gates manage the reconfiguration for the FRG extensions.

The number of gates and their order are identical for RFRG (I) and RFRG (II). Therefore the VHDL implementation of RFRG (I) is analogous to RFRG (II), except it contains smaller logical signal vectors *Ki* and *Ko* due to fewer key lines in the circuit (Fig. 2a). The VHDL code of both circuits is ready for programming on FPGA boards, such as a 5CSEMA4U23C6N of the Altera Cyclone V, and can be used to synthesize more complex reversible circuits.

```
1    architecture RTL of RFRG_II is
2      component CNOT is port (X0,X1: in STD_LOGIC; Y0,Y1: out STD_LOGIC);
3      | end component;
4      component TG is port (X0,X1,X2: in STD_LOGIC; Y0,Y1,Y2: out STD_LOGIC);
5      | end component;
6      signal s: STD_LOGIC_VECTOR(0 to 12)
7    begin
8      G0: CNOT port map(Ki(0), X(0),  s(0),  s(1));
9      G1: CNOT port map(Ki(1), X(1),  Ko(1), s(2));
10     G2: CNOT port map(Ki(2), X(2),  Ko(2), s(3));
11     G3: TG   port map(Ki(3), s(1),  s(2),  Ko(3), s(4),  s(5));
12     G4: TG   port map(Ki(4), s(4),  s(3),  Ko(4), s(6),  s(7));
13     G5: CNOT port map(s(7),  s(5),  s(8),  s(9));
14     G6: TG   port map(s(6),  s(9),  s(8),  s(10), s(11), s(12));
15     G7: CNOT port map(s(12), s(11), Y(2),  Y(1));
16     G8: CNOT port map(s(0),  s(10), Ko(0), Y(0));
17   end RTL;
```

Fig. 3. The VHDL implementation of the RFRG (II) architecture

A separate testing program has been created to verify the correctness of the reconfig-urable circuits and their VHDL description. The VHDL implementations of RFRG (I) and RFRG (II) are imported as components into the testing code. The best practice is to handle testing circuits as a black box knowing only the input and output data of the devices. The circuit assessments have been carried out to reach the following goals:

(1) Ensure that the key and control signals of the circuit are transmitted to the output unchanged.
(2) Check whether the key lines are utilized optimally, activating 8 gates for 3 key lines and 32 gates with 5 lines.
(3) Confirm that key signals correctly activate the FRG extensions by comparing factual output for RFRG (I) and RFRG (II) with the expected output shown in Tables 1 and 2, respectively.

The timing diagram of the RFRG (II) test simulation is presented in Fig. 4. It shows input signals $X[0..2]$, $Ki[0..4]$, output signals $Y[0..2]$, $Ko[0..4]$, and the circuit speci-fications such as the number of unique gates "unq" and the number of repeating gates

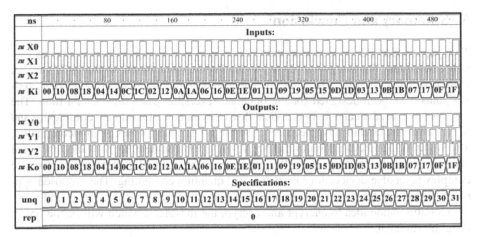

Fig. 4. The timing diagram of extended RFRG (II) in Active-HDL

"rep". Key signals are represented in hexadecimal values from 00 to 1F, where 00 corresponds to the key K[0..4] = "00000" and 1F to K[0..4] = "11111", respectively. The full simulation cycle is 514 ns, and the input data changes every two nanoseconds. Figure 4 confirms that the control signal and key signals are transmitted without changes to the outputs ($Y_0 = X_0$, $Ko = Ki$), and the key bits are optimally used as expected since "unq" = 32, "rep" = 0 at the end of the simulation.

For RFRG (I), the simulation cycle is almost 4 times shorter due to fewer key lines. The testing results show that RFRG (I) contains 8 unique gates and the output signals of these gates match with the output data of gates from RFGR (II), confirming that RFRG (II) fully contains the functionality of RFRG (I).

Table 3 shows the output values of the informational signal lines following the hexadecimal representation of the key signals. Since the control signals are transmitted unchanged ($Y_0 = X_0$), the output signals are represented in the following order $Y_1^0 Y_2^0 Y_1^1 Y_2^1$, where Y_1^0, Y_2^0 stand for target line values on the output while $X_0 = 0$, and Y_1^1, Y_2^1 are the values on the output while $X_0 = 1$. The arithmetic signs '+' and '−' indicate the polarity of the signals. Table 3 shows the correct key-gate activations of the extended Fredkin functions established in Table 2. Based on the simulation results, the proposed RFRG (II) activates 32 unique extended Fredkin gates without repeating activations, as expected.

The reconfigurable circuits for reversible extended Fredkin gates have been described in the VHDL code. The programmed models are fully functional and they have been implemented in a 5CSEMA4U23C6N chip of the Altera Cyclone V. This approach allows for the synthesis of new binary logic digital reversible devices in a more compact design and various applications, particularly for message encryption and decryption devices, which will be the subject of our further research.

5 Summary and Conclusion

This paper offers 32 variations of the Fredkin gate containing 8 known gates and 24 new extensions. The analysis of the extended Fredkin gate based on the NCT gate set with mixed polarity was carried out. Two new reconfigurable circuits of extended Fredkin gates have been designed for the obtained reversible gates for the first time. The devices have an advantage in operational flexibility in cost of hardware complexity and delay time. The circuits have been described in HDL code for FPGA implementation. The simulation and evaluation in the Active-HDL environment approve the functional correctness of the HDL models. The circuits can be utilized to design and describe complex reversible devices for low-power and quantum computing. The proposed approach allows synthesizing new extended Fredkin gates with N control lines and using the reconfiguration idea to encrypt and decrypt the messages, which will be the object of future research.

References

1. Chandrika, V.O., Kumar, S.M.: Design and analysis of SRAM cell using reversible logic gates towards smart computing. Supercomput **78**(2), 2287–2306 (2022)
2. Shafi, A., Bahar, A.N.: Fredkin circuit in nanoscale: a multilayer approach. Int. J. Inf. Technol. Comp. Sci. (IJITCS) **10**(10), 38–43 (2018)
3. Landauer, R.: Irreversibility and heat generation in the computing process. IBM J. Res. Dev. **5**(3), 183–191 (1961)
4. Bérut, A., Petrosyan, A., Ciliberto, S.: Information and thermodynamics: experimental verification of Landauer's erasure principle. Stat. Mech. Theory Exp. **6**, P06015 (2015)
5. Bennett, C.H.: Logical reversibility of computation. IBM J. Res. Dev. **17**(6), 525–532 (1973)
6. Abdessaied, N., Drechsler, R.: Reversible and Quantum Circuits. Optimization and Complexity Analysis. Springer, Cham (2016)
7. Sasamal, T.N., Gaur, H.M., Singh, A.K., Mohan, A.: Reversible circuit synthesis using evolutionary algorithms. Lect. Notes Electr. Eng. **577**, 115–128 (2020)
8. Ghosh, M., Dey, N., Mitra, D., Chakrabarti, A.: A novel quantum algorithm for ant colony optimization. IET Quantum. Commun. **3**(1), 13–29 (2022)
9. Shahidi, S.M., Borujeni, E.S.: A new method for reversible circuit synthesis using a simulated annealing algorithm and don't-cares. Comput. Electron. **20**(1), 718–734 (2021)
10. Deibuk, V., Biloshytskyi, A.: Design of a ternary reversible/quantum adder using genetic algorithm. Int. J. Inf. Technol. Comp. Sci. (IJITCS), **7**(9), 38–45 (2015)
11. Deibuk, V.G., Yuriychuk, I.M., Lemberski, I.: Fidelity of noisy multiple-control reversible gates. Semicond. Phys. Quantum Elec. Optoelec. **23**(4), 385–392 (2020)
12. Moraga, C.: OR-Toffoli and OR-peres reversible gates. In: Yamashita, S., Yokoyama, T. (eds.) Reversible Computation. RC 2021. Lecture Notes in Computer Science, vol. 12805. Springer, Cham (2021). https://doi.org/10.1007/978-3-030-79837-6_17
13. Fredkin, E., Toffoli, T.: Conservative logic. Int. J. Theor. Phys. **21**(3–4), 219–253 (1982)
14. Dovhaniuk, O., Deibuk, V.: Synthesis and implementation of reconfigurable reversible generalized Fredkin gate. In: Proceedings of 2021 IEEE 12th International Conference on Electronics and Information Technologies (ELIT), Ukraine, pp. 165–169 (2021)
15. Moraga, C., Hajam, F.Z.: The Fredkin gate in reversible and quantum environments. Facta Univ. Ser.: Electron. Energetics **36**(2), 253–266 (2023)

16. Murali Krishna, B., Sri Kavya, K.S., Sai Kumar, P.V.S., Karthik, K., Siva, N.Y.: FPGA implementation of image cryptology using reversible logic gates. Int. J. Adv. Trends in Comput. Sci. Eng. **9**(3), 2522–2526 (2020)
17. Rajesh, K., Umamaheswara, R.G.: FPGA implementation of encryption and decryption of a message using optimized reconfigurable reversible gate. Int. J. Recent Technol. Eng. **8**(2), 1654–1658 (2019)
18. Pawlowski, M., Szyprowski, Z.: Implementation of reversible gates in FPGA structure. Proc. SPIE **10445**, 445–455 (2017)
19. Dovhaniuk, O., Deibuk, V.: CMOS simulation of mixed-polarity generalized Fredkin gates. In: 2022 12th International Conference on Advanced Computer Information Technologies (ACIT), Slovakia, pp. 388–391 (2022)
20. Pedroni, V.A.: Circuit Design with VHDL. The MIT Press, Cambridge (2020)
21. Jamuna, S., Dinesha, P., Shashikala, K.P., Kishore Kumar, K.: Design and implementation of reliable encryption algorithms through soft error mitigation. Int. J. Comp. Netw. Inf. Secur. (IJCNIS) **12**(4), 41–50 (2020)

Applying Data Mining Techniques in People Analytics for Balancing Employees' Interests

Liana Maznyk[1], Zoriana Dvulit[2(✉)], Nadiia Seliuchenko[2], Marian Seliuchenko[2], and Olena Dragan[1]

[1] National University of Food Technologies, Volodymyrska Str. 68, Kyiv 01601, Ukraine
[2] Lviv Polytechnic National University, Bandera Str. 12, Lviv 79000, Ukraine
zoriana.p.dvulit@lpnu.ua

Abstract. In recent years the use of remote or hybrid formats of work organization in enterprises has become the world's tendency. The introduction of the severe quarantine limitations related to the pandemic of COVID-19 and the avalanche-type character of the digitalization of society activated the changes in the sphere of social-labor relations, where the interests of different groups of workers are asymmetric. Many researchers who studied these phenomena mark the presence of divergences in individual advantages concerning the different formats of a labor organization of employers and hired workers. The dynamic of changes predefines the necessity of further research, which is why this subject acquires greater relevance. Our research aims to substantiate the approaches to balancing the interests of different categories of workers regarding the labor organization formats by exposing disbalances in the conditions of full-scale war in Ukraine. For monitoring, diagnosis, and management of disbalances, we substantiate the use of Data Mining in People Analytics (PA), the necessity of their combination with the classic instruments of study of personnel behavior, the use of active (reactive) formats of study of persons' opinions in the extreme terms of running a business. The research results correlate with and confirm the world's tendencies of introducing a hybrid model of labor organization. The novelty of the executed research consists of substantiating the approaches to balancing the interests of the employees regarding different labor organization formats in the conditions of full-scale war in Ukraine. The presented study expands the scope of knowledge in the field of PA, as it takes into account new circumstances that have not existed since the end of World War II. It supplements data on work organization formats obtained by researchers of the impact of the COVID-19 pandemic with new data obtained as a result of a survey of employees and managers of various organizations in Ukraine, where there is a full-scale war. The presented study is one of the first that emphasizes the features of PA during wartime and explains why in such conditions it is necessary to combine Data Mining techniques with classic tools for studying personnel behavior. The authors substantiate possible approaches to balancing the interests of different categories of employees.

Keywords: People Analytics (PA) · Data Mining · Employers · Employees · Survey · Disbalance of interest · Formats of labor organization

Z. Hu et al. (Eds.): ICCSEEA 2023, LNDECT 181, pp. 106–115, 2023.
https://doi.org/10.1007/978-3-031-36118-0_10

1 Instruction

Long-term research conducted by HRM experts and scientists in personnel management proves that the high involvement (loyalty) of personnel directly affects their performance indicators such as the level of labor discipline, quality of performance of production tasks, compliance with deadlines for tasks, etc., which, in turn, affects the overall business performance [1]. Based on the data from the American institute of public opinion Gallup [2], in companies with a high level of personnel engagement the level of workers' absence in the workplace is 81% lower, the turnover of personnel is 18% lower, and the level of profitability is 23% higher. More than 50 years old research on employee engagement showed that engaged workers achieve better business results than other workers in any industry.

The study of personnel engagement problems shows that the most successful collaboration occurs when employees and the organization share common values. The values of the employee determine their behavior. The analysis of the personnel behavior helps to answer the question: «What behavior of worker is effective for the achievement of business goals and helps to react to any challenges faced by the organization?». These and other similar questions are the main subjects of People Analytics (PA) that get increasing interest from practicians and scientists [3–5]. With the introduction of PA, organizations began to actively implement technologies and methods of information collection about workers' behavior during business hours. The digitalization of the economy, the development of the digital labor market, and the COVID-19 pandemic which acted as the catalyst and activated the transition of the labor organization formats stipulated the necessity of processing large data volumes and the use of Data Mining techniques for PA [6–8]. The results of the use of PA for the last three years show that more workers are predisposed to remote work, and front-rank companies already prepare for it technically and technologically. However, on the whole, there is a significant divergence in the estimations of advantages of different formats of a labor organization of employers and hired workers [9, 10] that predetermines the necessity of social-labor relations harmonization for the corresponding direction.

The full-scale war in Ukraine with its negative consequences made adjustments to the processes of functioning of organizations on the territory of Ukraine. Research concerning the peculiarities of the interaction of managers and employees in such realities is limited, therefore the study of people's behavior in new socio-economic and political conditions is relevant.

The presented research targets to study the opinions of workers and leaders of Ukrainian organizations that have experience in different formats of labor not only during the pandemic of COVID-19 but in conditions of the full-scale war in Ukraine, as well as the results of identifying imbalances in their interests and finding possible ways to eliminate them. Hence, the requirement was defined to formulate the following research questions and find answers to them. First is how is the interest of workers and leaders balanced concerning the different formats of labor organization in the conditions of full-scale war with its consequences. Second, what is the organizations' technical and technological readiness level for remote work? Third, what approaches can organizations use to balance the interests of different categories of employees?

2 Method of Research

2.1 Methods of Collecting the Opinions of Employers and Employees Regarding Attitudes to Different Formats of Work Organization

The composition of the offered research considers the necessity of combining Data Mining for PA technologies with the classic methods of personal behavior study that are especially important in extreme terms (pandemics, military operations, blackouts). Since knowledge about people's behavior in the conditions of a full-scale war with all its negative consequences is limited, the proposed approach reduces the risks of using exclusively artificial intelligence in People Analytics. It makes it possible to take into account specific situations and the specifics of each person's reaction to them. Results of previously conducted monitoring confirm the constantly growing asymmetry of employers' and workers' answers concerning the advantages and disadvantages of the different formats of labor organization. The presented research was organized in the form of observational data analysis. A sample is formed with the use of the method of representative sampling. The choice of method of observational data analysis is based on the necessity to research the opinions of different employee groups that are included in the general population. The groups are formed based on the qualitative characteristics of "leaders" and "workers (specialists, professionals)".

To answer the research questions we formed two questionnaires that contained 13 questions of the closed type, one for the leaders and one for the workers. The survey envisaged the parallel study of leaders' and workers' opinions by an identical list of questions. We collected responses from 101 respondents who participated in the survey and worked in different industries of the economy, where 24 were leaders and 77 were workers (specialists, professionals). The survey took place in Ukraine from November 22 to December 02, 2022. To implement the questionnaires, we used the Google Forms tool and to collect responses we shared the questionnaires on the most popular social networks and messengers. Generalization of the survey results allows for determining the level of coordination of opinions of different groups of respondents. The sample is much smaller than the general population – the number of employees in Ukraine in 2020 was 9 948,1 thousand people, and the corresponding statistical data for 2021–2022 is missing. With a probability of 0,95, we can say that the margin of error of the survey results is 9,75%. Our experimental expectations were to diagnose imbalances in the interests of employees and managers of organizations regarding different formats of work organization. This made it possible to substantiate approaches to balancing their interests.

2.2 Literature Review

During the pandemic of COVID-19 and strict quarantine restrictions, when organizations passed to the remote format of work, HR managers faced several problems related to the functions of diagnosis, coordination, supervision, and motivation [11, 12]. It gave the impulse to the development and implementation of innovative and effective methods of employee engagement. The introduction of employee engagement measures based on digital technologies became especially important for the growth of organizations in crisis [13, 14].

The tendency of the increasing influence of personnel analytics teams, which was observed from 2020 to 2021, was stable in 2022. In 2021 the correlation between analytics team members and total employee headcount changed from 1: 4000 (2020) to 1: 2900. In 2022, this correlation remains at the level of 2021. Leaders of personnel analytics get a more significant influence on the top management of organizations: 21% of leaders report directly to the Chief Human Resources Officer comparatively with 13% in 2021 [15]. In regards to the increasing role of PA, the increasing popularity is acquired by research on the integration of artificial intelligence in managing human resources (HRM) in the context of different functions and practices in organizations. Experts in the field of HRM consider that possession of the relevant digital competencies by the workers in this field is the base condition of forming the modern labor market of HR [16]. The development of PA stipulated the introduction of the systems of HR analytics that are based on the analysis of large arrays of data and apply artificial intelligence to provide the possibility to get practical conclusions (unlike regular reports that give data).

Software that developers offer for PA is various: it can be concentrated only on one HR function (payroll management, management of talents, systems of watching of declarants, training and development, production management); offers a single window (HR platforms) that manages all processes from hiring and onboarding to engaging and evaluation of the personnel productivity. To monitor the workers, HR-tech companies offer packages that monitor everything, beginning with keypresses and ending with access to the external disk. The best software offered in the market in 2022 can monitor the visited web pages and used applications, block content and applications, make screenshots of users' activity, and generate detailed reports [17]. Digitalization of PA allows HR managers to control the workers' productivity and efficiency in the office and remotely. HR managers can regularly get information about workers that work remotely and, thus, estimate the influence of different formats of labor organization on the productivity, quality, efficiency, psychophysiological state of workers, levels of stress in a team, work difficulties, mutual relations in the team, informal leaders.

HR systems that collect anonymized data about employees using questioning or digital tracking (corporate messengers, email, and others), allow forecasting the future, e.g., to forecast personnel turnover or employee behavior. Artificial intelligence gave strong development to the management strategy of talents, as now, in the early stage, it is possible to estimate the potential of the candidate and choose the right moment for rotation and promotion at work. By studying employees' behavior, an organization can estimate the level of work satisfaction, psychical health, and prosperity of workers, timely reduce their professional burnout, determine the effective or toxic managers, etc. When using PA as an instrument for balancing the interests of leaders and workers, one should take into account its "shadow sides" [18], which are bringing about an illusion of control and reductionism; leading to estimated predictions and self-fulfilling prophecies; fostering path dependencies; impairing transparency and accountability; reducing employees' autonomy; marginalizing human reasoning; and eroding managerial competence.

3 Result and Discussion

The analysis of the survey's results showed that most of the polled workers (55,8%) and leaders (50%) represent service businesses. About 24,7% of workers and 16,7% of leaders represent the IT sector. The majority of workers (55,8%) represent organizations with over 100 people, and 41,7% of leaders represent organizations with under 50 people. The survey results showed that almost 45,3% of workers can perform their duties entirely remotely, 34,7% of workers can perform their duties in the hybrid format, and only 20% are not able to work remotely. Moreover, only every third leader gave a positive answer to the possibility of working remotely, 37,5% confirmed they can perform their duties in hybrid mode, and 29,2% of the polled leaders are not able to work remotely.

It is necessary to mark that among the polled workers some were in positions with functional duties that require a physical presence in the workplace. Thus, even the influence of extreme circumstances is not a substantial criterion for choosing the format of labor organization. To the question "Does your company regulate the use of different work formats in the regulatory documents?" almost 58% of workers gave a positive answer, 30,3% answered that work formats are not clearly regulated, and 11,8% pointed out the absence of such regulations.

Only 45,8% of leaders confirmed the presence of regulation of work formats in normative documents, 37,5% mark the present regulation as not clear enough, and about 17% state the absence of regulation of labor organization formats in their companies. The results testify that leaders have a more serious understanding of approaches to forming internal normative documents concerning corporate measures in extreme terms.

To the question "Does your company use the efficiency indexes (KPI, OKR) or other criteria to evaluate your job results?" almost 60% of respondents from the two groups gave positive answers, 37,5% of the polled leaders and 26,6% of workers are sometimes informed about kinds and controls over their professional duties performance and 16,8% of the polled workers expressed no interest in this question. The latter can be explained by the respondents that are confident about adequately implementing their duties. Some respondents (8,3% leaders and 7,8% workers) answered that such questions were not relevant due to the high level of personnel engagement in their organization.

According to the given answers, it is possible to mark that not many leaders and workers confirm the absence of information about the kinds and control methods. This is the evidence that Ukraine's system of social-labor relations is sufficiently transparent and "shadow sides" of PA are minimized.

About the work in the office, positive answers gave only 26% of workers, while this format of labor organization chose about 45,8% of leaders. Among the respondents, 55,8% of workers and 37,5% of leaders prefer the hybrid format of work organization. Fractions of workers and leaders that would like to work remotely are close (18,2% and 16,7%) which testifies to the presence of a balance of their interests concerning the organization of the remote work format (Table 1).

In the opinion of most workers (52,6%) and leaders (45,8%), the work in the office is characterized by higher productivity in performing their professional duties. Remote work is preferred by about 18,2% of workers and 16,7% of leaders. Interestingly, only 14,5% of workers and 12,5% of leaders consider productivity to be higher in terms of the remote format of work organization. Thus, the disbalance occurs between the

Table 1. The attitude of the employees toward the different formats of labor organization

Work format	Answers	Groups of employees	
		Leaders, %	Workers, %
In the office	Prefer	45,8	26,0
	Productivity is higher	45,8	52,6
Remote	Prefer	16,7	18,2
	Productivity is higher	12,5	14,5
Hybrid	Prefer	37,5	55,8
	Productivity is higher	41,7	32,9

answers of both polled groups concerning the advantages and productivity of working in a remote format: a particular part of respondents would like to work remotely despite the understanding that labor productivity will be lower. Fewer leaders (37,5%) prefer the hybrid format, although a more significant part (41,7%) asserts that such a format provides higher labor productivity. The more significant part of workers (55,8%) wants to work in the hybrid format, understanding that the labor productivity will be lower. Thus, leaders estimate labor productivity changes depending on the different work formats. More than half of polled workers consider working in the office more productive, but only a fourth of them prefer this format of work organization. For the management of companies, it must signal the presence of a large disbalance of interests, which is why corporate measures must include specific stimulation instruments. The choice of certain instruments will depend on the results of the PA.

Part of the survey questions focused on determining the factors that induce workers or leaders to prefer work in hybrid or remote formats. For this purpose, it was offered to define the priorities of the factors such as saving commuting time and expenses, saving time on complying with business style, and digitalizing the personal space. Summarizing the survey results showed that most leaders considered digitalization a priority in the ground of hybrid or distance formats. For workers, these priorities are saving time and travel expenses. The coordination of respondents' answers concerning their priorities was evaluated during the selection of remote or hybrid formats of work using the coefficient of mutual conjugating of Cramer. We will mark that the value of this coefficient can change in limits from 0 to 1. The closer the value to 0, the weaker the connection conditions and vice versa. Calculations showed that both polled groups were not homogeneous enough and that higher divergence of thoughts was among the leaders which equals to 0,1145. Concerning workers, the coordination in this group is not far but higher and equals to 0,1511. Thus, such values testify to the weak but noticeable coordination in each investigated group. Answers in the survey concerning existing technical and technological terms for remote work, level of satisfaction of hired workers by these terms, and reporting methods for the executed work allow evaluating the level of technical and technological readiness of organizations to remote or hybrid formats of work.

More than 60% of leaders confirm the presence of corresponding technical and technological terms for the remote and hybrid work formats. Only 49,3% of workers gave positive answers to this question. It can be explained by (1) comparatively, with the leaders, a more significant amount of workers have experience working in the remote and hybrid formats in the conditions of quarantine limitations, which allows them to estimate present terms more critically in war-time; (2) part of workers having experience in remote or hybrid work demonstrate the belated reaction to the situation of blackout in Ukraine; (3) among the polled leaders, 41,7% represent companies with under 50 employees, most characteristically for the IT sphere, electronic commerce, IT-marketing, where technical and technological terms are more adapted to the modern extreme terms of the functioning of Ukrainian business; (4) leaders perceive corresponding challenges more strategically and comprehensively approach to the evaluation of the created terms, which envisages not only the use of software facilities for remote and hybrid formats of work but also the necessity of the use of alternative methods of energy supply.

Regarding the level of satisfaction by the technical and technological terms for the realization of the remote and hybrid formats of work, fractions of answers of leaders and workers almost coincided, i.e., disbalance in this question is absent. The majority of respondents are totally or partly satisfied with the created terms. Answers of respondents to the question «What method do they use for evaluating the executed work»? were divided into 43,4% of workers and 45,8% of leaders reporting about the executed work by electronic reports; every third worker (30,3%) and every eighth leader (12,5%) reported using specialized software. This testifies that organizations use modern technologies for information collection about the workers. The results of the survey we conducted correlate with the conclusions obtained in other analogical surveys. However, such survey plans did not envisage the parallel study of leaders' and workers' opinions with an identical list of questions.

A Survey of leaders' opinions of leading companies of Ukraine, conducted from July to October 2022 by KPMG Ukraine [19], showed that 63% of business leaders of Ukrainian companies and 48% of leaders of the world companies marked that hybrid work positively influenced on investments in technology of workplace. At the same time, 5% of leaders in the world said that a hybrid format negatively influenced workers' maintenance. In contrast, business leaders in Ukraine admitted vice versa, that the possibility to work from home or the office had positively influenced the maintenance of talents. However, in the opinion of Ukrainian leaders, hybrid work negatively influences workers' productivity (42%) and worsens their morale state (37%). In the world, these indexes are far below the level of 13% and 12%, accordingly. Without regard to that, the leaders of Ukrainian companies in the future think about the work in a hybrid format (88%) that exceeds the expectation of world leaders (28%) three times. Only 5% of the Ukrainian leaders speak about the obligatoriness of traditional work in the office. At the same time, leaders in Ukraine (5%) and in the world (7%) do not plan to transform the working environment to the format of fully remote collaboration.

The results of our research practically coincided with the survey of KPMG Ukraine [11] regarding the leaders' that give the advantage to the hybrid format of work organization. However, a contradiction occurs concerning the results of the influence of such a format on labor productivity. From our data, in the opinion of 41,7% of leaders, the hybrid format of labor organization assists the increase of labor productivity, and according to the results of KPMG Ukraine, 42% of leaders think that such format will negatively influence labor productivity. Such divergences in the results of the survey can be explained by different times of their realization: KPMG Ukraine was conducted during July - October 2022, when the destruction of power infrastructure of Ukraine was not yet there; our research is undertaken in November - December 2022, when periodic partial and complete blackouts began to occur in the context of war. Thus, we took into account the change of terms of doing business in Ukrainian realities by using the instruments of Data Mining.

According to research [10], 95% of employers want workers to be in the office. At the existence of choice, 99% of workers expressed the desire to work remotely their whole life even in part-time jobs. However, Ukrainian realities showed quite another structure of such priority distribution in the choice of different work formats. About 45,8% of leaders and 26% of workers entirely give the advantage to working in the office, and among the supporters of remote and hybrid formats are 54,2% of leaders and 74% of workers. According to our research, such distribution of ideas does not conflict with the survey [19] results, where two-thirds of people want to work at home, and 80% of workers think about the remote format as the competitive edge of the enterprise at the labor market. According to AT&T research [20], the hybrid model of work in the USA was 42% in 2021, with the prospect of increasing to 81% in 2024. This format is the prevailing model of the organization of the working process, and most business leaders consider this as a new standard. In Ukraine, according to our research, 79,2% of leaders and 82,4% of workers have the experience to work in remote or hybrid formats. These numbers for 2022 in Ukraine correspond to the forecasts of researchers for the USA in 2024. Despite all advantages of hybrid and remote formats of work for the different groups of hired workers, some problems take place such as formal and informal communication in teams, work and personal life balance, level of trust in the team; inoperative reactions to current problems, etc.

To balance the interests of the hired workers, we recommend using such approaches: (i) combination of Data Mining, People Analytics and classical instruments of study of personnel behavior; (ii) the use of active (reactive) forms of studying people's opinions in the extreme terms of running a business; (iii) integration of the module of behavior analytics in any automated HRM system that is used in the enterprise; (iv) providing a transparent process of collecting information about workers using a personal electronic cabinet for every worker, where such individual information can be accessed by each worker in read-only mode as well as by the leader with the corresponding access level; (v) development of specific technical, technological, and organizational measures based on diagnostics of the hired workers attitude to implementing functional duties in the different formats of labor organization.

4 Summary and Conclusion

In the presented research, the disbalances of interests of separate categories of workers concerning the different formats of labor organization are exposed and approaches to harmonizing corresponding social-labor relations in the conditions of full-scale war in Ukraine. Our results showed imbalances that occured between the answers of both polled groups concerning the advantages and productivity of working in a remote format: a particular part of respondents would like to work remotely despite the understanding that labor productivity will be lower. Fewer leaders and workers prefer the hybrid format, although a more significant part asserts that such a format provides higher labor productivity. The majority of leaders confirm the presence of corresponding technical and technological terms for the remote and hybrid work formats. Less than half of workers gave positive answers to this question.

The presented study expands the scope of knowledge in the field of PA, as it takes into account new circumstances that have not existed since the end of World War II. Modern PA systems, for objective reasons, cannot use historical data on the behavior of employees in the conditions of a full-scale war in Ukraine. Therefore, to identify imbalances in the interests of employees, it is recommended to combine Data Mining, People Analytics and classical instruments of study of personnel behavior. The results of the study complement the already available data on the attitude of employees to various formats of work organization during the COVID-19 pandemic. Surveying employees through the distribution of Google Forms does not provide an opportunity to ensure the representativeness of the sample. However, in the difficult conditions of the war, with its help, the most acute problems in the field of labor organization can be identified. HR managers can further use the information obtained during the formation of a questionnaire for a regular survey within a specific organization and for making management decisions based on Data Mining techniques.

We expect that the results of the conducted research will contribute to the growth of scientific interest in the field of application of Data Mining in People Analytics in the conditions of global challenges of society (digital economy, pandemics, climate problems, armed conflicts, full-scale war in Ukraine, world food crisis, etc.).

References

1. Govender, P., Sukdeo, N., Ramdass, K.: Employee engagement and the impact on productivity in a South African FMCG company. In: Proceedings of the 7th North American International Conference on Industrial Engineering and Operations Management, Orlando, pp. 205–220 (2022)
2. Website Gallup. https://cutt.ly/p0KNlta
3. Huselid, M.: The science and practice of workforce analytics: introduction to the HRM special issue. Hum. Resour. Manage. **57**(3), 679–684 (2018)
4. Allaham, M.: Bibliometric analysis of HR analytics literature. Elec. J. Soc. Sci. **21**(83), 1147–1169 (2022)
5. McCartney, S., Fu, N.: Promise versus reality: a systematic review of the ongoing debates in people analytics. J. Organ. Effect. People Perform. **9**(2), 281–311 (2022)
6. Dahlbom, P., Siikanen, N., Sajasalo, P., Jarvenpää, M.: Big data and HR analytics in the digital era. Balt. J. Manag. **15**(1), 120–138 (2019)
7. Chamorro-Premuzic, T., Bailie, I.: Tech is transforming people analytics. Is that a good thing? Harvard Business Review (2020). https://cutt.ly/Z0KB23S
8. Repaso, J., Capariño, E., Hermogenes, M., Perez, J.: Determining factors resulting to employee attrition using data mining techniques. Int. J. Educ. Manag. Eng. **12**(3), 22–29 (2022)
9. Website Moneyzine. Martynas Pupkevicius. Revealing remote work statistics & facts for 2022. https://moneyzine.com/careers-resources/remote-work-statistics
10. Website KPMG. https://cutt.ly/E0KNyvc

11. Hussain, M., Mirza, T., Hassan, M.: Impact of COVID-19 pandemic on the human behavior. Int. J. Educ. Manag. Eng. **10**(5), 35–61 (2020)
12. Salehin, I., Tamim, D.S., Mohammad, T.I., Rayhan, I., Fatema, N.K.: Impact on human mental behavior after pass through a long time home quarantine using machine learning. Int. J. Educ. Manag. Eng. **11**(1), 41–50 (2021)
13. Nisha Chanana, S.: Employee engagement practices during COVID-19 lockdown. J. Publ. Affairs (2020)
14. Jack Linchuan Qiu: Humanizing the posthuman: digital labour, food delivery, and openings for the new human during the pandemic. Int. J. Cult. Stud. **25**(3–4), 445–461 (2022)
15. Ferrar, J., Verghese, N., González. N.: Impacting Business Value: Leading Companies in People Analytics. shorturl.at/uwD57
16. Shpak, N., Maznyk, L., Dvulit, Z., Seliuchenko, N., Dragan, O., Doroshkevych, K.: Influence of digital technologies on the labor market of HR specialists. In: Proceedings of the CEUR Workshop Proceedings, vol. 3171, pp. 1475–1487 (2022)
17. The Best Employee Monitoring Software for 2022. Business News Daily. https://www.businessnewsdaily.com/11143-best-employee-monitoring-software.html
18. Giermindl, L.M., Strich, F., Christ, O., Leicht-Deobald, U., Redzepi, A.: The dark sides of people analytics: reviewing the perils for organisations and employees. Eur. J. Inf. Syst. **31**(3), 410–435 (2022)
19. Global Workplace Analytics. Costs and Benefits. https://globalworkplaceanalytics.com/resources/costs-benefits
20. Is corporate America ready for The Future of Work? AT&T. shorturl.at/oqAG5

Object-Oriented Model of Fuzzy Knowledge Representation in Computer Training Systems

Nataliia Shybytska[✉]

National Aviation University, Kyiv 03058, Ukraine
nataliia.shybytska@npp.nau.edu.ua

Abstract. The situation in the world with covid and other historical events has accelerated the transition of the educational process to training remotely. That's why the task of developing models for the representation of didactic knowledge in a form suitable for computer processing is currently acquiring scientific and practical significance. The article shows one of the aspects of optimization of the learning environment by formalizing the process of representation and obtaining knowledge in computer-based training systems based on an object-oriented model of the representation of didactic knowledge. It is proposed to optimize the learning process by eliminating redundancy (procedural method), optimizing the execution of actions at a time (composition), and combining procedures to obtain one result (generalization). The object-oriented model of fuzzy knowledge representation in computer training systems is based on combining elementary semantic units into classes and identifying the corresponding relations between them by specifying a fuzzy membership function of the student's knowledge to the system's knowledge, obtained during the test. Thus, the proposed principles make it possible to create computer training systems with a flexible learning scenario by correcting the learning trajectory taking into account the results of intermediate testing.

Keywords: Knowledge representation · computer training systems · object-oriented technology · knowledge model · fuzzy set · fuzzy membership function

1 Introduction

Long-term didactic experience and the situation in the world with covid-19 have revealed a lot of disadvantages of the traditional classroom learning model [1–3]. The training process should be highly individual in speed and form, and take place in the location and time most suitable for a student.

The creation of intelligent learning systems - computer training courses with a flexible learning strategy and with multimedia elements of visualization, improves the efficiency of the learning process [2–4]. One of the aspects of the activation of the learning environment is the formalization of the process of presenting and acquiring knowledge in computer-based training systems.

The development of information technologies and the growth of the productivity of computer systems have predetermined the improvement of structural forms of presentation of information, which are characterized by a high level of formalization and abstraction. Thus, the task of developing models for the representation of didactic knowledge in a form suitable for computer processing is currently acquiring scientific and practical importance [5–7].

This article proposed the object-oriented model of fuzzy knowledge representation in computer training systems, where the student's knowledge is a subset of the knowledge of the system with some fuzzy membership function. This allows you to create of fully-connected and dynamically customizable computer-based training systems with a flexible learning scenario by correcting the learning trajectory using fuzzy function obtained as a result of intermediate test control students' knowledge.

2 The Task Analysis of Knowledge Representation Models

The representation of knowledge as a methodology for modeling and formalization of conceptual knowledge, focused on computer processing, is one of the most important sections of knowledge engineering [8–13]. This is due to the fact that the representation of knowledge ultimately determines the characteristics of computer training systems and depends on the nature and complexity of the tasks.

Let us analyze the existing knowledge representation models used to build computer training systems. In a broad sense, the learner model is understood as knowledge about the learner used to organize the learning process. The learner mode a set of accurately presented facts about the student, which describe various aspects of his condition: level of knowledge, personal characteristics, professional qualities, etc.

Computer training systems are developing in following directions [12]:

1) Creation of learner (student) models - is a structured representation of a learner's knowledge, misunderstandings, and difficulties, is constructed from learner data usually gathered by a system through the learner's interaction with a training system.
2) Creation of information structure subject area for obtaining a knowledge representation model which can be the basis of a computer-based training system.

2.1 Student Models

The student (learner) model is the core of a computer-based training system. For each user of the system, the student model maintains up-to-date information about his learning goals, path, experience, student knowledge, misunderstandings, mistakes, etc. It stores the history of his interactions with the system.

Thus, student model includes: objectives of a current training session; a target level of knowledge; a user's current knowledge level; a current training program; a system working parameters and options set by the user and/or automatically detected by the system; technical characteristics of the current client computer. The most famous student models are: overlay learner model, difference model, perturbative model.

The easiest model to implement is the overlay student model [15]. It is based on the assumption that the knowledge of the student and the knowledge of the system have a similar structure, while the knowledge of the student is a subset of the knowledge of the system. A numeric attribute is added to each topic to indicate the extent to which the learner understands the material on that topic. The value of this attribute will be determined during the student's survey.

Student difference model is the model of the student, in the construction of which the system analyzes the answers of the student and compares them with the knowledge that is embedded in the system and used by the expert when solving similar problems. The differences between this knowledge form the basis of the user model. This model allows taking into account not only the lack of knowledge of the student, but also their incorrect use.

Perturbation model of the student is the model of the student, which is built on the assumption that the knowledge of the student and the knowledge of the system may not partially coincide. In this case, an important prerequisite for constructing such a model is the identification of the reasons for the discrepancy, since without defining discrepancies, the learner model will be too fuzzy.

2.2 Knowledge Representation Models

Let's consider the most widely-used knowledge representation models: logical, production, semantic networks, frame and ontologies models [7, 13, 14].

The logical model is used to represent knowledge in the system of first-order predicate logic and formulate conclusions using syllogism. Such a model makes it possible to implement a system of formally accurate definitions and conclusions. To represent knowledge using predicate logic, it is necessary to choose constants that define objects in a given area, functional and predicate symbols that define functional dependence and relations between objects, and build logical formulas based on them. In the case when the choice of objects is difficult, you should choose a different model of knowledge.

In production models, knowledge is represented by a set of rules of the form "*IF - THAT*". These models are of two types: with direct and inverse conclusions and include three components: a rule base consisting of a set of products (rules of inference); a database containing many facts; interpreter to obtain a logical conclusion based on this knowledge. The rule base and the database form the knowledge base, and the interpreter corresponds to the inference mechanism. Direct output systems convert the contents of the database from the original to the target. However, such a conclusion is characterized by a large amount of data, as well as estimates of the state tree that are not directly related to the conclusion.

With the help of rules, a *AND/OR* tree is constructed in the production system with inverse conclusions, linking facts and conclusions. The logical conclusion is an assessment of this tree based on facts. Parts of the tree that are not relevant to the conclusion are not considered in this case. Thus, production systems, along with positive properties - the simplicity of creating and understanding individual rules and the simplicity of the inference mechanism - also have disadvantages: low processing efficiency, the complexity of the mutual relations of rules.

The next model for the representation of knowledge is semantic networks, representing the knowledge system in the form of a holistic image of the network, the nodes of which correspond to concepts and objects, and arcs - to the relations between objects.

Initially, the semantic network was invented as a model for representing the structure of long-term memory in psychology.

The hierarchical structure of concepts shows the relation of inclusion of concepts using relationship predicates: *IS-A* (inheritance of attributes in the hierarchical knowledge system) and *PART-OF* ("part-whole" relation). The advantage of the representation of knowledge by semantic networks is the simplicity of the description model based on the relationship between elements. However, with the increase in network size, the search time increases significantly. The methods of knowledge representation based on semantic network models have found application in natural language understanding systems, in particular, at the stage of structural analysis of the recognition of parts of speech, as well as highlighting keywords.

The next step in the formalization of the representation of knowledge was the creation of the Minsky frame theory [16]. A frame is a data structure designed to represent some standard situations. The frame representation model is a hierarchical structure of frames, each of which, in turn, consists of an arbitrary number of slots represented by a certain data structure, which is briefly described as a triplet "object-attribute-value". Frame systems are used not only in diagnostic expert systems, but also in the field of pattern recognition, natural language processing, and knowledge processing [16 18].

Ontology is a modern technique of knowledge representation that is grounded on a conceptualization. Conceptualization is an abstract, simplified view of some selected part of the world, containing the objects, concepts, and other entities that are assumed to exist in some area of interest and the relationships that hold among them [19, 20].

3   Object-Oriented Model of Knowledge Representation

The object-oriented model of presentation data is based on the concept of an object, which is a combination of data and methods for processing them. In object-oriented programming (OOP) environments, any object is defined by two characteristics - state and behavior. The state (attributes) of an object is determined by the current values of its parameters. Behavior (event) determines the ways of changing the object's own state and its interaction with others [21, 22].

The developed graphical interface of modern software allows you to process objects in the form of graphic images, which allows you to optimize not only the process of obtaining knowledge but also to simulate the process of training skills.

A perspective direction of using computer technologies in the learning process is the integration of computer capabilities and various methods of transmitting audiovisual information, which is made available using multimedia tools that allow you to create and use *3D* graphics, animation, and sound. At the same time, formalization of many different methods of assembling and analyzing information about the course of the learning process and the preparedness level of students becomes available.

The task of modeling and optimizing the process of managing the information flow in the training system, taking into account the individual level of training of the student, becomes relevant. The didactic principle of the formation of the scientific content of education can be formulated as follows: structuring the scientific content to indivisible (unable to be divided or separated) elements of knowledge within the discipline in order to identify interdisciplinary relationships and manage the process of declaring knowledge [23].

From a pedagogical viewpoint, setting goals for discipline training is to identify and logically strictly define the skill system that a learner has to master to solve professional problems. The objective learning function defines a system of skills, for the formation of which it is necessary to organize the process of mastering the elements $x_i \in X$ of the knowledge set X of the system.

To organize a learning process management system, firstly need to define goals - skills, then form the scientific content of the training, which is transformed into a skill system. To determine specific goals, it is necessary to structure the learning process and identify typical learning activities. As a result of this, the activity can be represented in the form of simpler actions and set to an adaptive level to the learner or to choose a higher level and to raise the speed of education. Convolution is the merging of elementary actions into a new action of a higher level.

The typical learning actions can be performed: simultaneously, sequentially, continuously, discretely, be combined into a linear, branching or cyclic structure.

Let us imagine the scientific content of training in the form of a hierarchical model (Fig. 1).

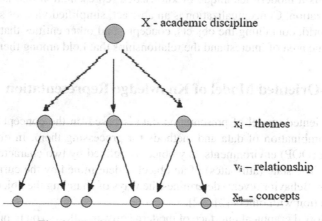

Fig. 1. The hierarchical structure of the discipline

The knowledge in this model is described in a modular fashion. The structure of the modules is determined by the relations of concepts typical of each module. The top-level x_i modules can be sections and topics of the studied subject area. At the lower level, the concepts x_n of the discipline under study are considered, while x_i, x_n are the modules of the upper and lower levels, respectively, v_i are the relations between the modules.

Optimization of the processes of knowledge formation is carried out on the basis of compilation mechanisms:

- The procedure method (Proceduralization) consists of the allocation of typical actions and the formation of procedures with the replacement of variables in the originally used universal rules by some special values suitable for solving the problem.

- The composition is a combination of independent rules and their transformation into one general.
- The mechanism of a knowledge compilation allows increasing the speed of assimilation and processing of information by the student while reducing the load on long-term memory, the learning process becomes more adaptive.

The stage of didactic knowledge coordination is carried out using three mechanisms:

1) Strengthening the rules is used when comparing them taking into account frequently used rules and is achieved through the organization of repetitions and visual-auditory effects.
2) Separation or specialization is used to create a new rule based on the information received in a previous attempt to apply known rules.
3) The mechanism of generalization allows you to expand the area of the rules and ensure their application in a wider context.

Thus, the optimization of the education process in computer-based training systems can be achieved using mechanisms of eliminating redundancy of didactic information (proceduralization), a mechanism that allows performing actions at a time (composition), and a mechanism for combining procedures that allow getting one result - (generalization). The mechanism of skill learning is that of compilation, which is a complex process consisting of proceduralization and composition.

4 Method of Describing Student's Knowledge as Fuzzy Subset of System's Knowledge

To manage the learning process and organize feedback from the student to the computer-based learning system, the object-oriented model of presenting didactic knowledge is proposed and shown in Fig. 2.

This model is based on the principles of encapsulation, inheritance, and polymorphism allows you to build a hierarchical model of the subject area by combining objects into classes and identifying the corresponding relations between them. Subsets of test indicators $t_i \in T$ are determined at the stage of test control of students' knowledge and form classes at the levels of hierarchy in the knowledge system with elements $x_i \in X$.

Today, there are no standards for estimating the processes of learning information; therefore, it is important to structure the scientific content of education to elementary semantic units - information objects. The task of creating semantic units of knowledge on the topic of lectures and a system of didactic knowledge is solved by the teacher subjectively. We do not have an objective definition of a knowledge unit. A knowledge unit is something that brings meaning to particular data.

The types of attributes of elements of these sets and test indicators are shown in Fig. 3. For creating computer learning systems, it must be really important to understand that each information object could represent a different level of detail (or abstraction) with different levels of information detail inside the structure.

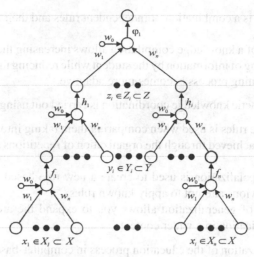

Fig. 2. The structure model of knowledge representation

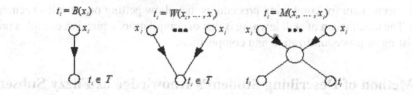

Fig. 3. Types of relations between elements

Concerning the system of content units of knowledge, which corresponds to the principle of information and logical completeness, test tasks are developed. Each question of the test task is of the following types:

– conceptual questions "who, what" and "define the concept;
– conceptual and analytical questions "define the analogy";
– conceptual and semantic tasks in order to determine the semantic equivalence or proximity of concepts and situations;
– factual questions "what - where - when" to determine the location of the object S and the time t of its existence. Let the studied object (phenomenon or situation) $S1$ be connected through the system-forming operator $\langle \psi 1 \rangle$ with the object S at time t, therefore, we can write:

$$S(t)\langle \psi \rangle S1(t).., Sn(t); \qquad (1)$$

– functional (target or causal) questions "for what - why" - to study the characteristic changes of an object in the chain of cause-and-effect relationships:

$$S \rightarrow S1(t-2) \rightarrow S1(t-1) \rightarrow S1(t) \rightarrow S(t+1) \rightarrow \ldots; \qquad (2)$$

– structural questions "of what" (or what parts the object consists of) - to analyze the appearance of the object $S1$, its structure and its components $S11, S12, \ldots, S1n$, which, using the operator χ, form a whole:

$$S\langle\chi\rangle S1, S2, \ldots, Sn; \tag{3}$$

- problem - operational tasks in order to solve the problem;
- quantitative questions "how much - how much" - to identify knowledge about the quantitative properties of objects and their application under various conditions and situations.

To assess the degree of matching of knowledge or skills of a group of student's subset $Y \subset X$ to the required knowledge subset $Z \subset X$, the concept of fuzzy function $\mu_Z(x_i)$ inheritance is proposed. The fuzzy function $\mu_Z(x_i)$ inheritance describes the process of accumulation by the student of a subset of $Y \subset X$ didactic knowledge or skills and evaluates the contribution of x_i elements from the set of X - descendants to the formation of the element of set Z of knowledge - the parent by the formula:

$$\mu_Z(x_i) = \sum_{i-1}^{N} \mu_y(x_i) \cdot \xi_V(x_i), \tag{4}$$

where $x_V(x_i)$ is the relative weight of the binary inheritance relation in a pair of elements (x_i, z_i), calculated by the formula:

$$\xi_V(x_i) = \frac{\mu_V(x_i)}{\sum_{i=1}^{N} \mu_V(x_i)}. \tag{5}$$

Let's look at these formulas in more detail. For a fixed element z of the set of knowledge - the ancestor will expertly set the structure of the inheritance relation in the form of the function:

$$v = H(x_i, z_i), \text{ where } i = 1, N. \tag{6}$$

We will assume that the set of v relations between the elements x_i, z_i knowledge of neighboring levels is represented by the weights of the values of the function $\mu_Z(x_i)$, given expertly at the stage of structuring knowledge and building a tree of goals (skills).

The values of the coefficients $\mu_Z(x_i)$ for elements $x_i \rightarrow z_i \in Z$ inheritance relations are given as follows:

- $\mu_Z(x_i) = 0$, if the parent-child relationship $x_i \rightarrow z_i \in Z$ is not defined;
- $0 < \mu_Z(x_i) <= 1$, parent-child relationship $x_i \rightarrow z_i \in Z$ determined.

Thus, the description of the process of knowledge accumulation by students using the fuzzy inheritance function of a subset of didactic knowledge or skills allows us to evaluate the contribution of the elements of the child set to the formation of an element of the parent knowledge set. In the process of repeating didactic material, the system allows the student to choose a shorter learning path, except knowledge elements for which $\mu_Z(x_i) \rightarrow 1$.

5 Summary and Conclusion

The described methodology of representation and structuration of didactic knowledge allows for building a hierarchical model of the subject area by combining objects into classes and identifying the corresponding relationships between them. Relations describe the way in which classes and instances can be related to each other.

The student's knowledge is considered as a subset of the system's knowledge with some fuzzy membership function obtained during the results of the intermediate test of the level of the student's knowledge. The proposed principles of representation of didactic knowledge will make it possible to create computer training systems with a flexible scenario and individual learning trajectory.

Creating computer-based training systems on the concept of applying didactic principles and an object-oriented approach allows to:

– establish a logical connection between the didactic elements of modules of different levels of complexity;
– inherit knowledge from modules of higher levels;
– limit the class of a search objects only to central thematic modules of knowledge and top-level modules from which didactic knowledge is inherited in the process of forming an individual trajectory of student learning;
– form modules of objects on the basis of the mechanism for eliminating redundancy (proceduralization) in the process of studying the sections of knowledge.
– perform actions simultaneously and create new rules using the composition mechanism;
– build a module structure for various purposes of using, indicating many relationships between moduless.

References

1. Syngene. Global E-Learning Market Analysis (2019). https://www.researchandmarkets.com/reports/4769385/global-e-learning-market-analysis-2019
2. Bullen, M., Morgan, T., Qayyum, A.: Digital learners in higher education: generation is not the issue. Can. J. Learn. Technol. 37(1), 1–24 (2011)
3. Ghavifekr, S., Rosdy, W.A.W.: Teaching and learning with technology: effectiveness of ICT integration in schools. Int. J. Res. Educ. Sci. 1(2), 175–191 (2015)
4. Floridi, L.: Philosophy of Computing and Information. Blackwell (2003). 388 p.
5. Borst, J.P., Anderson, J.R.: A step-by-step tutorial on using the cognitive architecture ACT-R in combination with fMRI Data. J. Math. Psychol. 94–103 (2017)
6. Benjamins, V.R., Fensel, D., et al.: Community is knowledge! In: Knowledge Acquisition Workshop on KAW 1998), Banff, Canada, pp. 18–23, April 1998
7. Lenko, V., Pasichnyk, V., Shcherbyna, Y.: Knowledge Representation Models. Lviv Polytechnic, pp.157–168 (2017)
8. Borst, J.P., Anderson, J.R.: A step-by-step tutorial on using the cognitive architecture ACT-R in combination with fMRI data. J. Math. Psychol. 76, 94–103 (2017)
9. Burgin, M.: Knowledge and data in computer systems. In: Proceedings of the ISCA 17th International Conference "Computers and Their Applications", San Francisco, California, pp. 307–310 (2002)

10. Mabel, V.H., Selwyn, J.: A review on the knowledge representation models and its implications. Int. J. Inf. Technol. Comput. Sci. **10**, 72–81 (2016)
11. Keith, L.: Theory of Knowledge, 2rd edn., p. 224 (2018)
12. Malhotra, M., Nair, T.R.G.: Evolution of knowledge representation and retrieval techniques. Int. J. Intell. Syst. Appl. **07**, 18–28 (2015)
13. Keserwani, P., Mishra, A.: Selecting integrated approach for knowledge representation by comparative study of knowledge representation schemes. Int. J. Sci. Res. Publ. **3**(2) (2013)
14. Bodily, R., et al.: Open learner models and learning analytics dashboards: a systematic review. In: Proceedings of the 8th International Conference on Learning Analytics and Knowledge (2018)
15. Sosnovsky, S.: Translation of overlay models of student knowledge for relative domains based on domain ontology mapping. In: Proceedings of the 2007 conference on Artificial Intelligence in Education: Building Technology Rich Learning Contexts That Work, pp. 289–296 (2007)
16. Minsky, M.: A Framework for Representing Knowledge. MIT-AI Laboratory Memo (1974)
17. Reimer, U., Schek, H.-J.: A frame-based knowledge representation model and its mapping to nested relations. Data Knowl. Eng. 321–352 (1989)
18. Hartono, S., Kosala, R., Supangkat, S.H., Ranti, B.: Smart hybrid learning framework based on three-layer architecture to bolster up education 4.0. In: 2018 IEEE international conference on ICT for smart society (ICISS), pp. 1–5 (2018)
19. Aminu, E.F., Oyefolahan, I.O., Abdullahi, M.B., Salaudeen, M.T.: A review on ontology development methodologies for developing ontological knowledge representation systems for various domains. Int. J. Inf. Eng. Electron. Bus. **2**, 28–39 (2020)
20. Davies, J., Fensel, D., Harmelen, F.: OnTo-Knowledge: Content-Driven Knowledge-Management through Evolving Ontologies. Wiley (2002). 312 p.
21. Hirschfeld, R., Costanza, P., Nierstrasz, O.: Context-oriented programming. J. Object Technol. **7**(3), 125–151 (2008)
22. Wegner, P.: Object-oriented programming (OOP). In: Encyclopedia of Computer Science, pp. 1279–1284, January 2003
23. Shybytska, N.: Logical-semantic model of structuring the content of learning. In: Modern International Relations: Current Problems of Theory and Practice/Collective Monograph, pp. 353–360. Naukowe Wyższej Szkoły Biznesu i Nauk o Zdrowiu, Łodz (2021)

Implementation of Optical Logic Gates Based on Color Filters

Victor Timchenko[1]([✉]), Yuriy Kondratenko[2,4], and Vladik Kreinovich[3]

[1] Admiral Makarov National University of Shipbuilding, Mykolaiv 54025, Ukraine
vl.timchenko58@gmail.com
[2] Petro Mohyla Black Sea National University, Mykolaiv 54003, Ukraine
[3] University of Texas at El Paso, El Paso, TX 79968, USA
[4] Institute of Artificial Intelligence Problems, Kyiv 01001, Ukraine

Abstract. This work is devoted to the creation of optical logic gates for the synthesis of efficient systems with a high processing speed of fuzzy data, as well as a wide possibility of parallel computing. Logical operations are based on the use of a light emitter of a certain color as a fuzzy variable - a carrier of logical information and the basis for constructing logical decisions by additive and subtractive transformation of the light emitter by the corresponding light color filters. Possible logical combinations of the proposed optical schemes for the implementation of disjunction and conjunction operations, as well as logical inference for various combinations of color filters, are given. The logical operations of negation and the search for a new decision required the creation of more complex optical circuits presented in the work, and are the basis for continuing scientific research in the field of creating optical circuits for the implementation of such logical operations as NOR, NAND, etc.

Keywords: Color information set · Fuzzy logical gates · Light color filters

1 Introduction

The development of optoelectronic technologies over the last 2–3 decades has aroused interest in the development of computing devices capable of performing binary operations and becoming an alternative to fully electrical semiconductor gates. It can be said that two main trends in the development of optical computing (logical) gates have been formed. The first trend is optoelectronic logic gates, which sought to combine the advantages of optical and semiconductor components, using the advantages of optical systems in the form of the possibility of parallel processing of large amounts of information, but at the same time, using semiconductor switches for binary encoding and calculations, for example, article [1]. The second direction focused on building all-optical logic gates [2–8]. The physical principles of creating optical switching devices are based on the properties of interference (for example, based on the combination of interference fringes, the use of a Mach-Zehnder interferometer [2], etc.), the polarization [3] and coherence [4] of a light beam, based on semiconductor optical amplifiers [5, 6], and also using the properties of diffraction gratings [7] and photonic crystals [8].

© The Author(s), under exclusive license to Springer Nature Switzerland AG 2023
Z. Hu et al. (Eds.): ICCSEEA 2023, LNDECT 181, pp. 126–136, 2023.
https://doi.org/10.1007/978-3-031-36118-0_12

CIE 1976 [9, 10] introduced the CIELAB and CIELUV color difference formulas as CIE XYZ space (and other tricolor spaces) with the limitations that equal differences in CIE XYZ trichromatic values (measured using Euclidean distance) do not correspond to equal perceived color differences [9]. This required further research in the field of transformation of color as a source of information [10, 11]. Such approaches provide color calculation models with which to perform logical operations simply by applying color-coded film to films, paper, or reflectors. Using spectroscopic analysis, the key optical properties of color codes for Boolean operations are highlighted [12, 13].

The first works were based on the representation of fuzzy logic operations for parallel computing [14] using the optical effect of anisotropic scattering [15], based on the method of areal spatial coding for the parallel implementation of optical fuzzy logic elements by changing the switching state of lenses and the corresponding threshold device [16], for the calibration of shadow diagrams, for example, an optical system based on the zone coding scheme and the shadow casting method [17]. The authors of work [18] develop the principles of using a spatial modulator of a Gaussian laser light source and a system of microprisms, paying special attention to applications with two dominant inputs, then the authors present [19] a real-time optical rules generator for a fuzzy inference mechanism with two inputs, and ways of optically implementing this procedure are shown. An optoelectronic fuzzy inference system for parallel processing of a large number of fuzzy rules based on a spatial light modulator with the implementation of various membership functions [20]. Fuzzy logic gates can also use optical-chemical, chemical and biological [21] construction principles. The use of optical logic gates in artificial intelligence systems or, somewhat already, in decision-making systems involves the processing of a large amount of data and a multi-level decision-making process (e.g., in various applications [22–25]).

The main advantage of optical logic systems is speed, compactness and unlimited possibility of parallel operations on fuzzy variables. The disadvantages of the above optical fuzzy logic gates are their significant complexity, and as a result, low technology and high energy consumption, because laser light sources with high requirements for the quality of light emitter, arrays of lenses and prisms, complex diffraction gratings, additional devices in the form of piezocrystalline elements and optoelectronic phase shifters, coding, holograms, shadow images, etc. are used.

At the same time, an analysis of the development of colorimetry shows that color carries an independent information load that does not require direct digital transformation. A certain positive or negative color rating is successfully used to assess the degree of danger (safety) of a situation in technical applications. For example, a negative color score: red R – a clear threat, yellow Yel – a likely threat, magenta M can be defined as the proximity of a threat; a positive color assessment: green G – the proximity of the absence of a threat, then you can continue: blue C – the probable absence of a threat, blue B – an absence of a threat. Basically, the white color W determines a positive assessment (for example, the existence of a decision), and black Blc - a negative one (for example, the absence of a decision). The use of color as a carrier of logical information can allow the creation of high-speed devices with performance based on calculations that use the speed of light to form an array of logical solutions. In the works of the authors [26–28], the main principles for constructing the architecture of innovative logical optical devices based on simple color filters and the formation of inference procedures for processing fuzzy information sets in the form of a given color are proposed. The main goal of this work is to continue efforts to expand the functionality and modeling of logical optical gates based on the additive and subtractive transformation of the light emitter, which allows further synthesizing of high-speed logic inference procedures for intelligent decision support systems.

2 Basic Optical Operations for Transformation of Color Information

It is well known that all visible colors can be obtained by an appropriate combination of three basic color sets: red $\{R\}$, green $\{G\}$, and blue $\{B\}$. When we do not have any color, we perceive it as black $\{Blc\}$. When we combine all three colors in equal proportion, we get white color $\{W\}$; when we combine red and blue, we get magenta $\{M\}$; when we combine red and green, we get yellow $\{Yel\}$, and when we combine green and blue, we get cyan $\{C\}$:

$$\{R\} + \{G\} + \{B\} = \{W\}; \ \{R\} + \{G\} = \{R, G\} = \{Yel\};$$
$$\{R\} + \{B\} = \{R, B\} = \{M\}; \ \{G\} + \{B\} = \{G, B\} = \{C\} \tag{1}$$

These are the colors that we will use in our proposal. We assume that we have ideal filters corresponding to all three basic colors (red, green, and blue) and all three combined colors (yellow, magenta, and cyan). An optical transformation of the form: $\{R\}+\{G\}+\{B\} = \{W\}$ can be defined as a simple (ordinary) decision under contradictory conditions (which can also be approximately attributed to estimate $\{G\}$). Let's imagine a light emitter and filter of a certain color in the form of a 3×3 diagonal matrix [25]:

$$\{R\} = \begin{pmatrix} R\,0\,0 \\ 0\,0\,0 \\ 0\,0\,0 \end{pmatrix}, \ \{G\} = \begin{pmatrix} 0\,0\,0 \\ 0\,G\,0 \\ 0\,0\,0 \end{pmatrix}, \ \{B\} = \begin{pmatrix} 0\,0\,0 \\ 0\,0\,0 \\ 0\,0\,B \end{pmatrix}, \ \{Yel\} = \begin{pmatrix} R\,0\,0 \\ 0\,G\,0 \\ 0\,0\,0 \end{pmatrix}$$
$$\{M\} = \begin{pmatrix} R\,0\,0 \\ 0\,0\,0 \\ 0\,0\,B \end{pmatrix}, \ \{C\} = \begin{pmatrix} 0\,0\,0 \\ 0\,G\,0 \\ 0\,0\,B \end{pmatrix}, \ \{W\} = \begin{pmatrix} R\,0\,0 \\ 0\,G\,0 \\ 0\,0\,B \end{pmatrix}, \ \{Blc\} = \begin{pmatrix} 0\,0\,0 \\ 0\,0\,0 \\ 0\,0\,0 \end{pmatrix} \tag{2}$$

We will distinguish between evaluation and, in fact, decisions in the decision-making process. The assessment will be determined in the accumulation or change of current information and take the values: $\{R\}, \{G\}, \{B\}, \{Yel\}, \{M\}, \{C\}$, positive and negative decision $\{W\}$ and $\{Blc\}$ will be defined as a logical conclusion made on the basis of estimates, but not further using the accumulated information. The operations of addition and subtraction of color are identical to the operations of union (disjunction) and intersection (conjunction) of sets (logical statements, operations). If necessary, each color can be assigned an appropriate numerical weight value for the interval $[0 \div 1]$. For example, $\{R\}$ (0); $\{Yel\}$ (0.25); $\{G\}$, (0.55); $\{C\}$ (0.75); $\{B\}$ (1); $\{M\}$ (0.45); $\{R\}$ (0), which corresponds to the location of the color on the inner hexagon of the circular spectrum.

3 Implementation of the Logical Operation

3.1 Implementation of the Logical Operation *OR* (Disjunction)

Let us assume that the light emitters used are spectral monochromatic. Light filters have primary (red, green, blue) and secondary (cyan, magenta, yellow) colors and do not attenuate the corresponding transmitted light emitters, i.e. are ideal. The main disjunction ∪ operations are carried out by adding color information sets. Suppose there are perfect filters corresponding to all three primary color (red, green and blue) and all three composite color (yellow, magenta and cyan). Of course, combining two or more lights of the same color does not change that color:

$$\{R\} \cup \{R\} = \{R\}; \{G\} \cup \{G\} = \{G\}; \{B\} \cup \{B\} = \{B\} \tag{3}$$

The total number of estimates and decisions at the output will be determined from the known expression for combinations with $n = 3, k = 3$:

$$A = (n + k - 1)!/k!(n - 1)! = 10 \tag{4}$$

Taking into account idempotency property (3), three repeated combinations are excluded. There are estimates: $\{R\}, \{G\}, \{B\}, \{Yel\}, \{M\}, \{C\}$ and decision $\{W\}$ The figure Fig. 1 shows the optical schemes for white and secondary colors at the output of coloroids. The results of the transformations shown in Table 1 demonstrate all possible combinations of logic operations of the proposed optical gates.

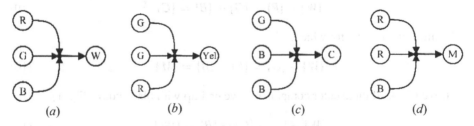

Fig. 1. Optical schemes of disjunction coloroid

3.2 Implementation of the Logical Operation *AND* (Conjunction)

Subtractive transformation of light emitters using light filters forms a blocking (subtraction) of the corresponding color. For example, a red filter blocks the green and blue components, allowing only the red to pass through; this can be described as:

$$\{R\} = \{W\} - \{G\} - \{B\} \tag{5}$$

We can also write similar expressions describing the blue filter:

$$\{B\} = \{W\} - \{R\} - \{G\} \tag{6}$$

Table 1. Logical operation for optical scheme

Logic form	Matrix form
$\{R\} \cup \{G\} \cup \{R\} = \{W\}$	$diag(R, 0, 0) + diag(0, G, 0) + diag(0, 0, B) = diag(R, G, B)$
$\{R\} \cup \{R\} \cup \{R\} = \{R\}$	$diag(R, 0, 0) + diag(R, 0, 0) + diag(R, 0, 0) = diag(R, 0, 0)$
$\{G\} \cup \{G\} \cup \{G\} = \{G\}$	$diag(0, G, 0) + diag(0, G, 0) + diag(0, G, 0) = diag(0, G, 0)$
$\{B\} \cup \{B\} \cup \{B\} = \{B\}$	$diag(0, 0, B) + diag(0, 0, B) + diag(0, 0, B) = diag(0, 0, B)$
$\{R\} \cup \{R\} \cup \{B\} = \{M\}$	$diag(R, 0, 0) + diag(R, 0, 0) + diag(0, 0, B) = diag(R, 0, B)$
$\{R\} \cup \{G\} \cup \{G\} = \{Yel\}$	$diag(R, 0, 0) + diag(0, G, 0) + diag(0, G, 0) = diag(R, G, 0)$
$\{G\} \cup \{B\} \cup \{B\} = \{C\}$	$diag(0, G, 0) + diag(0, 0, B) + diag(0, 0, B) = diag(0, G, B)$

and the green filter:

$$\{G\} = \{W\} - \{R\} - \{B\} \tag{7}$$

We can also have a yellow filter that blocks the blue components of the white light and keeps only the red and green components, which form the yellow light filter F_1:

$$\{W\} - \{B\} = \{R\} + \{G\} = \{Yel\} \tag{8}$$

we can similarly have a cyan filter F_2 for which:

$$\{W\} - \{R\} = \{G\} + \{B\} = \{C\} \tag{9}$$

and a magenta filter F_3 for which:

$$\{W\} - \{G\} = \{R\} + \{B\} = \{M\} \tag{10}$$

If we block all three color components, we end up with black color (Fig. 1):

$$\{W\} - \{R\} - \{G\} - \{B\} = \{Blc\} \tag{11}$$

When a yellow light emitter (for example) passes through a red filter, the green color is blocked, and the output is red:

$$\{Yel\} - \{G\} = \{R\} \tag{12}$$

through the green filter, the red color is blocked, and the output is green red:

$$\{Yel\} - \{R\} = \{G\} \tag{13}$$

through the blue filter, red and green are blocked, and the output is black (i.e., the absence of light emitter) color:

$$\{Yel\} - \{R\} - \{G\} = \{Blc\} \tag{14}$$

It is more convenient to describe the operations of subtracting a certain color using the conjunction operation ∩ corresponding to the multiplication of diagonal matrices (2), which are defined for the input information and the corresponding blocking filters, described by formulas (5–14) and other combinations for filters (for example, Fig. 2a–d). The total number of evaluations and decisions at the output will be determined from the formulas (4) for combinations with $n = 3$, $k = 3$ and 7 different input information sets $A = 70$ or excluding repetitions due to the formula (3) $A = 49$.

For a combination of color filters in the form $\{Yel\} \cap \{C\} \cap \{M\}$ for any input information as optical set $\{Q\}$ the output will always have the value $\{Blc\}$ or $diag(0, 0, 0)$. Therefore, this equation will not be reflected in the tables below. Considering all possible combinations and for the input set $\{Q\}$ we obtain the following formulas, which are presented in Tables 2, 3, 4, 5, 6, 7 and 8.

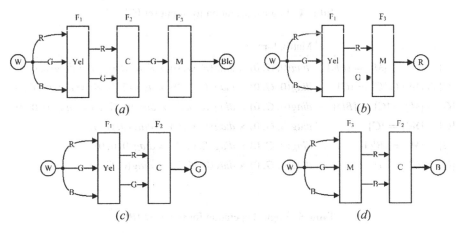

Fig. 2. Optical schemes of conjunction coloroid

Table 2. Logical operation for input set $\{W\}$

Logic form	Matrix form
$\{W\} \cap \{Yel\} \cap \{M\} = \{R\}$	$diag(R, G, B) \times diag(R, G, 0) \times diag(R, 0, B) = diag(R, 0, 0)$
$\{W\} \cap \{Yel\} \cap \{C\} = \{G\}$	$diag(R, G, B) \times diag(R, G, 0) \times diag(0, G, B) = diag(0, G, 0)$
$\{W\} \cap \{M\} \cap \{C\} = \{B\}$	$diag(R, G, B) \times diag(R, 0, B) \times diag(0, G, B) = diag(0, 0, B)$
$\{W\} \cap \{Yel\} = \{R\} \cup \{G\} = \{Yel\}$	$diag(R, G, B) \times diag(R, G, 0) = diag(R, G, 0)$
$\{W\} \cap \{M\} = \{R\} \cup \{B\} = \{M\}$	$diag(R, G, B) \times diag(R, 0, B) = diag(R, 0, B)$
$\{W\} \cap \{C\} = \{G\} \cup \{B\} = \{C\}$	$diag(R, G, B) \times diag(0, G, B) = diag(0, G, B)$

Table 3. Logical operation for input set $\{R\}$

Logic form	Matrix form
$\{R\} \cap \{Yel\} \cap \{M\} = \{R\}$	$diag(R, 0, 0) \times diag(R, G, 0) \times diag(R, 0, B) = diag(R, 0, 0)$
$\{R\} \cap \{Yel\} \cap \{C\} = \{Blc\}$	$diag(R, 0, 0) \times diag(R, G, 0) \times diag(0, G, B) = diag(0, 0, 0)$
$\{R\} \cap \{M\} \cap \{C\} = \{Blc\}$	$diag(R, 0, 0) \times diag(R, 0, B) \times diag(0, G, B) = diag(0, 0, 0)$
$\{R\} \cap \{Yel\} = \{R\}$	$diag(R, 0, 0) \times diag(R, G, 0) = diag(R, 0, 0)$
$\{R\} \cap \{M\} = \{R\}$	$diag(R, 0, 0) \times diag(R, 0, B) = diag(R, 0, 0)$
$\{R\} \cap \{C\} = \{Blc\}$	$diag(R, 0, 0) \times diag(0, G, B) = diag(0, 0, 0)$

Table 4. Logical operation for input set $\{G\}$

Logic form	Matrix form
$\{G\} \cap \{Yel\} \cap \{M\} = \{Blc\}$	$diag(0, G, 0) \times diag(R, G, 0) \times diag(R, 0, B) = diag(0, 0, 0)$
$\{G\} \cap \{Yel\} \cap \{C\} = \{G\}$	$diag(0, G, 0) \times diag(R, G, 0) \times diag(0, G, B) = diag(0, G, 0)$
$\{G\} \cap \{M\} \cap \{C\} = \{Blc\}$	$diag(0, G, 0) \times diag(R, 0, B) \times diag(0, G, B) = diag(0, 0, 0)$
$\{G\} \cap \{Yel\} = \{G\}$	$diag(0, G, 0) \times diag(R, G, 0) = diag(0, G, 0)$
$\{G\} \cap \{M\} = \{Blc\}$	$diag(0, G, 0) \times diag(R, 0, B) = diag(0, 0, 0)$
$\{G\} \cap \{C\} = \{G\}$	$diag(0, G, 0) \times diag(0, G, B) = diag(0, G, 0)$

Table 5. Logical operation for input set $\{B\}$

Logic form	Matrix form
$\{B\} \cap \{Yel\} \cap \{M\} = \{Blc\}$	$diag(0, 0, B) \times diag(R, G, 0) \times diag(R, 0, B) = diag(0, 0, 0)$
$\{B\} \cap \{Yel\} \cap \{C\} = \{Blc\}$	$diag(0, 0, B) \times diag(R, G, 0) \times diag(0, G, B) = diag(0, 0, 0)$
$\{B\} \cap \{M\} \cap \{C\} = \{B\}$	$diag(0, 0, B) \times diag(R, 0, B) \times diag(0, G, B) = diag(0, 0, B)$
$\{B\} \cap \{Yel\} = \{Blc\}$	$diag(0, 0, B) \times diag(R, G, 0) = diag(0, 0, 0)$
$\{B\} \cap \{M\} = \{B\}$	$diag(0, 0, B) \times diag(R, 0, B) = diag(0, 0, B)$
$\{B\} \cap \{C\} = \{B\}$	$diag(0, 0, B) \times diag(0, G, B) = diag(0, 0, B)$

3.3   Implementation of the Logical Operation *NOT* (*Negation*)

The negation operation is a more complex logical action, since the negative decision is based not on the level of the previous estimate (sum or product of sets), but on changing the original decision to the opposite decision, which in itself requires a separate decision on the application of the negation action. The first logical action in case of negation is a logical operation of the form: $\{Blc\} = -\{W\}, \{W\} = -\{Blc\}$ for which the optical scheme of coloroid can be used (Fig. 3). The difference between this scheme and the

Table 6. Logical operation for input set $\{Yel\}$

Logic form	Matrix
$\{Yel\} \cap \{Yel\} \cap \{M\} = \{R\}$	$diag(R, G, 0) \times diag(R, G, 0) \times diag(R, 0, B) = diag(R, 0, 0)$
$\{Yel\} \cap \{Yel\} \cap \{C\} = \{G\}$	$diag(R, G, 0) \times diag(R, G, 0) \times diag(0, G, B) = diag(0, G, 0)$
$\{Yel\} \cap \{M\} \cap \{C\} = \{Blc\}$	$diag(R, G, 0) \times diag(R, 0, B) \times diag(0, G, B) = diag(0, 0, 0)$
$\{Yel\} \cap \{Yel\} = \{Yel\}$	$diag(R, G, 0) \times diag(R, G, 0) = diag(R, G, 0)$
$\{Yel\} \cap \{M\} = \{R\}$	$diag(R, G, 0) \times diag(R, 0, B) = diag(R, 0, 0)$
$\{Yel\} \cap \{C\} = \{G\}$	$diag(R, G, 0) \times diag(0, G, B) = diag(0, G, 0)$

Table 7. Logical operation for input set $\{M\}$

Logic form	Matrix form
$\{M\} \cap \{Yel\} \cap \{M\} = \{R\}$	$diag(R, 0, B) \times diag(R, G, 0) \times diag(R, 0, B) = diag(R, 0, 0)$
$\{M\} \cap \{Yel\} \cap \{C\} = \{Blc\}$	$diag(R, 0, B) \times diag(R, G, 0) \times diag(0, G, B) = diag(0, 0, 0)$
$\{M\} \cap \{M\} \cap \{C\} = \{B\}$	$diag(R, 0, B) \times diag(R, 0, B) \times diag(0, G, B) = diag(0, 0, B)$
$\{M\} \cap \{Yel\} = \{R\}$	$diag(R, 0, B) \times diag(R, G, 0) = diag(R, 0, 0)$
$\{M\} \cap \{M\} = \{M\}$	$diag(R, 0, B) \times diag(R, 0, B) = diag(R, 0, B)$
$\{M\} \cap \{C\} = \{B\}$	$diag(R, 0, B) \times diag(0, G, B) = diag(0, 0, B)$

Table 8. Logical operation for input set $\{C\}$

Logic form	Matrix form
$\{C\} \cap \{Yel\} \cap \{M\} = \{Blc\}$	$diag(0, G, B) \times diag(R, G, 0) \times diag(R, 0, B) = diag(0, 0, 0)$
$\{C\} \cap \{Yel\} \cap \{C\} = \{G\}$	$diag(0, G, B) \times diag(R, G, 0) \times diag(0, G, B) = diag(0, G, 0)$
$\{C\} \cap \{M\} \cap \{C\} = \{B\}$	$diag(0, G, B) \times diag(R, 0, B) \times diag(0, G, B) = diag(0, 0, B)$
$\{C\} \cap \{M\} = \{G\}$	$diag(0, G, B) \times diag(R, G, 0) = diag(0, G, 0)$
$\{C\} \cap \{M\} = \{B\}$	$diag(0, G, B) \times diag(R, 0, B) = diag(0, 0, B)$
$\{C\} \cap \{C\} = \{C\}$	$diag(0, G, B) \times diag(0, G, B) = diag(0, G, B)$

universal coloroid (Fig. 2a) is that the filter values are fixed, and not a variable assessment (of experts or sensors). The inverse operation $\{W\} = -\{Blc\}$ defined as the search for a new decision, assumes that in the absence of light of any color $\{Blc\}$ in the optical information channel, the light sensor is triggered and a signal is given to turn on the light emitter $\{W\}$ to form a new decision. Further, considering the main information color set $\{R\}$, $\{G\}$, $\{B\}$ we understand that there are opposite estimates of the evaluation and decision-making process: $\{R\} = -\{B\}$, $-\{R\} = \{B\}$, at the same time, the negation operation for information color set $\{G\}$, $\{M\}$, is meaningless as it is the median value. For the informational color set $\{Yel\}$, $\{C\}$ approximate values $\{Yel\} \approx \{R\}$, $\{C\} \approx \{B\}$ can be taken.

The negation operation is formed by an optical scheme in Fig. 3. The logical coloroid includes color filters $F_1 - F_5$; Gate 1, Gate 2, Gate 3; light sources with switching on in the absence of light S_1L_1, S_2L_2, S_3L_3; blocking devices Bl_1, Bl_2, and functions as follows: with input information set $\{B\}$ in Gate 1 and light emitter S_1L_1 is triggered, and the light emitters S_2L_2, S_3L_3 are blocked by devices Bl_1, Bl_2. At the output of the coloroid we get $\{R\}$ when input information set $\{R\}$ in Gate 2, then light emitter S_2L_2 is triggered, and the light emitter S_3L_3 is blocked by device Bl_2. At the output of the coloroid we get $\{B\}$ When input information set $\{W\}$ in Gate 3 first we get $\{Blc\}$, and then light emitter S_3L_3 is triggered. At the output of the coloroid we again get $\{W\}$ Also note that in general all the gates of the coloroid are ready to operate when any input $\{W\}$, $\{B\}$, $\{R\}$ applied. To limit the information sets $\{G\}$, $\{Yel\}$ $\{M\}$, $\{C\}$ for which the negation operation is not applied, it is necessary to select the minimum and maximum levels of estimates, applying for an arbitrary light flux $\{Q\}$ respectively, the filters $\{R\}$ and $\{B\}$ highlighting the corresponding set: $\{R\} = min\{Q\}$, $\{B\} = max\{Q\}$, which are fed to the input of the optical negation gates. Based on the proposed optical schemes, it is also possible to implement neither optical logic operations *NOR*, *NAND* etc.

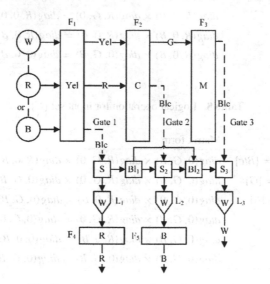

Fig. 3. Optical scheme of negation coloroid

4 Summary and Conclusion

The results of research in the field of creating high-performance logical color gates by converting optical fuzzy information to build a logical inference are presented. All possible variants of logical operations *OR* and *AND* for the proposed optical gates are given. The application of logical operations of negation *NOT* and search for a new decision is substantiated, a new optical scheme for their implementation is developed.

The advantages of logical optical gates (coloroid) are:

1) A much simpler design of logic gates, which ensures their high technological feasibility, in comparison with existing proposals.
2) The possibility of serial-parallel processing of a large amount of information with high performance, including through the processing of fuzzy information without the use of binary calculations.
3) High degree of visualization of the results of processing current and output information.
4) When processing and transmitting information, optical devices provide higher robustness in comparison with semiconductor elements.

The proposed optical logical gates are quite simple to implement and can be used as the basis for intelligent decision-making systems for objects with a large amount of processed information, for example, when managing the flights of large airports or the traffic of ships in conditions of high navigation intensity.

References

1. Azhigulov, D., Nakarmi, B., Ukaegbu, I.A.: High-speed thermally tuned electro-optical logic gates based on micro-ring resonators. Opt. Quant. Electron. **52**(9), 1–16 (2020). https://doi.org/10.1007/s11082-020-02526-y
2. Araujo, A., Oliveira, A., et al.: Two all-optical logic gates in a single photonic interferometer. Opt. Commun. **355**, 485–491 (2005)
3. Zaghloul, Y.A., Zaghloul, A.R.M.: Complete all-optical processing polarization based binary logic gates and optical processors. Opt. Express **14**(21), 9879–9895 (2006)
4. Zhu, Z., Yuan, J., Jiang, L.: Multifunctional and multichannels all-optical logic gates based on the in-plane coherent control of localized surface plasmons. Opt. Lett. **45**(23), 6362–6365 (2020)
5. Ma, S., Chen, Z., Dutta, N.K.: All-optical logic gates based on two-photon absorption in semiconductor optical amplifiers. Opt. Commun. **282**(23), 4508–4512 (2009)
6. Kotb, A.: Simulation of high quality factor all-optical logic gates based on quantum-dot semiconductor optical amplifier at 1 Tb/s. Optik **127**(1), 320–325 (2016)
7. Qian, C., Lin, X., Xu, J., et al.: Performing optical operators by a diffractive neural network. Light Sci. (2020). www.nature.com/Lsa2020/Open Access
8. Jung, Y.J., Yu, S., Koo, S., et al.: Reconfigurable all-optical logic AND, NAND, OR, NOR, XOR and XNOR by photonic crystal nonlinear cavities. In: Conference on Lasers and Electro-Optics, Pacific Rim (2009). Paper TuB4_3
9. Sharma, G., Vrhel, M.J., Russell, H.J.: Color imaging for multimedia. Process. IEEE **86**(6), 1088–1108 (1998)
10. Corletti, C., Toroba, R.: Color coded logic. Opt. Commun. **126**(4–6), 197–204 (1996)
11. Ohta, N., Robertson, A.R.: Colorimetry. Fundamentals and Applications. Wiley (2005)
12. Kawano, T.: Printable optical logic gates with CIELAB color coding system for Boolean, operation-mediated handling of colors. In: Genetic and Evolutionary Computing (ICGEC), pp. 270–275. IEEE (2012)

13. Moritaka, K., Kawano, T.: Spectroscopic analysis of the model color filters used for computation of CIELAB-based optical logic gates. ICIC Express Lett. Part B: Appl. **5**(6), 1715–1720 (2014)
14. Liu, L.: Optical implementation of parallel fuzzy logic. Opt. Commun. **73**(3), 183–187 (1989)
15. Lohmann, A.W., Weigelt, J.: Optical logic by anisotropic scattering. Opt. Commun. **54**(2), 81–86 (1985)
16. Lin, S., Kumazava, I., Zhang, S.: Optical fuzzy image processing based on shadow-casting. Opt. Commun. **94**(5), 397–405 (1992)
17. Zhang, S., Chen, C.: Parallel optical fuzzy logic gates based on spatial area-encoding technique. Opt. Commun. **107**(1–2), 11–16 (1994)
18. Gur, E., Mendlovic, D., Zalevsky, Z.: Optical implementation of fuzzy-logic controllers. Part I. Appl. Opt. **37**(29), 6937–6945 (1998)
19. Gur, E., Mendlovic, D., Zalevsky, Z.: Optical generation of fuzzy based rules. Appl. Opt. **41**(23), 4653–4761 (2002)
20. Alles, M., Sokolov, S.V., Kovalev, S.M.: Fuzzy logical control based on optical information. Autom. Control. Comput. Sci. **48**(3), 123–128 (2014)
21. Gentili, P.L.: The fundamental fuzzy logic operators and some complex Boolean logic circuits implemented by the chromogenism of a spirooxazine. Phys. Chem. **13**(45), 20335–20344 (2011)
22. Kunjir, A., Shah, J., Singh, N., Wadiwala, T.: Big data analytics and visualization for hospital recommendation using HCAHPS standardized patient survey. Int. J. Inf. Technol. Comput. Sci. (IJITCS) **11**(3), 1–9 (2019)
23. Omri, A., Omri, M.N.: Towards an efficient big data indexing approach under an uncertain environment. Int. J. Intell. Syst. Appl. (IJISA) **14**(2), 1–13 (2022)
24. Kondratenko, Y.P., Simon, D.: Structural and parametric optimization of fuzzy control and decision making systems. In: Zadeh, L.A., Yager, R.R., Shahbazova, S.N., Reformat, M.Z., Kreinovich, V. (eds.) Recent Developments and the New Direction in Soft-Computing Foundations and Applications. SFSC, vol. 361, pp. 273–289. Springer, Cham (2018). https://doi.org/10.1007/978-3-319-75408-6_22
25. Bhardwaj, S., Pandove, G., Dahiya, P.K.: A genesis of a meticulous fusion based color descriptor to analyze the supremacy between machine learning and deep learning. Int. J. Intell. Syst. Appl. (IJISA) **12**(2), 21–33 (2020)
26. Timchenko, V., Kondratenko, Y., Kreinovich, V.: Efficient optical approach to fuzzy data processing based on colors and light filter. Int. J. Probl. Control Inform. **52**(4), 89–105 (2022)
27. Timchenko, V.L., Kondratenko, Y.P., Kreinovich, V.: Decision support system for the safety of ship navigation based on optical color logic gates. In: CEUR Workshop Proceedings, vol. 3347, pp. 42–52 (2022)
28. Timchenko, V.L., Kondratenko, Y.P., Kreinovich, V.: Why color optical computing? In: Phuong, N.H., Kreinovich, V. (eds.) Deep Learning and Other Soft Computing Techniques. Studies in Computational Intelligence, vol. 1097, pp. 227–233. Springer, Cham (2023). https://doi.org/10.1007/978-3-031-29447-1_20

Enhancing Readability in Custom Templates for Displaying Semantically Marked Information

Viacheslav Zosimov ⓘ and Oleksandra Bulgakova(✉) ⓘ

Taras Shevchenko National University of Kyiv, Bohdan Hawrylyshyn str. 24, Kyiv 04116, Ukraine

`sashabulgakova2@gmail.com`

Abstract. The article presents new methods for displaying the information content of web resources and enhancing the effectiveness and quality of information search on the Internet. The new approach to information display utilizes client-side templates, resulting in a uniform styling of web resources of a particular type, customizable according to the user's personal template. The web resource itself only consists of data marked with relevant semantic markup meta tags. Furthermore, the quality of information search is improved through the additional multi-level processing of results from traditional search engines. The methods presented in the article are integrated into a comprehensive Internet data operating system (CIDO) that leverages the author's Web data operating language (WDOL) for data transformation and processing functions. The significance and value of this research lie in the provision of new and improved methods for displaying and searching for information on the Internet, ultimately enhancing the user's experience and efficiency.

Keywords: Information retrieval · Display templates · Semantic markup · Search results ranking

1 Introduction

The development of a comprehensive web data operation system is a crucial task, including methods, models, and software tools to operate web data, particularly on corporate web resources of the domestic segment of the worldwide web, on a semantic level in real-time, under conditions of weak structure, heterogeneity, and incomplete information, and also with increasing data sources. Noted scientists such as Anisimov A. V., McCarthy D., Glibovets M.M., Nuell A., Marchenko O.O., Lueger D. F., Foster D. M., Pospelev D. A., Rubashkin V. Sh., Vynograd T. V., Lande D. V., Popov E. V., Glibovets A.M., Pitts U., Osuga S., Lackoff Dzh., Hopcroft Dzh., Leont'ev N. N., Shemakin Yu. I., Galperin I. R., Kobozeva I. M., Melchuk I. A., Krongauz M. A., Graymas A. Zh., Sevbo I.P., and Shirokov V. A. and others have researched text information retrieval and analysis [1–7]. Despite their work, questions remain about methods for automatic information extraction from web resources, semantic markup integration in HTML code, alternative methods of content display without URL attachment, and search result personalization.

Z. Hu et al. (Eds.): ICCSEEA 2023, LNDECT 181, pp. 137–146, 2023.
https://doi.org/10.1007/978-3-031-36118-0_13

The exponential growth of the information sources number leads to the development of such areas as intelligent data analysis (Data mining), data extraction from web pages (Web mining), intelligent information search methods, machine learning, cluster and regression analysis. The concept of the semantic web development became the next step in the global network development. Semantic Web concept was designed to make information amenable to synthesis of inferences and automatic analysis [8–10].

In 2020, [11] a complex Internet data operating system (CIDO) was presented, which includes methods, models and software tools for real time web data operating at the semantic level, in conditions of weak structure, heterogeneity and incompleteness of information, with constantly growing number of sources. In the CIDO system, a new object-oriented language – WDOL (web data operating language) was proposed, for handling web data with a wide functionality for extracting, processing, saving and displaying web data [12]. All information processing capabilities are implemented as built-in language tools and are primarily focused on working with the object model of documents presented on the World Wide Web in the HTML code form. WDOL's main advantage is that it is implemented in a declarative style and allows to describe complex operations using simple inline functions. All language features are implemented with high level of abstraction, which allows changing individual internal mechanisms of data processing, such as the format of data storage, parsing the web pages structure methods, without the need to adjust the software code. The developed language became the core of the system for improving the web data processing efficiency, and the basis for the implementation of the platform for creating search agents based on the web resources semantic markup.

The paper proposes to consider the display templates mechanism of the web resources information content in the CIDO system, in which the display templates are applied on the client side, for all web sites of a given type.

2 CIDO System

The basic theory of enhancing readability in custom templates for displaying semantically marked information focuses on improving the user experience by optimizing the visual presentation of information. This is achieved through the use of customized templates that are designed specifically to display semantically marked information in a clear, concise, and easily understandable format. The templates use visual cues, such as color, font size, and layout, to highlight important information and improve the overall readability of the content. Additionally, the use of semantic markup allows for the automatic extraction and display of relevant information, reducing the need for manual interpretation and increasing the speed and accuracy of information delivery.

The CIDO system is a web data processing system that displays its results to end users in the form of web pages. This system is based on the theories of web resource semantic profiling and template-based data display.

The CIDO system has a modular structure. Each module can interact with others. The WDOL is the basis of the CIDO system kernel, which implements basic data transformation and processing functions of [12].

Modular architecture allows you to highlight the system's core as an abstraction level. It provides all system modules with the inline data operating functions, and hides the implementation details. Such architectural decisions are the basis of the I/O model implementation based on streams of bytes for the UNIX family operating systems.

Basic system modules add an abstraction level between functional modules and emulate system kernel functions. Such an architectural solution allows to implement a tiered approach to system design. Each level serves its interaction processes set. In such a structure, the joint work of functional and general modules becomes simpler, clearer and transparent.

Functional inline modules implement the applied system functions as the web applications form to meet the user information needs. Using the web applications we get the correct operation of the program irrespective of the user's OS.

Data display module in the CIDO system uses an approach based on visual templates. CIDO system generates two groups of results. Each group has its own display template type:

1) Templates based on the semantic profile of web site. This type of templates is the main one for the system and are used to display search results and web site content. The final web page is formed on the basis of the following components:

   – general data;
   – structure template (HTML), which includes the web page HTML structure and styling;
   – a web site semantic profile – is a HTML template with integrated semantic markup, which is necessary, for application of intelligent search methods and improvement of their correctness and effectiveness.

2) Display templates for the results of data processing and aggregation in the form of charts, lists, tables, etc.

3   Data Aggregation and Processing Results Display Templates

It is difficult, and sometimes impossible, to describe such data according to the semantic profile of corporate websites because they can have an abstract character not related to specific products or services. For example, research into weather change dynamics, formation of lists of art products such as films, books, audio recordings, etc.

The use of such templates is based on the approach implemented in FastTemplate templates in the PHP programming language. Any large page is composed of countless fragments, the smallest of which are ordinary text lines that are named and given a value.

Information display based on templates is one of the most widely used methods today. During the development of a website, a data display template is created, according to which the given website is displayed in the same way for all network users. Factors that affect the external appearance of the template:

The client's wishes, the developer's personal experience, and the current trends in web resource design are taken into account. It's obvious that one template cannot accommodate the unique ways in which different potential users may perceive information. This can lead to difficulties in searching for the necessary information among certain groups of users. The main difference in the approach implemented in the CODE system is that the display templates are not applied on the developer side of the web resource, but on the user side. In other words, all web resources created based on the semantic profile of corporate web resources presented in the work can be displayed consistently according to one user template. The scheme in Fig. 1 demonstrates the differences between the conventional approach to displaying web resource content and the one proposed in the work. The abbreviations in Fig. 1: $S_1,\ldots S_n$ - web resources, $T_1,\ldots T_n$ - display templates, $U_1, \ldots U_n$ - users.

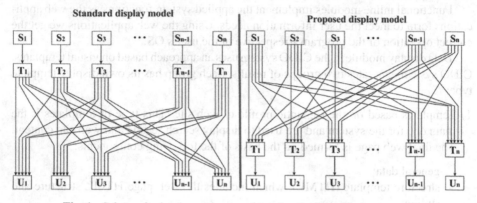

Fig. 1. Scheme for implementing templates for displaying web resources

4 Development of Web Resources with Integrated Semantic Markup

The semantic profile is a live structure. It can be expanded with new unique elements or by the schema.org dictionary base set elements [13].

The module architecture was developed using the positive experience of the Word Press CMS [14], which today is the most popular CMS for the web site development on the domestic and global market. The visual interface provides convenient addition of all actual data of the structure of web site, which is stored in the local database. The final web pages is generating according to the display template. Text data from the database is placed in the HTML blocks with the semantic markup included according to the actual semantic profile.

Figure 2 shows the scheme of the web pages semantic markup module components interaction.

5 Templates for Displaying the Informational Content of a Web Resource

To display the results of work in the CIDO system, a template mechanism is used. This module implements a new approach to display the web site data. Displayed templates are applied on the user's side, for all web sites of this type. It is an alternative approach to the traditional one, in which every single site has its own display template.

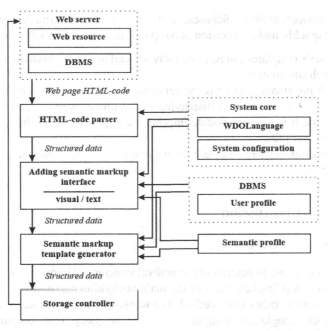

Fig. 2. The scheme of the module of integration the semantic markup into the existing web resources code

A new approach can be effectively implemented using the pre-designed display templates and semantic profile for web resources of a certain type.

The template includes:

- a web page structural scheme of the HTML blocks placement;
- stylistic design of the specified blocks;
- semantic markup of the HTML blocks.

Special attention must be paid to the HTML block concept with built-in semantic mauprk. It determines exactly which elements of semantic profile will be placed in a specific HTML blocks. The traditional approach includes the logical division of a web site into a set of web pages with unique URL within one domain (http://website.com/page1 … http://wesite.com/pageN). The result of a request to a specific URL shows the content of one web page on the screen. Proposed in the paper new approach provides two variants of the unique identification of information content:

1. Traditional one, when content is tied to a specific web site page. The URL of the page is used as an identifier.
2. Identification of information based on the semantic content, which can be used without binding to the URL of the higher level page. Such an approach makes it possible to display an aggregated content of several or all web resource pages on one screen. It can be displayed partial or full. For example, a visitor needs to view a shortened list of

company services, available licenses, certificates, contact information on one screen in order to quickly make a decision about possible cooperation with the company.

Custom user's templates can be effectively applied to display both the search engine results, and web site content.

Since each information unit has the semantic identifier which is unique, it makes it possible to form a template for displaying the results of web search in an arbitrary way. This approach is used as module important component for further processing of the search engine results.

The scheme of display template management module components interaction showed in Fig. 3.

6 Results and Discussion

6.1 Experimental Results

Goal of the experiment: to determine the general structure of corporate websites based on the analysis of top-level elements of the main navigation menu.

1000 corporate websites were studied. The websites for analysis were automatically selected from the Google search engine results for the query "Our company". In total, 1284 websites were processed, excluding duplicate links to websites, classifieds boards, and aggregator sites of services.

The elements of the navigation menu were extracted in the form of "Element name" → "Link" and stored in a database. Table 1 lists the structural elements ordered by decreasing frequency of occurrence in the websites. A total of 5371 unique navigation elements were obtained. Elements with fewer than 50 occurrences were ignored as uninformative.

As a result of the investigation, the top-level elements structure was formed. Structural elements with similar meanings were grouped into thematic groups for representation.

The overall structure of corporate web resources was built based on the results of the experiment, as well as on the results of research by leading specialists in the field of web design and usability [18–20]. The general scheme includes all possible options, and when developing each specific project, only the ones corresponding to the company's requirements will be selected.

Table 2 presents groups of structural elements and the overall frequency of occurrence of all elements in the group.

The number of occurrences exceeding the total number of analyzed websites is due to some websites having main navigation duplicated in the footer of the website. The navigation element extraction algorithm takes such items into account as separate.

Data from Table 2 was used as structural groups to which elements similar in meaning were added for convenient navigation through the website.

Based on the results of the study and the constructed overall structure of the top-level elements, a tree of structural elements of the corporate website was formed and a semantic profile of the corporate website was built.

Table 1. Fragment of the list of structural elements of corporate websites

#	NAME OF ELEMENT	NUMBER OF OCCURRENCES
1	Contacts	891
2	About company	637
3	Services	611
...	...	...
25	Partners	81
26	Clients about us	78
27	FAQ	75
...	...	...
44	Gallery of works	51
45	Our news	50

Table 2. Fragment of groups of structural elements of web resources

GROUPS	NUMBER OF OCCURRENCES
Articles, Information, FAQ, News, Our news, Press center, Promotions, Events	1041
Home, About us, About the company, Our company	1026
Contacts, Our contacts	953
Services, Our services, Service	865
...	...
Products, Our products, Production, shop, goods	463
Vacancies, Career	355
Reviews, Clients about us	220

6.2 Search Results Further Processing

The improving of the effectiveness task of information searching on the Internet by keywords can be solved in two directions:

– development of new autonomous search systems;
– additional processing of search output results of existing search systems [15].

It is important to note that in the first direction we are talking about the development of an independent search engines that do not use the results of other search engines at any stage of their work. At present, all implemented projects of domestic search systems do not have their own search engines and are based on the publication of google.com.ua [17].

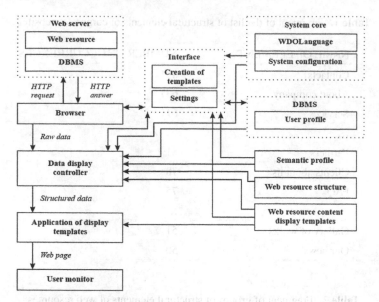

Fig. 3. The scheme of display template management module components interaction

Considering the Internet scale and the developed modern search systems infrastructure, it is obvious that the development of a new search system is an impossible task. The modern search systems quite effectively perform the task of collecting and indexing data, as well as searching for information in the index database according to a search query. To date, the undisputed world leader in the information search field on the Internet is the Google search engine [16]. Therefore, the system search output results are taken as the basis for further research. Based on this, the work focuses on additional processing of search results in three directions:

- development of alternative methods of search results ranking;
- creation of new approaches to displaying search results;
- improvement of intelligent search methods based on search agents.

As a new approach to displaying search results, the work uses custom templates based on the semantic profile of the web site.

The improvement of intelligent search methods is based on the semantic profile application as an information identifier for search agents. The search agents effective work becomes possible to the approaches and methods presented in the work complex application:

- web resource semantic profile;
- data mining method based on the WDOL language;
- data storage using version control systems, which provides ample opportunities for data aggregation and further analysis;
- custom templates for displaying search results.

The module provides programming and application of search agents directly in the process working with search results.

7 Summary and Conclusion

The use of custom templates for displaying web resource information is an innovative approach to information visualization. This approach redefines an information web site as a structured set of data based on a selected semantic profile, allowing the user to present the site content in a personalized and convenient manner. The quality of information search is improved through additional multi-level processing of traditional search engine results, solving problems such as search spam, better search results, and reduced search time. The effectiveness of this approach is increased through its integration into a single system, the Complex Internet Data Operating System (CIDO), with its core being the Web Data Operating Language (WDOL) developed by the author. This technology advances the field of knowledge by providing a new way to present and search for information on the web.

References

1. Manning, C D., Prabhakar, R., Hinrich, S.: Introduction to Information Retrieval, p. 469. Cambridge University Press, New York (2008)
2. Biro, I., Siklosi, D., Szabo, J., Benczur, A.A.: Linked latent Dirichlet allocation in Web spam filtering. In: Adversarial Information Retrieval on the Web: Proceedings of the 5th International Workshop, pp. 37–40. ACM, Madrid (2009)
3. Karpov, A., Kipyatkova, I., Ronzhin, A.: Very large vocabulary ASR for spoken Russian with syntactic and morphemic analysis. In: Proceedings of the 12th International Conference INTERSPEECH-2011. ISCA Association, Florence, Italy, pp. 3161–3164 (2011)
4. Maass, W., Kowatsch, T.: Semantic Technologies in Content Management Systems: Trends, Applications and Evaluations, p. 206. Springer, Heidelberg (2012). https://doi.org/10.1007/978-3-642-24960-0
5. Amerland, D.: Google Semantic Search: Search Engine Optimization (SEO) Techniques That Get Your Company More Traffic, Increase Brand Impact, and Amplify Your Online Presence. Que Publishing (2013)
6. Bernsen, N., Dybkjær, L. Multimodal Usability. Human–Computer Interaction Series. Springer, London (2009). https://doi.org/10.1007/978-1-84882-553-6
7. Castillo, C., Davison, B.: Adversarial Web Search, p. 126. Now Publishers Inc. (2011)
8. Structured Data Markup Wizard. https://www.google.com/webmasters/markup-helper/u/0/
9. Haag, S.: Management Information Systems for the Information Age, 9th edn., pp. 554. McGraw-Hill Higher Education (2012)
10. Ding, X.L.: The realization of the data mining technology and its application. Adv. Mater. Res. **971–973**, 1820–1823 (2014)
11. Zosimov, V., Bulgakova, O., Pozdeev, V.: Complex internet data management system. In: Babichev, S., Lytvynenko, V., Wójcik, W., Vyshemyrskaya, S. (eds.) ISDMCI 2020. AISC, vol. 1246, pp. 639–652. Springer, Cham (2021). https://doi.org/10.1007/978-3-030-54215-3_41
12. Zosimov, V., Bulgakova, O.: Development of domain-specific language for data processing on the internet. In: International Scientific and Technical Conference on Computer Sciences and Information Technologies, vol. 2, pp. 287–290 (2020)
13. Schema.org. https://schema.org/docs/schemas.html
14. Wordpress: Cms wordpress. https://wordpress.com/

15. Zosimov, V., Bulgakova, O., Stepashko, V.: Inductive building of search results ranking models to enhance the relevance of text information retrieval. In: International Workshop on Database and Expert Systems Applications-DEXA, Valencia, Spain, pp. 291–295 (2015)
16. Personalization and Google Search Results. https://support.google.com/
17. Zosimov, V., Bulgakova, O.: Application of personalized ranking models based on expert evaluations for sorting goods on e-commerce web resources. In: International Scientific and Technical Conference on Computer Sciences and Information Technologies, vol. 2, pp. 42–45 (2020)
18. Stephen, A., Promise, A.: A new query expansion approach for improving web search ranking. Inf. Technol. Comput. Sci. 1, 42–55 (2023)
19. Mithun, A.M., Bakar, Z.A.: Empowering information retrieval in semantic web. Int. J. Comput. Netw. Inf. Secur. 2, 41–48 (2020)
20. Yadav, U., Narula, G.S., Duhan, N., Jain, V.: Ontology engineering and development aspects: a survey. Int. J. Educ. Manag. Eng. 3, 9–19 (2016)

Metaheuristic Optimization Algorithms Usage in Recommendation System with User Psychological Portrait Generation

Mykhailo Vernik and Liubov Oleshchenko[✉]

National Technical University of Ukraine "Igor Sikorsky Kyiv Polytechnic Institute",
Kyiv 03056, Ukraine
oleshchenkoliubov@gmail.com

Abstract. Today, most Internet services are built using user recommendations: product recommendations (e-commerce), movies (Netflix), mobile applications (App Store, Play Market) and software code (GithubCopilot). The purpose of each of the software solutions is determined primarily by the business, which in turn is determined by the needs of the user. All of the above applications use classic neural network optimization methods for recommender systems development.

In this research classical methods of solving the problem of providing relevant recommendations were considered. The existing methods of building recommender systems were analyzed: classic and personalized models. This research proposes a new software method of creating a recommendation system using metaheuristic optimization algorithms with the use of generated user's psychological portrait, combination of the OCEAN model and 8 types of activities as hyperparameters, with its implementation in the "Entertainment Planner" mobile application for recreation sphere.

The proposed software method uses as a basis a neuro-collaborative filter with metaheuristic optimization methods, which ensures fast convergence and finding a way out of local traps of the function. Thanks to proposed software method, the result of improving the accuracy of the user's selection of events by 6–8% when using the full set of metaheuristic optimization algorithms in compared with existing methods. Implementation of the proposed software method is the mobile application, which has been developed using native technologies. The software product is implemented in the product-markets, which provides open access to the technology. Imperative technologies were used to implement a neural network for creating a psychological portrait of the recommendation system's user.

Keywords: Metaheuristic optimization · recommender system · user psychological portrait · neural networks · factorization matrix · mobile application

1 Introduction

In this research we're about to provide a proof of concept of usage and comparison metaheuristic optimization algorithms (MOA) in recommendation systems with classic optimization methods. Metaheuristic optimization methods in neural networks provide

© The Author(s), under exclusive license to Springer Nature Switzerland AG 2023
Z. Hu et al. (Eds.): ICCSEEA 2023, LNDECT 181, pp. 147–162, 2023.
https://doi.org/10.1007/978-3-031-36118-0_14

no worse results within a relative error of 10% or 1–2% more accurate than classical methods for recommender systems, which creates a new branch of recommender systems interaction with metaheuristic optimization methods. In an online store, there is a personalized recommendation algorithm for the customer. In a physical store personal recommendations are provided by consultants for an external assessment of the person needs based on which they offer recommendations. Despite this, most people make purchases through interaction with the person and person's emotional coloring, not with the recommendation itself. Thus, the development of software for a recommender system aimed at creating recommendations based on a new optimization method taking into account the psychological user component is relevant.

1.1 Analysis of Existing Solutions

One of the approaches to the design of recommender systems (RS), which is widely used, is joint filtering [1–5]. By finding users of the same age or items with a rating history similar to the current user or item, recommendations are generated using this neighborhood. Collaborative filtering methods are classified as memory-based and model-based [6, 7].

The following actions are used to collect RS user's data:

1. Ask the user to rate the item on a certain scale.
2. Ask the user to perform a search.
3. Ask the user to rate a collection of items from most favorite to least favorite.
4. Presentation of two items to the user and asking him to choose the best of them.

An example of implicit data collection is the analysis of product browsing time.

Content-based filtering methods use a set of discrete attributes and features to characterize an element in the system. The element representation algorithm is used to abstract the features of the elements in the system. A widely used algorithm is the TF-IDF representation (or vector space representation), and methods including text analysis, information retrieval, multimodal sentiment analysis, and deep learning are used. Disadvantages are cold start and overfitting [8, 9].

Session-based recommender systems use user interaction during a session to generate recommendations. Session-based recommender systems are used on YouTube and Amazon. They are particularly useful when a user's history (such as past clicks and purchases) is not available or does not match the user's current session. Methods of creating recommendations based on sessions mainly use generative sequential models. Disadvantages are high computing score processing large data and low accuracy in the existence of sparsity [10–13]. For reinforcement learning to build recommender systems, models or policies can be learned by rewarding the recommender agent. Recommendation methods based on reinforcement learning allow to train models and optimize based on indicators of involvement and user's interest of recommendation systems. The effectiveness of a recommendation system also depends on the degree to which it incorporates risk into the recommendation process. One of the options for solving this problem is a system that models context-sensitive recommendation and combines content-based techniques. Disadvantage is require big data [14–16].

Mobile data for mobile recommender systems is more complex than data, it requires spatial and temporal autocorrelation, and has validation problems [17]. Factors that can affect mobile recommender systems and the accuracy of prediction results are context, recommendation method, and privacy [18]. Mobile recommendations may not be applicable in all regions of the planet due to cultural and national characteristics of the population living in different territories. Uber and Lyft use this approach to generate driving routes for taxi drivers in a city using GPS data about the routes taken by taxi drivers during work, which includes location (latitude and longitude), timestamps and operational status [17]. This data is used to recommend a list of pick-up points on the route to optimize loading time and profit. Many recommender systems take a hybrid approach, using different types of filtering. Netflix is a good example of the use of hybrid recommender systems [19]. Some hybridization techniques include:

1. Weighted methods: numerical combination of evaluations of various components of recommendations.
2. Switching: choosing between recommendation components and applying the selected one.
3. Mixed methods: recommendations from different guidelines are presented together to give a recommendation.

Based on the conducted research, the following advantages and disadvantages of the considered methods were highlighted:

1. Personalized RS is more difficult to implement compared to classic RS, but at the same time, personalized RS provides more accurate results than classic RS.
2. The general trend of using RS is hybrid models.
3. The greatest concentration of development on risk-aware systems.

In previous studies, the classic model of building a psychological portrait of the user using a combination of the multilayer perceptron and back propagation algorithms was considered [20–25]. A conclusion novelty, which is made based on the existing solutions of the represented problem, demonstrates the disadvantages and ability of the proposed method to cover those objectives, as well as combining the worst approaches of the creation recommendation systems with a new method using combination the multilayer perceptron and MOA. The combination of the proposed method was not considered in the previous works, which could be used to demonstrate their accuracy and effectiveness in creating a recommender system with the use of generated user's psychological portrait.

1.2 Objective

The object of the research in this article is to compare classic optimization approach with a metaheuristic approach for a recommendation system in recreation sphere, by providing a proof of concept with increasing the accuracy of users' choice of recommender system events by using combinations of metaheuristic optimization algorithms and building a psychological portrait of the user using a neural network and a factorization matrix.

2 Research Method

2.1 Users Data Analysis

The main idea of the proposed method is to provide a proof of concept for metaheuristic optimization approach in recommendation systems, where technology will be demonstrated in the open-source mobile system with an implemented solution called "Entertainment Planner", so that each user of the "Entertainment Planner" system leaves data (for example, the event "Amusement Park", 2 people are invited, 2 min 13 s, free of charge, for 2 people, calmly, relaxingly):

1. Selected events.
2. Invited people.
3. Time spent on selection.
4. User selection parameters.
5. Psychological portrait of the user.

User and event data are represented as separate data vectors, where the number of active system users is 110 and the number of events is 5,370. More than 2,000 events have been added by users in total. Also more than 10,000 data on user actions, such as time spent on a particular page, invited people, selection parameters, psychological portrait. Thus, at the output we have more than 12,000 rows of data about users and their interaction in the system. This dataset can be presented in the form of 2 datasets for training and testing, where 10,000 rows are training, and 2,000 rows are testing. The basis of recommendations is user data, recommendations can be any characteristic (for example, a constructed neural network of the psychological portrait of the user to create a recommended list of invitations to the user), in our case it is a list of recommended events.

The proposed recommendation system uses the neuro-collaborative filter model with modification and optimization of the neural network learning process using 5 metaheuristic optimization algorithms: Gray Wolf Optimization (GWO), Particle Swarm Optimization (PSO), Whale Optimization Algorithm (WOA), Bee Algorithm (BA), Genetic algorithm (GA), which interacts with the neural network as follows:

1. The neural network is decomposed into a one-dimensional vector of levers of neural connections and levers of deviations.
2. The vector of the decomposed neural network will be called a generation.
3. Each of the 5 algorithms will generate a new generation.
4. Among the 5 generations, the best generation is selected based on the root mean square error of the final result.
5. The levers of neural connections and the levers of deviations are updated.

2.2 Recommender System Core Description

Figure 1 demonstrates the core of the proposed recommender system, which can be used in different scenarios of combining multiple neural networks and (or) arbitrary number of matrix factorizations as an additional parameters or different datasets, as well as its interaction with a set of metaheuristic optimization methods, could be extended with any new MOA algorithm, which in this case creates a new generation.

The best values are taken from the each used MOA algorithm and returned in one vector of weights and biases, for retraining / adjusting the weights and biases of the neural network.

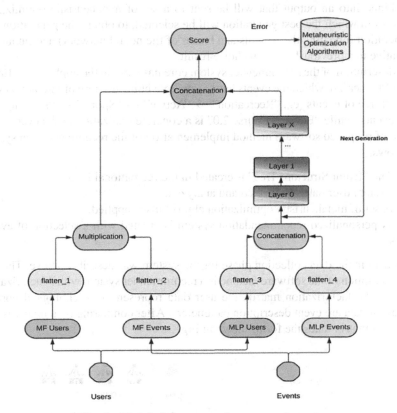

Fig. 1. Model of the proposed recommender system

The software system receives 2 vectors at the input: the "Users" and "Events" vectors, where the user vector is the data of psychological portraits, which is getting generated by using OCEAN model [26] for each user respectively and applying its values as a correction weights to the user's preselected choices, which consists of 8 additional parameters, such as "Sport", "Relax", "Romantic", "Extreme", "Calm", "Intellectual", "Fun", "Creative". The event vector is the data of event description parameters and applied 8 parameters from users' perspective. Accordingly, each of which creates vectors:

1. MF Users (user factorization matrix vector);
2. MF Events (event factorization matrix vector);
3. MLP Users (multilayer perceptron vector of users);

4. MLP Events (multilayer event perceptron vector).

After processing and decomposing the vectors into the factorization matrix (MF) will go "MF Users" and "MF Events", at the same time into the multilayer perceptron (MLP) will go the combined vectors "MLP Users" and "MLP Events".

At the stage of obtaining results from both the factorization matrix and the MLP are combined into an output that will be sent to a set of metaheuristic optimization methods, from which the best generation will be selected, to obtain the generation of a new generation (updating the weights and biases of the neural network) and finding an error relative to a previously known choice result.

The description of the recommender system core has a use in the application "Entertainment Planner", in which the events have a concrete built structure of interaction with an abstract tree of events (eg: "Recreation" → "Active" → "Sports" → "Dancing" are abstract events, while "Bachata Course 2.0" is a concrete realization of the event). The sequence of proposed software method implementation of the recommendation system is as follows:

Stage 1. The "Event Structure Tree" is created in the recreational field.

Stage 2. System user data is collected and analyzed.

Stage 3. A set of metaheuristic optimization algorithms is applied.

Stage 4. A personalized recommendation system is created with a selection of events and users.

User data in the data collection phase for the system was described above. The following description of the software method is a recommender system with a factorization matrix. For the factorization matrix, the user data represent a psychological portrait and the event data are event description parameters. After comparing both matrices, the factorization matrix takes the form, shown in Fig. 2.

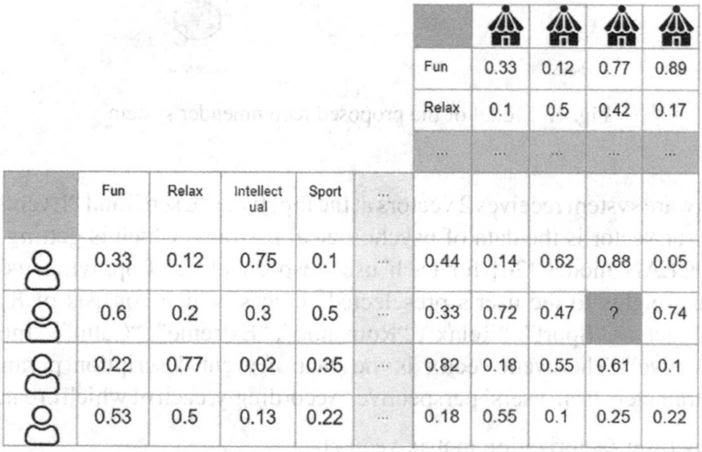

	🏛	🏛	🏛	🏛
Fun	0.33	0.12	0.77	0.89
Relax	0.1	0.5	0.42	0.17
...	...	...	...	...

	Fun	Relax	Intellect ual	Sport	...					
👤	0.33	0.12	0.75	0.1	...	0.44	0.14	0.62	0.88	0.05
👤	0.6	0.2	0.3	0.5	...	0.33	0.72	0.47	?	0.74
👤	0.22	0.77	0.02	0.35	...	0.82	0.18	0.55	0.61	0.1
👤	0.53	0.5	0.13	0.22	...	0.18	0.55	0.1	0.25	0.22

Fig. 2. Matrix of factorizations

Consider each component of the recommendation system core separately:

1. The "Users" component is a vector of system users.
2. The "Events" component is a vector of events visited by users.
3. Figure 3 shows the left part of the recommendation system represents the matrix factorization of Users and Events.

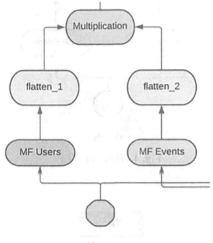

Fig. 3. Factorization matrix of Users and Events

4. Figure 4 shows the right part of the recommendation system. In this case, it represents the input to a multilayer perceptron.
5. Data sending from input to the MLP with the ReLU activation function (Fig. 5).
6. The next stage is to combine the results of 2 methods, MF and MLP (Fig. 6).
7. After merging, this result is normalized to the original data (Fig. 7).
8. To process the received error, metaheuristic optimization methods are used, thanks to which a new generation will be created to improve the results of the neural network.

The main idea of creating a single base of a package of metaheuristic algorithms is to use them to optimize a neural network. Since each of the proposed algorithms has its own advantages and disadvantages, when combining them into a single base according to the constructed generation in an arbitrary iteration, the best one will be chosen. The neural network will be decomposed into a vector, first the weights are written, and then the displacement weights, so all further operations will take place over this vector (Fig. 8).

Most metaheuristic algorithms are based on the swarming behavior of animals or biological interactions of living organisms. This is what gives metaheuristic algorithms simplicity in their software implementation, since they have the same implementation, which is based on the initialization of the initial generation, which is decomposed into a vector into which an objective function is applied, a stopping criterion is created.

The advantage of using proposed approach is that over time the researcher develops better and more advanced optimization methods, thus the base itself will be updated by

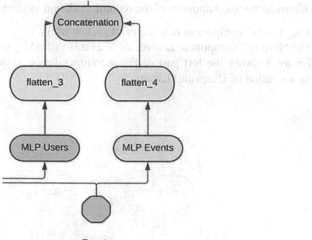

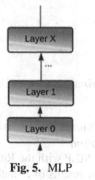

Fig. 4. Data for MLP

Fig. 5. MLP

adding new algorithms and, accordingly, the process of stagnation of the recommender system will not occur.

2.3 Software Implementation of the Proposed Method

To test the developed method, a corresponding software client part was created in the form of a mobile application. The following programming technologies for the client part of the proposed recommender system software development were used:

1. Flutter – for Google Play, App Store applications;
2. Firebase API – for notification message;
3. REST API – for communication with servers;
4. React JS – for creating a web application.

To develop the server part of the software, the following technologies were used:

1. .NET Core Web API is the main server;

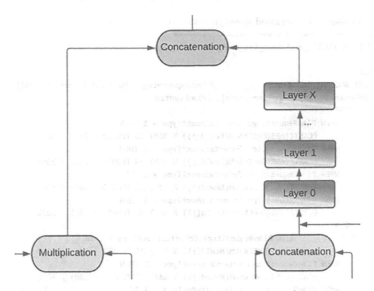

Fig. 6. The output combines the results of matrix factorization and multilayer perception

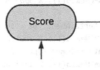

Fig. 7. Output result.

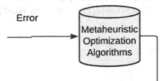

Fig. 8. The module for using the package of metaheuristic algorithms

2. AZURE MS SQL – for creation a database;
3. GOOGLE MAPS API – for locations of rest areas;
4. ASP.NET MVC – admin panel;
5. Python, Tensorflow, MOA, PyTorch, PySwarm – to create a psychological portrait of the recommender system user.

In Listing 1, the implementation fragment of one of the procedures for obtaining user moods (in general, Fun, Intellectual, Sport, Relax, Extreme, Calm, Creative, Romantic user moods were created) in the "Entertainment Planner" recommender system is given. The structure of the proposed software database consists of 45 tables and occupies ~500 mb on the hard disk of US-1 servers in the Azure cloud system, is given.

Listing 1. Get user mood stored procedure (code fragment)

```
============================================
ALTER PROCEDURE [dbo].[Get_All_User_Suppliers_Mood]
AS
BEGIN
    SET NOCOUNT ON SELECT          [PlanResponseLogs].[UserId],[PlanRequestId]
[PlanResponseLogs].[SupplierFirmId], [PlanEventId]
        CASE
            WHEN PlanRequestLogs.EntertainmentType = 1 THEN
                (CAST((ABS(CHECKSUM(NewId())) % 300) AS FLOAT) + 700) / 1000
            WHEN PlanRequestLogs.EntertainmentType = 2 THEN
                (CAST((ABS(CHECKSUM(NewId())) % 300) AS FLOAT) + 700) / 1000
            WHEN PlanRequestLogs.EntertainmentType = 3 THEN
                (CAST((ABS(CHECKSUM(NewId())) % 100) AS FLOAT) + 100) / 1000
            WHEN PlanRequestLogs.EntertainmentType = 4 THEN
                (CAST((ABS(CHECKSUM(NewId())) % 100) AS FLOAT) + 100) / 1000
        END as Sport,
        CASE        WHEN PlanRequestLogs.EntertainmentType = 1 THEN
                (CAST((ABS(CHECKSUM(NewId())) % 100) AS FLOAT) + 100) / 1000
            WHEN PlanRequestLogs.EntertainmentType = 2 THEN
                (CAST((ABS(CHECKSUM(NewId())) % 100) AS FLOAT) + 100) / 1000
            WHEN PlanRequestLogs.EntertainmentType = 3 THEN
                (CAST((ABS(CHECKSUM(NewId())) % 300) AS FLOAT) + 700) / 1000
            WHEN PlanRequestLogs.EntertainmentType = 4 THEN
                (CAST((ABS(CHECKSUM(NewId())) % 300) AS FLOAT) + 700) / 1000
        END as Relax,
CASE        WHEN PlanRequestLogs.EntertainmentType = 1 THEN
                (CAST((ABS(CHECKSUM(NewId())) % 300) AS FLOAT) + 700) / 1000
            WHEN PlanRequestLogs.EntertainmentType = 2 THEN
                (CAST((ABS(CHECKSUM(NewId())) % 100) AS FLOAT) + 100) / 1000
            WHEN PlanRequestLogs.EntertainmentType = 3 THEN
                (CAST((ABS(CHECKSUM(NewId())) % 100) AS FLOAT) + 100) / 1000
            WHEN PlanRequestLogs.EntertainmentType = 4 THEN
                (CAST((ABS(CHECKSUM(NewId())) % 100) AS FLOAT) + 100) / 1000
        END as Extreme
            …
    FROM [dbo].[PlanResponseLogs]
    INNER JOIN
dbo.PlanRequestLogs
ON   PlanRequestLogs.PlanRequestId = [PlanResponseLogs].PlanRequestId
    INNER JOIN
dbo.SupplierFirm_AbstractCategory
ON   SupplierFirm_AbstractCategory.SupplierFirmId=[PlanResponseLogs].SupplierFirmId
    INNER JOIN
dbo.MoodState
ON MoodState.AbstractCategoryId=SupplierFirm_AbstractCategory.AbstractCategoryId
        AND MoodState.SubCategoryId=SupplierFirm_AbstractCategory.SubCategoryId
        INNER JOIN
            dbo.Users
                ON Users.UserId = [PlanResponseLogs].UserId
        ORDER BY UserId, [SupplierFirmId]
END
```

This method provides an opportunity to receive an already generated user data file on the server, which is shown in Listing 2, which significantly speeds up the process of processing other calculation tasks.

Listing 2. Using a neuro-collaborative filter (code fragment)

```python
import torch
from layer import FeaturesEmbedding, MultiLayerPerceptron
class NeuralCollaborativeFiltering(torch.nn.Module):
    """

    A Pytorch implementation of Neural Collaborative Filtering.
    """

    def __init__(self, field_dims, user_field_idx, item_field_idx, embed_dim,
mlp_dims, dropout):          super().__init__()
        self.user_field_idx = user_field_idx
        self.item_field_idx = item_field_idx
        self.embedding = FeaturesEmbedding(field_dims, embed_dim)
        self.embed_output_dim = len(field_dims) * embed_dim
        self.mlp = MultiLayerPerceptron(self.embed_output_dim, mlp_dims, dropout,
output_layer=False)
        self.fc = torch.nn.Linear(mlp_dims[-1] + embed_dim, 1)
    def forward(self, x):
        """

        :param x: Long tensor of size ``(batch_size, num_user_fields)``
        """

        x = self.embedding(x)
        user_x = x[:, self.user_field_idx].squeeze(1)
        item_x = x[:, self.item_field_idx].squeeze(1)
        x = self.mlp(x.view(-1, self.embed_output_dim))
        gmf = user_x * item_x
        x = torch.cat([gmf, x], dim=1)
        x = self.fc(x).squeeze(1)
        return torch.sigmoid(x)
```

Customer support and reservations departments perform the main role at peak moments of software system load (for example, when an event is canceled).

Figure 9 shows the presentation of user data tables.

1. dbo.User table contains data on created users of the recommendation system.
2. dbo.UserLog table contains data on logging of all actions related to requests and user interaction with the recommendation system.
3. dbo.UserViolation table contains status data for problem users of the system.

This user data structure implements interaction with clustered and non-clustered keys, which allows to speed up the search for user records in the MSSQL database due to the fact that the clustered key is built on a B+ tree, a type of binary tree that has a logarithmic structure, which significantly speeds up the search process.

Such a key can be one, in our case there is also a non-cluster key, there can be an arbitrary number of such keys. Until we run out of hard disk space, each creation of a non-clustered key will map the table and turn it into a B+ tree.

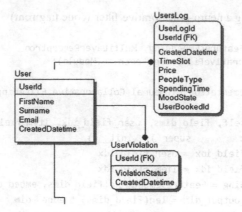

Fig. 9. User data structure

Figure 10 presents tables of abstract categories events:

1. dbo.AbstractCategory – the table corresponds to the hierarchical creation of types of entertainment and events (for example: category "Active recreation" - subcategory "Sports" - subcategory "Football").
2. dbo.AgeRanges – the table corresponds to which age ranges this category of entertainment is best suited for.
3. dbo.AgeRangesSexTypes – the table corresponds to which gender the age category is corresponding to the category of entertainment.
4. dbo.RepetitionRate – the table corresponds to the repetition rate of the event.
5. dbo.PeopleType – the table corresponds to which type of group of users the event is best suited for.
6. dbo.MoodState – the table corresponds to which user mood the event is suitable.

After adding the user into the recommendation system, the algorithm of forming a user psychological portrait and the algorithm of recommendations come into effect:

1. A basic psychological user's portrait is built on the basis of the initial choice.
2. The recommendation system takes the events visited by the user, user's choice of parameters, as well as the generated psychological portrait as parameters for generating recommended events. Each of the choices made by the user is stored in the system, which enables a learning process for the neural network, so the system can provide recommendations more accurately, as well as suggest people accordingly with a similar psychological profile.

3   Research Results

When compared with the classical method, the results of which are shown in Table 1, the proposed method gives an advantage in this case due to the fact that metaheuristic optimization algorithms allow the neural network to increase the search space and have the ability to get out of local traps, unlike classical methods such as stochastic gradient.

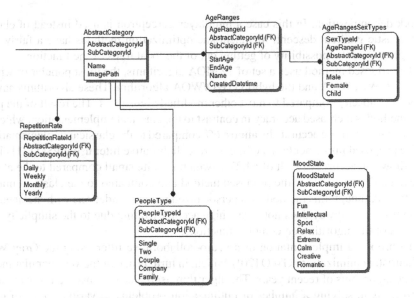

Fig. 10. The structure of the abstract category of events

The results of the tests show that the testing took place on a sample of data for Kyiv city (Ukraine) of in the amount of 6000 events, which were obtained using Google Maps API services and added >30 places and events through the created data administration system for "Entertainment Planner" recommendation system.

Table 1. Accuracy comparison of classic method for providing recommendations and MOA combinations of the proposed method for sampling ~6,000 recommended events

Method	Accuracy
Classic (MLP + BP)	76.68%
Proposed (MLP + MOA)	84.61%
Proposed (MLP + PSO)	44.27%
Proposed (MLP + GWO)	67.81%
Proposed (MLP + WOA)	65.52%
Proposed (MLP + GWO/WOA)	70.34%

The classic method implements a neuro-collaborative filter model using multilayer perceptron technology with back propagation method, the sample results showed high results compared to standard, non-personalized systems, such as the classic implementation of the collaborative filter, the subject filter. The accuracy of the method is 76.68%.

The proposed method implements a neuro-collaborative filter model with optimization modification using MOA set, which, in turn, use the vector technology of neural network decomposition. In this case a multilayer perceptron is used instead of classical stochastic gradient descent. Metaheuristic optimization methods have a fairly fast convergence and the possibility of getting out of the local traps of the function.

The proposed method uses a set of 12 MOA algorithms, the most popular of which are PSO, GWO, WOA and the hybrid GWO/WOA algorithm. These algorithms have a competent capacity compared to most other modified algorithms. The result of the proposed method is increased accuracy in contrast to the classical implementation, which is 84.61%, which is more accurate by almost 6% compared to the classical implementation.

The proposed implementation of the neuro-collaborative filter using PSO [21] algorithm shows an accuracy result of 44.27%, which is quite small compared to the above implementation methods of the proposed method and compared to the classical implementation. Although this method uses personalized recommendations with metaheuristic optimization anyway does not give a high accuracy rating due to the simplicity and weakness of the algorithm on complex functions.

The proposed implementation of a neuro-collaborative filter using the Gray Wolf metaheuristic optimization (GWO [20]), which, in turn, is one of the very popular metaheuristic algorithms of recent years. The algorithm itself has demonstrated considerable effectiveness in solving a number of optimization problems in various areas of constructive engineering. In this method, the accuracy result is 67.81%, which differs from the classical one by about 9%. This result has the right to live with the use of various types of modifications of the method, which can provide greater accuracy to the proposed method, such as the implementation of the hybrid Gray Wolf and Blue Whales algorithms (GWO/WOA [22]), which, in turn, confirmed the hypothesis of an increase in accuracy when using modifications to the GWO [20] algorithm up to 70.34%. The reliability and accuracy of the results have been provided by using a dataset of 6,000 real events, as a proof of concept of the proposed method, on the basis of a representative sample an 8% increase in accuracy was obtained from the classical method MLP + BP 76.68% to proposed MLP + MOA 84.61%.

Disadvantage of using the proposed software method is time costs, since the neural network is decomposed into a vector of weights and displacement weights, and processing and superimposition of activation functions and other matrix operations takes place over it, which in general have an $O(N^3)$ time estimate, which significantly slows down the processing process on large neural networks.

4   Summary and Conclusion

For the first time, it found further development, the superposition of methods proposed by the authors "MLP + MOA" improves accuracy for providing recommendations to users of the recommender system, which is better compared to existing software solutions in the recommendation systems field. The model for learning and optimizing neural networks is based on metaheuristic optimization algorithms, which, in turn, have advantages in solving specific tasks compared to accurate optimization methods. In particular, an advantage in the speed of convergence and the advantage of being able to avoid local traps (depending on the task, these can be both minima and maxima). Disadvantage of using the proposed software method is time costs for data processing.

The proposed software method uses as a basis a neuro-collaborative filter with meta-heuristic optimization methods, which ensures fast convergence and finding a way out of local traps of the function, thanks to which the result of improving the accuracy of the user's selection of events by 6–8% when using the full set of MOA algorithms compared to available methods.

Based on the received data, which, in turn, allows improving the accuracy of user analysis, the proposed method provides a reduction in the complexity of calculations and provides greater accuracy than the classical implementation. The modular structure of the proposed software, the modules of which implement the above-mentioned algorithms, greatly facilitates the implementation of the software product as a whole.

In the future, based on the obtained results, can be created universal approach of a self-developing recommendation system with super positioning optimization algorithms. Proposed method can change the classical realization of recommender systems development, as well as apply new technologies to optimize neural networks. The analyzed evaluation results and comparison of the efficiency of similar solutions can be used for various applied problems, where a choice of possible options is implemented.

References

1. Ghazanfar, M.A., Prügel-Bennett, A., Szedmak, S.: Kernel-mapping recommender system algorithms. Inf. Sci. **208**, 81–104 (2012). https://doi.org/10.1016/j.ins.2012.04.012. CiteSeerX 10.1.1.701.7729
2. Aggarwal, C.C.: Recommender Systems: The Textbook. Springer, Cham (2016). https://doi.org/10.1007/978-3-319-29659-3. ISBN 9783319296579
3. Brusilovsky, P.: The Adaptive Web, p. 325 (2007). ISBN 978-3-540-72078-2
4. Wang, D., Liang, Y., Xu, D., Feng, X., Guan, R.: A content-based recommender system for computer science publications. Knowl.-Based Syst. **157**, 1–9 (2018). https://doi.org/10.1016/j.knosys.2018.05.001
5. Blanda, S.: Online Recommender Systems – How Does a Website Know What I Want? American Mathematical Society. Accessed 31 Oct 2016
6. Feng, X.Y., et al.: The deep learning-based recommender system "Pubmender" for choosing a biomedical publication venue: development and validation study. J. Med. Internet Res. **21**(5), e12957 (2019)
7. Chen, M., Beutel, A., Covington, P., Jain, S., Belletti, F., Chi, E.: Top-K Off-Policy Correction for a REINFORCE Recommender System (2018). arXiv:1812.02353 [cs.LG]
8. Yifei, M., Narayanaswamy, B., Haibin, L., Hao, D.: Temporal-contextual recommendation in real-time. In: KDD 2020: Proceedings of the 26th ACM SIGKDD International Conference on Knowledge Discovery & Data Mining, pp. 2291–2299. Association for Computing Machinery (2020)
9. Hidasi, B., Karatzoglou, A.: Recurrent neural networks with top-k gains for session-based recommendations. In: Proceedings of the 27th ACM International Conference on Information and Knowledge Management, CIKM 2018, pp. 843–852. Association for Computing Machinery, Torino (2018). https://doi.org/10.1145/3269206.3271761. arXiv:1706.03847. ISBN 978-1-4503-6014-2. S2CID 1159769
10. Kang, W.-C., McAuley, J.: Self-Attentive Sequential Recommendation (2018). arXiv:1808.09781 [cs.IR]

11. Li, J., Ren, P., Chen, Z., Ren, Z., Lian, T, Ma, J.: Neural attentive session-based recommendation. In: Proceedings of the 2017 ACM on Conference on Information and Knowledge Management, CIKM 2017, pp. 1419–1428. Association for Computing Machinery, Singapore (2017). https://doi.org/10.1145/3132847.3132926. arXiv:1711.04725. ISBN 978-1-4503-4918-5. S2CID 21066930
12. Liu, Q., Zeng, Y., Mokhosi, R., Zhang, H.: STAMP: short-term attention/memory priority model for session-based recommendation. In: Proceedings of the 24th ACM SIGKDD International Conference on Knowledge Discovery & Data Mining, KDD 2018, pp. 1831–1839. ACM, London (2018)
13. Xin, X., Karatzoglou, A., Arapakis, I., Jose, J.: Self-Supervised Reinforcement Learning for Recommender Systems (2020). arXiv:2006.05779 [cs.LG]
14. Ie, E., et al.: SlateQ: a tractable decomposition for reinforcement learning with recommendation sets. In: Proceedings of the Twenty-Eighth International Joint Conference on Artificial Intelligence (IJCAI-19), pp. 2592–2599 (2019)
15. Zou, L., Xia, L., Ding, Z., Song, J., Liu, W., Yin, D.: Reinforcement learning to optimize long-term user engagement in recommender systems. In: KDD 2019: Proceedings of the 25th ACM SIGKDD International Conference on Knowledge Discovery & Data Mining, KDD 2019, pp. 2810–2818. https://doi.org/10.1145/3292500.3330668. arXiv:1902.05570. ISBN 9781450362016. S2CID 62903207
16. Lakiotaki, K., Matsatsinis, N.F., Tsoukias, A.: Multicriteria user modeling in recommender systems. IEEE Intell. Syst. 26(2), 64–76 (2011). https://doi.org/10.1109/mis.2011.33. S2CID 16752808. CiteSeerX 10.1.1.476.6726
17. Yong, G., Hui, X., Alexander, T., Keli, X., Marco, G., Pazzani, M.J.: An energy-efficient mobile recommender system (PDF). In: Proceedings of the 16th ACM SIGKDD International Conference on Knowledge Discovery and Data Mining, pp. 899–908. ACM, New York (2010). Accessed 17 Nov 2011
18. Pimenidis, E., Polatidis, N., Mouratidis, H.: Mobile recommender systems: identifying the major concepts. J. Inf. Sci. 45(3), 387–397 (2018). https://doi.org/10.1177/016555151879 2213. arXiv:1805.02276. S2CID 19209845
19. Gomez-Uribe, C.A., Hunt, N.: The Netflix recommender system. ACM Trans. Manag. Inf. Syst. 6(4), 1–19 (2015). https://doi.org/10.1145/2843948
20. Mirjalili, S., Mirjalili, S.M., Lewis, A.: Grey wolf optimizer. Adv. Eng. Softw. 69, 46–61 (2014). https://doi.org/10.1016/j.advengsoft.2013.12.007. ISSN 0965-9978
21. Trivedi, I.N., Jangir, P., Kumar, A., Jangir, N., Totlani, R.: A novel hybrid PSO–WOA algorithm for global numerical functions optimization. In: Bhatia, S.K., Mishra, K.K., Tiwari, S., Singh, V.K. (eds.) Advances in Computer and Computational Sciences. AISC, vol. 554, pp. 53–60. Springer, Singapore (2018). https://doi.org/10.1007/978-981-10-3773-3_6
22. Mirjalili, S., Lewis, A.: The whale optimization algorithm. Adv. Eng. Softw. 95, 51–67 (2016). https://doi.org/10.1016/j.advengsoft.2016.01.008. ISSN 0965-9978
23. Jadhav, S.L., Mali, M.P.: Pre-recommendation clustering and review based approach for collaborative filtering based movie recommendation. Int. J. Inf. Technol. Comput. Sci. (IJITCS) 8(7), 72–80 (2016). https://doi.org/10.5815/ijitcs.2016.07.10
24. Raghavendra, C.K., Srikantaiah, K.C., Venugopal, K.R.: Personalized recommendation systems (PRES): a comprehensive study and research issues. IJMECS 10(10), 11–21 (2018). https://doi.org/10.5815/ijmecs.2018.10.02
25. Neysiani, B.S., Soltani, N., Mofidi, R., Nadimi-Shahraki, M.H.: Improving performance of association rule-based collaborative filtering recommendation systems using genetic algorithm. IJITCS 11(2), 48–55 (2019). https://doi.org/10.5815/ijitcs.2019.02.06
26. Soto, C., Jackson, J.: Five-factor model of personality. In: Dunn, D.S. (ed.) Oxford Bibliographies in Psychology. Oxford, New York (2020). https://doi.org/10.1093/obo/978019982 8340-0120

Interest Point Detection at Digital Image Based on Averaging of Function

Iryna Yurchuk$^{(\boxtimes)}$ and Ivan Kosovan

Software Systems and Technologies Department, Taras Shevchenko National University of Kyiv, Kyiv, Ukraine

i.a.yurchuk@gmail.com

Abstract. Digital image processing is a fundamental task of computer vision. It can be realized with different purposes, in particular, images can be compared, certain segments are tracked, image quality changes, etc. There are many methods and approaches for processing, but methods that can make inferences about an image based on individual pixels, which are the individual card of an image, are valuable and useful. In particular, one of these concepts is interest pixels (points). The search for them is based on detectors at all. In this paper, the mathematical basis with further its application into the algorithm to research a pixel as an interest point of a digital image is obtained. It is based on the conditions of topological persistence of averaging of a function of two variables. Also, the authors implement this algorithm, test it on grey-scaled images, and compare it with the Harris detector.

Keywords: Parameter-dependent integral · Interest point · Digital image · Averaging · Topological persistence

1 Introduction

Nowadays, there are some points of view on conditions to pixel has to satisfy to be an interest point [1]. In general, it means any point in the image for which the signal changes two-dimensionally. It can be $X-$ corner ($T-$ corner, $Y-$ corner, etc.), a black dot on white, an ending of a branch, etc.

Also, there are three categories of methods to detect an interest point. They are based on contour (extract contours that have to be approximated, search for maximal curvature, etc., and search for intersection points), intensity (compute a measure that distinguishes a point from others based on calculations in many directions of discrete analogs of derivatives) or parametric model (fit a parametric intensity model to the signal). Each of them has its advantages and disadvantages, which are taken into account when applied in a certain subject area. For example, the parametric model is limited to specific types of interest points by the detected features sample such that each of them is closest to the observed signal.

Purpose of work – to propose the mathematical basis and its algorithmic realization for the detection of the point of interest based on the intensity model with providing

Z. Hu et al. (Eds.): ICCSEA 2023, LNDECT 181, pp. 163–172, 2023.
https://doi.org/10.1007/978-3-031-36118-0_15

approximated manifold with the further possibility of its persistent analysis. It facilitates universality due to a combination of both intensity and contour methods of interest point detection.

The aim of research – algorithm of detection of an interest point as a point that is persistent (stable) up to a translation and a scaling, taking into account the possibility of changing the value of its neighborhood, and detected interest point has to be distinct, stable, unique, interpreted and local.

Major research objectives:

- Formulate mathematical statements that provide the conditions for the persistence of a point up to a translation and a scaling.
- Based on the obtained results, propose an algorithm for finding a point of interest.
- Compare the obtained algorithm with the existing ready-made solution.

2 Literature Review

In the field of digital image processing, the averaging method is quite common. In addition, this method can be found by studying the processing of local pixels of one. For example, averaging the variance of neighboring pixels is used to increase the performance [2], and pixel-by-pixel averaging of variations of multiple images has increased the accuracy of predictions to change the color of the pixel in the original image [3].

The averaging of equalization and threshold made it possible to develop an algorithm that automatically corrects the white balance and increases the color contrast. The color range is not overstretched, and the nature of the original image does not change [4].

Averaging is also used to reduce noise, which greatly improves the quality of digital images. As an example, weighted averaging with overcomplete orthogonal transformation fights noise at the level of cutting-edge algorithms [5]. Another example is using the average values of neighboring non-noise pixels to detect and replace noisy pixels in the original image. This approach has shown high efficiency and provides low computational complexity, as it is implemented using iterative processes [6].

In modern realities, it is obvious that portable devices are an integral part of human life. However, such devices have limited memory and computing power, so they must use modern filtering methods, such as a spatial averaging filter [7]. The method of analysis and comparison of templates is also used to solve the problem of resource intensity. It can be used to recognize a particular person's face in a digital image with several different faces. For example, there is an approach that uses the averaging of half of the face, namely: divides the face in half, determines the features, and recognizes them in other images [8]. This method not only showed greater recognition accuracy but also improved resource efficiency.

It should also be noted that sufficient conditions have been derived for the topological stability of the averaging of piecewise smooth functions $f : \mathbb{R} \to \mathbb{R}$ with a finite number of extremes with respect to discrete measures with finite supports [9].

At the current stage of development of digital image processing, there are many algorithms to define the points of interest. As an example, we can take the use of the Dilation Operator [10]. First, the image is divided pixel by pixel into equidistant grids. As a stopping metric, the SpeedThreshold is used, which is calculated as critical values of two manually set speeds. The stopping points and their neighborhoods are the areas of further investigation - we select the area where the number of stopping points is greatest. First, we binarize this area to get rid of unnecessary noise. Using the Dilation operator, we reduce the studied area to the minimum value, which will be the point of interest.

Kumar and Pang [11] noted that a pixel of interest can be determined by examining its neighbors, namely, gray-scale grades. However, newer scientific work, such as that by Zhao et al. [12], departs from this principle and uses a multiscale Gabor filter to extract information from pixels in completely different directions. Information about the different scales of grayscale change is used to build an "entropy map" of each pixel. The pixels of interest are extracted directly from the map by entropy normalization. In this way, the work has paved the way for the study of pixels of interest not only for the contour itself but also for the grayscale, as well as increasing the robustness to noise and reducing the time-consuming execution with the right selection of the scale. The authors single out this method for finding pixels of interest as being more suitable for applications such as screening monitoring, sensing, and working with medical images.

In addition to working directly with images, local feature search is also used for video files. This is possible because a video is structured as a set of individual images. For example, Li et al. [13] describe a method for finding Spatio-Temporal Interest Points in video files, which is based on calculations of curvature and gradient points. The process of finding points of interest is divided into three steps: intensity calculation using a response function, obtaining candidate points by non-maximal suppression, and imposing constraints.

3 Background

Let consider points $A_1(x_1, y_1, z_1), A_2(x_2, y_2, z_2), A_3(x_3, y_3, z_3)$ and $O(0, 0, 0)$ such that for any i, $i = \overline{1,3}$, a value $z_i > 0$, $\overrightarrow{OA_1}$, $\overrightarrow{OA_2}$ and $\overrightarrow{OA_3}$ are pairwise non-collinear and non-coplanar. Also $O(0, 0, 0) \in \text{Int } A_1'A_2'A_3'$, where A_1', A_2' and A_3' are projections of A_1, A_2 and A_3 onto the xOy coordinate plane, respectively. We define 2-simplex formed by A_1, A_2, A_3 and O. It is easy to write the equations of planes α, β and γ such that $A_1, A_2, O \in \alpha$, $A_2, A_3, O \in \beta$ and $A_1, A_3, O \in \gamma$ by using the coordinates of corresponding points. So, we obtain $\alpha : a_1x + b_1y + c_1z = 0$, $\beta : a_2x + b_2y + c_2z = 0$ and $\gamma : a_3x + b_3y + c_3z = 0$.

Let define a function $f(x, y)$ by the following equalities:

$$f(x, y) = \begin{cases} a_1'x + b_1'y, & if\,(x, y) \in D_1; \\ a_2'x + b_2'y, & if\,(x, y) \in D_2; \\ a_3'x + b_3'y, & if\,(x, y) \in D_3, \end{cases} \tag{1}$$

where $a_i' = -\frac{a_i}{c_i}$, $b_i' = -\frac{b_i}{c_i}$ and $D_i = \text{pr}_{xOy}A_iOA_{i+1}$ where $A_4 = A_1$ for $i = \overline{1,3}$.

It is easy to see that $c_i \neq 0$. Since it is $z-$ coordinate of a vector $\overrightarrow{n_i}$ such that

$$\overrightarrow{n_i} \perp A_iOA_{i+1}, \text{ we can write } \overrightarrow{n_i} = \begin{vmatrix} \vec{i} & \vec{j} & \vec{k} \\ x_i & y_i & z_i \\ x_{i+1} & y_{i+1} & z_{i+1} \end{vmatrix} \text{ and } \begin{vmatrix} x_i & y_i \\ x_{i+1} & y_{i+1} \end{vmatrix} \neq 0.$$

By D we denote a closed domain that contains a projection of a triangle $A_1A_2A_3$ onto the xOy coordinate plane.

At a finite set of points $\{P_i(x_i; y_i)\}_{i=1}^m$ of D we define a function $\mu(x, y)$ such that $\mu(x_i, y_i) = p_i$, where $p_i \in \mathbb{R}_+$.

Let assume that $\{P_i(x_i; y_i)\}_{i=1}^m = P = P' \cup P'' \cup P'''$, $P' \subset D_1$, $P'' \subset D_2$ and $P''' \subset D_3$, also $|P'| = m_1$, $|P''| = m_2$ and $|P'''| = m_3$, where $m_1 + m_2 + m_3 = m$. We can renumber the points P_i: the first m_1 points belong to D_1, next m_2 points belong to D_2 and the last m_3 to D_3.

Let $f(x, y) : \mathbb{R} \times \mathbb{R} \to \mathbb{R}$ and $\alpha > 0$ be a number. Then a function $f_\alpha(x, y) : \mathbb{R} \times \mathbb{R} \to \mathbb{R}$ can be defined by the following formula:

$$f_\alpha(x, y) = \iint\limits_D f(x + \alpha t, y + \alpha s)\mu(t, s)dtds, \qquad (2)$$

where D is a closed domain, $D \subset \mathbb{R} \times \mathbb{R}$.

We have to remark that if $f(x, y)$ is defined into D, then $f_\alpha(x, y)$ is defined into $\tilde{D}$, $\tilde{D} \supset D$.

Two continuous functions $f(x, y) : \mathbb{R} \times \mathbb{R} \to \mathbb{R}$ and $g(x, y) : \mathbb{R} \times \mathbb{R} \to \mathbb{R}$ are called to be topologically equivalent if there exist the preserving orientation homeomorphisms $h_1 : \mathbb{R} \times \mathbb{R} \to \mathbb{R} \times \mathbb{R}$ and $h_2 : \mathbb{R} \to \mathbb{R}$ such that $h_2 \circ f = g \circ h_1$.

A function $f(x, y)$ is topologically persistent up to averaging to a function μ if there exists $\varepsilon > 0$ such that for all $\alpha \in (0, \varepsilon)$ the functions $f(x, y)$ and $f_\alpha(x, y)$ are topologically equivalent.

Let $\mu(t, s)$ be a discrete probabilistic measure with a finite support $P_1, P_2, \cdots, P_m$ at D. Set $\mu(P_i) = \mu(x_i, y_i) = p_i$, where $p_i \in \mathbb{R}_+$ and $\sum_{i=1}^m p_i = 1$. We assume that $f : D \to \mathbb{R}$ belongs to a class C^1 excepting a set of points $S = \{S_i(x_i^*; y_i^*)\}_{i=1}^\infty$ which is zero measurable and there exist finite limits $L_1^i = \lim\limits_{x \to x_i^* - 0} \frac{\partial f(x,y)}{\partial x}$, $R_1^i = \lim\limits_{x \to x_i^* + 0} \frac{\partial f(x,y)}{\partial x}$, $L_2^i = \lim\limits_{y \to y_i^* - 0} \frac{\partial f(x,y)}{\partial y}$, $R_2^i = \lim\limits_{y \to y_i^* + 0} \frac{\partial f(x,y)}{\partial y}$.

We remind that a point $M_0(x_0, y_0)$ is a local minimum (maximum) of function $f(x, y)$ if there exists its neighborhood $U(M_0)$ such that for any point (x, y), $M \in U(M_0)$, the inequality $f(M) > f(M_0)$ is true.

For formulating the conditions of function to be topologically persistent up to averaging we use Theorem 5.2., see [9].

Theorem. *Let assume that both of limits* $L_1 = \lim\limits_{x \to 0-0} \frac{\partial f(x,y)}{\partial x}$, $R_1 = \lim\limits_{x \to 0+0} \frac{\partial f(x,y)}{\partial x}$ $(L_2 = \lim\limits_{x \to 0-0} \frac{\partial f(x,y)}{\partial y}$, $R_2 = \lim\limits_{x \to 0+0} \frac{\partial f(x,y)}{\partial y})$ *are finite and* $X_j \neq 0$, *where* $X_j = L_1(p_1 + \ldots + p_j) + R_1(p_{j+1} + \ldots + p_m)(X_j = L_2(p_1 + \ldots + p_j) + R_2(p_{j+1} + \ldots + p_m))$, $j = 1, \ldots, m-1$. *Then* $f(x, y)$ *at* $(0;0)$ *is topologically persistent up to averaging to* μ.

We only provide the main argument for the proof and omit the technical details. Since $\frac{\partial f(x,y)}{\partial x}$ can be considered as a function of one variable, we can apply Theorem 5.2., which contains conditions to be topologically persistent up to averaging to μ for a function of one variable.

Corollary. *If* $f(x, y)$ *is piecewise linear, defined by the formula* (1) *and values* $X = L_1p_k + R_1p_k'$ *and* $Y = L_2p_k'' + R_2p_k'''$, *where* $k, k', k'', k''' \in \{1, 2, 3\}$ *and correspond* to D_i, are defined correctly. Then X and Y have the same substance as X_j in the Theorem and exhaust the whole set of X_j.

We have to remark before using Theorem for digital image processing in Sect. 4:

- any pixel has integer coordinates, all x_i and y_i are integer and z_i is also integer as its intensity and belongs to the interval $[0; 255]$;
- any pixel has to be translated along the two axes at the origin as a function of two variables.

4 Algorithm

An interest point in a digital image is a pixel such that its neighborhood is a blob, namely a region that is associated with at least one local extremum (either a maximum or a minimum), see [14].

In this paper, the authors consider the following conditions on a pixel to be an interest point of an image:

- The interest point position is accurate in the image domain;
- It is persistent up to local and global deformations in the image domain: according to definitions of $f_\alpha(x, y)$ and topologically persistent up to averaging, term stability includes a translation and a scaling;
- An interest point should include an attribute of scale so that the interest point can be obtained from the image even when the scale changes.

We have to remark that the last condition can be expanded to a concept of multi scaling, see [12, 15].

Let consider the algorithm of researching a pixel as an interest point in a digital image in denotations and terms of the previous section. A point O coincides with a researched pixel X. It is easy to hold a condition to be origin by using a translation. Points $A_1(x_1, y_1, z_1)$, $A_2(x_2, y_2, z_2)$ and $A_3(x_3, y_3, z_3)$ are pixels that belong to a neighborhood of X. These four points have to be checked on conditions of the possibility to construct 2-simplex on them. Then it needs to be checked whether a point O belongs to the internal area of the projection of 2-simplex on the xOy – plane. It is quite simple: first, we read the points, then we look for the vector product of the paired vectors, calculate the sum of the products and check whether it is equal to 3 or -3.

After obtaining the required construction, we proceed to the stage of integration with a parameter. In the course of the work, we have identified 16 possible arrangements of a triangle in which one of the vertices coincides with the origin of coordinates in the plane.

Then, we have to check the stability criterion theorem:

1. Obtain the coefficients at the variables in the three equations of the planes as parts of 2-simplex, see the formula (1);
2. Define the function $\mu(x_i, y_i) = p_i$, $i = 1, 2, 3$ by using pixels that belong to part D_i of a neighborhood of O. Let $\{K_s\}_{s=1}^{N_i}$ be a set of pixels that belong to D_i. Then, we

set for a grey-scaled image:

$$p_i = \frac{\sum_{s=1}^{N_i} I(K_s)}{255 \, N_i},\tag{3}$$

where $I(K_s)$ is an intensity of K_s and $i = 1, 2, 3$;

3. Calculate values X and Y, according to the corollary of the Theorem;

4. Check whether either $X \neq 0$ or $Y \neq 0$. If the statement is true, pixel O is stable.

The final check is to check for linearity preservation. The planes obtained by integration are passed through the construction check again.

We define those points that passed all the described steps as points of interest and plot them on the image.

A parameter α at formula (2) can be interpreted as a learning rate. As usually in machine learning it must be manually chosen for each given type of digital image (landscape, portrait, MRI, etc.) and may vary greatly depending on it.

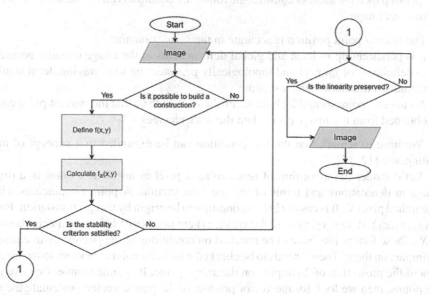

Fig. 1. Algorithm block diagram: the first condition is followed from pairwise non-collinear and non-coplanar of vectors which are the basis of the construction; $f(x, y)$ and $f_\alpha(x, y)$ are defined by formulas (1) and (2), respectively

In Fig. 1, there is a representation of the algorithm. It has a starting point in terms of the image to be processed. The processing consists of the alternate selection of three random neighbors of each pixel of the image, followed by checking the possibility of constructing a 2-simplex on the given points (see the first condition in the block diagram), as well as calculating the averages. The second condition is checking of the Corollary. If the conditions of the Corollary are satisfied, then a point is an interest point. At the output of the algorithm, we have an image with selected points of interest. We have to remark that the last condition checks the preservation of 2-simplex.

5   Illustrative Examples and Testing

Certain processing results were obtained for images of size 45×45.

For the first test, a handwritten symbol similar to the "+" sign was chosen. It can be seen that the number of points found is greater than the Harris detector (Fig. 2). We remind that the Harris detector extracts corners (a corner is a junction of two edges, where an edge is a sudden change in image brightness) and infers interest point. It is a first-order derivative-based interest point detection. According to the definition of interest point, the stability of constructed 2-simplex in obtained algorithm can be compared with the Harris detector correctly.

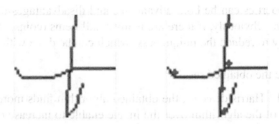

Fig. 2. The first comparison of the work of the algorithms. On the left is the author's one, and on the right is the Harris detector

Next, another handwritten symbol, "∀", was chosen. One pass over the image showed the same result as in the first test (Fig. 3). Much more interesting is the picture after several passes (Fig. 4). You can see that almost the whole figure is highlighted as pixels of interest. From this, we can conclude that more research needs to be done in the field of use as a contour detector.

Fig. 3. The second comparison of the work of the algorithms. On the left is the author's one, and on the right is the Harris detector

On the larger image, we got a slightly worse result because of the number of points found (Fig. 5). This can be explained by the fact that the Harris method examines the image through a sliding window, while our method goes through each pixel individually.

Let use a visual inspection to summarize distinct, stable, unique, interpreted and local properties of interest points: the more passes of the algorithm, the less distinct points are; the stability does not depend on passes and is defined by α as a parameter; the more passes of the algorithm, the less unique points are; the interpretability and locality are satisfied and do not depend on passes.

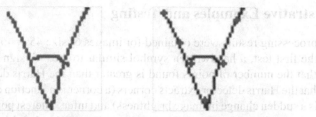

Fig. 4. Several passes of the algorithm over the image with the results saved. On the left is the author's one, and on the right is the Harris detector

The listed properties can be both advantages and disadvantages depending on the field of application. Obviously, if there are many small items (coins, etc.) in the scene, then it is necessary to reduce the uniqueness, which can be done with a larger number of passes.

Let summarize the obtained results:

1) Compared to the Harris detector, the obtained algorithm finds more interest points. Several passes of the algorithm over the image enable to increase of interest points set.
2) During the testing, there are no false positives, in particular, background points not specified as interest points.

To compare the performance of the algorithms, let summarize the obtained values in the table (Table 1).

Table 1. Algorithm performance comparison

Algorithm	Image size		
	45 × 45	255 × 255	1001 × 1001
The suggested algorithm	4.931[a]	13.392[a]	52.844[a]
Harris detector	0.022[a]	0.123[a]	0.482[a]

Fig. 5. Algorithm passing over the larger image. On the left is the author's one, and on the right is the Harris detector

6 Summary and Conclusion

The detection interest point algorithm based on topological persistence up to averaging to a two-variable function includes the possibility of constructing an approximation manifold in the neighborhood of the point using a 2-simplex, whose stability is respective up to the translation and scaling transformations. The basis of the algorithm is a criterion determined by partial derivatives. The author's algorithm was tested on grey-scaled images and compared with the Harris detector as a detection interest point algorithm based on the intensity model. The most advantage is a large enough amount of detected points. The most disadvantage of it is the long processing time.

In further research, the digital image processing based on the obtained algorithm will be adapted to video files with real-time processing. It needs additional research possibility to increase the number of considered neighbors (pixels) on which 2-simplex is constructed. In addition, the calculation of scaling coefficients which are realized as average values of the intensity of pixels in a neighborhood can be improved.

References

1. Jing, J., Gao, T., Zhang, W., Gao, Y., Sun, C.: Image Feature Information Extraction for Interest Point Detection: A Comprehensive Review. ArXiv, abs/2106.07929 (2021)
2. Sritapanu, S., Amornraksa, T.: Adaptive partial image watermarking based on the averaged variance of neighbor pixels. In: Proceedings of the International Symposium on Intelligent Signal Processing & Communication Systems, pp. 1–4. IEEE, Chengdu (2010). https://doi.org/10.1109/ISPACS.2010.5704661
3. Peungpenich, A., Amornraksa, T.: An improving method for image watermarking using image averaging and tuned pixels prediction. In: Proceedings of the 10th International Symposium on Communications & Information Technologies, pp. 755–760. IEEE, Tokyo (2010). https://doi.org/10.1109/ISCIT.2010.5665090
4. Tai, S.-C., Liao, T.-W., Chang, Y.-Y., Yeh, C.-P.: Automatic White Balance algorithm through the average equalization and threshold. In: Proceedings of the 8th International Conference on Information Science & Digital Content Technology (ICIDT 2012), pp. 571–576. IEEE, Jeju (2012)
5. Xu, J., Lin, F.: Hybrid multi-resolution analysis and weighted averaging of overcomplete orthogonal transform scheme for digital image denoising. In: Proceedings of the International Conference on Multimedia & Expo Workshops (ICMEW), pp. 1–6. IEEE, Chengdu (2014). https://doi.org/10.1109/ICMEW.2014.6890575
6. Konieczka, A., Balcerek, J., Dąbrowski, A.: Method of adaptive pixel averaging for impulse noise reduction in digital images. In: Proceedings of the Baltic URSI Symposium (URSI), pp. 221–224. IEEE, Poznan (2018). https://doi.org/10.23919/URSI.2018.8406738
7. Tsai, C.-M., Yeh, Z.-M.: Fast spatial averaging filter. In: Proceedings of the International Symposium on Computer, Consumer & Control, pp. 153–156. IEEE, Taichung (2012). https://doi.org/10.1109/IS3C.2012.47
8. Arora, S., Chawla, S.: An intensified approach to face recognition through average half face. In: IEEE India Conference (INDICON), pp. 1–6. IEEE, New Delhi (2015). https://doi.org/10.1109/INDICON.2015.7443802
9. Maksymenko, S., Marunkevych, O.: Topological stability of the averages of functions. Ukr. Math. J. **68**(5), 707–717 (2016)

10. Qing, W., Ding, C., Fu, K., Ren, W.: Interest point detection method based on dilation operation. Acta Armamentarii **38**(10), 2041–2047 (2017)
11. Kumar, A., Pang, G.K.H.: Defect detection in textured materials using optimized filters. IEEE Trans. Syst. Man Cybern. B Cybern. **32**(5), 553–570 (2002)
12. Zhao, Z., Li, B., Chen, L., Xin, M., Gao, F., Zhao, Q.: Interest point detection method based on multi-scale Gabor filters. IET Image Proc. **13**(12), 2098–2105 (2019)
13. Li, Y., Xia, R., Huang, Q., Xie, W., Li, X.: Survey of spatio-temporal interest point detection algorithms in video. IEEE Access **5**, 10323–10331 (2017). https://doi.org/10.1109/ACCESS.2017.2712789
14. Lindeberg, T.: Detecting salient blob-like image structures and their scales with a scale-space primal sketch: A method for focus-of-attention. Int. J. Comput. Vis. **11**(3), 283–318 (1993)
15. Satyabrata, S., Santosh, K.N., Tanushree, M.: Digital image texture classification and detection using Radon transform. Int. J. Image Graph. Sig. Process. (12), 38–48 (2013)

Analyzing Ukrainian Media Texts by Means of Support Vector Machines: Aspects of Language and Copyright

Maksym Lupei[1]([✉]), Oleksandr Mitsa[2], Vasyl Sharkan[3], Sabolch Vargha[2], and Nitsa Lupei[4]

[1] Department of Intelligent Information Technologies, Institute of Cybernetics, National Academy of Sciences of Ukraine, Kyiv, Ukraine
maxim.lupey@gmail.com

[2] Department of Informative and Operating Systems and Technologies, Uzhhorod National University, Uzhhorod, Ukraine

[3] Department of Journalism, Uzhhorod National University, Uzhhorod, Ukraine

[4] Department of Theory and History of State and Law, National Academy of Management, Kyiv, Ukraine

Abstract. The paper focuses on the possibilities of studying Ukrainian online media publications using the support vector machine. Through a database including about 95,000 texts of three Ukrainian online media («Censor.NET», «Ukrainska Pravda», «Zakarpattya Online») within the period from September 2021 to August 2022 there were traced possibilities for identifying the authorship of a media text (belonging to certain media). It was established that the accuracy of determining whether a text belongs to online media (regional, national) using SVM is significantly influenced by the presence in published texts keywords included in the names of the media: if these words remain in the dataset, the accuracy of identification of the media through the text fragment is 0.979, if these words are removed, the accuracy drops to 0.930. This means that the high frequency of use of certain words (primarily related to the name of the publication, which is typical for the texts of Ukrainian online media) significantly affects the results of the experiment. A more detailed analysis of language elements employing SVM makes it possible to conclude that the identification of the media occurs primarily on the basis of linguistic units of the lexical level. Therefore, for the analysis of texts written in the Ukrainian language, which has an inflectional system, the stemming procedure for variable words is essential.

The study also considered the legal aspects of authorship determination using the proposed machine learning method and highlighted new legal challenges when authorship determination is associated with automated journalism, that is, the creation of texts using artificial intelligence.

Keywords: Machine learning · support vector machine · definition of authorship · legal regulation · copyright protection · artificial intelligence · mass media

Z. Hu et al. (Eds.): ICCSEEA 2023, LNDECT 181, pp. 173–182, 2023.
https://doi.org/10.1007/978-3-031-36118-0_16

1 Introduction

In today's world, machine learning is intended to solve tasks related to large volumes of data (e.g. with large text corpora or all the words of a certain language in general) and for which there is no clear algorithm (e.g. the imitation of the work of the human brain, which has so far not yet been studied fully). Trying to solve such tasks with the help of algorithms, which are hard-coded into the program's source code, would make it too complex to both design and execute such a program even using a very powerful computer. It is in situations like these that machine learning methods are necessary.

In connection with the rapid development of artificial intelligence, the limits of its application are expanding. For instance, in the field of journalism, when creating various types of content and texts with the use of artificial intelligence, new legal discussions arise, since the perception of missing authorship can also lead to the perception of the disappearance of responsibility [1].

Modern studies of robotics in media news [2] distinguish three areas in which such robotics is used: content creation, news gathering (gathering, sorting and verifying information from sources) and news distribution (personalized news and advertising).

Journalistic materials are special from a copyright perspective, as some news items are not protected by copyright law. To illustrate, the Berne Convention for the Protection of Literary and Artistic Works (1979) does not protect «news of the day or various facts which are of the nature of press information only» [3]. The same applies in Ukrainian national legislation. The reason for this is to prevent the violations of copyright law to monopolize information. The mere nature of information or news cannot be considered «creative» and thus protected by copyright law. However, there are other legal tools that allow stakeholders to protect their investment in news production, such as protection against unfair competition or unjust enrichment. The development of systems for automated journalism can also be protected as a know-how by the trade secret.

The issue of copyright protection for works created with the help of computers appeared in legal literature as early as the 1960s [4]. Since then, quite a large amount of literature has explored this issue from different points of view. There currently exist two legislative models in the world: a) special protection of computer-generated works (adopted, e. g. by Great Britain), b) requirement for a person's original creative activity [5].

Copyright protection has traditionally proven itself as a valuable regulatory tool. However, when revolutionary technologies are challenged, unprecedented questions arise. Given the global and pervasive impact of artificial intelligence on our society, legislation now more than ever needs to be more supportive of ethical behavior.

Ethical design and ethical use of artificial intelligence systems are some of the priorities of the European Union. To remain competitive with the rest of the world, the EU aims to ensure user trust in artificial intelligence (AI) systems by guaranteeing that these systems developed in the EU provide their users with the safeguards for protection of human rights.

In particular, «the EU seeks to remain true to its cultural preferences and highest standards of protection against social risks posed by artificial intelligence, inter alia, those affecting privacy, data protection and discrimination rules, unlike other more lenient jurisdictions» [6].

Thus, in April 2021, the European Commission suggested new rules and actions aimed at turning Europe into a global center for reliable artificial intelligence. The combination of the first ever AI legislative framework and the new coordinated plan with the member states is meant to guarantee safety and fundamental human rights while enhancing the expansion of AI, investment and innovation across the European Union [7]. New technology regulations will complement this approach by adapting safety rules to increase user trust in a new, versatile generation of products.

Since copyright is not supposed to apply to materials that are of the nature of ordinary press information, in many cases the question of authorship of specific publications that belong to the genre of informative notes is not as relevant as it is with respect to authored materials of other genres. However, there is a separate category consisting of publications whose sources are the materials of news agencies, the use of which requires special permissions. For example, in many Ukrainian publications, particularly, on one of the most popular Ukrainian news websites «Ukrainska Pravda» («The Ukrainian Truth») and «Censor.NET», there is a warning about the prohibition of copying, reprinting information containing links to the agencies «Interfax-Ukraine», «Ukrainski Novyny» («The Ukrainian News»). To detect the unauthorized use of such materials, special technologies are required, and the usual identification of the sameness of text fragments may not be sufficient in this case.

The purpose of this study is to analyze the possibilities of SVM in determining the belonging of a Ukrainian publication to certain media based on language analysis and taking into account the legal aspects of authorship. To realize the goal, the following tasks must be completed:

1) create a dataset of Ukrainian media publications, which will allow us to draw conclusions about the reliability of using SVM to identify the text belonging to a certain media;
2) to determine with the help of a linear SVM model the accuracy of identification of text belonging to one or another media;
3) find out the prospects of using the proposed approach to solve the problem of determining the authorship of media publications.

2 Review of the Literature

Machine learning is effective for solving tasks related to analyzing data from the Internet – publications from social networks, articles in scientific journals, publications on news portals, in particular, with regard to determining their authorship [8]. Authorship identification is important in order to preserve intellectual property rights, prevent theft, and ensure that each article is referenced to its specific author. This enables institutions to provide identification data about the author [9]. It is also worth highlighting that the use of machine learning is relevant with regard to recognizing fake news, which is used in politics for various kinds of propaganda, in social networks such as Facebook, Twitter etc. [10]. Authorship identification programs have been recently developed in many fields, such as cybercrime law [11], opinion mining (also known as sentiment analysis) [12], etc. Artificial intelligence is also part of cryptographic detection, signature detection and intrusion detection. The biggest challenge of authorship identification is the selection of the most important features representing the author's writing style. The possibility to single out the most important characteristics can ensure an accurate determination of authorship.

The approaches to authorship identification can combine accumulated knowledge from the theory of image recognition, mathematical statistics and probability theory, neural networks, cluster analysis, Markov chains, and others. It should be noted that Support Vector Machines (SVM) gives good results when solving the problem of determining authorship. This is a popular and powerful tool in the field of machine learning. It is a type of supervised learning algorithm that can be used for both classification and regression problems. Support Vector Machines (SVM) was effective for sentiment analysis in social networks [13]. In the following research [14], SVM was used with a linear kernel using the One-Vrest strategy to classify blogs, tweets, and other documents. In the research work [15] a novel up-selling and appetency prediction scheme is presented based on a support vector machine (SVM) algorithm using linear and polynomial kernel functions. In the paper [16] new combined classification method is proposed using a bagging classifier in conjunction with a support vector machine as the base learner. It should be noted that the choice of the classifier for the bagging is problem-dependent. E.g., it is possible to employ neural networks with bithreshold activations [17] instead of SVM for some identification task. Sometimes this approach can provide better performance [18].

3 Dataset of Publications for the Analysis

Articles from three Ukrainian digital media were used for the analysis: «Ukrainska Pravda» (https://www.pravda.com.ua/), «Censor.NET» (https://censor.net/) and «Zakarpattya Online» (https://zakarpattya.net.ua/).

About 95,000 publications of these media for the period from September 2021 to August 2022 were stored in the local database (all publications were presented on the sites within this period):

- about 37,000 publications from «Ukrainska Pravda»,
- about 50,000 publications from «Censor.NET»,
- about 8,000 publications from «Zakarpattya Online».

To compare the results, there were selected two media of the all-Ukrainian level («Ukrainska Pravda», «Censor.NET») which use the materials of the information agencies «Interfax-Ukraine», «Ukrainski Novyny» (in total, the publications included in the dataset contain about 4,500 references to these news agencies), and one regional media («Zakarpattya Online») which does not use materials of news agencies as a direct source.

Since the vast majority of publications of these Internet media are published without specifying the authorship and authors are different persons, yet publications are often made on the basis of press releases, it is difficult (and sometimes impossible) to trace the peculiarities of the language style of the publications with the help of traditional analysis. In this regard, the data obtained with the help of SVM can be interesting for a deeper linguistic analysis – relating to the peculiarities of the language of certain media, as well as for analysis from the copyright point of view – concerning the use of news agencies as a source of materials.

The language of some of these media has already been the subject matter of the study conducted by linguists. Most of the publications deal with the functioning of the Ukrainian language units in the «Ukrainska Pravda» media, this topic has been systematically studied by researcher M. Navalna (publication [19] and others), certain aspects have been studied by Yu. Makarets [20], N. Polishchuk [21]. Various aspects of the functioning of the language of the Ukrainian mass media were analyzed in the collective monograph edited by M. Navalna [22]. These studies were conducted using traditional linguistic methods. The language of «Censor.NET», «Zakarpattya Online» has not yet been in the research spotlight, inter alia, in a comparative aspect. The study of the Ukrainian media «Ukrainska Pravda» and «Politeka» was carried out employing SVM through publication [23].

4 Study Methods

SVMs work by finding a hyperplane in a high-dimensional space that maximally separates different classes. This hyperplane is called the maximal margin hyperplane. The distance between the hyperplane and the closest data points of each class is called the margin. The purpose of SVM is to find a hyperplane with the maximum margin.

The support vector classification (SVC) is the type of support vector machine (SVM) which is employed for classification tasks. SVC aims to find a hyperplane in a high-dimensional space that maximally separates different classes. Once the hyperplane is found, new data points can be classified by assigning them to the class in which they are closest to the hyperplane.

One of the key advantages of SVC is its ability to process non-linearly partitioned data.

This is achieved by using kernel functions that map the data points into a higher dimensional space where they can be linearly separable. The common kernel functions include linear, polynomial and radial basis function (RBF). The choice of kernel function can significantly affect SVC performance, so it is important to carefully consider which kernel to use.

Another advantage of SVC is its noise resistance. Because it relies on only a subset of data points, called support vectors, to define the hyperplane, it is less sensitive to the presence of outliers. This makes SVC a good choice for tasks where the data may be noisy or have some degree of uncertainty.

In spite of its advantages, SVC has some shortcomings. One shortcoming is that it can be sensitive to the choice of parameters such as the cost parameter C and the kernel function. It is of importance to carefully set these parameters in order to obtain good performance. Moreover, SVC can be time-consuming, especially when working with large datasets.

Overall, SVC is a powerful and widely used machine learning tool that can be applied to a variety of classification tasks. Its ability to process non-linearly separable data and its noise resistance make it a perfect choice for many programs. However, it is essential to address its restrictions and to thoroughly set its parameters to achieve good performance.

The goal of SVC is to find a hyperplane in high-dimensional space which maximally separates different classes. Mathematically, it can be represented as follows:

$$w^T x + b = 0, \qquad (1)$$

where w is the vector of the normal to a hyperplane, x is a data point, b is the bias term. The bias term determines the position of the hyperplane along the normal vector.

To classify a new data point, we can use the following formula:

$$y = sign(w^T x + b), \qquad (2)$$

where y is the predicted class label (1 or -1). If the result of the equation is positive, the data point is classified as the one that belongs to class 1. If the result is negative, the data point is classified as the one that belongs to class -1.

To find the hyperplane of the maximum field, we need to maximize the distance between the hyperplane and the nearest data points from any class. This distance is called the margin. Mathematically, the margin can be represented as:

$$margin = \frac{2}{||w||}, \qquad (3)$$

where $||w||$ is the norm of the vector of the normal w.

In order to maximize the margin, we need to minimize the norm of the normal vector w. This can be achieved through the following optimization problem:

$$minimize \frac{1}{2} ||w||^2, \qquad (4)$$

$$y_i(w^T x_i + b) >= 1 \ for \ all \ i, \qquad (5)$$

where y_i is the true label of the data point class x_i.

This optimization problem can be solved by means of different methods such as quadratic programming or gradient descent. Once the optimal values of w and b are found, the maximum field hyperplane can be used to classify new data points.

The main parameters and toolsets employed in the computational experiments are listed in Table 1.

Table 1. Basic parameters and toolsets

Sample	Characters in a text	Converter	Model
95000	<3000	TfidfVectorizer	SVC

The main parameters and toolsets under study are constant in all experiments. All experiments use a sample of 95,000 texts, which are subsequently divided into two datasets – a training set and a test set. The dataset was divided into text fragments of less than 3000 characters each. From them there were formed 10000 dimensional vectors, used as an input of a neural network.

The main steps for preparation are:

- reading and preparing data,
- initialization and machine learning,
- data analysis.

5   Results of the Experiment

The dataset is divided into training and test parts. There were used classes to which the studied text can be referred – Pravda, Zakarpattya and Censor. By means of the Lime package, one can visualize working with texts and identify lexemes that allow you to determine whether the text belongs to the studied resource.

The experiment was conducted in two stages. At the first stage, it was revealed that the main words through which the program identified with a high degree of accuracy the belonging of the text to certain media (F1 Score = 0.9785; Accuracy Score = 0.9785; Precision Score = 0.9787; Recall Score = 0.9785), were the words that are in the names of these media («Zakarpattya», «censor», «pravda») (see Figs. 1, 2).

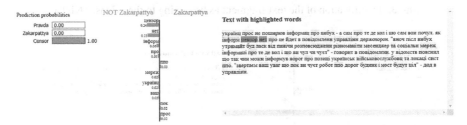

Fig. 1. Identification of the text fragment as belonging to the «Censor.NET»

Fig. 2. Identification of the text fragment as belonging to the «Ukrainska Pravda»

As seen from the above-suggested Figs. 1 and 2, the coefficient of keywords, used to determine the belonging to certain media, is different: «Censor.NET» was determined taking into account the words «censor» and «net» with a coefficient of 0.24 and 0.23; «Ukrainska Pravda» was identified taking into consideration the word «pravda» only with a coefficient of 0.05, yet other words were also determined as characteristic of this media (that, parliament, votes, European, party, bank, etc.).

At the second stage, the words, which are in the names of these media were excluded from the dataset («Zakarpattya», «censor», «pravda»). After that, the accuracy decreased be about 5%: F1 Score = 0.9304; Accuracy Score = 0.9311; Precision Score = 0.9305; Recall Score = 0.9310 (see Figs. 3–4). It should be noted that our SVM-based approach can be extended in the case of the network consisting of polynomial neural units and multithreshold neurons [24, 25].

Fig. 3. Identification of the text fragment as belonging to the «Censor.NET»

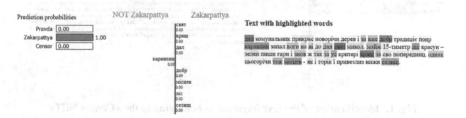

Fig. 4. Identification of the text fragment as belonging to the «Zakarpattya Online»

The program quite precisely identifies the belonging of the text to certain media based on the use of words. To exemplify, the belonging of the texts to «Ukrainska Pravda» was identified on the basis of employing such words as *parliament, source, details, reference, informs, the wounded, draft law*; the ones that belong to «Censor.NET» – based on the words *relations, reference, information, read, messages*; as those belonging to «Zakarpattya Online» – based on the lexemes *residents, regional, administration, local, village, customs, holiday, tells, housing, economy, currently, natural gas*. As noticed, in «Zakarpattya Online», there are clearly identified words related to regional issues, which are in the focus of this media.

6   Summary and Conclusion

The developed program for text parsing on Node JS enabled formation of a significant corpus of texts of publications of the all-Ukrainian media «Ukrainska Pravda» and «Censor.NET», as well as the regional media «Zakarpattya Online» (in total, approximately

95,000 articles for the period from September 2021 until August 2022). Many of the articles, published on these resources, were connected to social, economic and political topics, a large number of articles address the Russian Federation's War against Ukraine.

The analysis of these texts employing support vector machine (SVM) made it possible to obtain an average accuracy of 0.979, if the words included in the names of these media remain in the texts of the dataset («Zakarpattya», «pravda» and «censor»). Since it is characteristic of analyzed Ukrainian online media to indicate the names of these media in the text of publications, the retention of these keywords causes a significant error; if they are removed, the accuracy will drop by about 5% and will be 0.930. The texts, whose authorship was determined imprecisely, which is nearly 7%, require a separate in-depth analysis.

A more detailed exploration of language elements based on SVM enables the concluding that the identification of media «Ukrainska Pravda», «Censor.NET», «Zakarpattya Online» occurs primarily at the lexical level – through the detection of typical lexical markers and coefficients of the text belonging. This is also facilitated by the stemming procedure, carried out in relation to the texts, due to which inflections were not taken into account.

This methodology for analyzing and classifying texts possesses huge potential for using in the legal field, e. g. to determine authorship in copyright disputes. Nevertheless, there arise new legal challenges in identifying authorship, associated with automated journalism that is, generating texts by means of artificial intelligence which necessitate further research by interested parties.

The problem of determining the authorship of texts, the belonging of texts to one or another media remains relevant, especially with the development of such systems as Chat GPT. In this study, the linear SVM model was used, but other models should be tested in future studies and their capabilities should be tested on media texts and texts of other language styles written in Ukrainian.

References

1. Krausová, A., Moravec, V.: Disappearing authorship: ethical protection of AI-generated news from the perspective of copyright and other laws, 13, JIPITEC 132 para 1 (2022). https://www.jipitec.eu/issues/jipitec-13-2-2022/5540
2. Marconi, F.: Newsmakers. Columbia University Press, Artificial Intelligence and the Future of Journalism (2020)
3. Berne Convention for the Protection of Literary and Artistic Works, as amended on 28 September 1979
4. Lawlor, R.C.: Copyright Aspects of Computer Usage 11 Bulletin of the Copyright Society of the U.S.A. 380 (1964)
5. Gaudamuz, A.: Artificial intelligence and copyright, WIPO Magazine, October 2017 https://www.wipo.int/wipo_magazine/en/2017/05/article_0003.html
6. Madiega T. EU guidelines on ethics in artificial intelligence: Context and implementation, European Parliament, September 2019. https://www.europarl.europa.eu/RegData/etudes/BRIE/2019/640163/EPRSBRI(2019)640163_EN.pdf
7. European Commission official website. https://digital-strategy.ec.europa.eu/en/library/proposal-regulation-laying-don-harmonised-rules-artificial-intelligence

8. Khan, A.S., Ahmad, H., Asghar, M.Z., Saddozai, F.K., Arif, A., Khalid, H.A.: Personality classification from online text using machine learning approach. Int. J. Adv. Comput. Sci. Appl. **11**(3), 460–476 (2020)

9. Abbasi A., Javed A.R., Iqbal F. et al. Authorship identification using ensemble learning.Sci Rep 12, 2022: 9537

10. Bhargava M., Mehndiratta P., Asawa K. Stylometric Analysis for Authorship Attribution on Twitter. Lecture Notes in Computer Science, 2013: 37–47

11. Iqbal, F., Binsalleeh, H., Fung, B.C., Debbabi, M.A.: Unified data mining solution for authorship analysis in anonymous textual communications. Inf. Sci. **231**, 98–112 (2013)

12. Ziani, A., Azizi, N., Guiyassa, Y.T.: Combining random subspace algorithm and support vector machines classifier for Arabic opinions analysis. In: Le, Thi, H., Nguyen, N., Do, T. (eds.) Advanced Computational Methods for Knowledge Engineering, Springer, Cham, vol. 358, pp. 175–184 (2015). https://doi.org/10.1007/978-3-319-17996-4_16

13. Ahmad, M., Aftab, S.: Analyzing the performance of SVM for polarity detection with different datasets. Int. J. Mod. Educ. Comput. Sci. **9**(10), 29 (2017)

14. Chatterjee, S., Jose, P.G., Datta, D.: Text classification using SVM enhanced by multithreading and CUDA. Int. J. Mod. Educ. Comput. Sci. **11**(1), 11 (2019)

15. Lian-Ying, Z., Amoh, D.M., Boateng, L.K., Okine, A.A.: Combined appetency and upselling prediction scheme in telecommunication sector using support vector machines. Int. J. Mod. Educ. Comput. Sci. **11**(6), 1 (2019)

16. Govindarajan, M.: Empirical analysis of bagged SVM classifier for data mining applications. Int. J. Mod. Educ. Comput. Sci. **5**(10) (2013)

17. Kotsovsky, V., Batyuk, A., Mykoriak, I.: The computation power and capacity of bithreshold neurons. In: 2020 IEEE 15th International Scientific and Technical Conference on Computer Sciences and Information Technologies, CSIT 2020 **1**, 28–31 (2020)

18. Kotsovsky, V., Batyuk, A., Voityshyn, V.: On the size of weights for bithreshold neurons and networks. In: 16th IEEE International Conference on Computer Sciences and Information Technologies, CSIT 2021, 13–16 (2021)

19. Navalna, M.: New lexical processes in the language of online newspaper "Ukrainska Pravda" in the early XXI century, Studia Ukrainica Posnaniensia. ZESZYT III. Poznan, pp. 165–173 (2015). (In Ukrainian language)

20. Makarets, Yu.: On the issue of the deviation of Ukrainian Internet media texts (on the example of Internet edition "Ukrayinska Pravda"). Scientific Bulletin of the International Humanitarian University. Series: Philology **32**(3), 72–76 (2018). (In Ukrainian language)

21. Polishchuk, N.: Computer Terminology Vocabulary of Newspaper Text (based on the materials of the newspapers "Dzerkalo Tyzhnya", "Ukrayinska Pravda", "Hazeta po-ukrayinsky", "Vysokyj Zamok"). Scientific Bulletin of Lesya Ukrainka East European National University. Series: Philological sciences. Linguistics 1(250), 95–100 (2013). (In Ukrainian language)

22. Navalna, M. (ed.): Vocabulary of Ukrainian mass media: monograph. Pereyaslav-Khmelnytskyi: KSV, 2019: 212 (2019). (In Ukrainian language)

23. Lupei, M., Mitsa, O., Sharkan, V., Vargha, S., Gorbachuk, V.: The identification of mass media by text based on the analysis of vocabulary peculiarities using support vector machines. In: 2022 International Conference on Smart Information Systems and Technologies (SIST) (2022). https://doi.org/10.1109/sist54437.2022.9945774)

24. Kotsovsky, V., Geche, F., Batyuk, A.: Bithreshold neural network classifier. In: 2020 IEEE 15th International Scientific and Technical Conference on Computer Sciences and Information Technologies, CSIT 2020 **1**, 32–35 (2020)

25. Kotsovsky, V., Batyuk, A.: On-line relaxation versus off-line spectral algorithm in the learning of polynomial neural units. Commun. Comput. Inf. Sci. **1158**, 3–21 (2020)

Data Mesh as Distributed Data Platform for Large Enterprise Companies

Yevhenii Vlasiuk and Viktoriia Onyshchenko[✉]

National Technical University of Ukraine "Igor Sikorsky Kyiv Polytechnic Institute",
37 Peremohy Avenue, Kyiv 03056, Ukraine
v.onyshchenko@kpi.ua

Abstract. Rapid increase in data volume during the last decades promoted a lot of creation and evolution of data platforms and related technologies: data warehouses and data lakes, data fabric and data factory, lakehouses and cloud-native platforms allowed converting data into a valuable business asset. However, all these platforms are centralized and siloed which means that for large enterprise company they are growing to significant size and complexity level and becoming difficult to maintain and cost ineffective. To solve this problem, distributed data platform architecture and data mesh as the most widespread implementation of this architecture are proposed in this article. Firstly, analysis of siloed data platforms, their advancing features, implementation options and use cases are investigated. After that, analyzing reasons of large enterprises data platforms, characteristics of distributed data platform and data mesh are formed. Finally, data syndication and data quality process in data mesh platform architecture are identified as significant aspects of platform's adoption success and which require additional design, modeling and implementation research.

Keywords: Big data · data mesh · distributed data platform · data management processes

1 Introduction

Each day the amount of collected, stored, processed, analyzed and consumed data is growing. Earlier deep analysis was applied before start processing particular data points due to the high cost of storage, long processing time, absence of algorithms to process particular forms and formats of data. The rapid evolution of technologies allowed to mitigate the majority of mentioned limitations and brought new challenges of processing a large amount of data in different formats, with various velocities. Various types of data platforms were designed and built to handle increased volume, variety and velocity of data. Dedicated data management processes were introduced to make sure data flows from the initial state how it lives in the source system, towards the consumable form for the end business users. Such data management process includes but is not limited to the data reconciliation process, data quality assurance, data lineage, and data governance. These processes are adopted and implemented very well in a centralized data platform supported by modern database management systems and services provided by leading cloud providers allowing to support a majority of architecturally significant requirements from a business like scalability, performance, security, adaptability, flexibility, etc. However, what data & analytics industry is observing over the last two years is increasing demand for business domain-specific data platforms. From the author's practical experience working as Senior Data Delivery Manager at Epam Systems, a leading provider of digital platform engineering, and digital product design, large enterprises start lacking the flexibility of a centralized data platform for own various business units and lines of business. Demand for distributed data platforms starts being vocal.

More and more devices from our day-to-day life become "connected" or "smart". Nowadays it's the new normal when a robotic vacuum cleaner starts cleaning your apartment by schedule or by button click in the app on your smartphone. It's useful when it notifies you about the need to change the filter or sends you a message that usually you are doing cleaning once per two days so it's time to clean. Due to rapid development and reduced cost of production for microchips, sensors, storage devices, network speed and bandwidth, the amount of data that we can collect, store, process and analyze is increasing enormously. It launched a cyclic process when data produces more data and so on. The commercial value of the data played a definite role to push this process forward: producers got a powerful tool to understand better individual demand, react to it and get the customer to their list. Global concurrency increased the value of data and raised rates of in-time decision-making. As The Economist mentioned in 2017 "data is becoming new oil" [1] and building new areas of the economy [2]. More and more companies which historically were oriented on manufacturing, healthcare, and finance, now position themselves as data companies while other organizations and even whole industries start heavy expansion and adoption of data analytics [3].

As a consequence of this, a lot of small companies and large enterprises start investment in their own data platform. The main objective of this article is to explain the reason and provide the use cases for building distributed data platforms comparing to centralized and siloed platforms like data factory, data fabric, lakehouse or others and research a data mesh as a solution for distributed data platform. Siloed platforms are good solution for traditional data management use cases when data are logically connected, belongs to same business domain, validated and controlled by similar rules and policies. The problem starts when the number business domains starts growing and data is tightly coupled inside the domain and loosely cohesive across domains. Existing and popular solution for this problem is to build a number of data warehouses, data lakes or data fabric platforms each for particular domain. But the main limitation of this option is siloed nature of this solution: data becomes isolated in own platform and it's hard to build cross-domain analytics on a level of the whole enterprise. The main author's aspiration in this research is to propose how data mesh can solve the limitation of siloed platforms, discuss data syndication and data quality assurance process in data mesh platform as those are not yet sufficiently researched and explored.

2 Literature Review

Data fabric, data factory, lakehouse architectures of data platform focuses on describing processes that are executed inside the platform. By design such platforms contain all required layers to retrieve raw data from the source of truth system and based on it build consumable product by the end users. Majority of researches focuses on clarification of intra-platform processes: data ingestion, data hydration, data cleansing and enrichment, data reconciliation, data quality management, data governance, data lineage, data security and access management, data aggregation, data product build.

Hiren Dutta [4] approaches data governance process of the data lake with real time data ingestion. He proposes to use graph model to represent state and transitions of the data governance process in data lake. Suggested data governance architecture also allows to control data consumption by third-party platforms and applications and sets the ground for cross-platforms communication.

Anosh Fatima, Nosheen Nazir and Muhammad Gufran Khan [5] analyzed another data management process which is an integral part of data platform – data cleansing. While research data cleansing process, authors focus on missing data recovery. This process can be also considered as part of larger process of data quality assurance. Yuxin Wang, Joris Hulstijn & Yao-Hua Tan [6] also mentions other aspects of data quality assurance like data correctness check, data completeness validation, data consistency and data timeliness assurance.

Important process of data security and access is discussed in research from Isma Zulifqar1, Sadia Anayat2, Imtiaz Kharal [7]. Authors propose encryption as safer way to protect records in cloud-based data platform. It's definitely applicable solution however, if we analyze it from the perspective of data sharing, it introduces difficulties as third-party platforms and consumers will be struggling to interpret encrypted data and so separate mechanism of key sharing along with data sharing will be required.

3 Monolithic Data Architecture

A rapid increase in demand for data processing and analysis made a significant push into the development of corresponding architecture patterns, processes and technologies. A data lake, data warehouse, data mart, all these patterns were developed to be able to support a decent increase in capturing, storing and analyzing data. Leading technology companies like Google, Facebook, and Microsoft presented their implementations of these patterns, the majority of which became open-source solutions: Google File System which became a foundation for HDFS, a foundational storage platform for the Hadoop platform; Cassandra which is distributed database that ideally suits for processing time-series data and others.

Technology platforms were combined into comprehensive frameworks and architectural approaches. Big adoption got layered data architecture where each layer represents a defined stage of capturing, storing, processing or analyzing data. The combination of layers might vary depending on data sources, and the nature of data resulting in data products. A simplified version of a layered architecture blueprint designed for one financial enterprise is shown in Fig. 1

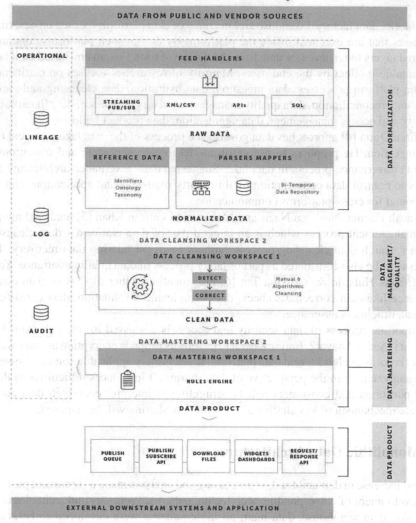

Fig. 1. Example of layered Data Architecture Blueprint [8]

This architecture focuses on the automation of the main processes of transforming the data from the raw to consumable form: data ingestion, data reconciliation, data quality assurance, data deduplication & enrichment, data lineage, data governance, data aggregation, and data exposure. Based on the architectural blueprint, the data platform was built utilizing PaaS and SaaS provided by the Google Cloud Platform. Implementation was done for the corporate services department of the company while in parallel similar platforms were built for other lines of business and other groups of consumers. All platforms lived in silos and operated with their own data within their own boundaries.

One of the variances of layered data architecture is data fabric. It's been a first time defined in the mid-2000s by Forrester analyst Noel Yuhanna but is still mentioned as a top trend for 2021 [9]. The focus of data fabric is on the high adoption of self-service capabilities based on metadata usage and management. Knowledge graphs and algorithms for insights generation as integral components of data fabric helps to minimize operational costs and efforts compared to traditional warehouse and lake-based architecture.

Mentioned architecture approaches were working well for the majority of cases. However, the rapid development of data-hungry AI and ML technologies started to demand a big amount of unstructured but at the same time validated data which occurred to be a problematic use case for both data lakes and data warehouses. It led to the development of the mixed model, an example of one of which is Lakehouse [10] developed by Databricks. The main aim of it is to combine SQL interface and enforced schema towards the data that resides in the data lake to provide high availability of data and avoid unnecessary data duplication along with minimizing operational efforts on supporting multiple platforms. Lakehouse concept being fresh has already got wide adoption among major players in the data analytics market including leading cloud providers with their services: Redshift Spectrum by AWS [11], BigQuery [12] and others.

Lakehouse architecture is aiming to simplify traditional layered data platforms by reducing the number of layers. It doesn't though cover the problem of sharing consumable data outside of the platform or between platforms. If multiple departments of a large enterprise build their own lakehouse platforms, they will end up keeping their data in siloes.

4   The Rationale for Distributed Data Platforms

Decoupling and restructuring complex systems into smaller and more manageable ones have already been applied to various organizational and technological platforms. Microservices architecture brought by Dr Peter Rodgers in 2005 can be one example of using a decoupling approach to split complex monolithic software platforms into a set of atomic services with well-defined boundaries following loosely coupled principles for microservices themselves and high cohesion for the components which are part of one microservice. This architecture pattern got high adoption across the software development industry as allowed to simplify solving of pressing problems: cost-effective application scalability, independent development and deployment of microservices, traceability of microservices-based platform to business domains and context and as a result ability to use suitable technologies to solve unique business challenges. This factor at the same time brings complexity to a microservices architecture: as each microservice is single-purposed, it can't cover the whole business process end-to-end. As a result, communication between microservices comes to a critical level and should not bring overhead. REST architectural pattern and its implementation on HTTP protocol are often used communication mechanisms for microservices.

If we assume data fabric architecture as monolith software architecture, then we would not refuse the benefits of applying the microservices approach to data fabric or layered data platform architecture. It would give simplicity and a more fine-grained approach to mapping our data persistence and data processing layers to business domains rather than a giant monolithic data lake or lakehouse, where finance data lives together with customer or purchase order data. At the same time, if we split our data into some sort of "micro-datalakes" or "micro-datawarehouses" the main idea of a unified data platform that allows building holistic cross-domain data views, will disappear. There should be an ability to keep existing platforms but establish new connections between them with defined communication contracts to gain added value.

A similar idea, however not on software architecture but rather on the organizational structure level, promotes value-creating networks (VCN). The first time term VCN was called by Kothandaraman and Wilson in 2001 with a focus on cross-firm interaction in the network to deliver value for the end user. Maila Herrala and Pekka Pakkala in their work [13] defined a value-creating network "to be a complex network structure where firms' core competencies are linked to each other through value exchanges such as flows of information, material, resources and money." This is another example of restructuring a monolithic entity into smaller and domain-centric parts and establishing connections between them to explore hidden added value. Herrala and Pakkala also provide an example of a value-creating network for travel services provider presented in Fig. 2.

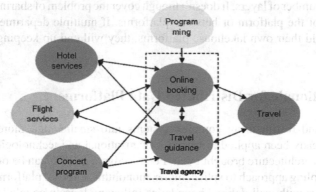

Fig. 2. Example of value-creating network for travel services provider [13]

If we think about large enterprises in the travel industry like Expedia, we will see they have firms and departments to cover all mentioned domains. Some of those firms were acquired, and some firms or departments were merged. Quite frequently such transformations raise the question of what to do with data platforms managed by these earlier independent companies or business units. Should those be integrated into one single data lake and data warehouse to get this holistic view of data views across the organization or should those be kept independent? When the second option is chosen, it immediately raises several questions on how to establish communications between multiple data platforms, how to enforce enterprise-level policies on standard data processes like data governance and data quality assurance and at the same time keep flexibility and independence of these business domains focused data platforms. One of the answers to these questions is a concept presented by Zhamak Dehghani in 2019 called "Data Mesh" [14].

5 Data Mesh as an Example of Distributed Data Platform

Inês Araújo Machado, Carlos Costa and Maribel Yasmina Santos "consolidate the background knowledge regarding the concept of Data Mesh and the current known approaches and features" in their work [15]. They also provide two implementations of data mesh architecture for Zalando and Netflix, however, they are not covering aspects of communications between distributed data platforms in data mesh as well as the evolution of data management processes like data governance and data quality assurance. Zhamak Deghani defined "data as product" as one of the four main principles of Data Mesh. This principle was actively used far before Data Mesh and got adoption in Data Factory Approach [16] as an "efficient way for organizations to focus on creating business value instead of just considering tools and technology". For Data Mesh though, data product gets exceptional importance as it becomes the foundation and contract for communication between platforms in the distributed mesh. Figure 3 visualizes that various domains use data products of others domains as data exchange unit. Analytics layer of the business domain uses data products from others domains to produce consumable data products in own scope. Furthermore, one of produced data products is used as input by another business domain for own data product build and analytics.

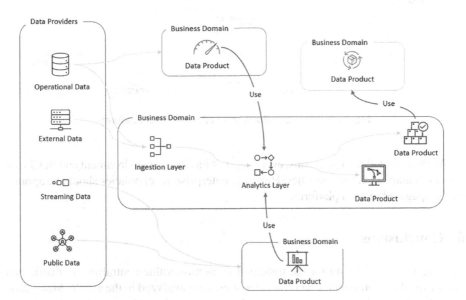

Fig. 3. Data Product as a contract for cross-domain communication in Data Mesh

Considering data product as a public API of your data platform raises several challenges related to product versioning, access management and governance, and data syndication. The last challenge becomes especially critical for business as it covers the mechanism of publishing and unpublishing data products to corporate marketplaces; notification of users of data products about new versions, and updates; requirement to provide backward compatibility of a data product.

Data Mesh changes the paradigm of data platforms from siloed or centralized to federated. Figure 4 visualizes differences between them: siloed platforms are isolated, data is not exchanged between each other or corresponding components; centralized platform is represented as one huge data platform where components from various domains are communicating with each other; finally federated data platform keeps isolation that is characteristic of siloed platform but at the same time defines clear contracts and mechanism of communication to break the data silos.

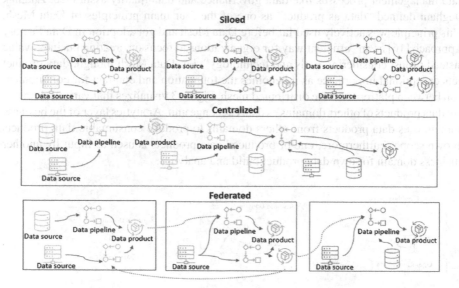

Fig. 4. Siloed, Centralized and Federated Data Platforms

With federated data platforms, every org unit has its own environment and flexibility of data management process rules. However, enterprise-level policies should be applied and propagated to each platform.

6   Conclusions

Reasons and justification for the transition from monolithic centralized to distributed platforms, their components and characteristics were analyzed in the article. Monolithic centralized data platforms are good for solving the problem of processing big volumes of varied data, but large enterprises start suffering with such platforms due to a couple of reasons:

1) data from various lines of business and departments are mixing, bringing unnecessary overhead for data owners;
2) difficulty in applying individual data management rules according to the business domain as processes are managed globally for the whole platform;

3) management of data platform product lifecycle and platform infrastructure is complicated as requires supporting priorities and needs from multiple lines of business.

Based on the analysis that was made, data mesh is proposed as a solution for distributed data platform architecture. In further research, the federated data governance pillar of data mesh will be explored in detail: how the data syndication process can be implemented, what is the mechanism of applying custom data quality assurance methods, specific to the business domain and at the same time enforce enterprise-level policies, what is the technological background for these processes and what architecture and execution algorithms are required.

References

1. The Economist. The world's most valuable resource is no longer oil, but data (2017). https://www.economist.com/leaders/2017/05/06/the-worlds-most-valuable-resource-is-no-longer-oil-but-data
2. The Economist. Data is giving rise to a new economy, 2017. Retrieved from https://www.economist.com/briefing/2017/05/06/data-is-giving-rise-to-a-new-economy
3. Anitha, P., Malini, M.P.: A review on data analytics for supply chain management: a case study. Int. J. Inf. Eng. Electron. Bus. (IJIEEB) 10(5), 30- 39 (2018). https://doi.org/10.5815/ijieeb.2018.05.05
4. Dutta, H.: Graph based data governance model for real time data ingestion. Int. J. Inf. Technol. Comput. Sci. (IJITCS) 8(10), 56–62 (2016). https://doi.org/10.5815/ijitcs.2016.10.07
5. Fatima, A., Nazir, N.: Muhammad Gufran Khan, "data cleaning in data warehouse: a survey of data pre-processing techniques and tools." Int. J. Inf. Technol. Comput. Sci. (IJITCS) 9(3), 50–61 (2017). https://doi.org/10.5815/ijitcs.2017.03.06
6. Wang, Y., Hulstijn, J., Tan, Y.-H.: Data quality assurance in international supply chains: an application of the value cycle approach to customs reporting. Int. J. Adv. Logist. 5(2), 76–85 (2016). https://doi.org/10.1080/2287108X.2016.1178501
7. Zulifqar, I., Anayat, S., Khara, I.: A review of data security challenges and their solutions in cloud computing. Int. J. Inf. Eng. Electron. Bus. (IJIEEB) 13(3), 30–38 (2021). https://doi.org/10.5815/ijieeb.2021.03.04
8. Gavrosnky, M.: Why Your Financial Services Company Should Consider Implementing a Data Factory (2020). https://www.epam.com/insights/blogs/why-your-financial-services-company-should-consider-a-data-factory
9. Gartner Press Release. Gartner Identifies Top 10 Data and Analytics Technology Trends for 2021 (2021). https://www.gartner.com/en/newsroom/press-releases/2021-03-16-gartner-identifies-top-10-data-and-analytics-technologies-trends-for-2021
10. Armbrust, M., Ghodsi, A., Xin, R., Zaharia, M.: Lakehouse: A new generation of open platforms that unify data warehousing and advanced analytics (2020). https://www.databricks.com/wp-content/uploads/2020/12/cidr_lakehouse.pdf
11. Kava, P., Gong, C.: Build a lake house architecture on AWS (2021). https://aws.amazon.com/blogs/big-data/build-a-lake-house-architecture-on-aws/
12. Levy, R., Thill, S., Tekiner, F.: Building a data lakehouse on Google Cloud Platform (2021). https://services.google.com/fh/files/misc/building-a-data-lakehouse.pdf
13. Herrala, M., Pakkala, P.: Value-creating networks – a conceptual model and analysis (2009). https://institute.eib.org/wp-content/uploads/2016/04/Final_Report_2009_Value-creating_networks_-_a_conceptual_model_and_analysis.pdf

14. Dehghani, Z.: How to move beyond a monolithic data lake to a distributed data mesh (2019). https://martinfowler.com/articles/data-monolith-to-mesh.html
15. Machado, I.A., Costa, C., Santos, M.Y.: Data mesh: concepts and principles of a paradigm shift in data architectures (2021). https://reader.elsevier.com/reader/sd/pii/S18770509210 22365?token=D16A531B8281668BA35608A988D318A68D6213FFDB2F7AE6600201 8933044CF03F2C5322EE58DD8DA7BD62ED11BBF6BC&originRegion=eu-west-1&ori ginCreation=20221111101328
16. Tsitlik, V.: How a Data Factory Approach Can Increase the Value of Your Data (2019). https://www.epam.com/insights/blogs/how-a-data-factory-approach-can-inc rease-the-value-of-your-data

Using the Strategy of Information Resistance to Improve Content in Virtual Communities Using the Example of the Facebook Social Network

Mariana Petryk, Oleksandr Marcovets, and Ruslana Pazderska[✉]

Lviv Polytechnic National University, Bandery str. 12, Lviv 79013, Ukraine
{mariana.petriv.mdkib.2021,oleksandr.v.markovets,
ruslana.s.pazderska}@lpnu.ua

Abstract. The conducted analysis of the relevance of the research topic is formulated on the basis of the goal, objectives, review of scientific developments and methods. They emphasize the importance of combating disinformation through the formulation of new types of false information, methods and means of its dissemination. For a better understanding of the process of organizing the use of strategies, a functional and detailed model of the subject area is presented, which displays data flows using the Data Flow Diagram (DFD) notation. The context diagram shows how the basic process of "Defining the role of social media users" based on feedback from "Researcher", content from "Social media user". As a result of this process, the "Researcher" will receive evaluated content, and the "Strategy Developer" in turn will be able to receive information about the role of the social media user. Based on the analysis of one of the communities, the use of the developed strategies was implemented. They will make it possible to improve the content of the virtual community, identify unscrupulous users and, in turn, increase the productivity of the community (attract more users, the content will be of high quality, it will be checked, conflict situations will not arise, etc.). That is, to implement measures that will contribute to the development of the community.

Keywords: Informational struggle · strategies · user · social community

1 Introduction

Today, a new way of waging war without weapons has appeared, which has been called information warfare, the main task of which is the destruction of humanity not as a material force, but as society as a whole.

We can observe the phenomenon of information-psychological effects in the mass media of those countries that are waging this "invisible" struggle among themselves. Information messages that each of us sees on TV screens, reads in newspapers or on the global Internet can sow suspicion and make a person hostile to the current government, to a certain nation or situation in the state. When such tension is created in a country

Z. Hu et al. (Eds.): ICCSEEA 2023, LNDECT 181, pp. 193–206, 2023.
https://doi.org/10.1007/978-3-031-36118-0_18

where there is an ongoing information struggle, there is a greater chance that the enemy will weaken the society so that it cannot resist. Such a situation can now be observed in Ukraine [10–12].

The aim of the study. To improve the content using strategies of conducting an information struggle in social networks.

The fulfillment of the purpose of the work is determined by the fulfillment of the following research tasks:

- analytical review of sources related to the research topic.
- determine the methods of information struggle in social networks.
- information model of the subject area.
- to obtain the results of the community analysis on the necessity of applying the developed strategies regarding informational resistance.

2   Review of Research Related to the Subject Area

Before conducting a detailed analysis of the influence of fake information on users of social networks during the information struggle, it is necessary to study in detail the literary sources related to the research topic. Sources include: articles, works and reports of famous people [3–5, 14, 15].

In the article V.Yu. Pravdenko and N.M. Tytova's "Influence of fake information on the consciousness of the individual" analyzed the essence of the concept of "fake". "Fake" is the presentation of certain information in a distorted form in order to mislead users. Fakes are false news, pages created by fictitious users, through which propaganda is spread. Also, the article states that the purpose of spreading fake information is to mislead people, sow doubts and convince them of the truth of the facts. Fake information is the main tool of the information struggle, therefore it is important to properly filter the information provided, and also not to spread propaganda through online networks. Users on social networks must be media literate, because with the help of spreading false information, you can not only manipulate people, but start a war [16].

In the article L.P. Makarenko "Evolution of forms and methods of information warfare" states that rumors are a special technology of information warfare. The main reason for the emergence of rumors is the lack of official information, or an incomplete presentation of information. When people lack information from their usual sources, they tend to create and accept rumours. Also, an important aspect of the generation of rumors is the incompetence of society, because when people do not understand the problem that has occurred, they "fantasize" and believe in what, in their opinion, is easily explained [9].

In the article M.M. Didenko and O.V. Zelinska "Informational-psychological influence on people during war" states that informational-psychological influence is a way of exerting influence on people with the aim of changing their behavior and views on a certain situation. The main methods of information-psychological influence are persuasion and suggestion. At the present time, the following actions are carried out during informational-psychological influence: manipulation of human consciousness, provoking conflict situations, inciting mistrust, creating an atmosphere of spiritlessness and immorality, intensifying the political struggle in the state and provoking various kinds of repression, etc. [2].

In the article by O. V. Kurban "Fakes in modern media: identification and neutralization", it is written that a fake is fake information that is distributed on online platforms. The author pointed out that the fake is quite a powerful weapon. It is quite difficult for an incompetent user to fight against false information, sometimes it is impossible at all, so it is important to develop specialized and simple tools for identifying and neutralizing fake news [6].

In the article by A.M. Peleshchyshyn and O.R. The thread "Determining the elements of socially oriented risks in the organization of the life cycle of a virtual community" describes the risks in the organization of the life cycle of a web community. In addition, the article proposes measures to avoid socially-oriented risks: prevention of users-provocateurs; suspension of flamers; preventing cyberbullying; improving content quality; improving the reliability of published information; avoiding plagiarism; appointment of a group leader, etc. The prescribed socially oriented risks will allow to ensure a high-quality process of creating and maintaining a web community [13].

In the article S.S. Fedushko and O.R. The thread "Determining the resilience of a virtual community against information attacks" indicates that the life cycle of a web community is the main process of creating and managing a group.

In the life cycle, the development of the web community takes place in such directions as:

- informative – updating content in the community, checking the truthfulness of information, creating various interesting surveys;
- user – analysis of personal information of group members, research of their activities, etc.;
- resource – technological support for managing a virtual community;
- reputational – increasing the rating of the web community and improving its effectiveness among other communities [18].

3 Research Methodology

The following general scientific methods were used to conduct this research: analysis, observation, scientific modeling.

The method of analysis was used to study the subject area, identifying features, signs, and relationships. Accordingly, it made it possible to formulate the purpose of the work and tasks. To analyze important trends of research in the specified direction – informational resistance in the social network.

The observation method was used in the work to study the specifics of the organization of the virtual community. In particular, its users, content, and their interaction with each other. This helped in tracking the effectiveness of the proposed strategies.

The method of scientific modeling was used to better understand the processes of using strategies, the possibility of exchanging certain information, the interaction of the main objects, etc.

4 Formal Problem Model

In order to better understand the entire process of developing strategies for information struggle in social networks, it is necessary to design a functional model.

An important aspect of a functional model is the representation of data flows using a Data Flow Diagram (DFD). The Gane-Sarson notation was used to construct the DFD of the diagram "Defining the role of a social network user".

The context diagram shows the main function – "Defining the role of a social network user" and three entities – "User", "Researcher", "Strategy Developer" (Fig. 1). In order to fully understand the work cycle of the main process, you need to think of clear ways of its implementation.

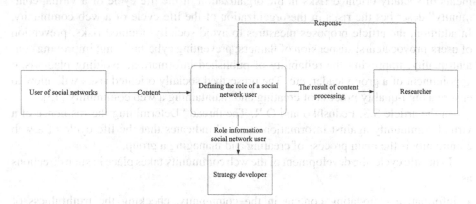

Fig. 1. Model of the context diagram "Definition of the role of the social community user"

So, first of all, the User of the social network sent the content to the system, the system processed it and sent the result of the analysis to the Researcher. In turn, the Researcher sent feedback to the system with full information about the user's content. The system processed the feedback and sent the generated social community user role to the Strategy Developer.

Next, you need to build a decomposition of the context diagram, namely, break the main process into sub-processes. A detailed diagram is built in order to clearly understand the stages of achieving the final result of the main function of the system. In Fig. 2 shows a detailed diagram of the main process "Defining the role of the social community user".

It is worth considering the process of building the first level of detail diagram.

When the content entered the system, you first need to:

- consider the content for processing – you need to choose the target content for its detailed analysis (process 1.1);
- analyze the content – you need to process the data from the post (process 1.2), namely:

analyze the text of the post – it is necessary to analyze the information from the post in detail and determine whether the publication is true or not (process 1.3);
analyze the metadata of the post – it is necessary to determine the location of the post (if specified); date and time of publication of the post (process 1.4);
analyze the author of the post – it is necessary to analyze the profile of the contributor: name; authenticity of acanthus; frequency of publications; user activity in communities (process 1.5).

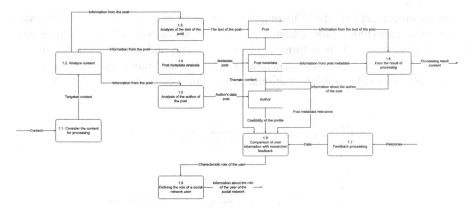

Fig. 2. Detailed diagram of the main process "Defining the role of a social network user"

After that, the processed information from the posts is recorded in specific data repositories: Analysis of the post text is recorded in the Post repository; Post metadata analysis is recorded in the Post Metadata repository; The analysis of the author of the post is recorded in the Author repository.

Later, the result of processing all the collected information from the post is formed (process 1.6). At the end, the result of content processing is obtained.

After the result of content processing by the system, the Researcher sends feedback. Feedback is processed (process 1.7), and the Researcher forms ready specific information: keywords (hashtags); short message (comment); photo and video confirmation.

Next, the user's information is compared with the researcher's feedback (process 1.8) according to certain rules.

Based on the analyzed information of the user and the researcher, in accordance with the Rules for recognizing fakes in social networks, the role of a social network user is determined (process 1.9), namely: Solver (distributes true information) or Slanderer (distributes false information).

As a result, information about the specific role of the social community user is sent to the strategy developer, and he makes a report on the completed tasks based on this data.

The context diagram (Fig. 3) shows the main process - "Development of strategies for information struggle in social networks" and two entities - "Social network", "Customer".

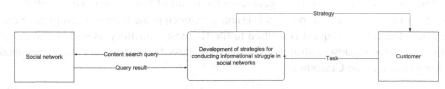

Fig. 3. Model of the context diagram "Development of strategies for conducting informational struggle in social networks"

In the process, at first, the "Customer" gave a specific task to the system, the system generated a search request for content. After that, the Social Network issued the result of the request, and the system, in turn, developed a strategy and sent it to the "Customer". Further, it is necessary to develop a detailed diagram of the main process "Development of strategies for conducting informational struggle in social networks" (Fig. 4).

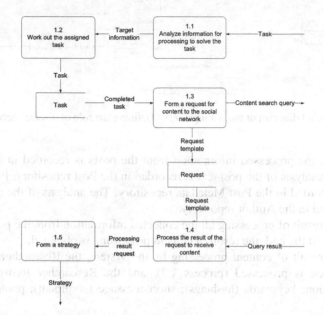

Fig. 4. Diagram detailing the main process "Development of strategies for conducting informational struggle in social networks"

It is appropriate to consider the process of developing a first-level decomposition diagram when a task from the Customer is entered into the system, first of all it is necessary: analyze the information for processing to solve the task - it is necessary to analyze the information by thematic direction (process 1.1); work out the assigned task – on the basis of the researched information, it is necessary to perform a specific assigned task (process 1.2);

After that, the completed task is recorded in the Task repository. Then, based on the completed task, a request for content to the social network was formed (process 1.3). As a result, a search request for content was received.

After that, the Social Network generated the result of the request. The result of the content request is processed (process 1.4) and developed in the form of a template. Next, the generated content request is written to the Request repository. Based on the result of the processed request, a strategy is formed (process 1.5). At the end, the developed strategy is sent to the Customer.

5 Strategies for Information Warfare in Social Networks

It is worth noting that guaranteeing information security in social networks is one of the main functions of the state. Therefore, a number of strategies for information warfare in social networks have been developed, which are aimed at protecting users' personal data.

Strategic goal №1. Anti-disinformation.

To do this, you need to take the following measures:

- to develop special groups of qualified specialists in the field of information security who will be able to quickly counteract various threats;
- to conduct information and psychological operations by the defense forces, which are aimed at deterring aggression from the Russian Federation;
- strengthen responsibility for spreading false information in social networks;
- prevent the spread of information products that support the position of the aggressor state;
- oppose information campaigns that deliberately involve Ukrainian citizens in actions that are not provided for by the Laws of Ukraine on military formations.

Strategic goal №2. Improving the level of media culture and media literacy of citizens.
To do this, you need to take the following measures:

- conduct training on media literacy of the population: develop critical thinking; improve abilities in checking various types of information for truthfulness, etc.;
- create comfortable conditions for the development of media specialists;
- strengthen responsibility for the dissemination of hidden information.

Strategic goal №3. Ensuring countermeasures against the spread of illegal content and protection of personal data of users in social networks.
To do this, you need to take the following measures:

- strengthen monitoring of information in social networks by special groups of specialists;
- to improve the legislation on the protection of personal data of users, as well as to create effective mechanisms for its implementation;
- protection of the country's information space from criminal content.

Strategic goal №4. Improving the cultural level of information society dialogue.
To do this, you need to take the following measures:

- publicly discuss current and socially important problems in social networks;
- implement innovations in the protection of information infrastructure;
- to ensure the comprehensive development of academic and analytical research in the field of information, in particular, the topic of the impact of online content of new mass media on the citizens of Ukraine needs to be studied in detail.

It should be noted that fake information in social networks spreads quickly and causes maximum damage to society. Modern technologies are an important factor in the fight against false news among the wide flow of information on the Internet, so the development of new algorithms for searching and verifying information is relevant today.

6 Practical Implementation

According to the strategies described in the previous section, an analysis of the community content of the public organization Varta1 in the Facebook social network was carried out. The analysis will allow us to assess the level of informational resistance in the community based on the identified features of each of the strategies.

The following important criteria for analyzing the compliance of strategy №1 were identified through research:

$Anti\_Disinformation(Community_i)$ – the importance of community misinformation;

$FakeMultimedia(Community_i)$ – amount of false multimedia content;

$GroupExperts(Community_i)$ – presence of a group of experts;

$ProvocativeAppeals(Community_i)$ – the number of provocative calls to promote information campaigns.

Let's create the following formula for calculation, where the value of the coefficients is determined according to the specifics of the research on the selection of the community strategy.

$$Anti\_Disinformation(Community_i) = k_1 * FakeMultimedia(Community_i)$$
$$+k_2 * GroupExperts(Community_i) + k_3 * ProvocativeAppeals(Community_i) \quad (1)$$

where, k_1 – coefficient importance factor in determining the value of community misinformation $FakeMultimedia(Community_i)$; k_2 – coefficient of importance of the presence of a group of experts in determining the value of community misinformation $GroupExperts(Community_i)$; k_3 – coefficient of importance of provocative appeals of users in determining the value of disinformation of the community $ProvocativeAppeals(Community_i)$;

The interpretation of determining the coefficients will look as follows $k_1 + k_2 + k_3 = 1$.

Considering the need to determine the value of the coefficient k we justify the following conditions for determining the level of disinformation $Anti\_Disinformation(Community_i)$: 1. If the value $0,5 < Anti\_Disinformation(Community_i) < 1$, then the content of the virtual community corresponds to a high level of compliance. 2. If the value $0,2 < Anti\_Disinformation(Community_i) < 0,5$, then the content of the virtual community meets the standards, but needs minor changes, is at an average level. 3. If the value $0 < Anti\_Disinformation(Community_i) < 0,2$, then the content of the virtual community corresponds to the low level.

The following important criteria for analyzing the compliance of strategy №2 were determined by research:

$User\_MediaLiteracy(Community_i)$ – the value of media literacy of community users;

$HiddenInformation(Community_i)$ – the amount of content involving the distribution of personal information;

$GrammaticalNorms(Community_i)$ – amount of content that does not meet grammatical standards;

$UncivilizedExpressions(Community_i)$ – the amount of content containing profanity.

$$
\begin{aligned}
&User\_MediaLiteracy(Community_i) \\
&= m_1 * HiddenInformation(Community_i) + m_2 \\
&*GrammaticalNorms(Community_i) + m_3* \\
&UncivilizedExpressions(Community_i)
\end{aligned}
\tag{2}
$$

where m_1 – coefficient of importance of the amount of personal (hidden) information when determining the value of media literacy $User\_MediaLiteracy(Community_i)$; m_2 – the coefficient of importance of content that corresponds to grammatical norms when determining the value of mediliteracy $User\_MediaLiteracy(Community_i)$; m_3 – coefficient of importance of users' non-cultural content when determining the value of media literacy $User\_MediaLiteracy(Community_i)$.

The interpretation of determining the coefficients will look as follows $m_1+m_2+m_3 = 1$.

Considering the need to determine the value of the coefficient $User\_MediaLiteracy(Community_i)$: 1. If the value $0.5 < User\_MediaLiteracy(Community_i) < 1$, then the media literacy of users is at a high level and does not require significant corrections. 2. If the value $0.2 < User\_MediaLiteracy(Community_i) < 0.5$, then the level of media literacy is at an average level. 3. If the value $0 < User\_MediaLiteracy(Community_i) < 0.2$, is low.

The following important criteria for analyzing the compliance of strategy №3 were determined by research:

$DataSecurity(Community_i)$ – the importance of protecting the community from illegal content;

$ViolationRights(Community_i)$ – the amount of content that violates the rights of users;

$Refutation(Community_i)$ – the number of rebuttals of such content;

$MonitoringGroup(Community_i)$ – the presence of a monitoring group for this type of content.

$$
\begin{aligned}
DataSecurity(Community_i) &= d_1 * ViolationRights(Community_i) \\
&+ d_2 * Refutation(Community_i) + d_3 * MonitoringGroup(Community_i)
\end{aligned}
\tag{3}
$$

where, d_1 – the coefficient of importance of content that violates the rights of users when determining the importance of ensuring the protection of the community from illegal content $DataSecurity(Community_i)$; d_2 – the coefficient of importance of content rebuttals in determining the value of ensuring the protection of the community from illegal content $DataSecurity(Community_i)$; d_3 – the coefficient of importance of the presence of a monitoring group that constantly monitors and regulates the appearance of such content when determining the importance of ensuring the protection of the community from illegal content $DataSecurity(Community_i)$.

The interpretation of determining the coefficients will look as follows $d_1+d_2+d_3 = 1$.

Taking into account the need to determine the value of the coefficient d, we justify the following conditions for determining the value of media literacy

$DataSecurity(Community_i)$: 1. If the value $0.5 < DataSecurity(Community_i) < 1$, then this indicates a high level of security and such a community does not need corrections. 2. If the value $0.2 < DataSecurity(Community_i) < 0.5$, then the security level is at an average level. 3. If the value $0 < DataSecurity(Community_i) < 0.2$, is low.

The following important criteria for analyzing the compliance of strategy №4 were identified through research:

$ContentAnalysis(Community_i)$ – the value of the relevance of the content to the direction it represents in the community;

$KeyWords(Community_i)$ – the number of keywords corresponding to the topics of information requests of community users;

$Reactions(Community_i)$ – the number of reactions to published content;

$Coments(Community_i)$ – the number of comments on published content.

$$ContentAnalysis(Community_i) = a_1 * KeyWords(Community_i)$$
$$+a_2 * Reactions(Community_i) + a_3 * Coments(Community_i) \tag{4}$$

where a_1 – the coefficient of importance of content that violates the rights of users when determining the importance of ensuring the protection of the community from illegal content $DataSecurity(Community_i)$; a_2 – the coefficient of importance of content rebuttals in determining the value of ensuring the protection of the community from illegal content $DataSecurity(Community_i)$; a_3 – the coefficient of importance of the presence of a monitoring group that constantly monitors and regulates the appearance of such content when determining the importance of ensuring the protection of the community from illegal content $DataSecurity(Community_i)$.

The value of the coefficient a is determined in accordance with the specifics of conducting research on community content analysis. Taking into account the factor of importance of keywords $KeyWords(Community_i)$, comments $Coments(Community_i)$ or reactions $Reactions(Community_i)$ it will be larger for one of them and smaller for others. For example, comments are more important to us than keywords, so a_1 for $Coments(Community_i)$ will be greater than other coefficients a_2, a_3. The interpretation of determining the coefficients will look as follows $a_1 + a_2 + a_3 = 1$.

Taking into account the need to determine the value of the coefficient a, we justify the following conditions of compliance with the content direction $ContentAnalysis(Community_i)$: 1. If the value $0.5 < ContentAnalysis(Community_i) < 1$, then the content of the virtual community corresponds to a high level of representation of this or that direction in the network. 2. If the value $0.2 < ContentAnalysis(Community_i) < 0.5$, then the content of the virtual community corresponds to the average level. 3. If the value $0 < ContentAnalysis(Community_i) < 0.2$, then the content of the virtual community corresponds to a low level.

Calculation Example. To calculate the relevance of strategy 1 to the value of community misinformation $Anti\_Disinformation(Community_i)$ the following coefficient values were assigned $FakeMultimedia(Community_i) - k_1 = 0, 6$, amount of false multimedia content, $GroupExperts(Community_i) - k_2 = 0, 3$, presence of a group of experts,, $ProvocativeAppeals(Community_i) - k_3 = 0, 1$, the number of provocative calls to promote information campaigns. That is, for this analysis, the presence of false multimedia content will be more important than the other two values.

Table 1. The results obtained

Post	k_1	k_2	k_3	Anti_Disinformation ($Community_i$)	Responds
Post 1	0	0	1	0.1	–
Post 2	1	1	0	0.9	+
Post 3	1	0	1	0.7	+
Post 4	1	1	1	1	+

According to Table 1, it can be seen that only the first post of this community requires the application of anti-disinformation measures. Other posts meet all requirements.

As a result, recommendations were provided for the development of the community, which will help the administration to better organize its activities in the future, in particular, to maintain a high level of combating disinformation, media literacy, communication culture, etc.

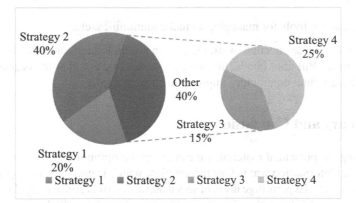

Fig. 5. Current strategies are necessary for community work

Also, based on the conducted research, we tried to analyze which of the proposed strategies are relevant for the work of the community (Fig. 5).

That is, based on the analysis of the content of the virtual community, we will be able to propose strategy №1 for the fight against disinformation in the virtual community, which is determined by the following indicators: the number of provocative and false information, groups of experts on the use of measures to counter disinformation.

The next auxiliary tool in protecting the community from illegal content is strategy №3. The main content of which is monitoring false content, its quick removal and protection of users' rights. Here we focus on checking all information submitted for publication. But the creation of not only a group of experts that can control these processes, but also resolve conflict situations.

When we analyze the user's content, namely the information that he broadcasts in the community, we take into account strategy №2 regarding users' media literacy. It will help not only to identify incorrect content, but also the user's role in the community, to investigate his behavior.

An important object in the activity of any virtual community is the user. Analysis of his reactions, comments and key words relevant to his topic, using strategy №4, will allow studying the direction of which he is a representative. It will open up new opportunities for interesting collaborations, access to interesting topics, development of new plans for the development of the virtual community.

As a result of the use of strategies, the administration of the virtual community will be able to:

- maintain the quality of the content, evaluate the suitability of this or that content offered for publication, correct errors;
- organize the processes of adding, blocking or deleting users;
- identify and avoid spreading false information;
- to support media literacy of users, through the formation of rules of behavior in the community;
- to unite people who strive to improve the quality of content in expert monitoring groups;
- practice modern tools for managing virtual communities, etc.

The result of the application of strategies is in principle unpredictable, as it is social communications. But after conducting several studies, we saw that the organization of the work of the virtual community improves with each strategy.

7 Summary and Conclusion

Summarizing the presented material, we can assert the opinion that the main aspect of conducting an information war is fake information, which is used to exert psychological pressure on society. Experts urge users of social networks to carefully study the published information, as misinformation is a powerful weapon today, which aims to mislead people.

Taking into account the high level of disinformation spread in the materials, a number of strategies for conducting informational struggle in social networks have been developed.

1. Strategic goal №1. Anti-disinformation.
2. Strategic goal №2. Improving the level of media culture and media literacy of citizens.
3. Strategic goal №3. Ensuring countermeasures against the spread of illegal content and protection of personal data of users in social networks.
4. Strategic goal №4. Improving the cultural level of information society dialogue.

Formal models of strategies have been developed, which will help in determining the necessity of their use. As a result, numerical data were obtained that show the content of one of the substantiated formal models for strategy. Possibilities of implementing strategies that will contribute to a higher level of organization of community activities in the network, namely increasing the level of resistance to information resistance, are presented.

That is, the developed strategies effectively organize the work of the virtual community, ensure the distribution of quality content, control users, constant monitoring and resolution of conflict situations, etc.

Today, we continue to constantly monitor research in this direction and try to make new adjustments to reduce the level of misinformation in the social media environment.

References

1. Meligy, A.M., Ibrahim, H.M., Torky, M.F.: Identity verification mechanism for detecting fake profiles in online social networks. Int. J. Comput. Netw. Inf. Secur. (IJCNIS) **9**(1), 31–39 (2017). https://doi.org/10.5815/ijcnis.2017.01.04
2. Didenko, M.M.: Information-psychological impact on people during the war. Bull. Stud. Sci. Soc. DonNU Named After Vasyl. Stus. **1**(14), 211–214 (2022)
3. Febrianita, R.: Information disorder & the online's gatekeeping mechanism struggle in post truth era. JCommsci – J. Med. Commun. Sci. **3**(3), 134. https://doi.org/10.29303/jcommsci.v3i3.78
4. Gururaj, H.L., Swathi, B.H., Ramesh, B.: Threats, consequences and issues of various attacks on online social networks. Int. J. Educ. Manag. Eng. (IJEME) **8**(4), 50–60 (2018). https://doi.org/10.5815/ijeme.2018.04.05
5. Ablazov, I., et al.: Information weapons within the interstate struggle in the XXI. Century Revista Amazonia Investiga **11**(52), 269–277. URL: https://doi.org/10.34069/ai/2022.52.04.29
6. Oleksandr, K.: Fakes in modern media: identification and neutralization. Sci. J. Libr. Sci. Doc. Sci. Informatol. **3**, 96–103 (2018)
7. Lord, W.P.: Designing for social connectivity (not everyone likes Webcams). eLearn **2021**(4). https://doi.org/10.1145/3462445.3457174
8. Luengo, M., García-Marín, D.: The performance of truth: politicians, fact-checking journalism, and the struggle to tackle COVID-19 misinformation. Am. J. Cult. Sociol. **8**(3), 405–427 (2020). https://doi.org/10.1057/s41290-020-00115-w
9. Makarenko, L.P.: Evolution of forms and methods of conducting information warfare. Int. Sci. J. Sci. Rev. **4**(3) (2014). https://naukajournal.org/index.php/article/view/185
10. Novorodovsky, V.: Information security of Ukraine in the conditions of Russian aggression. Soc. Doc. Commun. (9), 150–179. https://doi.org/10.31470/2518-7600-2020-9-150-1179
11. Onkovych A. Facebook social network and protection of the Ukrainian information space in the conditions of the Russian-Ukrainian conflict. Ukrainian information space. 2020. No. 1(5). P. 233–242. URL: https://doi.org/10.31866/2616-7948.1(5).2020.206131
12. Pazderska, R.S., Markovets, O.V.: Definition of content and strategies for increasing its effectiveness in virtual communities. Bull. Vinnytsia Polytech. Inst. **3**, 69–77 (2021)
13. Peleshchyshyn, A., Trach, O.: Determination of the elements of socially oriented risks in the organization of the life cycle of a virtual community. Inf. Secur. **23**(2), 130–135 (2017)
14. Limsaiprom, P., Praneetpolgrang, P., Subsermsri, P.: Security visualization analytics model in online social networks using data mining and graph-based structure algorithms. Int. J. Inf. Technol. Comput. Sci. (IJITCS) **6**(8), 1 (2014). https://doi.org/10.5815/ijitcs.2014.08.01
15. Limsaiprom, P., Praneetpolgrang, P.: Pilastpongs Subsermsri,"Visualization of Influencing Nodes in Online Social Networks". IJCNIS **6**(5), 9–20 (2014). https://doi.org/10.5815/ijcnis.2014.05.02

16. Pravdenko, V.Yu., Tytova, N.M.: The influence of fake information on the consciousness of the individual (2020). http://dspace.luguniv.edu.ua/123456789//5.pdf
17. Social networks as an environment of fakes: what you need to know about Facebook. https://explainer.ua/sotsialni-merezhi-yak-seredovishhe-fejkiv-shho-treba-znati-pro-facebook/
18. Trach, O., Fedushko, S.: Determination of the indicator of resistance of the virtual community in relation to information attacks. Inf. Secur. 22(1), 4–87 (2016)

A Modified Method and an Architecture of a Software for a Multivariate Polynomial Regression Building Based on the Results of a Conditional Active Experiment

Alexander Pavlov[✉], Maxim Holovchenko, Iryna Mukha, Kateryna Lishchuk, and Valeriia Drozd

National Technical University of Ukraine "Igor Sikorsky Kyiv Polytechnic Institute", Kyiv 03056, Ukraine

pavlov.fiot@gmail.com, {ma4ete25,lishchuk_kpi}@ukr.net, mip.kpi@gmail.com, drozdllera@gmail.com

Abstract. The application scope of the results presented in this paper is the creation of mathematical support for intelligent expert and diagnostic systems based on nonlinear regression models. The authors own an original method of estimating the coefficients at nonlinear terms of a multivariate polynomial regression given by its redundant representation based on the results of an active experiment. We reduce this problem solution to the sequential building of univariate polynomial regressions. The coefficients at nonlinear terms of the regressions are estimated by the original algorithmic procedure. For each univariate polynomial regression, we successively formulate nondegenerate systems of linear equations in which variables are the coefficients at nonlinear terms of the multivariate polynomial regression given by the redundant representation. We proposed to estimate the coefficients at linear terms by known methods. The main disadvantages of the proposed method were the need for the existence of a common range of possible values of all deterministic input variables and independence of the range of possible values of a part of the input variables at fixed values of all other input variables. These are significant drawbacks when designing, in particular, medical diagnostic systems. For example, if we fix the patient's age as a constant parameter, then this age will determine the range of changes to other input characteristics according to which the disease is diagnosed. This work presents a modification of the method that allows extending its scope to the cases formulated above, that is, when the common range of possible values of the input variables is an empty set or when the range of possible values of the input variables depends on the values of the fixed input variables in all tests of an active experiment. Also, with the help of an original fact of the general theory of the least squares method that we mention here, we have achieved an increase in the efficiency of estimating the coefficients at linear terms of a multivariate polynomial regression.

Keywords: Multivariate polynomial regression · Conditional active experiment · Multivariate linear regression · Software architecture · Diagnostic systems

© The Author(s), under exclusive license to Springer Nature Switzerland AG 2023
Z. Hu et al. (Eds.): ICCSEEA 2023, LNDECT 181, pp. 207–222, 2023.
https://doi.org/10.1007/978-3-031-36118-0_19

1 Introduction

1.1 Related Works Review Substantiating the Theoretical and Practical Relevance of the Results We Present

The practical significance of the ability to efficiently build multivariate nonlinear regressions is evident, for example, from the works [1–13] which show the fields of their application such as economics, finance, agriculture, medicine, forensics, image filtering, etc.

The method of estimating coefficients at nonlinear terms of a multivariate polynomial regression (MPR) was fully published in the paper [14]. Section 1.1 of [14] was fully devoted to a critical review of known universal methods of building nonlinear regressions [10, 15–17] such as a stepwise technique for adding and removing polynomial terms, group method of data handling, hierarchical approximation method, and genetic algorithms. We have concluded that:

1. None of the known universal methods of nonlinear regressions building can be claimed more efficient than others, i.e., the problem of creating an efficient universal method of building an MPR given by a redundant representation is still relevant both in theoretical and in applied aspects.
2. The method of building an MPR given by a redundant representation described in [14] is completely original and is not a modification of any other method. Indeed, this method reduces the estimating of nonlinear terms of the MPR to the sequential building of univariate polynomial regressions (UPRs). Such reducing allows constructing, based on the structure of the redundant representation of the MPR and the estimates found for the nonlinear terms of the UPRs, a sequence of nondegenerate systems of linear equations which variables are estimates of the coefficients at nonlinear terms of the MPR given by the redundant representation. We have achieved simplification of estimating the coefficients at nonlinear terms of UPRs based on an original theoretical result [14, 18] that has allowed us to estimate these coefficients in an arbitrary active experiment using only a single set of normalized orthogonal polynomials of Forsythe (NOPFs) built in advance with a given accuracy.
3. The statistical studies' results also presented in [14] have substantiated the practical efficiency of the proposed method.
4. In addition to its obvious advantages, the method described in [14] has the following significant drawbacks.

The first drawback: necessarily should exist a common range (e.g., a $[a, b]$ segment) in which each of the deterministic scalar variables can take arbitrary values.

The second drawback: the method [14] (that allows reducing the estimating of the coefficients at nonlinear terms of an MPR given by a redundant representation to the building of UPRs and systems of linear nondegenerate equations) requires that, in the general case, an arbitrary subset of the input variables which take fixed values in the active experiment in all tests did not affect the ranges in which other input variables can take arbitrary values. The second drawback is essential when designing, in particular, medical diagnostic systems. For example, if we fix the patient's age as a constant parameter, then this age will determine the range of changes to other input characteristics according to which the disease is diagnosed.

The main goal of this work is to extend the application scope of the method of building an MPR given by a redundant representation [14] to the cases when:

1. There is no common range of possible values of the input variables.
2. Fixed values of input variables in all tests of an active experiment change the range of possible values of other input variables.

The authors believe that this extension qualitatively increases the applied value of the method [14]. In this paper, we present modified algorithmic procedures that remove the above formulated restrictions when using the method of building an MPR given by a redundant representation [14]. The estimates of coefficients at nonlinear terms of the MPR given by a redundant representation were proposed to be found by any known universal methods. This work presents the result of the general theory of the least squares method (LSM) that significantly increases the efficiency of any universal method of estimating the coefficients of a multivariate linear regression (MLR).

1.2 Methodology for Extending the Application Scope of the Method of Building an MPR Given by a Redundant Representation [14]

The methodology of modification of the method [14] is based on the implementation of a theoretical result given in [18]. Namely, we utilize the possibility of reducing the solution of the UPR problem based on the results of an active experiment in an arbitrary segment $[c, d]$ to the construction of a UPR based on the results of a virtual active experiment with an input variable that takes arbitrary values in a predetermined segment of the number line. Using this result allows us to remove the above formulated restrictions from the method [14].

Remark 1. This has been the result that allowed us in [19] to use only a single set of NOPFs built with a given accuracy to find estimates of a UPR coefficients based on the results of an arbitrary active experiment.

1.3 The Structure of the Paper

In Sect. 2, we give the formal problem statement and outline the main provisions of the method [14] which allow us to reasonably present in Sect. 3 modified algorithmic procedures that remove the restrictions we talked about in Subsect. 1.1. In Sect. 4, we give the theoretical properties of a repeated active experiment which significantly simplify the algorithmic procedures for estimating an MLR coefficients when using the LSM. Section 5 presents an architecture of a software that in general implements the method of building an MPR given by a redundant representation based on the results of a limited active experiment. Section 6 contains the conclusions of the work.

2 General Theoretical Provisions

Paper [14] gives the following formulation of the problem of building an MPR given by a redundant representation. An MPR given by a redundant representation has the form

$$y(\overline{x}) = \sum_{\forall (i_1,...i_t) \in K} \sum_{\forall (j_1,...,j_t) \in K(i_1,...,i_t)} b_{i_1...i_t}^{j_1...j_t} (x_{i_1})^{j_1} \cdot (x_{i_2})^{j_2} \cdot .... \cdot (x_{i_t})^{j_t} + E \quad (1)$$

where $\bar{x} = (x_1 \ldots x_m)^\top$ is a deterministic vector of input variables, E is a random variable with mean $ME = 0$ and variance $\text{Var}(E) = \sigma^2 < \infty$. The values of $\forall b_{i_1 \ldots i_t}^{j_1 \ldots j_t}$ are unknown, some of them may equal to zero. The values of the unknown coefficients of the MPR (1) are estimated based on the results of an active experiment that must follow two restrictions. The first one is:

$$\bigcap_{j=1}^{m} [x_{j\,min}, x_{j\,max}] = [c, d] \neq \{\varnothing\} \tag{2}$$

where $[x_{j\,min}, x_{j\,max}]$, $j = \overline{1, m}$, are the ranges of arbitrary values of x_j in the active experiment. For all input variables, the active experiment is conducted within the range (2).

The second restriction: during the tests in the active experiment, an arbitrary number of input variables can take fixed values that do not change the range of arbitrary values of other input variables in the active experiment. That is, the segment $[c, d]$ in all tests is the range of arbitrary values of all input variables in the tests of the active experiment.

When the first and second conditions are met, the problem of estimating the coefficients at nonlinear terms of the MPR (1) is reduced to building a sequence of UPRs and the corresponding nondegenerate systems of linear equations. The variables of the equations are the coefficient estimates for the nonlinear terms of the MPR (1), and their right sides are the estimates for the nonlinear terms of UPRs. This sequence of steps is implemented by the algorithm [14] consisting of the following two sub-algorithms.

2.1 The l-th Step of the First Sub-algorithm

The l-th step of the first sub-algorithm is performed for a subsequent nonlinear term of the MPR (1) which value was not evaluated in the previous steps of the first sub-algorithm and which contains at least one input variable in a degree greater than or equal to two. Let x_{j_l} be such a variable. The first active experiment at the l-th step is performed for the following values of the input variables: $x_{j_l i} = x_i \in [c, d]$, $i = \overline{1, n}$; $x_{ji} = x_j^{F_1} \forall j \neq j_l, i = \overline{1, n}$. For this active experiment, the MPR (1) is transformed into a UPR. The estimates for the nonlinear terms of the UPR become the right side of linear equation(s); the variables of the equations are the estimates of the unknown coefficients for the nonlinear terms of the MPR (1).

Remark 2. We give the detailed description of the l-th step of the first sub-algorithm in [14] and illustrate it by a corresponding example. If at least one of the equations contains two or more unknowns, we implement at the l-th step of the first sub-algorithm the procedure for building the corresponding number (l_k) of UPRs at the other fixed values of the input variables: $x_{ji} = x_j^{F_p}$, $i = \overline{1, n}, p = \overline{2, l_k}, j \neq j_l$. Here, the number l_k of UPRs and the values of $x_j^{F_p}, p = \overline{1, l_k}$, are chosen such that as a result we get a quadratic nondegenerate system(s) of linear equations containing unknown coefficients for the nonlinear terms of the MPR (1), the right parts of the equations are the estimates of the coefficients for the nonlinear terms of the l_k UPRs built at the first step. The upper estimate to the number of steps of the first sub-algorithm is the number of terms of the MPR (1) containing at least one variable in a degree greater than or equal to two.

2.2   The *l*-th Step of the Second Sub-algorithm

The *l*-th step of the second sub-algorithm is performed for a subsequent nonlinear term of the MPR (1) which value was not evaluated in the previous steps of the second sub-algorithm and which contains input variables in the first degree, that is, has the form $b_{i_1 \ldots i_t}^{1 \ldots 1} \prod_{j=1}^{t} x_{ij}$. Then, in order to obtain a UPR in the active experiment, the variables $x_{i_j}, j = \overline{1,t}$, take the values of $x_{iji} = x_i \in [c,d], i = \overline{1,n}; x_{ji} = x_j^{F_{/1}} \forall j \notin \{i_1, \ldots, i_t\}, i = \overline{1,n}$. As in the first sub-algorithm, we build the corresponding number of UPRs and nondegenerate systems of linear equations. In each of the equations, the variables are the estimates of unknown coefficients of the MPR (1), and the right sides are the estimates of coefficients at nonlinear terms of the UPRs.

Remark 3. If a nonlinear term containing the maximum number of coefficients is chosen at the *l*-th step, then the estimate of its coefficient is the solution of a single equation with a single unknown.

Remark 4. According to [14, 18, 19], we find the estimates for nonlinear terms of all UPRs in the first and second sub-algorithms using only a single set of NOPFs built in advance with a given accuracy.

Remark 5. We use repeated active experiments [14] to obtain the given values of variances of the estimates for nonlinear terms of the UPRs.

The above brief overview of the main algorithmic procedures of the method [14] allows us to formulate in the most compact form and to substantiate the modified algorithmic procedures which significantly extend the application scope of the method [14].

3   A Modified Method of Estimating Coefficients at Nonlinear Terms of the MPR (1) Based on the Results of a Limited Active Experiment

3.1   The Problem Statement

An MPR given by a redundant representation has the form (1). We assume that the following restrictions on an active experiment may be satisfied. The first one is

$$\bigcap_{j=1}^{m} [x_{j\,min}, x_{j\,max}] = \{\varnothing\}. \tag{3}$$

The second restriction: if in all tests of the active experiment an arbitrary subset of the input variables takes fixed values,

$$x_{j_1 i} = x_{j_1}^F, \ldots, x_{j_{k_1} i} = x_{j_{k_1}}^F, i = \overline{1,n}, \tag{4}$$

then other input variables can take values only from the ranges which depend on the values (4):

$$x_{i_l} \in \left[c_{x_{j_1}^F \ldots x_{j_{k_1}}^F}, d_{x_{j_1}^F \ldots x_{j_{k_1}}^F} \right], l = \overline{1, k_2},$$

$$\forall l \ i_l \notin \left\{ j_l, l = \overline{1, k_1} \right\}, k_1 + k_2 = m. \tag{5}$$

We implement the consideration of conditions (3)–(5) in the modified method using the following result [18].

Let us consider a UPR

$$Y(x) = \sum_{j=0}^{r} \theta_j x^j + E, \tag{6}$$

the input scalar variable x can take values in the range $x \in [c, d]$. For the virtual input variable z at the values $z_i, i = \overline{1, n}, \forall i \ z_i < z_{i+1}$, we have built NOPFs

$$Q_j(z), j = \overline{0, r} \tag{7}$$

with a given accuracy. Here,

$$r = \max \left\{ \max \{ j_1, \ldots, j_t \}, \forall i_1, \ldots, i_t \in K, \forall (j_1, \ldots j_t) \in K(i_1, \ldots i_t); t, \forall b_{i_1 \ldots i_t}^{1 \ldots 1} \right\}. \tag{8}$$

That is, the number of NOPFs (7) allows us to estimate the coefficients at nonlinear terms of the UPR in the first and second sub-algorithms (see Sect. 2). Let

$$a = \frac{d - c}{z_n - z_1} > 0, b = c - \frac{d - c}{z_n - z_1} z_1. \tag{9}$$

We substitute $az + b$ for the input variable x in (6). Then,

$$\sum_{j=0}^{r} \theta_j (az + b)^j = \sum_{j=0}^{r} \gamma_j z^j. \tag{10}$$

Having set

$$x_1 = c, x_j = az_j + b, j = \overline{2, n}, \tag{11}$$

we get

$$x_1 = c < x_2 < \ldots < x_n = d, \sum_{j=0}^{r} \theta_j (x_i)^j = \sum_{j=0}^{r} \gamma_j (z_i)^j. \tag{12}$$

Remark 6. It is shown in [18] that the coefficients θ_j and $\gamma_j, j = \overline{0, r}$, are interrelated through a nondegenerate system of linear equations with an upper triangular matrix of constraints. This result allowed us [18, 19] to estimate the coefficients $\theta_j, j = \overline{0, r}$, at arbitrary numbers $c < d$ in the UPR problem using only a single set of NOPFs (7).

3.2 Modified Algorithmic Procedures of the Method for Estimating the Coefficients at Nonlinear Terms of an MPR (1)

We illustrate the meaning of the outlined below modified algorithmic procedures using the following example:

$$Y(x_1, x_2, x_3) = \theta_0 + \theta_1 x_1 + \theta_2 x_2 + \theta_3 x_3 + \theta_4 x_1 x_2 + \theta_5 x_1 x_2^2$$

$$+ \theta_6 x_1^4 x_2 + \theta_7 x_1 x_2 x_3 + \theta_8 x_1 x_2^2 x_3^2 + E. \tag{13}$$

According to (8), $r = 4$. We find NOPFs $Q_j(z), j = \overline{0, 4}$, with a given accuracy [14] for $z_1 < z_2 < \ldots < z_n, n > 4$.

3.3 Modified Algorithmic Procedure of the First Sub-algorithm of Sect. 2

The modification in the first sub-algorithm of Sect. 2 consists in replacement of the segment $[c, d]$ with

$$\left[c_{x_{j_1}^F \ldots x_{j_{m-1}}^F}, d_{x_{j_1}^F \ldots x_{j_{m-1}}^F} \right], i_l \notin \{j_1, \ldots, j_{m-1}\} \tag{14}$$

where we change only the input variable x_{i_l} in the active experiment, and for each of its steps we choose a nonlinear term of the MPR (1) that has the input variable in the largest degree.

For the example (13), the first sub-algorithm implements the first step for the nonlinear term $\theta_6 x_1^4 x_2$. The corresponding virtual UPR has the form

$$Y(z) = \theta_0 + \theta_1(az + b) + \theta_2 x_2^{F_1} + \theta_3 x_3^{F_1} + \theta_4(az + b)x_2^{F_1}$$

$$+ \theta_5(az + b)\left(x_2^{F_1}\right)^2 + \theta_6(az + b)^4 x_2^{F_1} + \theta_7(az + b)x_2^{F_1} x_3^{F_1}$$

$$+ \theta_8(az + b)\left(x_2^{F_1}\right)^2 \left(x_3^{F_1}\right)^2 + E, \tag{15}$$

the coefficients a and b are found by (9) for $c = c_{x_2^{F_1} x_3^{F_1}}, d = d_{x_2^{F_1} x_3^{F_1}}$. The values of the input variable x_1 in the active experiment for the real UPR

$$y\left(x_{1i}, x_2^{F_1}, x_3^{F_1}\right) = \theta_0 + \theta_1 x_{1i} + \theta_2 x_2^{F_1} + \theta_3 x_3^{F_1} + \theta_4 x_{1i} x_2^{F_1} + \theta_5 x_{1i}\left(x_2^{F_1}\right)^2$$

$$+ \theta_6 x_{1i}^4 x_2^{F_1} + \theta_7 x_{1i} x_2^{F_1} x_3^{F_1} + \theta_8 x_{1i}\left(x_2^{F_1}\right)^2 \left(x_3^{F_1}\right)^2 + \varepsilon_i$$

$$= \theta_0^* + \theta_1^* x_{1i} + \theta_4^* x_{1i}^4 + \varepsilon_i \tag{16}$$

are found according to (11). In (16), ε_i is the realization of the random variable E in the i-th test. Depending on the upper estimate of $\text{Var}(E) = \sigma^2$, to find the estimates

$\hat{\theta}_j^*, j \geq 2$, with a given variance, the active experiment can be repeated [14]. That is, the input variable value sequence is $x_1, \ldots, x_n, x_1, \ldots, x_n, \ldots, x_1, \ldots, x_n$ and the results of the active experiment have the form $\left(x_i \to y_{j+i}, i = \overline{1, n}, j = \overline{0, p-1}\right)$. The results of a real active experiment are used to design a virtual active experiment based on the input values of the virtual input variable z: $z_1, \ldots, z_n, \ldots, z_1, \ldots, z_n$ for the virtual UPR $\sum_{j=0}^r \gamma_j(z_i)^j + E$ (15). Taking [14] into account, the experiment is implemented as

$$\left(z_i \to \frac{y_i + y_{n+i} + \cdots + y_{(p-1)n+i}}{p}, i = \overline{1, n}\right). \tag{17}$$

In this case, the variance of estimates $\hat{\gamma}_j, j = \overline{0, 4}$, is reduced by p times compared to the estimates obtained as the result of the virtual active experiment $\left(z_i \to y_i, i = \overline{1, n}\right)$. The estimate of the coefficient θ_6 based on the results of the virtual active experiment is equal to $\hat{\theta}_6 = \hat{\gamma}_4 / \left(a^4 x_2^{F_1}\right)$. Due to [19], the estimates of the coefficients $\theta_j^*, j = \overline{0, 4}$, obtained using the NOPF built directly from the results of the real active experiment, and those obtained from the virtual active experiment are the same.

The second step of the first sub-algorithm is implemented for the nonlinear term $\theta_8 x_1 x_2^2 x_3^2$ of the MPR (13) where the variable in the active experiment is the input variable x_2. The input variables x_1 and x_3 take fixed values. At the second step of the first sub-algorithm, we estimate the coefficients for the input variable in the second degree of two UPRs (for different fixed values of the input variables x_1, x_3, i.e., for $x_1^{F_1} x_3^{F_1}, x_1^{F_2} x_3^{F_2}$) and form a nondegenerate system of two equations, where the variables are the estimates of the coefficients θ_5 and θ_8.

Remark 7. To avoid solving the system of two equations, we could set the variable x_3 in the second step and perform the third step of the first sub-algorithm for the term $\theta_5 x_1 x_2^2$ of the MPR (13).

3.4 Modified Algorithmic Procedure of the Second Sub-algorithm of Sect. 2

The l-th step of the second sub-algorithm is performed for a subsequent term of the MPR (1) which coefficient was not evaluated in the previous steps of the second sub-algorithm and which contains the maximum number of input variables only in the first degree.

Remark 8. Such a choice of a nonlinear term of the MPR (1) leads to the fact that the estimate of its coefficient is the solution of a single equation with a single variable.

Let this nonlinear term have the form

$$b_{i_1 \ldots i_{t_l}}^{1 \ldots 1} \prod_{j=1}^{t_l} x_{i_j}. \tag{18}$$

In this case, the conditional active experiment is implemented under the following restrictions. If $x_{j_p i} = x_{j_p}^F, i = \overline{1, n}, p = \overline{1, m - t_l}, \{i_1, \ldots, i_{t_l}\} \cap \{j_1, \ldots, j_{m-t_l}\} =$

$\varnothing$, $\{i_1, \ldots, i_{t_l}\} \cup \{j_1, \ldots, j_{m-t_l}\} = \{1, \ldots, m\}$, then the input variables can take arbitrary values within the ranges

$$x_{i_p i} \in \left[c^{i_p}_{x^F_{j_1} \ldots x^F_{j_{m-t_l}}}, d^{i_p}_{x^F_{j_1} \ldots x^F_{j_{m-t_l}}} \right]. \tag{19}$$

The ranges (19) may have an empty intersection. In this case, we replace each input variable x_{i_p} with

$$x_{i_p} = a_{i_p} z + b_{i_p}, p = \overline{1, t_l}, x_{i_p i} = a_{i_p} z_i + b_{i_p}, i = \overline{1, n} \tag{20}$$

where a_{i_p}, b_{i_p} are found according to (9) where

$$c = c^{i_p}_{x^F_{j_1} \ldots x^F_{j_{m-t_l}}}, d = d^{i_p}_{x^F_{j_1} \ldots x^F_{j_{m-t_l}}}, \tag{21}$$

and the real experiment for the UPR has the form

$$\left(x_{i_1 i}, \ldots, x_{i_{t_l} i}, x_{j_1 i} = x^F_{j_1}, \ldots, x_{j_{m-t_l} i} = x^F_{j_{m-t_l}} \to y_i, i = \overline{1, n} \right). \tag{22}$$

Remark 9. In a repeated active experiment with p repetitions, the sequence of input values (22) is repeated p times. We determine the value of p by the restriction on the variance of estimate of the UPR's coefficient b_{t_l} at the variable z to the power of t_l.

$$\hat{b}^{1 \ldots 1}_{i_1 \ldots i_{t_l}} = \frac{1}{\prod_{j=1}^{t_l} a_{i_p}} \left(\hat{b}_{t_l} + \text{const} \right) \tag{23}$$

where the known value of the constant in (23) is not zero if some nonlinear terms of (1) containing at least one input variable in a degree not lower than two, were transformed in the UPR into terms at z to the power of t_l.

Remark 10. For the example (13), the UPR at the first step of the second sub-algorithm has the form

$$\begin{aligned} Y(z) &= \theta_0 + \theta_1(a_1 z + b_1) + \theta_2(a_2 z + b_2) + \theta_3(a_3 z + b_3) \\ &+ \theta_4(a_1 z + b_1)(a_2 z + b_2) + \theta_5(a_1 z + b_1)(a_2 z + b_2)^2 \\ &+ \theta_6(a_1 z + b_1)^4(a_2 z + b_2) + \theta_7(a_1 z + b_1)(a_2 z + b_2)(a_3 z + b_3) \\ &+ \theta_8(a_1 z + b_1)(a_2 z + b_2)^2(a_3 z + b_3)^2 + E, \end{aligned} \tag{24}$$

the constant in (23) is uniquely determined by the parameters $\hat{\theta}_5, \hat{\theta}_6, \hat{\theta}_8, a_i, b_i, i = \overline{1, 3}$.

Remark 11. The number of steps of the second sub-algorithm given by the number of different terms of an MPR (1) in the form (18) can be reduced by evaluating more than one coefficient of the MPR (1) at some steps by building more than one UPR.

Thus, the implementation of the modified sub-algorithms of paragraphs 3.3, 3.4 is guaranteed to allow estimation, at a limited number of steps, of the coefficients at nonlinear terms of an MPR given by a redundant representation for the case when restrictions (3)–(5) are met. This conclusion is based on the fact that the use of the result [18] (see formulas (6)–(11)) in the modified sub-algorithms reduces their theoretical substantiation to the theoretical substantiation (given in [14]) of the sub-algorithms of paragraphs 2.1, 2.2. This is clearly illustrated by the example (13) for which the modified sub-algorithms of paragraphs 3.3, 3.4 are applied in Sect. 3.

4   Variances of Estimates of MLR Coefficients in the Repeated Experiment

As shown in [14], when the estimates for nonlinear terms of an MPR (1) are found sufficiently accurately and the terms with sufficiently small values of coefficient estimates are excluded, the problem of building an MPR given by a redundant representation (1) is reduced to building an MLR given by a redundant representation. We proposed [14] to solve this problem using one of the well-known universal methods. This section presents a theoretical result that can be useful for the implementation of any method of estimating the coefficients of an MLR given by a redundant representation using the general LSM procedure.

Suppose we have an MLR in the form

$$Y(\overline{x}) = \theta^\top x + E \tag{25}$$

where $\theta = (\theta_0, \theta_1, \ldots, \theta_m)^\top, x = (1, x_1, \ldots, x_m)^\top, x_j, j = \overline{1, m}$, are input deterministic variables, E is a random variable with $ME = 0, \mathrm{Var}(E) = \sigma^2 < \infty$. The active experiment has the form $(\overline{x}_i \to y_i, i = \overline{1, n})$ where

$$y_i = \overline{x}_i^\top \theta + \varepsilon_i, \tag{26}$$

ε_i is the realization of E in the i-th test, $\overline{x}_i = (1, x_{1i}, \ldots, x_{mi})^\top$ is the vector of input variables in the i-th test. We can consider y_i, ε_i to be realizations of independent virtual random variables $Y_i, E_i, i = \overline{1, n}$ where E_i are independent virtual copies of E. Then $Y_i = \overline{x}_i^\top \theta + E_i, i = \overline{1, n}$, or in vector form

$$Y = A\theta + \overline{E} \tag{27}$$

where $\overline{E} = (E_1, \ldots, E_n)^\top, Y = (Y_1, \ldots, Y_n)^\top$. Then $MY = A\theta$, the vector of estimates $\hat{\theta} = \left(\hat{\theta}_0, \hat{\theta}_1, \ldots, \hat{\theta}_m\right)^\top$ obtained by the LSM has the form

$$\hat{\theta} = \left(A^\top A\right)^{-1} A^\top Y, M\hat{\theta} = \theta. \tag{28}$$

Remark 11. In the MPR (1) building problem, the vector Y is replaced by the vector.

$$Y^* = (Y_1 - f(\overline{x}_1), \ldots, Y_n - f(\overline{x}_n))^\top \tag{29}$$

where $f(\overline{x}_i)$ is the known value of the nonlinear part of the MPR considering the found coefficient estimates for its nonlinear terms and excluding the nonlinear terms that are considered redundant.

Suppose that we conduct a repeated active experiment

$$(\overline{x}_i \to y_i, i = \overline{1, n}, \overline{x}_i \to y_{n+i}, i = \overline{1, n}, \ldots, \overline{x}_i \to y_{n(p-1)+i}, i = \overline{1, n}),$$

i.e., we have conducted np tests in which the sequence of input variables $\overline{x}_i, i = \overline{1, n}$ is repeated p times. Then the following statement holds.

Statement 1. The Following is True:

1. The vector $\hat{\theta}$ of estimates of unknown coefficients $\theta_0, \theta_1, \ldots, \theta_m$ obtained by the LSM has the form

$$\hat{\theta} = \left(A^\top A\right)^{-1} A^\top \left(\frac{1}{p} \sum_{j=1}^{p} Y^j\right) \tag{30}$$

where the random vectors $Y^j, j = \overline{2, p}$, are virtual independent copies of the virtual random vector Y (27) and $Y^1 = Y$. Realizations of their components are the numbers $y_{(j-1)n+i}, i = \overline{1, n}, j = \overline{1, p}$.

2. The estimate of the random vector $\hat{\theta}$ (30) is unbiased, and the variance of its components is p times less than that of the estimates vector (28).

Proof. Formula (30) follows from the vector equality

$$\left(\left(A^\top \cdots A^\top\right)\begin{pmatrix} A \\ \vdots \\ A \end{pmatrix}\right)^{-1} \left(A^\top \cdots A^\top\right)\begin{pmatrix} Y^1 \\ \vdots \\ Y^p \end{pmatrix} = \frac{1}{p}\left(A^\top A\right)^{-1} A^\top \sum_{j=1}^{p} Y^j. \tag{31}$$

The unbiasedness of the random vector $\hat{\theta}$ (30) follows from the equality

$$M\left(A^\top A\right)^{-1} A^\top Y = M\left(A^\top A\right)^{-1} A^\top Y^j = \theta, j = \overline{1, p}, \tag{32}$$

and the property of variances of its components from the equalities

$$\mathrm{Var}\left[\left(A^\top A\right)^{-1} A^\top \frac{1}{p} \sum_{j=1}^{p} Y^j\right] = \frac{1}{p^2} \sum_{j=1}^{p} \mathrm{Var}\left[\left(A^\top A\right)^{-1} A^\top Y^j\right],$$

$$\mathrm{Var}\left[\left(A^\top A\right)^{-1} A^\top Y^j\right] = \mathrm{Var}\left[\left(A^\top A\right)^{-1} A^\top Y\right], j = \overline{1, p}. \tag{33}$$

Corollary 1. Knowing the variance of the components of the estimates vector $\hat{\theta}$ (28), we can find such the number of repetitions p of the active experiment that the largest variance of the estimates $\hat{\theta}_0, \ldots, \hat{\theta}_m$ has an acceptable value.

Remark 13. The variances of the components of the vector $\hat{\theta}$ (28) are easily found because its components Y_i are virtual independent random variables, and $\mathrm{Var}(Y_i) = \sigma^2, i = \overline{1, n}$.

Corollary 2. We do not need to additionally compute the matrix inverse in the general formula of the LSM at arbitrary p, by virtue of (31).

We have used a partial case of the presented theoretical result in the method of estimating the linear terms of an MPR given by a redundant representation [21]. The software architecture presented in Sect. 5 includes algorithmic procedures for estimating all the coefficients of an MPR given by a redundant representation. It can be considered an original model of the software we are now designing. We publish it for the first time in this form.

5 Software Architecture

Figure 1 shows the architecture of a software that implements the mathematical methods for estimating the values of unknown coefficients of an MPR (its nonlinear and linear parts) described in this paper. It is presented in the form of a universal library with a high-level component-based software architecture.

In this library, we use a set of independent components for an MPR coefficients estimation. The components can be used to estimate unknown coefficients of other types of regressions, e.g., an MLR. The library itself is implemented as a separate component, but of a higher level, with the possibility of embedding in a third-party software.

We briefly describe below the components that make up the library:

- MPR is the main component, in fact the library itself, that implements the mathematical apparatus for an MPR coefficients estimation described in this work.
- RA_to_rational: provides the input data of the MPR problem in the form of rational fractions. This is necessary to ensure the calculation accuracy (eliminate rounding errors during calculations).
- RA_redundant_represent: builds a redundant representation for an MPR building problem.
- RA_calculation_MPR_evalt: finds estimates at nonlinear MPR terms implementing modified versions of the first and second sub-algorithms.
- MLR: finds coefficient estimates for linear parts of an MPR.
- RA_multivariate_linear: implements the LSM for finding estimates during multivariate linear regression analysis. To increase its performance, we use parallel calculations during the matrix multiplication and inversion operations in the general formula of the LSM [20].
- RA_cluster_analysis: significantly reduces the number of partial representations by splitting all the coefficients into two classes. The set of partial representations definitely contains the sought solution.
- RA_search_structure_MLR: finds the sought structure of the MLR (in interaction with RA_multivariate_linear component). To increase its performance, we use parallel calculations in the part of estimating the coefficients from the partial representations of MLR and further calculations of the LSM [20].
- RA_calculation_epsilon_evalt: finds estimates of the realizations of E.

- RA_chi_square: contains a modified χ^2 criterion to determine the reliability degree of the found estimates for an MLR.
- UPR: finds estimates of a UPR coefficients (is auxiliary for finding the coefficient estimates for the nonlinear part of an MPR).
- RA_Forsythe_uniform: implements an algorithm for finding the coefficients of NOPFs of a given degree with the required accuracy based on the use of symbolic calculations in a segment with a uniform distribution of a deterministic variable.
- RA_detx: finds the value of a deterministic variable x for the given starting and ending points of an active experiment.
- RA_detx_real_to_virtual: finds the value of a deterministic variable x for the given starting and ending points of a real active experiment based on the previously conducted virtual experiment.

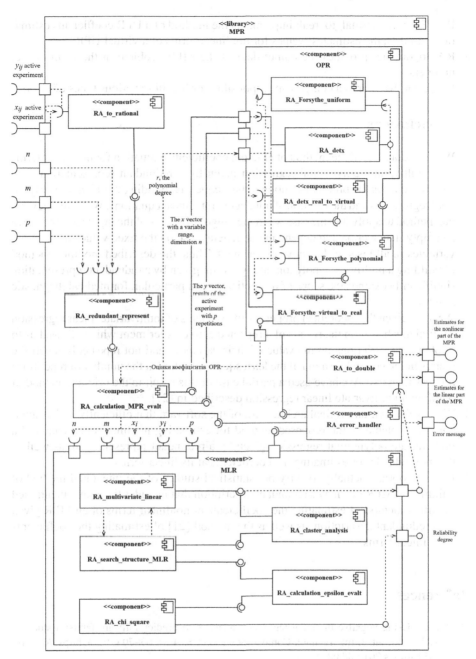

Fig. 1. The component-based software architecture of the MPR library

- RA_Forsythe_polynomial: uses NOPFs to implement the method for estimating the coefficients at nonlinear terms of a UPR using the output data of an active experiment.

- RA_Forsythe_virtual_to_real: implements the method of a UPR coefficients estimation based on the estimates' values for nonlinear terms of a virtual UPR.
- RA_to_double: provides the output data of the MPR problem in the form of real numbers.
- RA_error_handler is a component responsible for handling various types of errors.

6 Conclusions

1. We have considered the method of efficient coefficient estimation for nonlinear terms of a multivariate polynomial regression given by a redundant representation. The method reduces the problem solution to the sequential building of univariate polynomial regressions and the solution of nondegenerate linear equations. We have extended the method to apply it to the case when the ranges of values of the input variables have an empty intersection and depend in the general case on the fixed values of the input variables in all tests of the active experiment. Thus, the described modified method of building a multivariate polynomial regression given by a redundant representation significantly extends the scope of its application, in particular, for medical diagnostic systems.
2. We have given the theoretical properties of estimates of multivariate linear regression coefficients based on the results of a repeated active experiment which allowed us to implement restrictions on the value of their variances and not repeatedly invert the matrix in the general formula of the least squares method at the number of repetitions not less than two. We have used a partial case of this result in the original method of building a multivariate linear regression described in [21].
3. We have shown an original architecture of the software that efficiently and systematically implements the proposed method for estimating nonlinear coefficients of a multivariate polynomial regression given by a redundant representation, as well as the method [21] for estimating the coefficients at its linear terms.
4. It is appropriate actually to carry out statistical studies of the integrated method of building an MPR given by a redundant description that includes the above-mentioned modified method of estimating the coefficients at nonlinear terms of an MPR given by a redundant description, as well as the method [21] of estimating the coefficients at its linear terms.

References

1. Yu, L.: Using negative binomial regression analysis to predict software faults: a study of Apache Ant. Int. J. Inf. Technol. Comput. Sci. (IJITCS) **4**(8), 63–70 (2012). https://doi.org/10.5815/ijitcs.2012.08.08
2. Shahrel, M.Z., Mutalib, S., Abdul-Rahman, S.: PriceCop–price monitor and prediction using linear regression and LSVM-ABC methods for e-commerce platform. Int. J. Inf. Eng. Electron. Bus. (IJIEEB) **13**(1), 1–14 (2021). https://doi.org/10.5815/ijieeb.2021.01.01
3. Satter, A., Ibtehaz, N.: A regression based sensor data prediction technique to analyze data trustworthiness in cyber-physical system. Int. J. Inf. Eng. Electron. Bus. (IJIEEB) **10**(3), 15–22 (2018). https://doi.org/10.5815/ijieeb.2018.03.03

4. Isabona, J., Ojuh, D.O.: Machine learning based on kernel function controlled Gaussian process regression method for in-depth extrapolative analysis of Covid-19 daily cases drift rates. Int. J. Math. Sci. Comput. (IJMSC) **7**(2), 14–23 (2021). https://doi.org/10.5815/ijmsc. 2021.02.02

5. Sinha, P.: Multivariate polynomial regression in data mining: methodology, problems and solutions. Int. J. Sci. Eng. Res. **4**(12), 962–965 (2013)

6. Kalivas, J.H.: Interrelationships of multivariate regression methods using eigenvector basis sets. J. Chemom. **13**(2), 111–132 (1999). https://doi.org/10.1002/(SICI)1099-128X(199903/ 04)13:2%3c111::AID-CEM532%3e3.0.CO;2-N

7. Ortiz-Herrero, L., Maguregui, M.I., Bartolomé, L.: Multivariate (O)PLS regression methods in forensic dating. TrAC Trends Anal. Chem. **141**, 116278 (2021). https://doi.org/10.1016/j. trac.2021.116278

8. Guo, G., Niu, G., Shi, Q., Lin, Q., Tian, D., Duan, Y.: Multi-element quantitative analysis of soils by laser induced breakdown spectroscopy (LIBS) coupled with univariate and multivariate regression methods. Anal. Methods **11**(23), 3006–3013 (2019). https://doi.org/10.1039/ C9AY00890J

9. Nastenko, E.A, Pavlov, V.A, Boyko, A.L., Nosovets, O.K: Mnogokriterialnyi algoritm shagovoi regressii (Multi-criterion step-regression algorithm). Biomedychna inzheneriya i tekhnolohiya **3**, 48–53 (2020). https://doi.org/10.20535/2617-8974.2020.3.195661 (in Russian)

10. Sergeev, V.V., Kopenkov, V.N., Chernov, A.V.: Comparative analysis of function approximation methods in image processing tasks. Comput. Opt. **26**, 119–122 (2004). (in Russian)

11. Babatunde, G., Emmanuel, A.A., Oluwaseun, O.R., Bunmi, O.B., Precious, A.E.: Impact of climatic change on agricultural product yield using k-means and multiple linear regressions. Int. J. Educ. Manag. Eng. (IJEME) **9**(3), 16–26 (2019). https://doi.org/10.5815/ijeme.2019. 03.02

12. Bolshakov, A.A., Karimov, R.N.: Metody obrabotki mnogomernykh dannykh i vremennykh riadov (Methods of multidimensional data and time series processing). Goriachaia liniia–Telekom, Moscow (2007). (in Russian)

13. Zgurovsky, M.Z., Pavlov, A.A.: The Four-Level Model of Planning and Decision Making. In: Zgurovsky, M.Z., Pavlov, A.A. (eds.) Combinatorial Optimization Problems in Planning and Decision Making. SSDC, vol. 173, pp. 347–406. Springer, Cham (2019). https://doi.org/10. 1007/978-3-319-98977-8_8

14. Pavlov, A., Holovchenko, M., Mukha, I., Lishchuk, K.: Mathematics and software for building nonlinear polynomial regressions using estimates for univariate polynomial regressions coefficients with a given (small) variance. In: Hu Z., Dychka I., Petoukhov S., He M. (eds) Advances in Computer Science for Engineering and Education V. ICCSEEA 2022. Lecture Notes on Data Engineering and Communications Technologies, vol. 134, pp. 288–303 (2022). https://doi.org/10.1007/978-3-031-04812-8_25

15. Vaccari, D.A., Wang, H.K.: Multivariate polynomial regression for identification of chaotic time series. Math. Comput. Model. Dyn. Syst. **13**(4), 395–412 (2007). https://doi.org/10. 1080/13873950600883691

16. Ivahnenko, A.G.: Modelirovanie Slojnyh Sistem. Informacionnyi Podhod (Complex Systems Modeling. Informational Approach). Vyshcha shkola, Kyiv (1987). (in Russian)

17. Jackson, E.C., Hughes, J.A., Daley, M.: On the generalizability of linear and non-linear region of interest-based multivariate regression models for fMRI data. IEEE Conf. Comput. Intell. Bioinform. Comput. Biol. (CIBCB) **2018**, 1–8 (2018). https://doi.org/10.1109/CIBCB.2018. 8404973

18. Pavlov, A.A.: Estimating with a given accuracy of the coefficients at nonlinear terms of univariate polynomial regression using a small number of tests in an arbitrary limited active experiment. Bulletin of Natl. Tech. Univ. "KhPI". Seri.: Syst. Anal. Control Inf. Technol. **2**(6), 3–7 (2021). https://doi.org/10.20998/2079-0023.2021.02.01

19. Pavlov, A.A., Holovchenko, M.N., Drozd, V.V.: Construction of a multivariate polynomial given by a redundant description in stochastic and deterministic formulations using an active experiment. Bull. Natl. Tech. Univ. "KhPI". Ser.: Syst. Anal. Control Inf. Technol. **1**(7), 3–8 (2022). https://doi.org/10.20998/2079-0023.2022.01.01

20. Pavlov, A.A., Holovchenko, M.N., Drozd, V.V., Revych, M.M.: Doslidzhennya efektyvnosti metodu pobudovy bahatovymirnoyi liniynoyi rehresiyi, zadanoyi nadlyshkovym opysom (Дослідження ефективності методу побудови багатовимірної лінійної регресії, заданої надлишковим описом; Study of the efficiency of a method of building a multivariate linear regression given by a redundant representation). In: Materialy Vseukrayins'koyi naukovo-praktychnoyi konferentsiyi molodykh vchenykh ta studentiv "Inzheneriya prohramnoho zabezpechennya i peredovi informatsiyni tekhnolohiyi" (SoftTech-2022), Kyiv, 22–26 May and 22–25 October 2022, pp. 10–13. NTUU "KPI", Kyiv (2022). https://drive.goo gle.com/file/d/1CP9EaBTT_rJAXsINbanSVGnP2jkg9FJ0/view. Accessed 30 Dec 2022. (in Ukrainian)

21. Pavlov, A.A., Holovchenko, M.N.: Modified method of constructing a multivariate linear regression given by a redundant description. Bull. Natl. Tech. Univ. "KhPI". Ser.: Syst. Anal. Control Inf. Technol. **2**(8), 3–8 (2022). https://doi.org/10.20998/2079-0023.2022.02.01

Research and Development of Bilinear QoS Routing Model over Disjoint Paths with Bandwidth Guarantees in SDN

Oleksandr Lemeshko ⓘ, Oleksandra Yeremenko(✉) ⓘ, Maryna Yevdokymenko ⓘ, and Batoul Sleiman ⓘ

Kharkiv National University of Radio Electronics, Nauky Ave. 14, Kharkiv 61166, Ukraine
{oleksandr.lemeshko.ua,oleksandra.yeremenko.ua,
maryna.yevdokymenko}@ieee.org

Abstract. The work proposes developing a bilinear Quality of Service routing model over disjoint paths with bandwidth guarantees in the Software-Defined Network data plane. The advantage of the solution is formulating the routing problem as optimization under bilinear conditions aimed at bandwidth guarantee over the set of disjoint paths (multipath). Solving the routing problem allows obtaining disjoint routes by Boolean variables calculation responsible for the link belonging to the collection of network paths. Also, the model introduces conditions of the disjoint routes connectivity and the intersection absence. The model's novelty is the bilinear conditions for a given Quality of Service level guarantee. Fulfilling such bilinear requirements provides that the total bandwidth of the disjoint routing multipath will not be lower than the demanded level. Also, the optimality criterion is defined by the Quality of Service requirements. A numerical study confirmed the improved model's workability in providing a set level of obtained disjoint paths bandwidth.

Keywords: Multipath routing · Disjoint Paths · Bandwidth · Load Balancing · QoS · Software-Defined Network

1 Introduction

The significant interest in Software-Defined Networks (SDN) is due to many reasons and advantages compared to traditional networks. The use of softwarized networking approaches makes it possible to make traffic management more flexible and use optimization techniques for different purposes [1–3]. The practical application of SDN solutions is directly related to increasing the efficiency of providing a given level of Quality of Service (QoS) [4]. Since the control plane is separated from the data plane, containing network equipment, the so-called router's programming based on table-based forwarding becomes possible [1–5].

Due to various reasons and requirements for Quality of Service, fault tolerance, and security, different types of SDN architectures are used [3–8]. Figure 1 represents the primary differences between conventional (traditional) networks and SDN [1, 3]. At the

© The Author(s), under exclusive license to Springer Nature Switzerland AG 2023
Z. Hu et al. (Eds.): ICCSEEA 2023, LNDECT 181, pp. 223–235, 2023.
https://doi.org/10.1007/978-3-031-36118-0_20

same time, one should consider the existing deployment of communication networks, which include both SDN devices and traditional network equipment. In this way, hybrid SDNs are organized [3, 5].

It is noted that appropriate routing tools are actively used to ensure the Quality of Service, network security, and fault tolerance under established requirements [9–12]. One of the well-known directions is disjoint routing solutions [13]. Therefore, the presented work is devoted to improving the mathematical model of routing by disjoint paths, which can serve as a basis for QoS routing protocols in the SDN data plane.

The paper structure is as follows:

1. Section 2 is devoted to related work analysis in the field of disjoint routes calculation and its application to QoS routing in SDN.
2. Section 3 describes the mathematical model of disjoint paths calculation that provides multipath guaranteed bandwidth.
3. Section 4 contains the numerical research results of the developed model.
4. Section 5 discusses the proposed QoS routing model.
5. Section 6 concludes the work and suggests further development directions in advanced performance-based disjoint QoS routing.

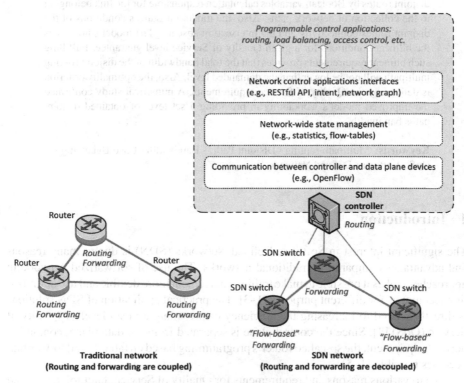

Fig. 1. Difference between traditional networks and SDN [1, 3]

2   Related Work Analysis

The analysis [10–24] shows that multipath routing over disjoint paths is an effective solution for better resource allocation, scalability, network resilience, and security. Moreover, such an approach for multipath routing is often utilized for Quality of Service and Quality of Experience improvement. Consider in more detail the results of some recent research in this area.

Thus, in [14] the need to develop search algorithms for multiple disjoint paths is substantiated. Furthermore, the corresponding algorithms must satisfy the demands of computational complexity to achieve scalability. Having such a set of routes adds more value to the multipath. In addition, it is noted that a set consisting of only two disjoint paths is usually used. This approach is widely used in fault-tolerant routing, where one route is used as the primary and the other as a backup. Thus, in [14] an efficient algorithm for calculating multiple disjoint paths with acceptable computational cost and ensuring scalability is proposed and investigated. Moreover, in [15], the authors propose One-Shot Multiple Disjoint Path Discovery Protocol (1S-MDP) that can provide both node-disjoint or link-disjoint paths.

The widespread use and application of disjoint paths routing are found in Mobile Ad-hoc Networks (MANETs). The specifics of this type of network impose certain limitations when developing routing algorithms. First, you should consider the dynamics of the MANET topology, error-less data broadcasting, and the need to ensure a high level of fault tolerance and security. For example, in [16], a link-disjoint multipath routing method was proposed to choose the shortest path from multiple paths in MANET. Moreover, the simulation results proved the possibility of using and efficiency in the traffic load of the proposed method in a dynamic environment.

Also, many current research works relate to improving the Quality of Service during QoS routing and implementing the strategy of disjoint multipath routing [17–22], including applying the Traffic Engineering concept principles [19]. Additionally, in [20], the authors consider adaptive multipath routing over both shortest and non-shortest disjoint paths.

While in [22], a novel multipath transport scheme for real-time multimedia using disjoint multipath and segment routing in Software-Defined Networks is proposed. It is noted that satisfactory Quality of Experience (QoE) level currently remains an urgent task. Thus, in the solution [22], the SDN controller centrally calculates multiple disjoint paths meeting bandwidth requirements and load balancing in the network. In this case, sub-flows are transmitted over disjoint paths to reduce the end-to-end delay and improve QoE.

Therefore, the basic model proposed in [10] is further used to modify and develop an improved model of disjoint multipath routing to improve the overall Quality of Service level by guaranteeing bandwidth for multipath flow transmission in application to SDN.

In contrast to the solutions presented in previous studies, primary attention is paid to the task of disjoint paths set calculation in this work. The load balancing processes in the SDN data plane are not explicitly considered, but the multipath routing strategy is used. The problem of calculating disjoint paths in the proposed mathematical model is presented in an optimization form. Mixed integer linear and nonlinear programming methods are used to solve the optimization problem formulated in work.

3 Mathematical Model of Disjoint Paths Calculation that Provides Multipath Guaranteed Bandwidth

Work [10] presents a model we will modify further and calculate the maximum number of disjoint paths and their bandwidth. In the modified model, the network structure and its functional parameters we described as follows:

$G = (R, E)$	graph presenting the network structure, in which $R = \{R_i;\ i = \overline{1, m}\}$ is a set of vertices simulating routers, and $E = \{E_{i,j};\ i, j = \overline{1, m};\ i \neq j\}$ is a set of arcs representing links;
s_k	source node;
d_k	destination node;
K	set of flows for transmitting in the network ($k \in K$);
$a_{i,j}^k$	control variables, each of which determines whether the link $E_{i,j} \in E$ belongs to the set of calculated disjoint paths for transmission of the k th flow;
$\varphi_{i,j}$	link $E_{i,j} \in E$ capacity (measured in packets per second, pps);
M^k	integer variable characterizing the number of disjoint paths;
$w_{i,j}$	weighting coefficients related to the capacity of a link $E_{i,j} \in E$

As a result of solving the problem of calculating the maximum number of disjoint paths for transmission of the k th flow, it is necessary to calculate the set of variables $a_{i,j}^k$, where the set of calculated disjoint paths corresponds to maximum multipath bandwidth.

The routing variables $a_{i,j}^k$ have the constraints of type:

$$a_{i,j}^k \in \{0; 1\}. \tag{1}$$

In addition, the following conditions for every distinct pair of source and destination nodes must be fulfilled [10]:

$$\sum_{j:E_{i,j}\in E} a_{i,j}^k = M^k;\ k \in K,\quad R_i = s_k; \tag{2}$$

$$\sum_{j:E_{j,i}\in E} a_{j,i}^k = M^k;\ k \in K,\quad R_i = d_k \tag{3}$$

At the same time, for the transit nodes in the network ($R_i \neq s_k, d_k$), the following restrictions are introduced in the basic model [10]:

$$\begin{cases} \sum_{j:E_{i,j}\in E} a_{i,j}^k \leq 1,\ k \in K; \\[2mm] \sum_{j:E_{j,i}\in E} a_{j,i}^k \leq 1,\ k \in K; \\[2mm] \sum_{j:E_{i,j}\in E} a_{i,j}^k - \sum_{j:E_{j,i}\in E} a_{j,i}^k = 0,\ k \in K. \end{cases} \tag{4}$$

The physical meaning of the first condition in the system (4) is that no more than one path can outgo from the transit node R_i. Fulfillment of the second condition in the system (4) must ensure that the transit node R_i cannot belong to more than one path from the calculated set. Implementing the third condition from the system (4) is responsible for the fact that the path can outgo from the transit router R_i only if it has income previously to this node. For the telecommunication network in general, the implementation of the restrictions system (4) must guarantee compliance with the following conditions:

1. Involved communication links ensure the connectivity of each specific calculated path.
2. Calculated paths will not intersect, i.e., they can share only source s_k and destination d_k nodes.

However, in any case, the following condition takes place:

$$M^k \geq 1. \tag{5}$$

In the general case, the range of available values M^k directly depends on the network topology, the network nodes' connectivity level, and the degree of the graph G vertices, which simulate the source and destination routers.

The basic mathematical model (1)–(5) can also be modified in QoS routing to provide maximum or specified bandwidth using the calculated set of disjoint paths. Consequently, within the frame of the basic model, it is necessary to introduce additional conditions to ensure a given Quality of Service level relative to bandwidth. Therefore, we denote by β_{path}^k the minimum bound value for the bandwidth of any of the disjoint paths set used to transmit the kth packet flow. Then the following condition can be introduced into the framework of the routing model (by analogy with [10]):

$$a_{i,j}^k \varphi_{i,j} + W(1 - a_{i,j}^k) \geq \beta_{path}^k, \tag{6}$$

where the weighting coefficients W take values higher than the maximum bandwidth of the communication network links, fulfillment of condition (6) ensures that each route belonging to the set of disjoint paths calculated for the kth flow has a bandwidth of not less than β_{path}^k.

Next, we should choose the following objective function J for maximization as the optimality criterion for the QoS-routing over disjoint paths problem solution:

$$J = c_M M^k + c_\beta \beta_{path}^k - c_v \sum_{E_{i,j} \in E} v_{i,j} a_{i,j}^k, \tag{7}$$

where weighting coefficients c_M, c_β, and c_v determine the importance of each component in expression (7).

Introducing the first term in the objective function (7) is associated with maximizing the number of disjoint paths used. The second term is responsible for maximizing the lower bound bandwidth of the calculated paths. If we limit ourselves to using only these two terms, then, on the one hand, the least efficient path will have a bandwidth equal to the value β_{path}^k. However, such a solution may not always contribute to the inclusion in the calculated set of the highest bandwidth routes. Therefore, the novelty of the proposed model is the use of the third term in criterion (7), which is introduced by analogy with the metrics of the OSPF and EIGRP protocols to include in the calculated disjoint paths high-bandwidth links. Thus, it is proposed that in the objective function (7) the weighting coefficients $v_{i,j}$, taking into account the bandwidth $\varphi_{i,j}$ of the corresponding communication link $E_{i,j} \in E$, are defined as:

$$v_{i,j} = 10/\varphi_{i,j}. \tag{8}$$

It is experimentally established that to implement QoS routing with the calculation of the maximum number of disjoint paths with maximum bandwidth, the weighting coefficients in expression (7) must meet the following condition:

$$M >> c_\beta >> c_v. \tag{9}$$

The peculiarity of using the proposed mathematical model (1)–(9) is that it focuses on choosing paths in the network that do not intersect and provide maximum bandwidth. Increasing the bandwidth allocated to a particular flow allows for improving average delay and packet loss. However, in practice, to save network resources when servicing the traffic of network users, there are requirements for the boundary value of the network bandwidth available for use by a particular flow. Let us denote, for example, for the kth packet flow this boundary value through β^k. Therefore, in the model (1)–(9), it is proposed to introduce the following condition:

$$M^k \beta_{path}^k \geq \beta^k. \tag{10}$$

Therefore, in the general case, the left part of inequality (10) is a bilinear form of two types of control variables M^k and β_{path}^k that characterize the lower bound of bandwidth, which in total provides the use of the calculated paths. The bound is lowest, as each of these disjoint paths, according to conditions (6), has a capacity not lower but may be higher β_{path}^k. Fulfillment of condition (10), depending on the form of the selected optimality criterion, can be achieved either by increasing the number of involved disjoint routes M^k or by raising the boundary value relative to their minimum bandwidth β_{path}^k.

Given the use of the optimality criterion of routing solutions (7), the priority of increasing the control variables that are included in the bilinear form in (10) will be determined by the hierarchy of values of weights in (7). In the case of introducing conditions (10), the optimization problem with criterion (7) and the supplemented set of constraints (1)–(6), (8)–(10) will belong to the class of problems of mixed-integer nonlinear programming (MINLP). However, for example, if the number of paths to be calculated is known in advance, i.e., in criterion (7), the first term also becomes a constant, condition (10) becomes linear, and the optimization problem remains of the MILP class.

4   Numerical Research

Consider several numerical examples for calculating disjoint paths and providing a given bandwidth on the network structure presented in Fig. 2. Communication link gaps for this network structure (Fig. 2) indicate their bandwidth (pps). Additionally, Table 1 shows the set of available paths and their bandwidth.

Then Table 2 shows the calculations to determine for one flow transmitted in the network from the first to the ninth router (Fig. 2) a set of disjoint paths and guarantees the lower bound of total bandwidth.

Table 2 demonstrates possible solutions to the QoS-routing problem using disjoint paths that provide a given bandwidth. Depending on the ratio of weighting coefficients M, c_β, and c_v in (7), the use of the model (1)–(6), (10) allows for obtaining a different set of paths.

If condition (9) was satisfied, then the fulfillment of condition (10) was achieved, as a rule, based on increasing the number of involved disjoint routes M^k. However, as shown in Table 2, it is better to use the network resource economically to satisfy the constraint $M \gg c_\beta \gg c_v$. Then condition (10) was fulfilled based on raising the boundary value relative to the minimum bandwidth (β_{path}^k) of the calculated paths. If ρ^k is equal to 1200 pps, then in Table 2, only the solutions at $M^k = 3$ (third row) satisfy (Fig. 3).

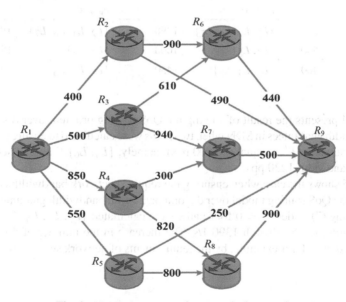

Fig. 2. Network structure for numerical research.

The introduction of the third term in the objective function (7) under $c_v \neq 0$ allowed for including more productive communication links in the set of calculated paths. Such a result was accompanied by an increase in the total bandwidth of disjoint paths used in QoS routing (Table 2, Fig. 3).

Table 1. Characteristics of available disjoint paths for the network structure (Fig. 2).

Path #	Designation	The set of links that form the path	Path Bandwidth, pps
1	L_1	$\{E_{1,2}, E_{2,6}, E_{6,9}\}$	400
2	L_2	$\{E_{1,2}, E_{2,9}\}$	400
3	L_3	$\{E_{1,3}, E_{3,6}, E_{6,9}\}$	440
4	L_4	$\{E_{1,3}, E_{3,7}, E_{7,9}\}$	500
5	L_5	$\{E_{1,4}, E_{4,7}, E_{7,9}\}$	300
6	L_6	$\{E_{1,4}, E_{4,8}, E_{8,9}\}$	820
7	L_7	$\{E_{1,5}, E_{5,9}\}$	250
8	L_8	$\{E_{1,5}, E_{5,8}, E_{8,9}\}$	550

Table 2. The results of calculations to determine the set of disjoint paths and provide guarantees for their lower bound of total bandwidth (k = 1).

β^k, pps	M^k	β^k_{path}, pps	The set of paths that are calculated using (7)	Total bandwidth, pps	The set of paths that are calculated using (7)	Total bandwidth, pps
			$c_v = 0$		$c_v \neq 0$	
1000	4	250	$\{L_2, L_3, L_5, L_7\}$	1390	$\{L_2, L_4, L_6, L_7\}$	1970
	2	500	$\{L_4, L_8\}$	1000	$\{L_4, L_6\}$	1220
	3	400	$\{L_2, L_4, L_8\}$	1450	$\{L_2, L_4, L_6\}$	1720

Figure 4 presents the result of solving the QoS routing problem over disjoint paths with bandwidth guarantees in SDN when two paths were calculated to ensure the required network bandwidth (β^k is equal to 1000 pps), namely, $\{L_4, L_6\}$ (Table 2, second row) with total bandwidth 1220 pps.

Figure 5 shows the case when ensuring the required network bandwidth ($\beta^k = 1000$ pps) using the QoS routing model over disjoint paths with bandwidth guarantees in SDN (1)–(10) using (7) under $c_v = 0$ four paths were calculated $\{L_2, L_3, L_5, L_7\}$ (Table 2, first row) with total bandwidth 1390 1/s. The increase in the number of disjoint paths used can be dictated, for example, by the requirements of network security policies [14].

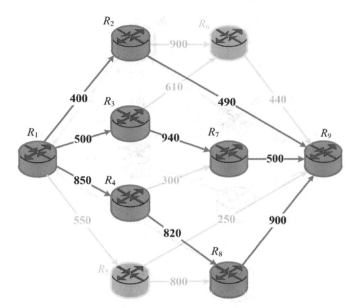

Fig. 3. The set of paths $\{L_2, L_4, L_6\}$ calculated using (7) under $c_v \neq 0$ with total bandwidth 1720 pps ($M^k = 3$).

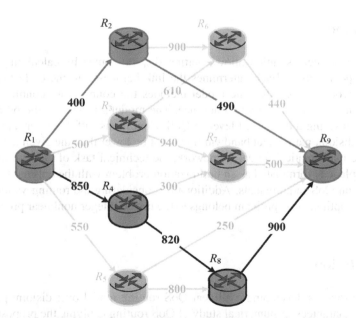

Fig. 4. The set of paths $\{L_4, L_6\}$ calculated using (7) under $c_v \neq 0$ with total bandwidth of 1220 pps ($M^k = 2$).

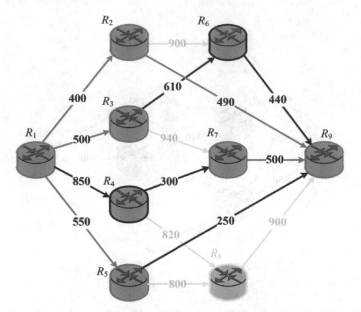

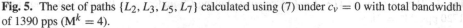

Fig. 5. The set of paths $\{L_2, L_3, L_5, L_7\}$ calculated using (7) under $c_v = 0$ with total bandwidth of 1390 pps ($M^k = 4$).

5   Discussion

The routing problem solution allows getting disjoint routes by calculating a set of Boolean-type variables. Each determines the link belonging to the collection of calculated network routes. Also, the model includes the connectivity conditions of the obtained paths and the intersection absence. The model's novelty is the bilinear conditions for ensuring a demanded level of QoS. Conditions fulfillment ensures that the calculated disjoint paths' total bandwidth will not be lower than the given level. Therefore, in the mathematical model framework, the technical task of QoS routing in the SDN data plane is formulated as an optimization problem with the optimality criterion depending on QoS requirements. Additionally, considering the routing variables constraints, the optimization problem belongs to the mixed integer nonlinear programming class.

6   Conclusion

The work proposes developing a bilinear QoS routing model over disjoint paths with bandwidth guarantees. A numerical study of QoS routing applying the proposed model on a set of examples proved the improved model's workability and effectiveness in providing a demanded level of obtained disjoint paths bandwidth.

The basic mathematical routing model (1)–(6) is proposed for further improvement. Thus, improved model (1)–(10) allows for calculating the corresponding multipath, namely, a set of disjoint paths under Quality of Service conditions. A graph in the model

(1)–(6) describes the network structure where the vertices set represents network routers, and the arcs set represents communication links.

The application of the routing model proposed in the work is, first of all, softwarized telecommunication networks (SDN data plane) where the functions of route calculation are assigned to SDN controllers. The linear nature of the expressions that form the basis of the developed QoS routing model, criteria, and limitations of the formulated optimization problem regarding the calculation of disjoint paths should contribute to the acceptable computational complexity. Therefore, their technological implementation in SDN could be the basis of the algorithmic and software background of promising QoS routing protocols.

We see further development of an advanced performance-based disjoint path QoS routing model in implementing fault-tolerant and secure routing. Moreover, complementary protection schemes will support additional QoS indicators such as bandwidth, average delay, packet loss probability, and network security parameters, which are crucial when transmitting mission-critical applications' traffic.

References

1. Ahmad, S., Mir, A.H.: SDN interfaces: protocols, taxonomy and challenges. Int. J. Wirel. Microwave Technol. (IJWMT) **12**(2), 11–32 (2022). https://doi.org/10.5815/ijwmt.2022. 02.02

2. Barabash, O., Kravchenko, Y., Mukhin, V., Kornaga, Y., Leshchenko, O.: Optimization of parameters at SDN technologie networks. Int. J. Intell. Syst. Appl. **9**(9), 1–9 (2017). https://doi.org/10.5815/ijisa.2017.09.01

3. Al Mtawa, Y., Haque, A., Lutfiyya, H.: Migrating from legacy to software defined networks: a network reliability perspective. IEEE Trans. Reliab. **70**(4), 1525–1541 (2021). https://doi.org/10.1109/TR.2021.3066526

4. Islam, N., Rahman, H., Nasir, M.K.: A comprehensive analysis of QoS in wired and wireless SDN based on mobile IP. Int. J. Comput. Netw. Inf. Secur. **13**(5), 18–28 (2021). https://doi.org/10.5815/ijcnis.2021.05.02

5. Rohitaksha, K., Rajendra, A.B., Mohan, J.: Routing in hybrid software defined networking using hop-count approach. Int. J. Wirel. Microwave Technol. (IJWMT) **12**(3), 54–60 (2022). https://doi.org/10.5815/ijwmt.2022.03.04

6. Strelkovskaya, I., Solovskaya, I., Makoganiuk, A.: Predicting self-similar traffic using cubic B-splines. In: 2019 3rd International Conference on Advanced Information and Communications Technologies (AICT) Proceedings, pp. 153–156. IEEE (2019). https://doi.org/10.1109/AIACT.2019.8847761

7. Bielievtsov, S., Ruban, I., Smelyakov, K., Sumtsov, D.: Network technology for transmission of visual information. selected papers of the XVIII international scientific and practical conference "information technologies and security" (ITS 2018). In: CEUR Workshop Proceedings, vol. 2318, pp. 160–175 (2018). https://ceur-ws.org/Vol-2318/paper14.pdf

8. Smelyakov, K., Smelyakov, S., Chupryna, A.: Adaptive edge detection models and algorithms. In: Mashtalir, V., Ruban, I., Levashenko, V. (eds.) Advances in Spatio-Temporal Segmentation of Visual Data. Studies in Computational Intelligence, vol. 876, pp. 1–51. Springer, Cham (2020). https://doi.org/10.1007/978-3-030-35480-0_1

9. Lemeshko, O., Hu, Z., Shapovalova, A., Yeremenko, O., Yevdokymenko, M.: Research of the influence of compromise probability in secure based traffic engineering model in SDN. In: Hu, Z., Petoukhov, S., Dychka, I., He, M. (eds.) Advances in Computer Science for Engineering and Education IV. ICCSEEA 2021. Lecture Notes on Data Engineering and Communications Technologies, vol. 83, pp. 47–55. Springer, Cham (2021). https://doi.org/10.1007/978-3-030-80472-5_5

10. Lemeshko, O., Yeremenko, O., Yevdokymenko, M., Sleiman, B.: System of solutions the maximum number of disjoint paths computation under quality of service and security parameters. In: Ilchenko, M., Uryvsky, L., Globa, L. (eds.) Advances in Information and Communication Technology and Systems. MCT 2019. Lecture Notes in Networks and Systems, vol. 152, pp. 191–205. Springer, Cham (2021). https://doi.org/10.1007/978-3-030-58359-0_10

11. Lemeshko, O., Yeremenko, O., Yevdokymenko, M., Shapovalova, A., Hailan, A.M., Mersni, A.: Cyber resilience approach based on traffic engineering fast reroute with policing. In: 2019 10th IEEE International Conference on Intelligent Data Acquisition and Advanced Computing Systems: Technology and Applications (IDAACS) Proceedings, vol. 1, pp. 117–122. IEEE (2019). https://doi.org/10.1109/IDAACS.2019.8924294

12. Lemeshko, O., Yevdokymenko, M., Yeremenko, O., Hailan, A.M., Segeč, P., Papán, J.: Design of the fast reroute QoS protection scheme for bandwidth and probability of packet loss in software-defined WAN. In: 2019 IEEE 15th International Conference on the Experience of Designing and Application of CAD Systems (CADSM), pp. 1–5. IEEE (2019). https://doi.org/10.1109/CADSM.2019.8779321

13. Gomes, T., Jorge, L., Girão-Silva, R., Yallouz, J., Babarczi, P., Rak, J.: Fundamental schemes to determine disjoint paths for multiple failure scenarios. In: Rak, J., Hutchison, D. (eds.) Guide to Disaster-Resilient Communication Networks. Computer Communications and Networks, pp. 429–453. Springer, Cham (2020). https://doi.org/10.1007/978-3-030-44685-7_17

14. Lopez-Pajares, D., Rojas, E., Carral, J.A., Martinez-Yelmo, I., Alvarez-Horcajo, J.: The disjoint multipath challenge: multiple disjoint paths guaranteeing scalability. IEEE Access 9, 74422–74436 (2021). https://doi.org/10.1109/ACCESS.2021.3080931

15. Lopez-Pajares, D., Alvarez-Horcajo, J., Rojas, E., Carral, J.A., Martinez-Yelmo, I.: One-shot multiple disjoint path discovery protocol (1S-MDP). IEEE Communications Letters 24(8), 1660–1663 (2020). https://doi.org/10.1109/LCOMM.2020.2990885

16. Robinson, Y.H., Julie, E.G., Saravanan, K., Kumar, R., Abdel-Basset, M., Thong, P.H.: Link-disjoint multipath routing for network traffic overload handling in mobile ad-hoc networks. IEEE Access 7, 143312–143323 (2019). https://doi.org/10.1109/ACCESS.2019.2943145

17. Sreeram, K., Unnisa, A. and Poornima, V.: QoS aware multi-constrained node disjoint multi-path routing for wireless sensor networks. In: 2019 5th International Conference on Advanced Computing & Communication Systems (ICACCS) Proceedings, pp. 382–385. IEEE (2019). https://doi.org/10.1109/ICACCS.2019.8728475

18. Sarma, H.K.D., Dutta, M.P., Dutta, M.P.: A quality of service aware routing protocol for mesh networks based on congestion prediction. In: 2019 International Conference on Information Technology (ICIT) Proceedings, pp. 430–435. IEEE (2019). https://doi.org/10.1109/ICIT48102.2019.00082

19. Zhang, C., et al.: Scalable traffic engineering for higher throughput in heavily-loaded software defined networks. In: NOMS 2020–2020 IEEE/IFIP Network Operations and Management Symposium Proceedings, pp. 1–7. IEEE (2020). https://doi.org/10.1109/NOMS47738.2020.9110259

20. Besta, M., et al.: High-performance routing with multipathing and path diversity in ethernet and hpc networks. IEEE Trans. Parallel Distrib. Syst. 32(4), 943–959 (2020). https://doi.org/10.1109/TPDS.2020.3035761

21. Hou, A., Wu, C.Q., Zuo, L., Zhang, X., Wang, T., Fang, D.: Bandwidth scheduling for big data transfer with two variable node-disjoint paths. J. Commun. Netw. **22**(2), 130–144 (2020). https://doi.org/10.1109/JCN.2020.000004
22. Zhang, W., Lei, W., Zhang, S.: A multipath transport scheme for real-time multimedia services based on software-defined networking and segment routing. IEEE Access **8**, 93962–93977 (2020). https://doi.org/10.1109/ACCESS.2020.2994346
23. Zeng, G., Zhan, Y., Pan, X.: Failure-tolerant and low-latency telecommand in mega-constellations: the redundant multi-path routing. IEEE Access **9**, 34975–34985 (2021). https://doi.org/10.1109/ACCESS.2021.3061736
24. Kaneko, K., Van Nguyen, S., Binh, H.T.T.: Pairwise disjoint paths routing in Tori. IEEE Access **8**, 192206–192217 (2020). https://doi.org/10.1109/ACCESS.2020.3032684

Improved GL-Model of Behavior of Complex Multiprocessor Systems in Failure Flow

Alexei Romankevitch[1], Kostiantyn Morozov[1(✉)], Vitaliy Romankevich[1], Daniil Halytskyi[1], and Zacharioudakis Eleftherios[2]

[1] National Technical University of Ukraine "Igor Sikorsky Kyiv Polytechnic Institute", 37 Peremohy Avenue, Kyiv 03056, Ukraine
romankev@scs.kpi.ua, mcng@ukr.net

[2] Neapolis University Paphos, 2 Danais Avenue, 8042 Paphos, Cyprus

Abstract. GL-models can be used as behavioral models of fault-tolerant multiprocessor systems in the flow of failures, to evaluate their reliability parameters by conducting statistical experiments. Estimation of the reliability parameters of fault-tolerant multiprocessor systems is an extremely complex and important theoretical and practical problem. To build models of complex systems, in particular those consisting of several subsystems, simpler basic GL-models can be combined. To build models of systems, between processors of different subsystems of which task exchange is possible for the purpose of restoring performance, a so-called dual model can be applied, built by replacing the edge functions of the original GL-model with corresponding dual functions. In addition, the work describes a special auxiliary model that reflects the limitation of the volume of tasks that can be transferred between subsystems.

Keywords: GL-model · fault-tolerant multiprocessor system · dual function

1 Introduction

In modern world, automation is becoming more and more important. This trend has several reasons. Yes, on the one hand, it frees a person from monotonous and uncreative work. On the other hand, it allows you to reduce the probability of errors caused by the human factor, as well as bypass the physical limitations of the human body, such as reaction speed, amount of calculation capabilities, etc.

To ensure automatic and automated functioning of various systems and complexes, so-called control systems (CS) are used [1, 2]. Today such systems are often created based on microprocessors (MP). However, if one MP is usually enough to build a simple CS, then more complex and responsible systems may require the use of many processors.

This especially applies to the so-called mission critical systems (MCS) [3], the failure of which can lead to significant material losses, threaten people's lives and health, etc. In particular, such systems often include energy facilities (nuclear power plants, thermal power plants, hydroelectric power plants), transport systems (aircraft, trains, railway hubs), large industrial enterprises, space vehicles and systems, complex military

Z. Hu et al. (Eds.): ICCSEEA 2023, LNDECT 181, pp. 236–245, 2023.
https://doi.org/10.1007/978-3-031-36118-0_21

equipment. Recently, this is beginning to apply to more commonly used objects, in particular, such as automatically piloted cars. The control system for such objects should usually be very reliable. In addition, such systems often must perform a large number of calculations per unit of time. Therefore, the developer of such an CS must solve the problem of ensuring both a high level of its reliability and a high level of its productivity.

Both problems can be solved in the case of building a CS based on the so-called fault-tolerant multiprocessor systems (FTMS) [4–6]. FTMS can continue full functioning in the event of failure of a certain number or subset of its processors. Such systems may include processors of various types with a complex system of connections between them.

The FTMS developer faces the task of assessing the reliability parameters of the developed system [7, 8], in particular, the probability of its trouble-free operation. The calculation of these values can occur in different ways, which can be divided into two categories [9].

The first category includes approaches based on the construction of certain analytical expressions for the sought parameters of system reliability. This allows you to get their values with great accuracy. However, for complex and heterogeneous systems, the formation of such expressions becomes a difficult task. For each of the systems, it may be necessary to develop a separate specialized method. That is, approaches from this category are usually non-universal [10–14].

Approaches from the second category make it possible to evaluate the reliability parameters of the system with a given accuracy, by performing statistical experiments with models of its behavior in the flow of failures [15]. Such a model takes as input the so-called system state vector and returns a certain value characterizing the state of the system as a whole, which corresponds to it (for example, 1 – the system is operational and 0 – the system has failed). The system state vector is a vector whose elements correspond to the state of each of the system's processors (for example, 1 – the processor is operational, 0 – the processor has failed).

By applying the methods of mathematical statistics to the values obtained with the help of such models on sets of random system state vectors, an estimate of its reliability parameters can be obtained with a certain accuracy. Accuracy can be increased, in particular, by increasing the number of analyzed vectors, and/or by using specialized procedures for their formation.

2 GL-Models and Their Properties

The so-called GL-models [16], which combine the properties of graphs and Boolean functions, can be used as behavioral models of FTMS in the flow of failures. Such a model is an undirected graph, each edge of which corresponds to a certain Boolean edge function that depends on all or some of the elements of the system state vector. If the edge function takes a zero value on a certain vector, the corresponding edge is removed from the graph. The connectivity of the graph corresponds to the state of the system: if the graph remains connected, the system is operational, and if not, it fails.

A number of methods can be used to build GL-models of different types of systems. In particular, the method described in [17] allows building models based on cyclic graphs. In addition, these models have a number of additional properties.

A system that remains operational when any m of its n processors fail and fails in the event of a failure of a larger number of processors (any combination of them), as well as the GL-model corresponding to it, are called basic and denoted by $K(m, n)$.

Model $K(m, n)$, built according to [17], will have exactly $\varphi(m, n)$ edges, where

$$\varphi(m, n) = n - m + 1. \tag{1}$$

In addition, on vectors with l zeros it will lose exactly $\psi(m, l)$ edges, where

$$\psi(m, l) = \begin{cases} 0, & \text{if } l < m \\ l - m + 1, & \text{if } l \geq m \end{cases}. \tag{2}$$

Such models are called MLE-models (minimum of lost edges), because on vectors with $m + 1$ zeros, they always lose a minimum sufficient number of edges, in other words, 2 edges.

Unfortunately, real systems are not always basic. In particular, this may apply to so called multi-module systems consisting of several subsystems. In some cases, processors of one subsystem can perform part of the tasks of another subsystem. Methods of building GL-models for some of these subsystems are proposed in [18, 19]. However, for cases of more complex interaction between subsystems, these methods may not be effective enough.

3 Dual GL-Model

Consider a multimodule system in which the processors of one subsystem can be used to perform the tasks of others. Let the subsystem contain n processors and be resistant to failure of m of them. In other words, $n - m$ capable processors are enough to perform tasks of the subsystem. The rest of the processors can be used to perform tasks of other subsystems.

To build a GL-model of the system, it will be useful to be able to obtain a vector containing exactly as many ones as the processors of the subsystem can be involved in solving external problems. For the aforementioned subsystem, this number can be calculated using the formula:

$$\mu(m, l) = \begin{cases} 0, & \text{if } l \geq m \\ m - l, & \text{if } l < m \end{cases}. \tag{3}$$

Let the model $K'(m, n)$ contain edge functions that are dual with respect to the functions of the MLE-model $K(m, n)$. We will call such a model a *dual* model. Using de Morgan's rules, it is easy to show that the expressions of the edge functions of the dual model can be obtained from the expressions of the edge functions of the basic model by replacing conjunctions with disjunctions, and disjunctions with conjunctions. It is obvious that in the dual model $K'(m, n)$ on vectors with l ones exactly $\psi(m, l)$ functions will take a value equal to one.

Suppose that in a basic m-fault-tolerant system consisting of n processors, l of them failed. Accordingly, the number of working processors is $n - l$. In the dual model $K'(n -$

$m + 1, n)$, which accepts as an input the system state vector of exactly $\psi(n - m + 1, n - l)$ functions will take a value equal to one. According to (2) and (3) and taking into account that n, m and l are integers, we can conclude that $\psi(n - m + 1, n - l) \equiv \mu(m, l)$. Thus, the vector consisting of the values of the edge functions of the dual model $K'(n - m + 1, n)$ will contain exactly $\mu(m, l)$ ones. In addition, according to (1), such a vector will have a length equal to m. We will call such a model a *redundancy model* and, to simplify notation, denote it as $R(m, n)$, i.e.,

$$R(m, n) \triangleq K'(n - m + 1, n) \tag{4}$$

4 Use of the Dual GL-Model

4.1 Model of a Pair of a "Donor" - "Recipient" Subsystems

Consider the situation when "redundant" processors of one subsystem can be used to restore the performance of another subsystem. That is, we can say that the first subsystem is a "donor", and the second – a "recipient".

Let the first of the subsystems contain n_1 processors and be resistant to the failure of any m_1 of them, and the second – n_2 processors and be resistant to the failure of m_2 of them. At the same time, the "excess" resources of the first subsystem can be used to solve part of the problems of the second subsystem, if its own resources are not enough. That is, the first subsystem acts as a "donor" and the second as a "recipient".

For the first subsystem, it is enough to build the usual basic model $K(m_1, n_1)$, which will reflect its behavior in the flow of failures. In addition, let's build a dual redundancy model $R(m_1, n_1)$, which accepts as input the state vector of the first subsystem, and with the help of which the auxiliary vector w, length m_1, will be obtained. By combining this vector (by concatenation) with the state vector of the second subsystem, we will get a vector (denoted as v) containing as many ones as there are processors available to solve the problems of the second subsystem. Its length will be equal to $m_1 + n_2$.

Recall that the second subsystem is resistant to the failure of no more than $m2$ of $n2$ of its own processors, i.e., to solve its problems, no less than $n_2 - m_2$ of operational processors are required. Considering the processors that can be provided by the first subsystem (the number of which can be up to m_1), we can say that the system is resistant to the failure of no more than $m_1 + m_2$ of $m_1 + n_2$ processors in total, that is, it will remain operational exactly as long as in the vector v there will be no more than $m_1 + m_2$ zeros. Thus, the second subsystem will correspond to the model $K(m_1 + m_2, m_1 + n_2)$, which receives the vector v described above as an input.

The model of the entire system can subsequently be built on the basis of the models of its subsystems obtained above, in particular, by the methods proposed in [19].

4.2 Any Number of "Donor" and "Recipient" Subsystems

In general, the number of subsystems corresponding to the behavior of the "donor", as well as the number of subsystems corresponding to the behavior of the "recipient", can be arbitrary.

In the case of several "donors", all their redundant processors can be used to restore the functionality of the "recipient" subsystems. Let the system have s "donor" subsystems, each of which contains n_i processors and is resistant to the failure of m_i of them ($i = 1, 2, \ldots, s$). For each of them, we will construct *dual redundancy models* $R(m_i, n_i)$, with the help of which the vectors w_i will be obtained. Let's combine these vectors by concatenation and get the vector w. This vector, which will have a length of $m = m_1 + m_2 + \ldots + m_s$, will contain exactly as many ones as there are reserve processors in all "donor" subsystems in general. Further, this vector can be used similarly to the case of one "donor" system.

In the case of multiple "recipient" subsystems, each of which can use available donor processors, the approach described in [18] can be applied. In this case, the available reserved processors of the "donor" subsystems can be considered as a sliding redundancy. Accordingly, the vector w will correspond to the part of the state vector that characterizes the processors of the sliding redundancy in [18].

4.3 Model of a "Symmetric" Pair of "Donor"-"Recipient" Subsystems

The case was considered above when one subsystem can contribute to the restoration of the performance of the second subsystem. In turn, it is possible to organize the system in such a way that the second subsystem can also contribute to the restoration of the first subsystem. That is, depending on the situation, each of the subsystems can become both a "donor" and a "recipient".

In this case, to build a model of the first subsystem, it is enough, by analogy with the previous sections, to build a dual redundancy model $R(m_2, n_2)$, which accepts the state vector of the second subsystem as an input. With the help of this model, an auxiliary vector of length m_2 will be obtained. By combining this vector (by concatenation) with the state vector of the first subsystem, we will get a vector containing as many ones as the number of processors that can be used to solve the problems of the first subsystem. Its length will be equal to $m_2 + n_1$. The first subsystem will correspond to the model $K(m_2 + m_1, m_2 + n_1)$, which accepts this vector as an input. The model of the second subsystem is built in the same way as in the previous case.

Note that for both dual models, parts of state vectors containing elements corresponding to processors that are contained only in one or another subsystem are used. Indeed, these models reflect the amount of "excess" resources directly in the subsystems, without taking into account the resources that can be obtained from other subsystems.

We also note that in a simple case (in particular, in the case considered in this section), the behavior of such a system corresponds to the basic model $K(m_1 + m_2, n_1 + n_2)$. However, the proposed approach can be used for more complex cases that cannot be reduced to basic models, in particular, when there are more than two systems and the relationships between them are more complex. For example, some of the subsystems cannot act as "donors" for others, or the number of processors that can participate in restoring the subsystem's performance is limited (which is discussed in the next section). Also, the proposed model can be useful in cases where it is necessary to obtain separate models for each of the subsystems, and not only the entire system.

4.4 Limiting the Amount of the "Donor" Processors

In a real system, the possibilities of using "excess" resources of one subsystem to restore the operability of another may be limited, for example, by bus bandwidth. In this case, no more than a certain number of available "redundant" processors can be used. This feature should also be reflected in the model.

To do this, we will solve the following problem. Let there be a vector of length n containing l ones. We need to get a vector of length h, $h < n$, containing $k = \min(l, h)$ ones. It is obvious that the vector considered above contains $n - l$ zeros. The $K(n - h + 1, n)$ model that accepts it as an input will lose $\psi(n - h + 1, n - l)$ edges, i.e. exactly the same number of its edge functions will take zero value. The total number of edge functions of this model according to (1):

$$n - (n - h + 1) + 1 = h. \tag{5}$$

Thus, the vector consisting of the values of the edge functions of this model will have a length of h, will contain $\psi(n - h + 1, n - l)$ zeros and, accordingly, $\rho(l) = h - \psi(n - h + 1, n - l)$ ones. So, according to (2):

$$\rho(l) = \begin{cases} h, & \text{if } l > h - 1 \\ l, & \text{if } l \leq h - 1 \end{cases}. \tag{6}$$

Taking into account that l and h are integers, $\rho(l) \equiv \min(l, h)$.

Thus, in the event that the number of processors of the subsystem that can be used to solve external problems is limited to some number h, it is sufficient to apply the auxiliary vector obtained using the dual model to the model $K(n - h + 1, n)$, where n is the length of this vector. The vector built on the basis of the values of the edge functions of such a model will contain the number of ones equal to the number of "redundant" processors of the subsystem, taking into account the above limitation. This vector can be used in the same way as the vector obtained by the auxiliary dual model was used in the previous cases.

Thus, in the event that the number of processors of the subsystem that can be used to solve external problems is limited to some number h, it is sufficient to apply the auxiliary vector obtained using the dual model to the model $K(n - h + 1, n)$, where n is the length of this vector. The vector built based on the values of the edge functions of such a model will contain the number of ones equal to the number of "redundant" processors of the subsystem, taking into account the above limitation. This vector can be used in the same way as the vector obtained by the auxiliary dual model was used in the previous cases.

We will also call such a model a *constraint model* and, to simplify notation, denote it as $L(h, n)$, i.e.,

$$L(h, n) \triangleq K(n - h + 1, n) \tag{7}$$

5 Examples

5.1 An Example of a Model of a Pair of "Donor"-"Recipient" Subsystems

Let the system consist of two subsystems. The first subsystem contains 10 processors and is tolerant to the failure of no more than 5 of them, and the second subsystem contains 7 processors and is tolerant to the failure of no more than 3 of them. At the same time, "redundant" processors of the first subsystem can be used to restore the performance of the second. The structure of this system is presented on Fig. 1 (A).

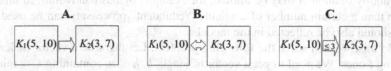

Fig. 1. System structure: A – a pair of "donor"-"recipient" subsystems, B – a "symmetric" pair of "donor"-"recipient" subsystems, C – a system with a limit on the number of "donor" processors

Let's build models of each of the subsystems. Let the states of the processors of the first subsystem correspond to the variables $x_1, x_2, \ldots, x_{10}$, and the variables $x_{11}, x_{12}, \ldots, x_{17}$ of the second.

The first subsystem will correspond to the model $K_1(5, 10)$ with 6 edge functions, which we denote by $f_1^1, f_2^1, \ldots, f_6^1$. We will also construct an auxiliary dual redundancy model $R_1(5, 10) = K_1'(6, 10)$ with 5 edge functions, which we denote by $g_1^1, g_2^1, \ldots, g_5^1$.

We will form the auxiliary vector v_1, which will contain the values of the elements of the system state vector that correspond to the states of the processors of the second subsystem, as well as the values of the edge functions of the auxiliary model:

$$v_1 = \left\langle x_{11}, x_{12}, x_{13}, x_{14}, x_{15}, x_{16}, x_{17}, g_1^1, g_2^1, g_3^1, g_4^1, g_5^1 \right\rangle.$$

The second subsystem will correspond to the model $K_2(8, 12)$, which accepts v_1 as an input vector. This model will have 5 edge functions.

In order to save space, the cumbersome expressions of the edge functions are not given [17]. Structure of the model is shown on Fig. 2 (A).

5.2   An Example of a Model of a "Symmetric" Pair of "Donor"-"Recipient" Subsystems

Consider a system similar to the one considered in the previous section, but in which the second subsystem can also use its "excess" resources to restore the performance of the first subsystem. The structure of the system is presented on Fig. 1 (B).

Note that the model of the second subsystem will be exactly the same as in the previous example, so we will limit ourselves to building the model of the first subsystem.

Let's build an auxiliary dual redundancy model $R_2(3, 7) = K_2'(5, 7)$, which will contain 3 edge functions: g_1^2, g_2^2, g_3^2. Let's form the auxiliary vector v_2, which will contain the values of the elements of the system state vector that correspond to the states of the processors of the first subsystem, as well as the values of the edge functions of the auxiliary model:

$$v_2 = \left\langle x_1, x_2, x_3, x_4, x_5, x_6, x_7, x_8, x_9, x_{10}, g_1^2, g_2^2, g_3^2 \right\rangle.$$

The first subsystem will correspond to the model $K_1(8, 13)$, which accepts the vector v_2 as an input vector. This model will contain 6 edge features.

In order to save space, the cumbersome expressions of the edge functions are not given [17]. Structure of the model is depicted at Fig. 2 (B).

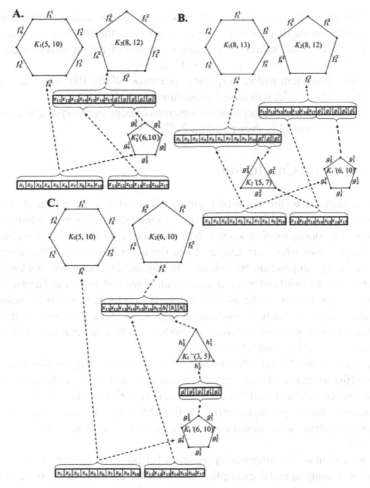

Fig. 2. System models: A – a pair of "donor"-"recipient" subsystems, B – a "symmetric" pair of "donor"-"recipient" subsystems, C – a system with a limit on the number of "donor" processors

5.3 An Example of Limiting the Number of "Donor" Processors

For example, consider a system similar to the one considered in Sect. 5.1, but in which the number of "redundant" processors of the first subsystem that can be used to restore the performance of the second subsystem is limited to 3. The structure of the system is presented on Fig. 1 (C).

Note that in this case only the model of the second subsystem will be changed. At the same time, the auxiliary dual redundancy model will remain the same as in the previous example.

Let's build an auxiliary constraint model that will limit the number of "redundant" processors used. This model $L(3, 5) = K_1''(3, 5)$, which will accept the input vector $u_1 = \langle g_1^1, g_2^1, g_3^1, g_4^1, g_5^1 \rangle$ will contain three edge functions: h_1^1, h_2^1, h_3^1.

Let's form an auxiliary vector u_2, which will contain the values of the elements of the system state vector that correspond to the states of the processors of the second subsystem, as well as the values of the edge functions of the auxiliary model formed above: $u_2 = \langle x_{11}, x_{12}, x_{13}, x_{14}, x_{15}, x_{16}, x_{17}, h_1^1, h_2^1, h_3^1 \rangle$.

The second subsystem will correspond to the model $K_2(6, 10)$, which accepts u_2 as an input vector. This model will have 5 edge functions.

In order to save space, cumbersome expressions of edge functions are not given [17]. The structure of the model is shown in Fig. 2 (C).

6 Summary and Conclusion

The work is devoted to the issue of building GL-models of complex multi-module systems, in which it is possible to transfer some of the tasks of one subsystem to another in order to restore operational efficiency. It is proposed to use the so-called *dual model*, the edge functions of which are formed from the edge functions of the original GL-model by replacing conjunction operations with disjunction operations and vice versa. It is shown that such a model can be used as a so-called *redundancy model* and indicate the number of processors that can be used to restore the performance of other subsystems. It is also shown the possibility of introducing into the model a limitation on the volume of tasks that can be transferred between subsystems, which is achieved by introducing an additional so-called *constraint model*.

Subsystems of the "donor"-"recipient" type are considered, in which one of the subsystems (the so-called "donor") can use its processors to restore the performance of another (the so-called "recipient"), as well as their variations (with an arbitrary number of subsystems of each type, "symmetric", with restrictions). For each case, methods of constructing a system model based on a combination of dual and ordinary GL-models are proposed.

The application of the proposed approaches for models of some of the considered types is shown using specific examples based on the experiments carried out by the authors.

References

1. Yalcinkaya, E., Maffei, A., Onori, M.: Application of attribute based access control model for industrial control systems. IJCNIS **2**, 12–21 (2017)
2. Gokhale, S., Dalvi, A., Siddavatam, I.: Industrial control systems honeypot: a formal analysis of conpot. IJCNIS **6**, 44–56 (2020)
3. Drozd, O.V., Rucinski, A., Zashcholkin, K.V., Drozd, M.O., Sulima, Y.Y.: Improving FPGA components of critical systems based on natural version redundancy. Appl. Asp. Inf. Technol. **4**(2), 168–177 (2021)
4. Yizhong, F., Yuting, L., Tuo, H., et al.: Design of missile incremental adaptive fault tolerant control system. J. Beijing Univ. Aeronaut. Astronaut. **48**(5), 920–928 (2022). (in Chinese)
5. Abbaspour, A., Mokhtari, S., Sargolzaei, A., Yen, K.K.: A survey on active fault-tolerant control systems. Electron. **9**(9), 1513 (2020)
6. Arfat, Y., Eassa, F.E.: A survey on fault tolerant multi agent system. IJITCS **9**, 39–48 (2016)

7. Wason, R., Soni, A.K., Qasim Rafiq, M.: Estimating software reliability by monitoring software execution through OpCode. IJITCS **7**(9), 23–30 (2015)
8. Thomas, M.O., Rad, B.B.: Reliability evaluation metrics for internet of things, car tracking system: a review. IJITCS **9**(2), 1–10 (2017)
9. Kuo, W., Zuo, M.J.: Optimal Reliability Modeling: Principles and Applications, p. 560. Willey, Hoboken (2003)
10. Belfore, L.A.: An O(n(log2(n))2) algorithm for computing the reliability of k-out-of-n:G & k-to-l-out-of-n:G systems. IEEE Trans. Reliab. **R-44**, 132–136 (1995)
11. Rushdi, A.M.: Utilization of symmetric switching function in the computation of k-out-of-n system reliability Microelectron. Reliab. **26**(5), 973–987 (1986)
12. Chang, G.J., Cui, L., Hwang, F.K.: Reliabilities for (n, f, k) systems. Stat. Probab. Lett. **43**, 237–242 (1999)
13. Sfakianakis, M., Kounias, S., Hillaris, A.: Reliability of a consecutive k-out-of-r-from-n: F system. IEEE Trans. Reliab. **41**(3), 442–447 (1992)
14. Hwang, F.K.: Simplified reliabilities for consecutive k-out-of-n systems. SIAM J. Algebraic Discrete Methods **7**(2), 258–264 (1986)
15. Lee, W.S., Grosh, D.L, Tillman, F.A., Lie, C.H.: Fault tree analysis, methods and applications: a review. IEEE Trans. Reliab. **34**(3), 194–203 (1985)
16. Romankevich, A., Feseniuk, A., Maidaniuk, I., Romankevich, V.: Fault-tolerant multiprocessor systems reliability estimation using statistical experiments with GL-models. Adv. Intell. Syst. Comput. **754**, 186–193 (2019)
17. Romankevich, V.A., Potapova, E.R., Hedayatollah, B., Nazarenko, V.V.: GL-model of behavior of fault-tolerant multiprocessor systems with a minimum number of lost edges. Visnyk NTUU KPI Inform. Control Comput. **45**, 93–100 (2006). (In Russian)
18. Romankevich, A., Romankevich, V., Morozov, K.: On one GL-model of a system with a sliding redundancy. Radio Electron. Comput. Syst. **5**, 333–336 (2013). (in Russian)
19. Romankevich, A.M., Morozov, K.V., Romankevich, V.A.: Graph-logic models of hierarchical fault-tolerant multiprocessor systems. IJCSNS **19**(7), 151–156 (2019)

A Method for Evaluating the Efficiency of the Semantic Kernel in the Internet Promotion Channel

Sergey Orekhov[✉], Andrii Kopp, Dmytro Orlovskyi, and Tetiana Goncharenko

Intelligent Information Technologies and Software Engineering Department, National Technical University Kharkiv Polytechnic Institute, Kharkiv 61002, Ukraine
sergey.v.orekhov@gmail.com

Abstract. Over the past twelve years, search engine optimization has become a classic technology for product promotion. Its tools have also become classics, such as Google AdWords advertising, HTML code optimization and building the semantic core of a WEB resource. However, recently they no longer give significant results, and their efficiency drops. We completed several WEB projects in 2020–2021, the results of which show a catastrophic drop in the efficiency of classic SEO tools by at least 50%. The main reason for this is that the online market is constantly changing due to competition. In our study, we propose to look for ways to improve efficiency based on the reengineering of existing tools. To implement the reengineering procedure, it is necessary to evaluate the effectiveness of existing SEO tools. Therefore, in this paper, we focused on the problem of evaluating the effectiveness of the semantic kernel of digital content as the most common approach in SEO. It is based on the method of analyzing the current situation and its automatic evaluation. By the situation, we mean a set of WEB documents that contain a semantic kernel. Therefore, the evaluation of the efficiency of the kernel must be performed both from the point of view of the document and from the point of view of the keyword. Further, such an assessment should also include a measure of the quality of the semantic kernel from the position of the search server. Such a server offers two evaluation tools. The first is WEB statistics, that is, how many visits this kernel attracted. Another way is the structure of the kernel itself, that is, what keywords are included in its composition. And the third component of the assessment is the budget or financial resources that need to be spent on the synthesis and implementation of the kernel on the Internet. Based on the results of this modeling, the article shows a real example (WEB project in the American online services market), in which this evaluation procedure was applied.

Keywords: Semantic Kernel · Virtual promotion · Search engine optimization · Customer journey map

1 Introduction

For five years, we have been researching the effectiveness of classic SEO tools, in particular, methods for synthesizing semantic kernels. Completed WEB projects concerned the market of Ukraine and the USA. The results obtained show a catastrophic decrease in

Z. Hu et al. (Eds.): ICCSEEA 2023, LNDECT 181, pp. 246–259, 2023.
https://doi.org/10.1007/978-3-031-36118-0_22

the effectiveness of methods for implementing the semantic kernel. For example, a WEB project for the wood products market in Ukraine showed that a simple implementation of a semantic core in parallel with code optimization did not give a single online order for half a year. These techniques increased the attendance of the WEB resource, but without result. At the moment, there is no alternative to SEO. Therefore, the only way to improve the efficiency of product promotion on the Internet is to reengineer classic SEO techniques.

It is known that reengineering includes three stages: assessment, change and planning. In this paper, we have focused our attention on the assessment stage. As an object of study, we have chosen the process of introducing the semantic core of digital content. Our goal is to create a methodology for evaluating the effectiveness of the semantic core of digital content. Consider a review of scientific publications on this issue and start with the concept of digital content.

2   Related Works

Nowadays, digital content has become a real product that is actively bought and sold. Digital content describes real goods on the Internet and exists in parallel with them. Digital content performs several tasks. All of them are divided into two groups depending on the two roles played by the end user on the Internet. The first role of the user is the seller of digital content, and the second role is the buyer. In both roles, digital content is an intermediary. In turn, according to the classical theory of marketing, digital content plays the role of a message in the marketing channel [1–4]. That is, it is a message between the buyer and the seller. Such a message informs the buyer that a digital product or service is available to satisfy a particular need.

The reason for the creation of both digital content and digital marketing was a significant change in the scheme of selling goods and services. Earlier we lived in the conditions of a post-industrial society. The schema of business was "goods – money – goods" and it existed because there was a limitation in the quantity of goods. However, with the advent of globalization and the concept of open markets, the goods are in abundance, and now it is difficult to find a buyer for these numerous goods and services. Thus, the main problem was finding a solvent buyer. That is, a new paradigm was formed: attract – earn – spend. Under new conditions, a digital product (content) becomes a combination of three elements: an audience, a communication channel (promotion) and a target message (digital content).

The following restrictions are imposed on digital content by the Internet:

- First, the digital description of the product is a set of keywords that are scanned by the search engine. The server forms answers according to a certain algorithm (Page Rank). Therefore, an important factor is the first position in the answers of a search server, for example, Google [5].
- Secondly, the format of the digital product description becomes the HTML language. It allows you to select keywords that the search server responds to.
- Third, the search server registers the number of links to the web page where our digital content is hosted.

– Fourthly, end users work massively in social networks. They have a technological connection with the search server and built-in search algorithms.
– Fifth, mobile applications allow you to connect to social networks.

To manage digital content a new business process scheme called virtual promotion was proposed (Fig. 1) [6, 7].

The analysis of the new process of virtual promotion shows that the digital content moving in the channel was called the semantic kernel (SK) [8, 9]. This is a new object for research. Let us consider its properties.

The first step in the implementation of virtual promotion is the synthesis of the semantic kernel. The success of virtual promotion depends on its effectiveness.

The works [8, 9] define the semantic kernel as a set of keywords, phrases and morphological forms that most accurately characterize the type of activity, goods and services offered for sale.

For the first time, the term "semantic kernel" was voiced in the methodology of search optimization [10]. Here, too, the semantic core is defined as a set of keywords that should be included in digital content to improve the search engine side of a web page. However, in the methodology of search optimization, the semantic kernel is considered as a duplicate of a potential request from a user in the form of a request to a search server. In other words, the semantic kernel is a targeted set of keywords that tells the search server what page to display to the user in response to his query. Moreover, the query must also contain a similar set of keywords. That is, the semantic kernel is a potential request from a user on the Internet.

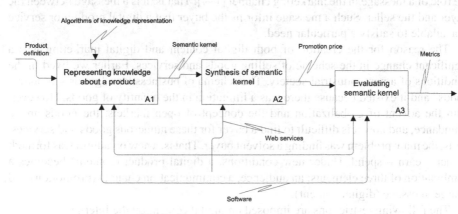

Fig. 1. The process of virtual promotion

The research conducted in [11] expands the concept of the semantic kernel. Here, based on many years of research into various forms of interaction between users and the search server, it is proposed to consider the semantic core as a concise description of a product or need circulating on the Internet. If you correctly create a core and place it in the right points (nodes) of the Internet, you can create a larger flow of customers for the enterprise. In addition, it has already been proven that digital content hides marketing information that is useful not only for the buyer, but also for the seller and his competitors [12].

In this way, SJ begins to play a much larger role than just copies of the user's request in the network. It becomes a message in the format of an HTML document, which can potentially be responded to by a buyer on the Internet. The effectiveness of the entire promotion process on the Internet becomes dependent on SK.

As a result, the task of evaluating the effectiveness of the semantic kernel, as a concise annotation of knowledge about the product and the need it covers, becomes relevant.

In which case is it necessary to solve this problem? First, the performance evaluation model allows you to manage the semantic core, that is, to change it if its performance falls or is insufficient. This is shown in work [13], where the assessment of the quality of the SI is the basis for solving the management problem.

The second case, where the semantic core efficiency model is appropriate, is contextual advertising [3, 4, 14]. Here, the effect of the kernel on the advertising message is analyzed: it attracts the required number of buyers of the product.

The third case, where the efficiency of the kernel is analyzed, is the thematic search or thematic traffic on the Internet [15]. Here SK is embedded in the WEB page and the quality of the core depends on whether the page will be in the TOP or not.

In this case, the terms quality and effectiveness of the SK should be considered as synonyms.

Thus, the assessment of effectiveness is the central essence of various channels of promotion of goods on the Internet which is confirmed by the data (Fig. 1).

To determine the quality or effectiveness of a message in a marketing channel, the following indicators are identified that affect its use [16]:

– message display frequency;
– demand or the number of effective actions of the buyer in response to the message;
– competitiveness of the request, where there is a message;
– rejection rate (the number of visitors who did not want to view the message);
– seasonality of the request, which affects the display of the message;
– geographical dependence of the request, where the message is displayed.

As can be seen from the indicators listed above, they all describe the functioning of the message in the promotion channel. Let us assume that SK is embedded in the message in the promotion channel. Then we will consider the metrics that are already available for evaluating the message in the channel.

The classic definition in the marketing theory says that the goal of a product promotion is to increase the efficiency of demand and sales [1]. The content of promotion consists in communication, that is, to inform potential customers about the product, its packaging, brand, advertising exhibitions, demonstrations, relation to the enterprise, etc. As a result, it is clear that the effectiveness of sales depends on the effectiveness or attractiveness of the message about the product that reaches potential customers (buyers). That is, the effectiveness of promotion is determined by the effectiveness of communication in the communication channel between the company and buyers.

Currently, the main key performance indicator that can give an answer about the effectiveness of a message on the Internet will be traffic [3]. It is the number of visits to a WEB page with a semantic kernel or the number of visits to a WEB site through an advertising message with a semantic kernel that forms the number of unique users per unit of time.

Based on the "traffic" metric, the first metric is formed - the traffic indicator. It determines the conditional efficiency of the channel and message [3]:

$$Efficiency = \frac{traffic^2}{budget}. \tag{1}$$

Another classic indicator of the effectiveness of the promotion channel is the return on investment indicator [3]:

$$ROI = \frac{profit - \exp enses}{\exp enses}. \tag{2}$$

The next important marketing funnel metric is conversion:

$$Conversion = \frac{buyers}{visitors}. \tag{3}$$

That is, it reflects the achievement of the goal - to attract the required number of buyers of the product.

In parallel with this, other indicators can be found in the literature, for example: the number of refusals, the cost of attracting one buyer, the average check, the number of repeated visits. But all the above indicators are only metrics, unfortunately, no stable efficiency criterion has been defined among them. The modern theory of Internet promotion is based only on the metric approach.

Thus, there is an urgent task of evaluating the effectiveness of the semantic kernel in the promotion channel created on the Internet. Let us consider its statement.

3 Problem Statement

To formulate the research problem, it is necessary to analyze the requirements for functioning and quality from the end user. In other words, it is required to understand how, when and with what result the user will apply this technique for evaluating the semantic kernel. To do this, we will construct a diagram of use cases (Fig. 2) in the UML language [17, 18].

The evaluation process will be realized thanks to the implementation of two roles: the developer and the user. It can be seen that the product promotion cycle must be managed (activated and stopped). It is also necessary to adjust the digital content, SY and a set of metrics during the entire time of its execution. It is also clear that management is based on evaluations.

It is clear that the promotion cycle is generally unlimited [6] because competitors also have every chance to use either search engine optimization techniques based on the semantic kernel or virtual promotion.

As a result, there is a situation of competition of semantic cores in the promotion channel on the Internet. Moreover, the victory is won by SK with the highest efficiency, which means that the company itself wins over its competitors. Thus, as a result of the

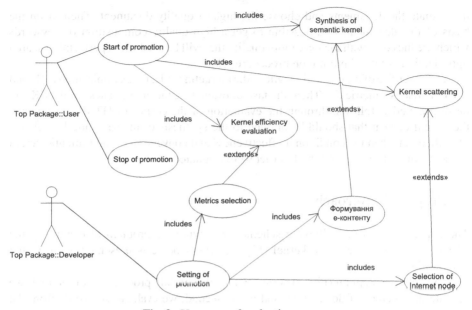

Fig. 2. Use cases of evaluation process

high efficiency of SK, there will be a maximum of calls and online payments. Now it is possible to formulate the statement of the research problem.

According to the definition of SK, its essence is a set of keywords contained in an HTML document. Therefore, efficiency should also be determined on the basis of keywords and document evaluation.

The classic metric for evaluating the quality of a keyword is the C-value metric:

$$C - value = \begin{cases} \log_2|a|f(a), & not\ included \\ \log_2|a|(f(a) - \frac{1}{P(T_a)}\sum_{b \in T_a}f(a)), & included \end{cases} \tag{4}$$

where a is the keyword, $|a|$ is the length of the keyword, $f(a)$ is the frequency of occurrence of the keyword a, T_a is the set of elements that include the word a, $P(T_a)$ is the number of elements in T_a, $\sum f(a)$ is the sum of frequency elements $b \in T_a$, including a. That is, a is part of the sentence b.

This metric (4) is based on the fact that the keyword exists within the text corpus. A text corpus is a set of HTML documents describing a given digital content. That is, the input information for evaluation is a text corpus containing a semantic kernel.

We will assume that the text corpus is a set of documents, and each document, in turn, is a set of sentences and the sentences themselves contain key words. Thus, the first assessment of effectiveness will be on the side of the quality of the semantic kernel, that is, on the side of the validity of the semantic kernel in terms of the structure and composition of keywords. In this case, metric (4) is the most suitable.

Thus, the task of evaluating the effectiveness of the semantic kernel consists in step-by-step evaluation first of HTML documents from the text corpus, and then of keywords from sentences. Then our evaluation should be carried out in two stages. First, you need to evaluate the documents and choose the highest quality document. Then, if on the basis of this document, it is possible to get a high-quality composition of keywords which are the cores with the maximum quality that will be used when performing search optimization or virtual promotion procedures [6].

The other side of the evaluation will be the marketing side, for example, metric (1) and WEB statistics – metric (2). Then, the task of evaluating the effectiveness of the SK can be formulated as follows: through the evaluation of the corpus of HTML documents, the digital content that should be applied to the synthesis of the semantic kernel was identified, and then the conditional effectiveness and conversion of the semantic kernel in the promotion channel on the Internet were evaluated.

4 Proposed Approach

The paper proposes the following scheme for solving the problem of evaluating the effectiveness of the semantic kernel (Fig. 3). The scheme is presented as an action diagram in UML.

The solution diagram (Fig. 3) shows two cycles of data processing. In the first, we operate on a corpus of documents, and in the second, we evaluate keywords from the highest quality document.

Based on the algorithm Page Rank [19], we assume that a collocation describes an ideal Internet query that will lead a potential buyer to the desired product. That is, collocation describes the ideal semantic kernel of a product on the Internet. Then the presence of collocation in the document indicates the high efficiency of this digital content and, accordingly, the semantic kernel in the promotion channel.

Therefore, initially, we propose to select documents with the highest quality of keywords, and then check whether it is possible to build at least a third-order collocation based on the selected documents. Why collocations are of the third order? In work [12], it was shown that in order to build an ideal query on the Internet, a user needs a query that consists of three keywords. Moreover, each word answers one of the questions asked. Only such a query structure leads to the maximum efficiency of Page Rank algorithm [5]. Therefore, if a collocation or an ideal query is hidden in our document, then Page Rank algorithm will work with maximum efficiency.

Let us consider the SK model for building an evaluation method.

Taking into account the proposal to evaluate the SK from three sides: document and keyword evaluation, conditional efficiency evaluation, and conversion evaluation, let us first consider the modeling of the "document-keyword" based on the text model shown in [20].

We will assume that at the input we have a text corpus of documents D, presented as HTML text:

$$D = \{s_1, ..., s_L\}, \tag{5}$$

where s_i, $i = \overline{1, L}$ is a sentence ending with a comma or other end-of-sentence sign. Each sentence s_i is a combination of three sets: a set of keywords to describe the main

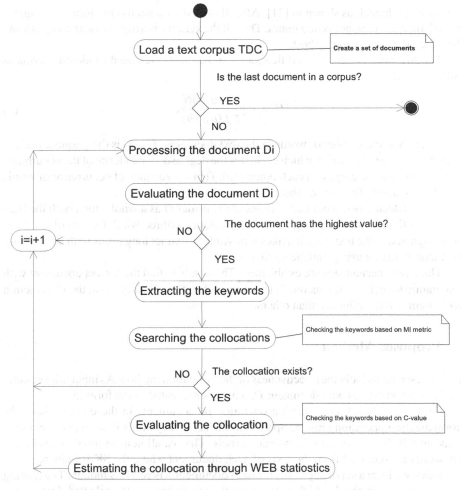

Fig. 3. Schema of the solution

content, a set of sentence-ending characters, and a set of keywords that expresses the general meaning [20, 21]:

$$s_i = \langle K, G, E \rangle \; i = \overline{1, L} \tag{6}$$

where K are keywords that form the product description on the Internet, G are keywords of general content, E are end-of-sentence symbols.

Next, we introduce the operation of metric evaluation of documents: $M : D \to D_M$. M is a set of metrics to evaluate a document D.

Let's introduce the normalization operation N, when we exclude the set E, and transform the other keywords to the infinitive: $N : D_M \to D_M^N$.

In work [12], when the user's request should include the answer to three questions (what, where, when) to describe the product. Then trigram models can be used to describe the semantic kernel, as shown in [21]. After that, if the kernel has the form of a trigram model, then its efficiency is maximum. Thus, if the kernel has a trigram in its composition, then the efficiency is the highest.

The MI measure and its modifications are the most often used to identify terms as collocations:

$$MI = \log_2 \frac{f(a, b)N}{f(a)f(b)}, \tag{7}$$

where N is the number of words in the text document. $f(a, b)$ is the frequency of co-occurrence of words a and b, which evaluates the degree of dependence of the occurrence of two words in the corpus on each other. $f(a), f(b)$ – frequency of occurrence of words a and b separately from each other.

If the detected two-word collocations are considered as a whole, then with the help of the mentioned measures, longer word combinations (three-word, four-word, etc.) can be recognized in the text, which makes it possible to extract long terms with an arbitrary syntactic structure using statistical criteria.

Thus, two mechanisms are established. The first is to find the correct document with a semantic kernel, and the second is to detect collocations. If they exist, then interaction with them is more effective than others.

5 Proposed Method

Let us describe verbally the effectiveness of the evaluation method. As input information, we have a text corpus with document D, which is presented in the form (5).

First, we conduct an evaluation operation of documents in the corpus. Next, the normalization operation must be applied to the set of keywords in selected document. This set will be the source of semantic kernels. That is, all non-essential symbols and keywords are removed from the texts of each document where the SK is hidden.

Next, we form a set of keywords for each document where SK is hidden. Then, using the metric (7), we check which order the collocations have in the selected documents. If there are collocations of the second and third order in the documents, then the effectiveness of such a system will be the highest from the point of view of the ideology of linguistic analysis of texts [12].

The verbal description of the method coincides with the scheme for solving the research problem shown in Fig. 3.

Thus, evaluating the document and finding the second- and third-order collocations will be the first evaluation cycle. Next, it is proposed to make an assessment from the side of marketing theory and WEB statistics. Then, if the semantic kernel passes this evaluation, it has the highest level of efficiency.

To implement the second evaluation cycle, we suggest using metrics (1)–(4).

The specified algorithm makes it possible to evaluate SK both from the side of the theory of texts, and from the side of the theory of marketing and WEB statistics. In addition, our evaluation is based on both the document evaluation and the SK. This is an important point, because the search server operates with the concept of document and keyword, and the user builds his queries on the basis of collocations and WEB statistics.

Each buyer analyzes the number of visits to the page of the product that he wants to buy, as well as the number of reviews, i.e. refusals and so on.

The advantage of this method is that it is the implementation of the first stage of reengineering of the classic SEO tool – the semantic kernel. At the same time, it takes into account both the aspect of evaluating digital content from where the kernel is extracted, and the aspect of the influence of the search engine on this SEO tool.

Let us consider an example of using the specified evaluation method to understand the effectiveness of the real semantic kernel in the process of product promotion on the Internet.

6 Experiment

To test the method of evaluating the effectiveness, the paper considered a WEB project to promote online services in the USA: www.celestialtiming.com. Currently, the project is fully operational and is constantly online.

According to the solution scheme (Fig. 3), at the first stage we analyze the text corpus of our WEB project. The corpus includes N documents. To do this, all documents must first be uploaded and evaluated. The following document metrics $M = \{A, R, S, Q, Z\}$ were selected for evaluation [20, 21]:

$$A = \frac{Verbs}{N}, \tag{8}$$

$$R = \frac{Verbs}{Nouns + Adjs + \text{Pr}\,onouns}, \tag{9}$$

$$S = \frac{Nouns + \text{Pr}\,onouns}{Verbs + Adjs}, \tag{10}$$

$$Q = \frac{Adjs + Advs}{Nouns + Verbs}, \tag{11}$$

$$Z = \frac{\text{Pr}\,epositions}{Sentences}. \tag{12}$$

Based on metrics (8)–(12), the quality of each document in the corpus was calculated (Table 1). The results show that the semantic kernel of this web project exists on the "contacts" page.

It is necessary to check now whether it is possible to extract the kernel from the content of the "contacts" page. An example of content is presented in Table 2.

We have found that in this digital content there are collocations of the second order (Fig. 4). This means that the semantic kernels that can be synthesized will include collocations of the second order. It is also seen that it is potentially possible to obtain collocations of the third order.

Now we need to compare how the kernels behave in terms of WEB statistics. To do this, we use the data of the WEB service – Google Analytics (Fig. 5).

Table 1. Evaluating the documents in a text corpus

No	Link	A	R	S	Q	Z	Sum
1	home	1.336	0.265	0.127	0.347	0.417	2.491
2	celebrity	1.196	0.339	0.138	0.415	2.125	4.213
3	myself	0.778	0.455	0.152	0.455	2.0	3.838
4	another	0.636	0.5	0.182	0.5	2.0	3.818
5	saved	0.83	0.451	0.152	0.403	1.609	3.445
6	applications	1.071	0.287	0.13	0.477	2.764	4.728
7	free	1.263	0.407	0.125	0.303	3.4	5.499
8	**contact**	**1.5**	**0.469**	**0.057**	**0.111**	**7.667**	**9.803**
9	account	0.667	0.455	0.15	0.471	2.25	3.991

Table 2. The fragment of digital content of tested WEB project

No	Link	Content
1	contact	CelestialTiming is the product of international collaboration with a goal of providing site users with an accurate and easy tool to bring them in touch with their natural rhythms of the universe and enable them to make more effective personal and business decisions Members of the CelestialTiming team have extensive experience and advanced degrees in the fields of physics, engineering, computer science, psychology, and education. They have developed professional psychological and astrological software, taught in colleges and universities, and published books and articles on a variety of topics

	A	B	C	D	E	F	G	H	I	J	K	L	M
1	**Metrics MI**	CelestialTiming	product	collaboration	goal	site	user	tool	rhythm	universe	team	member	experience
2	CelestialTiming	-	4.61	-	-	-	-	-	-	-	4.61	4.61	-
3	product	4.61	-	5.61	-	-	-	-	-	-	-	-	-
4	collaboration	-	5.61	-	5.61	-	-	-	-	-	-	-	-
5	goal	-	-	5.61	-	-	-	-	-	-	-	-	-
6	site	-	-	-	-	-	5.61	-	-	-	-	-	-
7	user	-	-	-	-	5.61	-	5.61	-	-	-	-	-
8	tool	-	-	-	-	-	5.61	-	-	-	-	-	-
9	rhythm	-	-	-	-	-	-	-	-	5.61	-	-	-
10	universe	-	-	-	-	-	-	-	5.61	-	-	-	-
11	team	4.61	-	-	-	-	-	-	-	-	-	-	5.61
12	member	4.61	-	-	-	-	-	-	-	-	-	-	-
13	experience	-	-	-	-	-	-	-	-	-	5.61	-	-

Fig. 4. Searching second order collocations

There are three spikes on the visit chart. They correspond to three variants of the semantic kernel that have been launched on the Internet. Thanks to our evaluation method, three points of decrease in the efficiency of the kernel were identified.

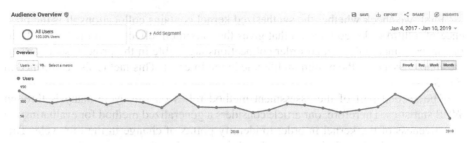

Fig. 5. WEB statistics of tested WEB project

Three variants of the semantic kernel were successively launched on the Internet. However, all these variants were synthesized on the basis of second-order collocations. This is confirmed by the table data (Table 2). Also, these data show that it is difficult to build third-order collocations for this case of digital content. This means that the efficiency of such cores is low. Therefore, if we take into account the effect of aging of nuclei, which was described in [8], then such kernels are unreliable and quickly age. This fact for this WEB project is also confirmed by WEB statistics (Fig. 5).

Therefore, we recommended the owners of this WEB project to change the digital content so that it is dominated by collocations of the third and fourth order. What kind of collocations these should be can be seen from the data in Fig. 4.

To ensure the accuracy and reliability of the results obtained, three WEB projects were analyzed in this work. Two projects prove the need to develop this assessment method. The third project, as the most complete and versatile, is shown above. All projects initially used the classical method of semantic kernel synthesis, which led to long inefficient cycles of product promotion on the Internet.

But after applying a new method for evaluating the efficiency of the kernel, noticeable improvements were obtained. The accuracy and validity of the new method is proven by the fact that all three projects were carried out for different markets and market segments, both in terms of geography and in terms of products to be promoted.

7 Conclusion

Among the main tools of search engine optimization on the Internet, work with semantic kernels occupies a central place. However, it is not enough to simply synthesize the kernel based on WEB services. The state and efficiency of the synthesized kernel must be constantly monitored.

The fact is that we discovered the effect of aging of the semantic kernel [20, 21]. It lies in the fact that the kernel repeats the product life cycle model in the Internet according to the classical marketing theory. Therefore, the problem becomes determining at what stage of this life cycle the kernel is. An example of the kernel life cycle is perfectly visible in Fig. 5. Therefore, to reveal changes in the cycle we need to check the efficiency of the kernel. The paper proposes an approach that is based on evaluating the efficiency of the kernel both from the standpoint of the Page Rank algorithm and from the standpoint of WEB statistics.

First, we check whether the synthesized kernel contains collocations. It is the presence of a third-order collocation that gives the maximum effect from the kernel. If, as in our test, since only second-order collocations are stable in the project, such kernels are characterized by the minimum lifetime in the Internet. This fact is also confirmed in Fig. 5.

The second part of the assessment method relies on life cycle analysis through WEB statistics. Therefore, our article considers a generalized method for evaluating the effectiveness of the kernel in order to identify points of change in the life cycle. This is an urgent task, the solution of which will increase the effectiveness of search engine optimization.

Based on the results obtained, the following conclusions can be drawn:

1) The practical application of the new evaluation method lies in the fact that it can be used to recognize the moment when the semantic kernel changes either to a new one or to a changed one.
2) The implementation of this method helps the content manager to understand the relationship between the life cycle of the semantic core, WEB document and WEB statistics. Until now no such relationship has been found in scientific publications. And since competition on the Internet is high, without noticing the fact that competitors use third-order collocations, we risk to lose in this market segment.
3) The development of a new evaluation method opens up prospects for applying the theory of situational management. This is a smart control class. Therefore, it is advisable to implement this control method in automatic mode. Then you can manage the process of promotion on the Internet using artificial intelligence systems [9].

The purpose of our further work is the software implementation of a special component of the semantic kernel. Its creation is planned on the platform NodeJS. Such a component will work as a machine learning system that will accumulate information about successful kernels in order to automatically generate new ones. At the same time, without a method for evaluating the effectiveness of the core, which works at the level of a keyword, document, and metric, it is not possible.

References

1. Kotler, P., Keller, K.L.: Marketing Management. Prentice Hall, USA (2012). 812 p.
2. Sihare, S.R.: Image-based digital marketing. Int. J. Inf. Eng. Electron. Bus. 5, 10–17 (2017)
3. Hashimova, K.K.: Analysis method of internet advertising-marketing information's dynamic changes. Int. J. Inf. Eng. Electron. Bus. 5, 28–33 (2017)
4. Hashimova, K.K.: Development of an effective method of data collection for advertising and marketing on the internet. Int. J. Math. Sci. Comput. 3, 1–11 (2021)
5. Ziakis, C., Vlachopoulou, M., Kyrkoudis, T., Karagkiozidou, M.: Important factors for improving Google search rank. Future Internet 11(32), 3–14 (2019)
6. Orekhov, S.: Analysis of virtual promotion of a product. In: Hu, Z., Petoukhov, S., Yanovsky, F., He, M. (eds.) ISEM 2021. LNNS, vol. 463, pp. 3–13. Springer, Cham (2022). https://doi.org/10.1007/978-3-031-03877-8_1
7. Orekhov, S., Malyhon, H.: Metrics of virtual promotion of a product. Bulletin of the National Technical University "KhPI". Series: System analysis, control and information technology. NTU «KPI», Kharkiv, vol. 2, no. 6, pp. 23–26 (2021)

8. Orekhov, S., Malyhon, H.: Method for synthesizing the semantic kernel of web content. In: CEUR-WS, vol. 3171, pp. 127–137 (2022)
9. Orekhov, S., Malyhon, H.: Method of solving the problem of situational control of semantic kernel of web content. Comput. Syst. Inf. Technol. **2**, 6–13 (2022)
10. Clarke, A.: SEO 2016. Learn search optimization with smart internet marketing strategies. DigitalBook.Guru, USA (2016). 140 p.
11. Orekhov, S.: An alternative way to search engine optimization. Inf. Technol. Sci. Eng. Technol. Educ. Health (2022). Abstracts of reports of the XXX international scientific and practical conference Micro-CAD-2022 (19–21 October 2022). Kharkiv. Ukraine: NTU «KPI», 2022. – P. 804
12. Orekhov, S., Malyhon, H., Liutenko, I., Goncharenko, T.: Using internet news flows as marketing data component. In: CEUR Workshop Proceedings, vol. 2604, pp. 358–373 (2020)
13. Godlevsky, M., Orekhov, S., Orekhova, E.: Theoretical fundamentals of search engine optimization based on machine learning. In: CEUR Workshop Proceedings, no. 1844, pp. 23–32 (2017)
14. Mezzi, M., Benblidia, N.: Study of context modelling criteria in information retrieval. Int. J. Inf. Technol. Comput. Sci. **3**, 28–39 (2017)
15. Pawade, D.Y.: Analyzing the impact of search engine optimization techniques on web development using experiential and collaborative learning techniques. Int. J. Mod. Educ. Comput. Sci. **2**, 1–10 (2021)
16. Bendle, N.T., Farris, P.W., Pfeifer, P.E., Reibstein, D.J.: Marketing Metrics. The Manager's Guide to Measuring Marketing Performance, 3rd edn. Pearson Education, Inc., London (2016). 456 p.
17. Zaretska, I., Kulankhina, O., Mykhailenko, H., Butenko, T.: Consistency of UML design. Int. J. Inf. Technol. Comput. Sci. **9**, 47–56 (2018)
18. Pathak, N.: An empirical perspective of roundtrip engineering for the development of secure web application using UML 2.0. Int. J. Intell. Syst. Appl. **5**, 43–54 (2017)
19. Srivastava, A.K., Garg, R., Mishra, P.K.: Discussion on damping factor value in PageRank computation. Int. J. Intell. Syst. Appl. **9**, 19–28 (2017)
20. Orekhov, S.: Advanced method of synthesis of semantic Kernel of E-content. In: CEUR-WS, vol. 3312, pp. 87–97 (2022)
21. Orekhov, S., Malyhon, H., Goncharenko, T.: Mathematical model of semantic Kernel of WEB site. In: CEUR Workshop Proceedings, vol. 2917, pp. 273–282 (2021)

Automated Dating of Galaktion Tabidze's Handwritten Texts

Tea Tvalavadze[1], Ketevan Gigashvili[2], Esma Mania[3], and Maksim Iavich[4](✉)

[1] Giorgi Leonidze State Museum of Georgian Literature, 8 Gia Chanturia Street, 0108 Tbilisi, Georgia
[2] Iakob Gogebashvili Telavi State University, 1 Georgian University Street, 2202 Telavi, Georgia
[3] Korneli Kekelidze Georgian National Centre of Manuscripts, 1/3 Merab Aleksidze Street, 1093 Tbilisi, Georgia
[4] School of Technology, Caucasus University, 1 Paata Saakadze Street, 0102 Tbilisi, Georgia
miavich@cu.edu.ge

Abstract. The study of Georgian history greatly involves the fundamental investigation of the handwritten texts. The scientists use different methods for this investigation, but they face different problems. It is related to the fact that the time of the text creation of manuscript in the most cases differs from the date of the manuscript creation and the identification of each of them needs special approach. The automatic dating approaches of the manuscripts is now very common in practice, and it involves the work of computer scientists. The paper studies the approaches of the dating the manuscripts and offers the novel methodology for dating Georgian ones using neural networks. The authors offer the neural network and train it with the existing data. The system is trained using the classified writings of Galaktion Tabidze, which is the greatest Georgian poet of XX century. The system is needed for dating his manuscripts, which are not dated yet. It is very important for the study of Georgian history. The paper also offers the trained model, which can classify Georgian handwritings of Galaktion Tabidze. The received system has the accuracy score, which is considered good in comparison with the related research in this field. The offered approach can be used to date the handwritings of different authors.

Keywords: Automatic dating approach · Neural network · Automated handwriting dating

1 Instruction

Georgia has 16-century-old written heritage and as far as any kind of historical research, be it social, political, cultural etc. is based on written sources, our awareness of Georgian history greatly depends on the fundamental investigation of the handwritten texts. One of the most basic aspects of the researches is dating of the manuscript sources and scholars use different methods for this purpose. The time of the text creation often differs from the date of the manuscript creation and identification of each of them needs special approach. e.g. dating of the text based on the information extracted from its contents cannot always be applied to the manuscript dating. On the other hand, in cases when manuscripts are the first original copies of the texts, the dates of the text and the manuscript are the same. Dating of such handwritings also means the text datation and it is very important. When we identify the time of the text creation, we specify at least the upper time limit of all the happenings described in it. It is one of the most challenging issues of historical studies as without knowing the time of the events we can't appoint them to their particular historical context or define their exact place in the biographical timeline.

While working on the archival materials of different writers and historical figures we have often come to the dead end with the dating of undated holographs. It generally happens when the contents of the texts do not contain any information that could be appointed to the particular period of time. In such cases, the only way to solve the problem is the manuscript analysis. In order to identify specific features of one and the same author's handwriting according to the years we need to analyze a great number of his/her manuscripts belonging to each chronological year (paradigmatic row). Georgian manuscript repositories are full of undated documents that contain important information but as they are undated, they can't be appointed to the particular period of time and can't be relied on in the scientific researches. Of course, we would be glad to date each of them, but the automated dating is based on statistics and therefore the great amount of samples in the datasets that should be investigated (including training sets, validation sets and testing sets) is very important. In order to use this method for manuscript dating we had to choose such an author whose archive could provide enough material needed for such researches. The largest Georgian manuscript repositories: Historical Archives of Georgia, Korneli Kekelidze National Institute of Manuscripts and Giorgi Leonidze National Museum of Georgian Literature include a great number of personal archives among which one of the richest is the archive of the XX century greatest Georgian poet Galaktion Tabidze. He had been writing poems for more than half a century and at the same time was very productive. His archive includes dated manuscripts for quite a long period from 1908 to 1959 and there is a great number of undated manuscripts too. The main objective of the research is to offer the methodology of the dating of Georgian handwritings, to offer the corresponding model and to train it according the dated samples. The most existing solutions offer the manual dating, which is very time consuming. Therefore, the automated solution must be offered. In the papers we offer the automated dating solution using Neural Network. The main limitation of our approach is that the accuracy score is not high enough, but is still considered good in the comparison with the related researches, but in the different languages. In the paper the authors try to achieve the maximal possible accuracy score.

2 Related Works

Dating of the literary pieces is extremely important for literary studies as it often helps us to identify the main idea of the text. e.g. dating of Galaktion Tabidze's poem "Let the banners wave on high" proved that it couldn't be dedicated to the 1917 October revolution in Russia as it was written before it. It became clear that the idea of its connection with the revolution was promoted by the soviet ideologized critics. It was supposed that the first variant copy of Galaktion Tabidze's another poem "Aspindza" was dedicated to a woman (in some researches to Meri Shervashidze and in the others – to Olia Okudjava) and the feelings represented in it were romantic but dating of the text made it clear that the poem was patriotic from the very first variant copies and was dedicated to the Georgians' victory over the Ottoman Empire in the Battle of Aspindza [1, 2].

Nowadays the use of digital technologies has greatly enhanced the possibilities of textual and manuscript studies. Using the grapheme-based method, scientists of Groningen University have dated Dead Sea Scrolls. Computer technologies and software enable automatic dating of handwritten manuscripts but, of course, there should be the data, chosen by scholars as the basis for such an analysis. Gustave Flaubert's undated manuscripts have been dated with an accuracy of about 62% and as the scholars say, "incorporation of newer methods of handwriting analysis can improve the accuracy of automatic datation" [3, 4]. Samson Ullmann used the grapheme-based method to date Matthew Arnold's manuscript of the poem "Dover Beach". He considers that there should be at least three thousand words for each year obtained from dated manuscripts and at least two lines for each letter of the alphabet to get accurate results. Tara Gilliam in her research "Writer identification in medieval and modern handwriting" suggested "three novel methodological approaches to the analysis of writer identification data" [5–7]. Miroslava Bozekova made experiments on 100 images from 40 different writers and achieved 96.5% accuracy in writer verification. Manoj Kumar Sharm and Vanshika Chanderiya created a writer identification framework which measures "the hand pressures during script writing using identical grapheme and writing strokes and then generates the pressure descriptors". They came to the conclusion that this method "gives the encouraging results and average success rate of the descriptors with the distinct alphabets is 98%". Kalthoum Adam, Somaya Al-Maadeed and Younes Akbari suggest a novel approach that improves classical dating methods by applying feature-level hierarchical fusion" and shows that "that applying a hierarchical fusion based on subsets of multi features in the KERTAS dataset can obtain promising results" [8–11]. There are lots of other projects, publications and novelties that prove effectiveness of digital technologies in these kinds of researches. There are some new approaches of dating the data using different machine learning techniques [12–15].

3 Automated Handwriting Dating

Georgia is making its first steps in this direction and has to do a lot to catch up with the processes. Automated handwriting dating as a method has never been used by our scientists before but being aware of those very interesting projects carried out in different countries for the author identification and handwriting dating we always desired to practice similar approaches on Georgian data. These two aspects of research are based on the same principles and both of them need voluminous databases with lots of high-resolution binarized images (manuscript images should be cleaned of the background noise). For the authorship identification scholars use algorithms and training sets to train the network investigate holographs of a great number of authors and by means of feature vectors, codebooks and self-organizing feature maps (SOFM) identify as many between-writer variations or so called writer-informative features in them as possible. The more discriminative these classifiers are the higher is the verification ratio of the results. After having trained the system on this data scholars run in the system the query sample – an anonymous handwriting in order to discover which are its closest-matching samples, or which are the manuscripts that have the greatest resemblance. As the manuscripts of the training sets are authorized, finding the resemblance between the anonymous manuscript with any of the images from the training set means that the author is identified.

For the automated dating of manuscripts scholars use different training sets – handwritings belonging to one and the same person and arranged according to the years. They use special algorithms to make computer identify and group their classifiers. These features are also mentioned as within-writer variations and date-informative features. Then scholars run in the system the query sample – an undated handwriting to find out manuscripts of which year in the paradigmatic row resemble it most of all ("nearest-neighbour-based verification system"). The received answer will be the date of the undated manuscript.

In both automated authorship investigation and automated dating researches that are based on handwriting studies there are two main approaches that mean two different objects of investigation. The first one is graphematic (based on allograph studies) and the other – textural. Both of the methods are applied to make the computer identify differences and similarities and detect the differentiating features but in the first case, the object of the investigation is: how different writers (writer-specific information) or one and the same writer in different years (date-specific information) write graphemes. The similarity and the difference both are based on the study of the grapheme forms. In the other case scholars do not consider graphemes, text contents and even which alphabet is used. They are oriented only at the text-independent statistics, which they call "textural". These are: "spacing between graphemes, words and lines, margins, proportions of the grapheme parts, ligatures, links, ties, slant, deviation from the baseline, the relation between height and width of letters, paper and ink specifications, height of the text line, hand pressure, pace of writing, roundness, curvature, contour-direction, Contour-hinge, Direction co-occurrence, horizontal run length, vertical run length, white run length, thickness of the ink trace etc.

Georgian alphabet consists of 33 letters and is very specific. Our handwritings have never been analysed, structured and clustered for automated processing. We have never used manuscript segmentation, feature extraction or pattern recognition techniques. Therefore, at this stage graphematic approach would be much more complicated and take a lot of time. So, in order to reach the higher level of performance we decided to apply textural method but at the same time to prepare some foundation for the usage of the graphematic method in future.

Every new project of this kind besides carrying out particular investigation and receiving particular results supports the further development of the Humanities as the algorithms and the databases created for these researches can be used for a great number of other investigations as well.

4 Offered Methodology

The image classification problem is considered to be rather complicated problem in computer vision. It is considered to identify the presence of different visual structures in the uploaded image. The algorithm of image classification has to analyze the needed numerical properties of the different image features and afterwards has to classify them into different categories. Many different classification methods, such as artificial fuzzy-sets, expert systems and neural networks can be used to classify the images, all of them have different advantages and disadvantages, but their main problem is the low accuracy score.

Deep convolutional neural network (CNN) is considered to be the modern and efficient approach of image classification, it can obtain rather good results when solving complicated machine learning problems such as classification [16–18]. In deep learning, CNN is considered as a class of artificial neural network. It is mostly used to analyze the visual images. CNNs can be mentioned as Space Invariant Artificial Neural Networks or Shift Invariant networks, they are based on the architecture, which is shared-weight of the convolution kernels or filters that process input features and output the responses, which are known as feature maps [19, 20].

Therefore, to date our image samples we decided to create the algorithm using image classification, by means of creating Deep Neural Network model. As the label, we have used the year of handwriting. We have used Keras library in order to implement deep convolutional neural network algorithm.

We had the labeled images of 54 different years. Each image we have saved with the resolution 300 dpi. The whole size of the images dataset was 16 GB. We have divided our dataset into the validation and test datasets, the best result we have got for 75% for training and 25% for validation.

We have processed the following procedures:

Image Scaling: The images we have resized to the size of: img_height = 180 and img_width = 180. It must be mentioned that this scaling was got after the experiments, the algorithm could classify the writings. It is illustrated on Fig. 1.

Mean/Standard Deviation of the Images: We considered the 'mean image', which we got by calculating the mean values for every pixel in our train datasets. It gave us the opportunity to understand the underlying structure in input datasets. For the concrete images we chose to augment the information with the perturbed images, because we considered the corresponding structure inappropriate.

Data Normalization: Data normalization is obligatory in order to ensure that all input pixels have the similar distribution of data. It makes the algorithm to process the information much faster during the training procedure. We normalized our data by means of subtracting the mean values from every pixel and afterwards divided the obtained result by the values of the standard deviation. This gave us opportunity to get the curve similar to a Gaussian one, which is centered at zero. For the input images the pixel values must be positive, therefore we scaled the normalized data in 0–255 range.

It must be also mentioned that the writer was also drawing on the handwritings and he was writing on the handwritings with the already printed text. It caused the reduction of the accuracy score; therefore, we cleaned the images in Photoshop before applying them to the input.

As our model has faced the overfitting, when the accuracy was reaching 52–58% the validation accuracy began to decrease.

Fig. 1. Scaled images.

Therefore, we performed the following procedures:

Data Augmentation: We used the pre-processing technique, it means that we augmented the input data with the perturbed versions of the input data. Rotation gave us very good results to mitigate the overfitting. By means of this, we exposed CNN at the much more variations of the data. It is rather interesting techniques because in this case it is much more complicated for CNN to recognize the not needed characteristics in data.

As it was mentioned above, we had 54 folders of data, after the literature review we have decided to keep accuracy score not less than 74%, as it the good result in the related works, which consider the hand writing classification, but for the different languages.

We decided to add the folders to training and validation data sets one by one. If the adding concrete folders caused the accuracy decrease, we continued to clean the image using Photoshop. Finally, we succeeded to add 50 folders. That folders which contained the small quantity of images in them decreased the accuracy score and caused the overfitting. The proposed gives us the possibility to achieve the main objective of the research, to date Georgian handwritings automatically using artificial intelligence.

5 Experiments

We have trained our model and the validation accuracy score 74.8%. We used Google Collab for it. The attributes of the final model are shown on the Fig. 2.

```
Model: "sequential_3"

Layer (type)                    Output Shape             Param #
=================================================================
rescaling_3 (Rescaling)         (None, 180, 180, 3)      0

sequential_2 (Sequential)       (None, 180, 180, 3)      0

conv2d_3 (Conv2D)               (None, 178, 178, 32)     896

max_pooling2d_3 (MaxPooling     (None, 89, 89, 32)       0
2D)

conv2d_4 (Conv2D)               (None, 87, 87, 64)       18496

max_pooling2d_4 (MaxPooling     (None, 43, 43, 64)       0
2D)

conv2d_5 (Conv2D)               (None, 41, 41, 128)      73856

max_pooling2d_5 (MaxPooling     (None, 20, 20, 128)      0
2D)

flatten_1 (Flatten)             (None, 51200)            0

dense_2 (Dense)                 (None, 128)              6553728

dense_3 (Dense)                 (None, 48)               6192

=================================================================
Total params: 6,653,168
Trainable params: 6,653,168
Non-trainable params: 0
```

Fig. 2. Model attributes

We have tested our model and it worked fine in the most cases, but it mixed the years with the previous and next ones, which can be considered logical, because the write could write the manuscript in the beginning or in the end of the year. In the experiment we used 100 not dated handwriting. We dated the using our model and using the manual dating. We have got the same output is 78 handwritings, which confirms our accuracy score.

6 Conclusion and Future Work

Our research shows the automated dating method of handwritten texts. Our research show that the best was for this is to use neural networks. We offer the methodology of training the system using neural network with the corresponding pre-processing procedures and techniques. As the result we have designed our model and made the experiments on

Galaktion Tabidze's handwritten texts, written in Georgian. Our system has the accuracy score 74.8%, which is good in comparison with the related research, but still is low to be used in practice. In our future work, we will work on improving the accuracy score and we think to combine manual and automated techniques.

The results of the work advance the process of the study of Georgian history. It must be mentioned that mostly the existing approaches are manual and time consuming.

The offered approach can date the handwritings much more efficiently. The offered methodology can also be using to date the handwritings of different writes, which will work with much higher accuracy.

Acknowledgment. This work was supported by Shota Rustaveli National Science Foundation of Georgia under grant [No. FR-21-7997] – Graphematic research and methodology of dating manuscripts.

References

1. Ninidze, M.: Creative history of Gelaktion Tabidze's poem (once more about "Aspindza"), Galaktionology, IX, Tbilisi, pp. 265–266 (1922)
2. Dhali, M.A., Jansen, C.N., de Wit, J.W., Schomaker, L.: Feature-extraction methods for historical manuscript dating based on writing style development. Pattern Recognit. Lett. **131**, 413–420 (2020)
3. Garain, U., Parui, S.K., Paquet, T., Heutte, L.: Machine dating of manuscripts written by an individual. J. Electron. Imaging **17**(1) (2008)
4. Ullmann, S.O.A.: Dating through calligraphy: the example of "Dover Beach." Stud. Bibliogr. **26**, 20–25 (1973)
5. Gilliam, T.: Writer identification in medieval and modern handwriting, p. 160. University of York Department of Computer Science, September 2011
6. Bozekova, M.: Comparison of handwritings, physics and informatics, pp. 7–8. Comenius University Bratislava, Slovak Republic (2008)
7. Sharma, M.K., Chanderiya, V.: Writer identification using graphemes, Department of Computer and Communication Engineering, School of Computer Science & Information Technology, Manipal University Jaipur, p. 14. (2019)
8. Adam, K., Al-Maadeed, S., Akbari, Y.: Hierarchical fusion using subsets of multi-features for historical arabic manuscript dating. J. Imaging **8**(60), 10 (2022). https://doi.org/10.3390/jimaging8030060
9. Ninidze, M.: Modern Research Technologies and the Electronic Scholarly Edition, Tbilisi, p. 10 (2016)
10. Bulacu, M., Schomaker, L.: Automatic handwriting identification on medieval documents. In: Proceedings of 14th International Conference on Image Analysis and Processing (ICIAP 2007), p. 279. IEEE Computer Society (2007)
11. Tvalavadze, T.: Grapheme-based method of handwriting dating and its use in archival studies
12. van der Lubbe, M.F.J.A., Vaidyanathan, A., de Wit, M., et al.: A non-invasive, automated diagnosis of Menière's disease using radiomics and machine learning on conventional magnetic resonance imaging: a multicentric, case-controlled feasibility study. Radiol. Med. **127**, 72–82 (2022). https://doi.org/10.1007/s11547-021-01425-w
13. Goswami, A., et al.: Change detection in remote sensing image data comparing algebraic and machine learning methods. Electronics **11**, 431 (2022). https://doi.org/10.3390/electronics11030431

14. Yadav, S.S., Jadhav, S.M.: Deep convolutional neural network based medical image classification for disease diagnosis. J. Big Data 6(1), 1–18 (2019). https://doi.org/10.1186/s40537-019-0276-2

15. Deep Learning: Challenges, Methods, Benchmarks, and Opportunities. IEEE J. Sel. Top. Appl. Earth Obs. Remote Sens. 13, 3735–3756 (2020). https://doi.org/10.1109/JSTARS.2020.3005403

16. Bocu, R., Iavich, M.: Real-time intrusion detection and prevention system for 5G and beyond software-defined networks. Symmetry 15, 110 (2023). https://doi.org/10.3390/sym15010110

17. Bocu, R., Bocu, D., Iavich, M.: An extended review concerning the relevance of deep learning and privacy techniques for data-driven soft sensors. Sensors 23, 294 (2023). https://doi.org/10.3390/s23010294

18. Sarvamangala, D.R., Kulkarni, R.V.: Convolutional neural networks in medical image understanding: a survey. Evol. Intell. 15(1), 1–22 (2021). https://doi.org/10.1007/s12065-020-00540-3

19. Maduranga, M.W.P., Nandasena, D.: Mobile-based skin disease diagnosis system using convolutional neural networks (CNN). Int. J. Image Graph. Signal Process. (IJIGSP) 14(3), 47–57 (2022). https://doi.org/10.5815/ijigsp.2022.03.05

20. Zaman, S., Rabiul Islam, S.Md.: Classification of FNIRS using Wigner-ville distribution and CNN. Int. J. Image Graph. Signal Process. (IJIGSP) 13(5), 1–13 (2021). https://doi.org/10.5815/ijigsp.2021.05.01

Artificial Neural Network and Random Forest Machine Learning Algorithm Based TCSC Controllers for Mitigating Power Problems

Niharika Agrawal[1]([✉]), Faheem Ahmed Khan[1], and Mamatha Gowda[2]

[1] Department of Electrical and Electronics Engineering, Ghousia College of Engineering, Ramanagaram, Karnataka 562 159, India
niharika.svits@gmail.com, faheemahmedkhan11@gmail.com
[2] Department of Artificial Intelligence and Data Science, BGS College of Engineering and Technology, Mahalakshmi Puram, Bengaluru, Karnataka 560 086, India
mahesh.mamatha@gmail.com

Abstract. The demand for electrical energy is rising every year. There is a need for more generation and transmission facilities which require huge investment in the erection of new power stations and lines. The conventional sources of energy are coal, oil, fossil fuel, and nuclear energy-based power plants which are fast depleting. The coal-based plants release different gases like nitrogen dioxide, sulfur dioxide, particulate matter (PM), mercury, and other substances which are very harmful to human life as these gases pollute the atmosphere. The erection of new power plants is costly and affects the environemnt too. The existing transmission line was overloaded which created stability problems in the system. Fixed capacitors (FC) were used in the transmission lines to meet this problem but there were problems like series resonance, wear and tear and slow response. Here comes the role of Power -Electronics based FACTS device Thyristor Controlled Series Capacitor (TCSC) to provide variable compensation and to mitigate these power problems and challenges without the need for investment in the construction of new lines. Traditionally Proportional Integral (PI) controllers have been used for power flow control by TCSC. In the present work the Artificial Neural Network (ANN) technique and Random Forest Machine Learning Algorithm (RFMLA) which comes under the umbrella of Artificial Intelligence have been applied to enhance the power flow transfer capacity of the Kanpur Ballabhgarh transmission line. Power obtained at the receiving end is more with an ANN-based controller. Power obtained is furthermore by the use of RFMLA. The existing power system is utilized more effectively, the efficiency of the system is enhanced, the losses are reduced with modern technology enabled/based TCSC. Thus the objective of meeting the shortage of power without harming the environment is achieved with ANN and RFMLA-based TCSCs. The TCSC based on ANN and RFMLA is successful in meeting various other power challenges and problems also like stability, power quality, low voltage profile of the system. The results of power flow are checked with MATLAB simulation.

Keywords: oscillations · power flow · power system · stability · TCSC

© The Author(s), under exclusive license to Springer Nature Switzerland AG 2023
Z. Hu et al. (Eds.): ICCSEEA 2023, LNDECT 181, pp. 269–288, 2023.
https://doi.org/10.1007/978-3-031-36118-0_24

1 Introduction

The objective of **Sustainable Development Goals** (**SDGs**) is the development of prosperity and peace for people and the planet now and also into the future. SDGs were formulated in 2015 by the United Nations General Assembly (UNGA). There are 17 SDGs and for each SDGs there are 8 to 12 targets and between one and four indicators for each target. The indicators are for measuring the progress towards reaching the targets. From the 17 different goals, the goal 7 is to provide an affordable, reliable sustainable and modern energy to all. Some of the targets to achieve this goal are to ensure the availability of affordable, reliable, and modern energy services by 2030, to increase the share of renewable energy by 2030, to make the global rate of energy efficiency improvement rate double in 2030, to use advanced and cleaner fossil fuel technology. In order to meet the above goal and target of SDGs the solution is to meet the rising demand of power through efficiently using the existing generation and transmission system without deteriorating/harming the environment. The installation of new plants and lines for meeting the rising demand is costly and impact the environment negatively. Fixed series capacitors were used for improving the power capacity but there were problems of wear and tear, slow response and sub synchronous resonance leading to damage of shaft. This challenge/limitation is overcome in this paper by the installation of a Series FACTS device -TCSC based on Machine Learning Algorithms (MLAs) which are ANN and RFMLA. The MLAs have extensive computational capabilities to solve complex Power System Engineering problems and produce accurate and fast results. The simulation results showed an improvement in transmission line power transfer capacity with MLA based TCSCs. The accuracy of decision tree, ANN and RFMLA is checked and the results are shown.

Fig. 1. Affordable and clean energy

The objective of the proposed work which is meeting the goal of providing clean and affordable energy to all is achieved successfully. Increased power flow by MLA based TCSC will help in the technical and economical development of country. The TCSC behaviour has been discussed for the two operating modes of TCSC and the results are analysed. The different errors are plotted for the inductive and capacitve modes of TCSC. Figure 1 shows the goal 7 and Fig. 2 shows the TCSC.

Fig. 2. TCSC (hitachy energy)

2 Literature Review

Various techniques have been used to control the power flow by TCSC. The PI controller was implemnted in the system for controlling the power flow. The simulation diagram consisted of various MATLAB blocks like AC Source, transmision line, three phase load, measurement block. A switch was provided for changing the different modes of TCSC. The results showed an increase in power transfer. For the generation of appropriate firing pulses SR flip flop was used [1]. To reduce the voltage sag using TCSC PWM Generator Pulse Controller was developed. The programmable voltage source was used to vary the voltage with time. The testing with TCSC was done with connections at different locations. The different locations were testing at sending end, testing at middle end, and testing at receiving end of the transmission line, with varying lengths of the transmission line i.e., 100 km, 200 km, 300 km and 400 km long. The power transfer capacity was found to be better at the sending end as compared to other ends [2]. The Pulse Generator block was implemented in the system for giving the firing pulses for analysing the thyristor, capacitor current, capacitor voltage. For proper working of TCSC in capacitive and inductive mode appropriate pulses were generated. In the inductive mode the TCSC acted as a controlled inductor. Power flow decreased in inductive mode and increased in capacitive mode [4].

The efficiency and voltage regulation of the transmission line was improved on using TCSC. The results of power flow improvement with different values of firing angle were plotted. The power was increased, the stability and damping profile of the system was improved using a decentralised synergetic TCSC Controller [3]. The TCSC was implemented using Arduino uno and the laboratory model was prepared. The improvement in power flow, voltage regulation was found when the TCSC system was implemented in series with the pi section of transmission line [5].The TCSC was used in multimachine bus system for improving stability and power system oscillations. The TCSC is effective in fault current limiting. TCSC is also used to mitigate the various power quality issues and problems like voltage sag and swell which were created due to disturbances or faults [6].The closed loop model of TCSC was found to be effective than open loop model for power flow control. The TCSC can be used in constant current and constant power mode. The simulation was done without TCSC and with TCSC and various waveforms for power, impedance and firing angle were plotted for both the modes of TCSC [7]. FACTS devices are used to control power flow in the system by governing the various parameters like impedance, voltage, and phase angle. The TCSC lead lag structure and impedance firing angle relation was nonlinear and the nonlinear parameter/element was given to the neural network using ANN because of non-linear activation layers. The TCSC was used as a fault current limiter by taking care of impedance value to large inductive range depending on the design of TCSC. Using TCSC in the system there was enhancement of voltage sag at the time of disturbances [8].

The split TCSC is used for meeting the small change in power demand using many modules of TCSC with different or same values of inductor and capacitor. There is possibility of giving different firing angles with split TCSC. Many reactance characteristic curves are now possible and fine tuning of line reactance is done. The dimension of triggering matrix now is 92 * 92 which gives a huge range of triggering choices increasing the flexibility of the system [12].

Due to faults, there were losses to the company, losses to the consumers. The correct location and repair of faults was significant. For the different values of distances, the simulation of fault in different modes of TCSC was done [9]. The result showed the plot of Fault section indicator FSI to locate the exact position of fault in the line. The identification of faults at the time of power swing was analysed [10]. A robust and unique fault detection algorithm was proposed. The two FACTS devices TCSC and Static Var Compensator (SVC) were used for the control of power flow and improvement of voltage profile. The control parameters of the devices were estimated based on ANN. The objective of the study was the enhancement of available transfer capacity with FACTS devices based on ANN [11].

3 Test System

The Kanpur Ballabhgarh transmission line in India and its Project Data is taken as for testing the PI, ANN and RFMLA techniques [12]. The compensation chosen was 75%. The inductance of the line is 1.044 mH /km. Total inductance of line of length 400 km is 0.4176 H. Total reactance of line 131.1929 ohms. For 75% series compensation the value of TCSC capacitor is 32.3503 μF and TCSC inductor is 0.07828 H. The resonance factor is 2 and resonance angle is 45 degrees. The resonance region is between 35 to 55 degrees. The ratio of TCR reactor inductance and the capacitance of series capacitor in this work is 0.25 which is in the desired range of 0.1 to 0.3. For 90% compensation the values of TCSC Capacitor and Inductor are 26.9722 μF and 0.09400 H respectively. This is 400 kV and 400 km long line. In this work the firing angle in capacitive mode of TCSC is chosen. The TCSC can be operated in blocked, bypassed and vernier mode. The vernier modes are the capacitive and the inductive mode. There is a resonance region between the inductive and capacitive vernier regions. In the present system the TCSC is worked in the capacitive mode to increase the power flow capacity of the system.

4 Proposed Methodology and Simulation Diagram

4.1 Proposed Methodology

In the present work the power transfer capacity of the system is enhanced using Artificial Intelligence based techniques like ANN and RFMLA. The Fig. 3. Shows the four main types of AI approaches. The major branches were developed and modified by Kalogirou, Dounis, Smolensky. Applications of AI are increasing in all domains at a fast rate because AI has the remarkable capacity to mimic the intelligence of human brain. The first category of AI that is Symbolic AI was based on symbols and has wide applications in process engineering. ANN are very popular and special machine learning algorithms that learn from the data and provide output/predictions. The heuristic methods solve the problem faster and in an efficient manner [13, 14].The entire solution space is properly searched to find the best solution. Some studies require Hybrid method that combine ANN and Fuzzy Logic. In the present work the TCSC is connected between source and infinite bus with transmission line. The different types of controllers selected can be PI, ANN and RFMLA. The PI is the traditional controller used in power systems. The key contribution of the paper is the use of ANN and RFMLA controllers to improve the power flow. ANN and RFMLA based controllers are very fast and are excellent computing systems. Figure 4 shows the methodology of power flow control.

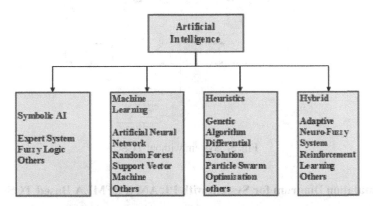

Fig. 3. The major branches of AI

4.2 Simulation Diagram Without TCSC in the System

This simulation diagram consists of different blocks in MATLAB like source block, the load block, the transmission line which is represented by the three phase series RLC branch block, RMS, Scope, display, VI measurement, Power measurement, from and go to blocks. The scope block is used to plot the variation of voltage and the variation of active power in the line. Then the simulation diagrams are developed including TCSC with different controller techniques like the traditional PI, the ANN method and the novel RFMLA. The results for power flow enhancement using different methods are tabulated and the waveforms are plotted. Fig. 5 shows the system without TCSC.

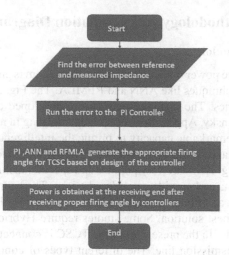

Fig. 4. Flow chart of Methodology

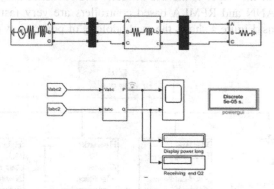

Fig. 5. System without TCSC

4.3 Simulation Diagram for System with PI, ANN, RFMLA Based TCSC

The simulation diagram with PI, ANN and RFMLA based TCSC controller is shown in Fig. 6. The systen consists of transmission line, TCSC block, source, load, control system and firing unit, active and reactive power blocks. The internal details of control subsystem are different for the three types of controller. The TCSC block is added in this system. The TCSC is bypassed for the first 0.5 s using the circuit breaker[20].The simulation is run and the different waveforms are seen on various scopes. The details of the simulation diagram is discussed in Sect. 4.3.1.

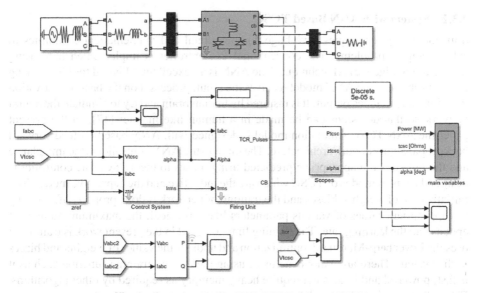

Fig. 6. System with PI, ANN, RFMLA based TCSC

4.3.1 System with PI Based TCSC Controller

The traditional controllers used in the power system are P, PI, PD and PID controllers. In the present work the tuning of firing angles of SCR is done first by PI controller. By properly tuning the controller the impedance error is reduced. The impedance errror is the difference betweeen the measured impedance and the reference impedance given to the system. Properly tuning the system is important to improve the performance and stablity. The simulation system consists of a PI controller block, the impedance measurementt block, firing unit subsystem. In the firing unit subsystem block there are blocks for firing unit of all the three phases A,B and C. There are three PLLs in the firing circuit subsystem which are synchronised with the line current. The PI controller adjust the firing angle of TCSC properly according to the change in impedance. There is a toggle switch for the change of modes which are the manual, inductive and the capacitive modes. In the present work the capacitive mode is used at firing angle of 78 degrees for the PI based TCSC controller. Fig. 7 shows the PI based Controller.

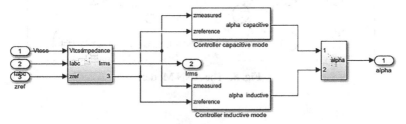

Fig. 7. PI based Controller.

4.3.2 System with ANN Based TCSC

Soft computing techniques have huge computational and mathematical capabilites to solve complex problems. The conventional PI controller is implemented here using (Artficial Intelligence) AI technique. The ANN is an excellent AI based tool for solving Power problems. The ANN model receives the input, process it on the basis of activation function and produces output. It is inspired by human brain and try to simulate the neural network so that decisions can be made in a human like manner [17]. In the present paper the ANN TCSC simulation model is designed with ANN based on feed-forward backpropagation network architecture. The input to the ANN controller is the impedance and the output is the appropriate/predicted firing pulses to the TCSC. The controller is designed for both modes of TCSC which are the inductive and the capacitive modes. The initialization of weights, biases and the training of network is done properly after taking the appropriate values of various parameters like error goal, the maximum number of epochs and the learning rate. The training function used in the present work is trainlm. It uses the Levenberg-Marquardt optimization and updates the values of weights and biases in the system. There are many benefits of using the trainlm training function such as it is fast, powerful and it does not require heavy memory as required by other algorithms. Figure 8 shows the ANN model.

Proper selection of weights and input output function/transfer function affect the working of ANN. The transfer functions here is tansig and purelin functions. The MATLAB command newff is used in the program which is used to initialize the network and creating a network object. The random number generator is used by newff and the initial weights for the network are generated. The parameters of trainlm are used to train the network. This ANN model trains the data of input power and angle. The final value of firing angle is generated based on the values of various ANN parameters fed into the system[18, 19].The ANN TCSC in the capacitive mode finally improved the system's power transfer capacity. The Fig. 9 shows the manual alpha control, the ANN controls (Custom designed Capacitive and Inductive Controls) for the two operating modes of TCSC.

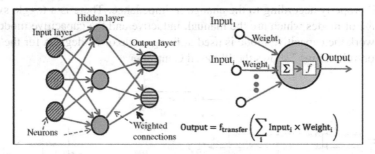

Fig. 8. The ANN Model

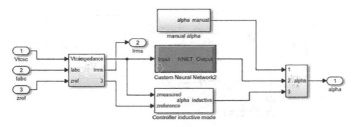

Fig. 9. ANN based Controller

4.3.3 System with Random Forest Machine Learning Algorithm

RFMLA is very popular and having high computational capabilities supervised machine learning method for classification and regression problems. The different types of classification algorithms are the decision tree, random forest-nearest neighbour, Naïve Bayes. As the name suggests RFMLA randomly creates a forest with trees. More the number of trees in the forest the more robust and accurate results are. Multiple decision trees are created here from the dataset and then they are combined together to get and more stable and accurate prediction of results. This is called ensemble learning in RFMLA. Here the predictions are done on the combined results of individual sample models Earlier decision tree method was used which worked effectively with the trained data, but this method captured the results of the data which is then reflected in the test data. Hence the results are influenced by the trained data of decision tree method. As against this is the random forest method, the training of data set is done by the bagging method. This method is used to decrease the variations in the predictions. It combined the results from many decision trees built on different samples of the given data set. The entire data set is divided into a different random subset and then these random subsets are used to split the tree. Hence the drawbacks of the decision tree method are removed here. The representation of Random Forest by Breiman, 2001; Geurts et al., 2006 is as given in Fig. 10 [15]. The internal detail of the system is shown in Fig. 11 below:

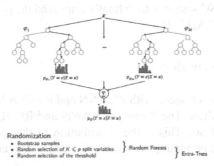

Fig. 10. Random Forest Machine Learning Algorithm (RFMLA) [15]

A call function is created. The data of power and angle is loaded in the workspace from mydata. Mat file. The variable X is chosen for power in MW and variable Y is for the output firing angle in degrees. The values of power are chosen according to the

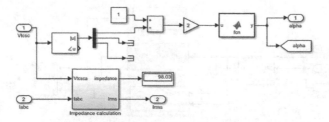

Fig. 11. RFMLA based Controller

TCSC characteristics that power flow decrease in inductive vernier region and increase in capacitive vernier region such that Power in MW = [390 380 360 340 320 300 590 580 570 560] and Firing angle in degrees = [0 10 20 30 40 50 60 70 80 90]. This input data is loaded into the system and the classification method of RFMLA is performed. The entire data set is used for training the classification trees based on bagged ensemble learning method. The tree bagger class is used in this function to create a bag of decision trees. The Baggedensemle. Trees stored the 60-by-1 cell vector which are trained classification trees. Using tree bagger, the trees are grown deeper in size. Hence it is used in the work. The number of bags is chosen for bootstrapping is 60. The different samples are taken from the data and many decision trees are prepared. As the classification is used in the above function hence the majority of the votes is taken for decision. The variable X and variable Y was converted into a [30 × 4] matrix and [30 × 1] matrix of power and angle respectively to plot the errors. The Num Predictors and Num Predictors to Sample are 4 and 2 respectively. The minimum leaf size by default is 1. The in-bag fraction, and the sample with replacement are kept both 1. The compute OOB Prediction command was kept 1 to show the error of out of bag. The class names obtained after optimization are all the angles in the variable Y. The firing angle by RFMLA is found to be 60 degrees. This firing angle is assigned to the simulation file of the system with TCSC using the MATLAB function. This function used EVALIN and ASSIGNIN commands and the data from the workspace. A for loop is created inside the function which sets the appropriate firing pulses to the TCSC system which finally improved the power transfer capacity to the maximum by RFMLA[15, 16].

5 Results and Discussion

Table 1 shows the power obtained with PI, ANN and RFMLA based TCSC. The power is maximum with RFMLA. The Power with ANN and RFMLA is more than that of traditional PI based TCSC. This is the contribution of this paper. RFMLA properly tuned the system based on input power and firing angle data and the line's capacity is significantly enhanced.

From the Table 2.the highest accuracy is found with RFMLA. This work is further extended for checking the errors with RFMLA based TCSC for the two operating modes of TCSC. Two test cases are generated by RFMLA for building the decision trees and various errors are plotted or two cases.

Table 1. Power and Voltage (Phase to Ground) at receiving end of the line

S.No	Controller/Device	Power (MW)	Voltage (kV)
1	Without TCSC	339.44	212.70
2	With PI based TCSC	380.04	225.10
3	With ANN based TCSC	382.43	225.50
4	With RFMLA based TCSC	386.00	227.00

Table 2. Accuracy of different algorithms

S.No	Method	Accuracy (%)
1	Decision Tree	94.20
2	Artificial Neural Network(ANN)	95.00
3	Random Forest Machine Learning Algorithm	97.04

5.1 Case 1: Test Case for Capacitive Mode of TCSC

5.1.1 Building Decision Tree

The decision tree is a popular tool in machine learning for making decisions and is commonly used in operational research and other studies The Random Forest creates many decision trees and then reaches to the solution based on input data The Pseudo code for the decision tree by Breiman et al., 1984 is shown in Fig. 12 [15, 16].

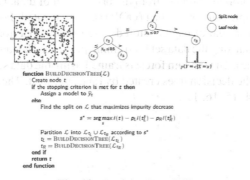

```
function BUILDDECISIONTREE(L)
  Create node t
  if the stopping criterion is met for t then
    Assign a model to ŷₜ
  else
    Find the split on L that maximizes impurity decrease
        s* = arg max i(t) − pₗi(tₗ*) − pᵣi(tᵣ*)
             s
    Partition L into Lₜₗ ∪ Lₜᵣ according to s*
        tₗ = BUILDDECISIONTREE(Lₜₗ)
        tᵣ = BUILDDECISIONTREE(Lₜᵣ)
  end if
  return t
end function
```

Fig. 12. Decision Tree [15]

For the capacitve region of TCSC the power value chosen is 590 MW and the desired firing angle is predicted for this power using predict (Bagged Ensemble, [590]). The prediction is based on the trained input data of power and angle. The result shown by the decision tree based on RFMLA is 60 degrees which lies in the capacitive range of this TCSC. The following decision trees for classification are generated by MATLAB simulation and are shown in Figs. 13a and b.

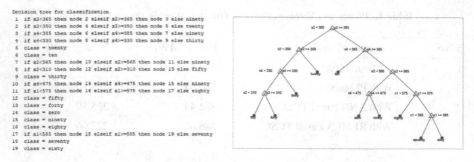

```
Decision tree for classification
 1  if x2<365 then node 2 elseif x2>=365 then node 3 else ninety
 2  if x2<350 then node 4 elseif x2>=350 then node 5 else twenty
 3  if x4<385 then node 6 elseif x4>=385 then node 7 else ninety
 4  if x4<330 then node 8 elseif x4>=330 then node 9 else thirty
 5  class = twenty
 6  class = ten
 7  if x2<565 then node 10 elseif x2>=565 then node 11 else ninety
 8  if x2<310 then node 12 elseif x2>=310 then node 13 else fifty
 9  class = thirty
10  if x4<475 then node 14 elseif x4>=475 then node 15 else ninety
11  if x1<575 then node 16 elseif x1>=575 then node 17 else eighty
12  class = fifty
13  class = forty
14  class = zero
15  class = ninety
16  class = eighty
17  if x1<585 then node 18 elseif x1>=585 then node 19 else seventy
18  class = seventy
19  class = sixty
```

Fig. 13. a. Decision tree (DT) of type 1 **b.** Decision Tree (DT) of type 2

Many decision trees are created by the RFMLA and the desired result for firing angle is generated by MATLAB. The classification method is chosen by the tree bagger. For splitting two variables are selected at random. The decision tree is plotted for getting outcomes of all the possible choices of angles for the given input of power demand. The node of the tree is used for classification and a condition is created on the features of the input to divide the data set according to the conditions chosen. The first decision tree shown in Fig. 13a is in the form of algorithm and the second is shown in Fig. 13b is in the form of a tree by RFMLA.

5.1.2 Plotting Various Errors

On applying bootstrap aggregation two independent sets of bags which are in the bag and out of the bag are created. In the bootstrap sample using sampling with replacement method the data chosen is "in-the-bag". The out-of-bag set of data is that data which is not selected in the sampling process from the original set of data. Using Random Forest many bootstrap samples and out of bag (OOB) sets are created. For each of the tree the observations which are out of bag are stored. The pseudo code for algorithm for OOB error consists of initialising the data set for training purpose. The random forest is created with k number of trees. The random forest is trained using the data set. The bootstrapping is done for each of the decision trees created by random forest. The predictions are done on data set (Figs. 14, 15, 16, 17).

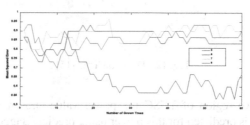

Fig. 14. MSE with no of trees grown

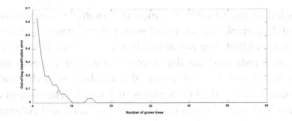

Fig. 15. Out of the Bag (OOB) Classification error (CE)

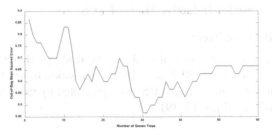

Fig. 16. Out of the Bag (OOB) MSE with trees

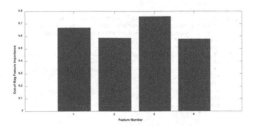

Fig. 17. Out of the Bag (OOB) feature importance

Then the errors are calculated using the difference between the actual value and predicted value. The MSE give how far the classifications are from the true values. It is given by the following equation:

$$\widehat{y_t} = mean(y|t), \, i(t) = \frac{1}{Nt}\sum_{x,y\in}\mathcal{L}_t\left(y - \widehat{y_t}\right)^2$$

The error is plotted for out of the bag (OOB) observations and the number of trees grown using OOB Error Bagged Ensemble. From the results of above figures, it is seen that all the errors which are mean square error (MSE), out of the bag (OOB) classification error (CE) and out of the bag (OOB) mean square error (MSE) are decreasing with the number of trees grown. The MSE is plotted for different leaf sizes. The minimum leaf size by default is 1. The error was calculated for leaf sizes 5, 6, 7, 8 also. The error is minimum in leaf size 5 and the optimal leaf size found by algorithm is also 5. If the size of leaf is increased, it will lead to the problem of overfitting. The out of bag feature importance is calculated using the command OOBPermutedVarDeltaError. The power values are included in the variable X which is a matrix of dimension 30 by 4. In all the 4 columns of X variable the values of power flow corresponding to angles from 0 degree to 90 degree is filled. The out of bag feature is found to be column no 2 and 4 of the variable X which indicated that the values of power in that column of dataset is out of bag. The columns 2 and 4 of the matrix of variable X is having values which is not in range and match with the data of columns in 1 and 3. The data in column no 1 and column no 3 have high probabilities than the column no 2 and column no 4 for getting the appropriate values of power flow.

5.2   Case 2: Test Case for Inductive Mode of TCSC

5.2.1   Building Decision Tree:

For predicting the angle for power flow 340 MW using predict (Bagged Ensemble, [340]) the angle predicted was 30 degrees by RFOA which lies in the inductive range of this TCSC. The following decision tree was generated by running the algorithm on simulation in MATLAB (Figs. 18, 19).

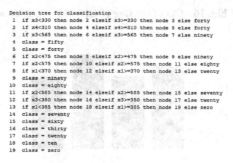

Fig. 18. Decision tree type 1

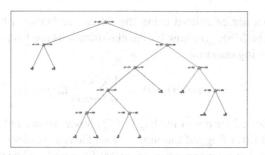

Fig. 19. Decision Tree type 2

For the inductive region also the treebagger is used which provided the ensemble with 60 bagged decision trees. The tree bagger was created with the method of classification. The number of predictors were 4 and the number of predictors to sample was 2. This decision tree is created for 340 MW power. Firing angle obtained from the decision tree corresponding to this power is 30 degrees. This is passed to the simulation file using the MATLAB function.

5.2.2   Plotting Various Errors

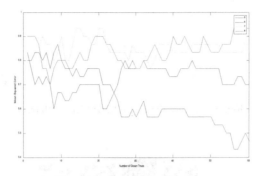

Fig. 20. MSE with grown trees

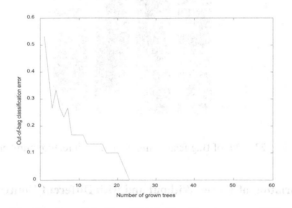

Fig. 21. Out of the Bag (OOB) Classification error (CE)

The compute OOB prediction was set to 1 for calculating the error. The plot of various errors like MSE, out of the bag (OOB) classification (CE) error and out of the bag (OOB) mean square error (MSE) with trees is done for the TCSC in inductive mode also. From the various Figs. 20, 21, 22 it is seen that all these errors are decreasing with the number of grown trees. The plot of out of bag feature importance is shown in Fig. 23. The out of the (OOB) bag feature result showed that the column no 3 and 4 of the data set have identical features and that of column 1 and column 2 dataset have small difference.

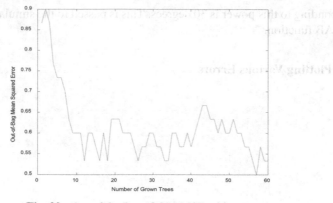

Fig. 22. Out of the Bag (OOB) MSE with trees

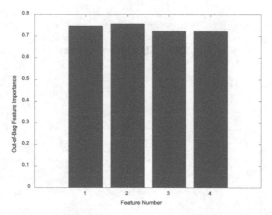

Fig. 23. Out of Bag feature importance with feature number

5.3 Plot of Variation of Power Without and with Different Controllers

Figure 24 shows the plot of variation of power without using any controller and then Figs. 25, 26, 27 shows the power with all the three different types of controllers.During the first 0.5 s the TCSC is bypassed. At 0.5 s the TCSC starts working and the transfer of power is increased. There is more power transfer with ANN method than with PI method. The transfer of power is maximum with RFMLA based controller.

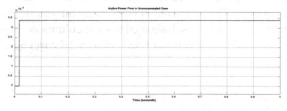

Fig. 24. Power without any Controller

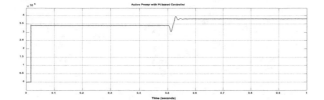

Fig. 25. Power with PI based Controller

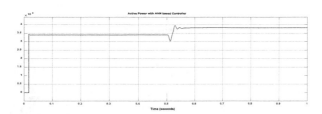

Fig. 26. Power with ANN based Controller

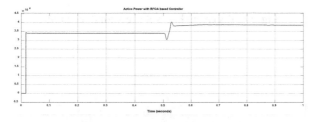

Fig. 27. Power with RFMLA based Controller

5.4 Plot of Variation in Voltage at the Receiving End Without TCSC and with TCSC Based on Three Types of Controllers.

The receiving end voltage is plotted for the system without any controller and then the system with different controllers. The TCSC is bypassed for the first 0.5 s. The voltage by ANN is better than the PI based TCSC. The voltage is maximum with RFOA based TCSC controller which shows that RFMLA is an excellent machine learning algorithm (Figs. 28, 29, 30, 31).

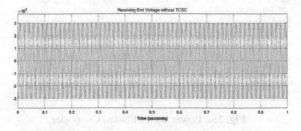

Fig. 28. Receiving End Voltage without TCSC

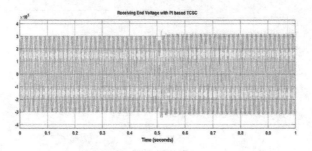

Fig. 29. Receiving End Voltage with PI controller

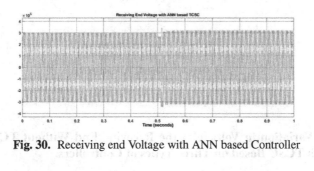

Fig. 30. Receiving end Voltage with ANN based Controller

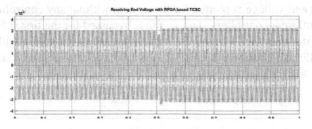

Fig. 31. Receiving End Voltage with RFMLA based controller

6 Conclusion and Future Work

Normally PI controllers are used for power flow enhancement by TCSC. In the proposed research work the use of AI based techniques which are the ANN, and the Random Forest Machine Learning Algorithm is done for power flow enhancement and voltage profile improvement. These ANN and RFMLA are intelligent algorithms and are very fast with excellent computational capabilities. From the results of Table 1 it is seen that using these AI techniques the power and voltage obtained at the receiving end are increased. These AI based techniques properly tuned the system and improved the stability of the system also by increasing the power transfer. The values of power and voltage at the receiving end are obtained more by ANN method than that received by traditional PI controller. These values of power and voltage are further enhanced by RFMLA based model. In this way the power is utilized more efficiently and effectively without the need for investment in new lines and stations. The Power system is now robust due to integration with AI enable technology is and capable of meeting the power challenges due to faults and other disturbances.

The equation showing the relationship between the impedance and firing angle of thyristor is non-linear and the ANN has very good ability to model this non linear and complex relationship. It is successfully applied to the present system. As the number of trees are higher in random forest there is higher accuracy in the results with RFMLA than the decision tree method. The results with RFMLA agrees with the results of [15]. The decision trees are plotted for both the modes of TCSC. The different types of errors are also plotted with the number of trees and the error is found decreasing in both the operating modes of TCSC. In this way ANN and RFMLA based controllers are successfully applied in the present work to meet the various power problems and challenges. The work can be extended on different systems with different loading or operating conditions and the accuracy and performance of the controller can be checked for those conditions. The work on load flow analysis and THD can be done. Thus, TCSC based on ANN and RFMLA is found to be a versatile device for meeting the challenge/objective of various power problems, providing clean, affordable energy to all and for the economic and all-round development of the country.

Acknowledgement. The authors would like to pay a vote of thanks to Electrical and Electronics Engineering Department, Principal of Ghousia College of Engineering for providing warm support.

References

1. Arora, K., Agarwal, S.K., Kumar, N., Vir, D.: Analysis of power flow control in power system model using TCSC. Int. J. Eng. Res. Appl. **3**(3), 821–826 (2013)
2. Kamel, S., Abokrisha, M., Selim, A., Jurado, F.: Power flow control of power systems based on a simple TCSC model. Ain Shams Eng. J. **12**, 2781–2788 (2021)
3. Fathollahi, A., Kargar, A., Derakhshandeh, S. Y.: Enhancement of power system transient stability and voltage regulation performance with decentralized synergetic TCSC controller. Int. J. Electr. Power Energy Syst. 1–11 (2021)

4. Hemeida, A.M., Hamada, M.M., Mobarak, Y.A., El-Bahnasawy, A., Ashmawy, M.G., Senjyu, T.: TCSC with auxiliary controls-based voltage and reactive power controls on grid power system. Ain Shams Eng. J. **11**, 587–609 (2020)
5. Ibrahim, D.K., Abo-Hamad, G.M., Zahab, E.E. D.M.A., Zobaa, A.F.: Comprehensive analysis of the impact of the TCSC on distance relays in interconnected transmission networks **8**, 228315–228325 (2020)
6. Fasihipour, H., Seyedtabaii, S.: Fault detection and faulty phase(s) identification in TCSC compensated transmission lines. IET Gener. Transm. Distrib. **14**(6), 1042–1050 (2020)
7. Cephas, O., Pandya, A.: Analysis of fault location in a transmission line with and without TCSC. In: International Conference on Advances in Power Generation from Renewable Energy Sources, pp. 1–12 (2019)
8. Pandiyan, M.K., Idhayaselvi, V.A., Danalakshmi, D., Sheela, A.: Online estimation of control parameters of FACTS devices for ATC enhancement using artificial neural network. In: IOP Conference Series: Materials Science and Engineering, IOP Publishing, pp. 1–11 (2021)
9. Ruhi, Z.M., Jahan, S., Uddin, J.: A hybrid signal decomposition technique for transfer learning based industrial fault diagnosis. Ann. Emerg. Technol. Comput. **5**(4), 37–53 (2022)
10. Dodangeh, M., Ghaffarzadeh, N.: An intelligent machine learning-based protection of ac microgrids using dynamic mode decomposition. Iranian J. Electr. Electron. Eng. **4**, 1–9 (2022)
11. Gandotra, R., Pal, K.: FACTS technology: a comprehensive review on FACTS optimal placement and application in power system. Iranian J. Electr. Electron. Eng. **18**(3), 1–14 (2022)
12. Meikandasivam, S., Vijayakumar, D., Nema, R.K., Jain, S.K.: Investigation of split TCSC on Kanpur-Ballabhgarh transmission system. Int. J. Electr. Comput. Eng. **8**, 76–83 (2014)
13. Liao, M., Yao, Y.: Applications of artificial intelligence-based modelling for bioenergy systems: a review. Wiley On Line Library (2021)
14. Keshkeh, K., Jantan, A., Alieyan, K.: A machine learning classification approach to detect TLS-based malware using entropy-based flow set features. J. ICT **21**(3), 279–313 (2022)
15. Akash, N.S., Rouf, S., Jahan, S., Chowdhury, A., Uddin, J.: Botnet detection in IoT devices using random forest classifier with independent component analysis. J. ICT **21**(2), 201–232 (2022)
16. Niva Mohapatra, K., Shreya, A.C.: Optimization of the random forest algorithm. In: Borah, S., Balas, V.E., Polkowski, Z. (eds.) Advances in Data Science and Management: Proceedings of ICDSM 2019, pp. 201–208. Springer Singapore, Singapore (2020). https://doi.org/10.1007/978-981-15-0978-0_19
17. Santra, A., Dutta, A.: A comprehensive review of machine learning techniques for predicting the outbreak of Covid-19 cases. Int. J. Intell. Syst. Appl. **14**(3), 40–53 (2022)
18. Panda, D., Mukhopadhyay, S., Bachhar, A.K., Roy, M.: Multi criteria decision making based approach to assist marketers for targeting BoPs regarding packaging influenced purchase during Covid-19. Int. J. Intel. Syst. Appl. **14**(3), 1–17 (2022)
19. Hagshenas, N., Mojarad, M., Arfaeinia, H.: A fuzzy approach to fault tolerant in cloud using the checkpoint migration technique. Int. J. Intel. Syst. Appl. **14**(3), 18–26 (2022)
20. Jovcic, D., Pillai, G.N.: Analytical modeling of TCSC dynamics. IEEE Trans. Power Delivery **20**(2), 1097–1104 (2005)

Child Access Control Based on Age and Personality Traits

Alguliyev M. Rasim, Fargana J. Abdullayeva, and Sabira S. Ojagverdiyeva[✉]

Institute of Information Technology, Baku, Azerbaijan
allahverdiyevasabira@gmail.com

Abstract. Exposure to harmful information on the Internet and constant use of digital devices (computer, phone, tablet, etc.) harm children's psychology and health. The use of methods that control access to the Internet and filter web content considered harmful to children is an effective means of solving this problem. The article proposes a method to control access to the Internet, taking into account several personality traits of the user-child (age, eye diseases, heart diseases, neurological and psychological conditions, etc.) using Mamdani-based fuzzy logic inference system. The values of the input parameters of this system are described by five linguistic parameters, which are "very low", "low", "medium", "high" and "very high" and with a triangular membership function. This approach is focused on the individual user and the main advantage is that parameters are used in vector form. This research is significant for the use of parents, guardians and those responsible for children.

Keywords: fuzzy logic extraction system · malicious information · Mamdani model · screen time · children's personality traits

1 Instruction

Modern children use digital devices from a younger age, and digital technology has become a key component of their lives [1]. Using smartphones and tablets has already become an easy activity for children. According to research, 97% of families have at least one smartphone, 75% have a tablet or a computer, and 44% of young children have a personal tablet and a computer [2].

Content posted on the web pages affects children's intellectual level, emotional state, mental and physical development, personality formation, and in short, psychology. These factors require a serious concern for children's online safety and finding effective solutions to protect them in cyberspace [3, 4].

There are current works on this feature, but in these cases one or two privacy features are used to block harmful internet content. (age, voice, etc.) The approach we propose is multi-criteria: age, psychological state, heart diseases, eye diseases, etc. includes such personality traits.

Some studies suggest an approach to protecting children's use of digital devices by implementing built-in safety and privacy controls, such as screen time or parental controls. Here are some age-appropriate safety precautions for children when using smart devices like the iPhone [4]. Cyber grooming is a threat that can cause more damage to children's psychology. [5] proposes a model that classifies cyber grooming attacks based on artificial intelligence to organize online protection of children.

In [6], children's access to harmful content is prevented by creating age-restricted profiles, and a method based on machine learning that automatically detects their age and filters out any suitable content by applying suitable restriction programs is proposed. Another point here is related to the proper regulation of the time children and adolescents spend in front of the computer. In many cases, users get certain diseases by spending a lot of time in front of a computer [7].

It should be noted that a standard approach to protecting children from the Internet or digital devices, a set of official rules, etc. are not developed. In this case, experts offer a variety of options to monitor the time a child spends on the computer. The purpose of the article is to prevent children from being exposed to harmful information while using Internet resources and to organize screen time control. User-child health, psychology, etc. when certain problems arise, the time spent in front of the monitor is limited and decisions are made by applying a control system based on Mamdani.This method is applied to each child individually, and as a result, decisions are made according to the child's situation.

1.1 Related Work

In order to protect minors from harmful web content and age-inappropriate information on websites, Youtube and other virtual spaces, methods that detect web content and implement web filtering are used. Systems that control users' access to web content are based more on biometric methods. These systems use many personality traits: age, face recognition, gender recognition, keystroke dynamics, voice recognition, etc. uses various functions such as [8, 9].

The methods used to detect web pages that are considered harmful content to children are based on the use of "blacklists". In the field of computer science, "blacklists" are a mechanism that controls access. [10] suggests a hybrid approach using a blacklisted URL database and the Naive Bayes algorithm. By blocking a link to a malicious website, only positive results are stored in the list and presented to the user. This content is considered accessible only for children aged 5–12.

Considering that pornographic information (images, audio, video, text containing pornographic content) is not relevant for child audience, using the data sanitization method, the contents of the malicious multimedia are cleaned and presented to the user [11]. The literature has developed a new Deep Neuron Network architecture called ChildNet, which filters out harmful image content that is age-inappropriate for children [12].

Another study uses machine learning methods to detect erotic texts on social media. Minors are prevented from exposure to the content in the form of obscene, aggressive, erotic or rude comments on the Internet [13]. Based on the fact that more and more unwanted and harmful information is spread among the users of the social network environment, a unified system for monitoring the harmful effects of social networks is introduced [14].

Some approaches offer methods to protect children online using biometric technologies. When using biometric methods, the privacy of the individual is more reliably ensured. The security condition applied here requires the security of sensitive information belonging to individuals, services or means. Such security systems provide access only to authorized persons, the access to the children's profiles by other persons, as well as criminals (sexual exploiters, cybercriminals, etc.) is restricted.

Access control is a key element of security systems that is responsible for assessing whether a user is allowed to run a particular program. The literature [15] introduces an access control system called BiometricAccessFilter. This system restricts the access of malicious people to the virtual world of children and adolescents and is important in preventing violence or aggression against them and in ensuring security. Here the age of the user is determined by voice recognition by the system.

A lot of research is being done with IoT applications on child protection issues [16]. A GSM-based security system is proposed. Another approach is child safety through wearables [17]. An integrated Care4Student in-system notification system is offered, which includes a server-based as well as a mobile-based module for security [18].

According to some of the approaches mentioned above, it can be concluded that existing research designed to protect children from harmful content and habits acquired from the Internet has taken into account mostly one biometric trait of the users. However, in our proposed work, screen time is set for each user, taking into account several individual characteristics of children and adolescents (age, heart and eye disease, psychological condition) and their access to information is controlled.

2   The Architecture of the Proposed System

Parental control programs can prevent children from accessing content that is not suitable for their age, manage the applications they use, check their log files, virtual friends they connect with, view their messages, and more. is used for.

The architecture of the system providing individual protection of children is described below (Fig. 1).

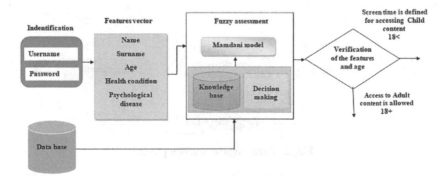

Fig. 1. Children's access control system on the Internet

According to the given architecture, the information about the child's age, health and psychological condition is considered as a criterion. The system is installed by the parent on the child's personal computer, and the process is performed sequentially when the child wants to access the computer. The elements that make up this system are described below.

Identification: the process of recognition of a child or adolescent to enter the system.

Features vector: includes information about the child's personal traits (name, surname, age, health and psychological condition).

Database: this database contains information about the diseases and psychological state of children from an early age.

Fuzzy assessment: in this section, the knowledge base is formed by determining the conditions for access to the content, taking into account the age, health and psychological condition of the child, and decisions are made based on this knowledge.

Verification of signs: according to the age of the child, his access to Child content or Adult content is determined. As shown in Fig. 1, if a child is under the age of 18, he or she is allowed to access Child content or a screen time is set. Otherwise, i.e. if the child is older than 18 years, his access to Adult content is determined.

3   Fuzzy Logic Extraction System

The expert's knowledge and experience are used in building the system. Rules based on the IF-THEN condition are given between the input and output data of the process [19]. The information processed in the fuzzy logic extraction model consists of two parts. The first part defines the membership functions of the input and output variables. The second part is based on rules.

Fuzzy variable is a set in the form $< \alpha, X, A >$, here we have:

α is the name of the fuzzy variable;

X is the domain of the function;

A is a fuzzy set in the space X.

The knowledge base is covered by general knowledge related to this domain, with the conditions and results of the problem. FLES generally consists of 3 blocks (Fig. 2) [19]:

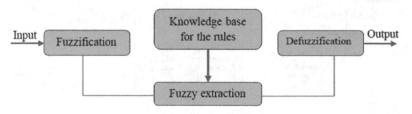

Fig. 2. Fuzzy logic extraction process

1. *Fuzzification block.* Converts fuzzy input data into fuzzy data (creates a fuzzy subset).
2. *Fuzzy extract block.*

3. *Defuzzification block.* Converts fuzzy output values to non-fuzzy values.

The structure of Mamdani algorithm consists of several successive stages. At the same time, each subsequent step takes the values obtained in the previous step as input. Here, the Rules consist of conditions and conclusions, which create fuzzy expressions.

3.1 Application of Mamdani Type Fuzzy Logic Extraction System

In the proposed work, a fuzzy logic extraction algorithm of the Mamdani type is used. In this model, the evaluation stages consist of the following steps [19]:

Database of Rules. It is designed to formally represent the empirical knowledge of experts in a particular subject area in the form of fuzzy extraction rules. This stage is the initial stage, a set of rules where a specific weight factor is assigned to each sub-result. In this study, a knowledge base on the topic is established first to determine security and screen time.

Fuzzification. Evaluation is carried out on the basis of 6 rules in the implementation of this task and each of these rules depends on whether the entry belongs to a different linguistic fuzzy set. In the process of fuzzification, regardless of what the input variables describe, the output variables are rated in the range 0 to 1. This step shows the basic parameters needed to determine security. The dimensions of each parameter are given in linguistic terms and converted to the corresponding fuzzy number. The research uses the triangular membership function. The triangular membership function is defined by 3 parameters {a, b, c} and is written as in the (1) formula:

$$f(x : a, b, c) = \begin{cases} 0, & x \leq a \\ \dfrac{x-a}{b-a}, & a \leq x \leq b \\ \dfrac{c-x}{c-b}, & b \leq x \leq c \\ 0, & c \leq x \end{cases} \tag{1}$$

Establishing Fuzzy Rules. Fuzzy rules are defined using the conditional operators If and Then. The result of the If-Then rule is determined by a fuzzy set.

Condition 1: If x = A, then y = B and z = C.

Here x and y are linguistic variables, and A and B are linguistic values defined by fuzzy sets. The values of the input parameters of the fuzzy logic extraction system are described by five fuzzy sets "very low", "low", "medium", "high", and "very high". Three parameters are taken as input parameters: the user's age, health and psychological condition. These fuzzy sets define the form and state of the membership function and are described in Table 1, Table 2 shows the values of the Internet content extraction block.

Extraction. At this stage, decisions are made on the basis of fuzzy rules. The output parameters are calculated for the set rules. Some of these rules are structured as follows:

1. If (child's age is 5, very low) and (health condition is medium) and (psychological condition is medium) then childcontent can be accessed no more than 10 min.

Table 1. Values of the fuzzy triangular number

Output parameters	Linguistic values	Interval values for the fuzzy triangular number
Adult content (no access)	Very low	[−0.4 0 0.2474]
Child content (no more than 10 min)	Low	[0.0127 0.2447 0.491]
Child content (no more than 20 min)	Medium	[0.251 0.504 0.742]
Child content (no more than 1.5 h)	High	[0.497 0.718 0.996]
Adult content (accessible)	Very high	[0.747 0.9987 1.05]

Table 2. Linguistic values of the Internet content extraction block

Linguistic values	Child's age
Very low	4 years old
Low	6 years old
Medium	12 years old
High	16 years old
Very high	18 years old

2. If (child's age is 4, very low) and (health condition is medium) and (psychological condition is medium) then childcontent cannot be accessed.
3. If (child's age is 6, low) and (health condition is medium) and (psychological condition is medium) then childcontent can be accessed for 15 min.
4. If (child's age is 12, medium) and (health condition is low) and (psychological condition is low) then childcontent cannot be accessed.
5. If (child's age is 14, medium) and (health condition is low) and (psychological condition is medium) then childnet can be accessed for 20 min.
6. If (child's age is 16, high) and (health condition is low) and (psychological condition is low) then childnet cannot be accessed.

Aggregation. This step uses the values of the functions obtained in the fuzzification stage. For each of the fuzzy extraction system rules, the conditions of the fuzzy extraction system and the situation of the validity of the conditions are calculated. The maximum or sum of the output of the rules assigned is calculated to check the accuracy.

Defuzzification. In this step, the fuzzy number is converted to a non-fuzzy value using the defuzzification method. Here, a decision algorithm is implemented that selects the best non-fuzzy value based on a fuzzy set and the value of the i-th output variable is calculated as in formula (2) using the central gravity method. Here, the i-th output variable is included in $E_i (i = 1...n)$.

$$y_i = \frac{\int_{min}^{max} x \cdot \mu_i(x) dx}{\int_{min}^{max} \mu_i(x) dx} \tag{2}$$

Here y_i is the value obtained as a result of defuzzification,

  μ_i is a membership function corresponding to E_i fuzzy set.
  min, max are the boundaries of the fuzzy set.

3.2 Results

The presentation of graphic data in the proposed expert system is processed in the fuzzy extraction system of the Matlab program. As one of the results of the experiment, a surface model based on certain dependencies is presented in the following style. Here, the child's access to harmful content is visualized, taking into account parameters suitable for the child's age and neurological condition (Fig. 3).

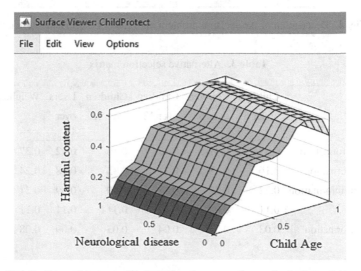

Fig. 3. Dependence graph of children by age and neurological condition

In the figure below, the child's access to harmful content is visualized, taking into account the parameters appropriate to the child's age and neurological condition.

3.3 Comparison of the Proposed Method with the AHP Method

Comparison of the proposed method with AHP (Analytic Hierarchy Process) can be done as follows. In the approach proposed using the AHP method, the type of information is determined using only the age category. In the model below, this idea is depicted and its decomposition is given [20] (Fig. 4).

Here, the purpose of the analysis is described in the first layer, the set of criteria in the second layer, and the alternatives to be selected in the third layer. As can be seen from the calculations made according to Table 3. Ranking of children's access to Internet content is determined based on age category.

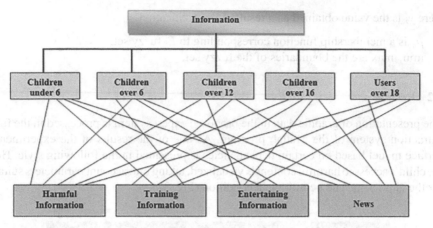

Fig. 4. Description of the decision problem in terms of criteria and alternatives

Table 3. Alternative selection matrix

Alternatives	Children under 6	Children over 6	Children over 12	Children over 12	Users over 18	Weightage	Score
Harmless information	0.47	0.47	0.37	0.10	0.12	0.37	0.38
Training information	0.16	0.18	0.31	0.35	0.28	0.24	0.23
Entertaininginformation	0.22	0.22	0.18	0.40	0.39	0.21	0.25
News	0.11	0.09	0.10	0.11	0.11	0.11	0.10
Harmful information	0.03	0.03	0.04	0.03	0.09	0.08	0.04

As can be seen from the table, when using this method, information for children is determined by age. However, in our proposed Mamdani method, the number of personal traits is more.

4   Conclusion and Discussion

Experiments showed that different surface models were obtained. The various parameters of these surface models are interdependent and visually different. As the parameters we choose change, so does the convexity of the model we build.

In today's age, it is necessary for children and adolescents to have computer skills and develop digital skills. But protecting their health is more important.

1) The article proposes an approach to protect children from harmful content, to ensure proper use of digital devices without harming their health and psychology, using a Mamdani-type fuzzy logic extraction system.
2) If-then rules are defined in this approach. A fuzzy output method was developed in the Matlab software package.

3) The results obtained in the output of the fuzzy logic inference system were satisfactory.

References

1. Konca, A.S.: Digital technology usage of young children: screen time and families. Early Childhood Educ. J. **50**(7), 1097–1108 (2022)
2. Rideout V., Robb M.B.: The common sense census: media use by kids age zero to eight **2020**, 2020 (2020)
3. Alguliyev, R., Ojagverdieva, S.: Conceptual model of national intellectucal system for children safety in internet environment. Int. J. Comput. Network Inf. Secur. **10**(3), 40–47 (2019)
4. Siddiqui, Z., Zeeshan, N.: A survey on cybersecurity challenges and awareness for children of all ages. In: 2020 International Conference on Computing, Electronics & Communications Engineering (iCCECE), Southend, UK, 131–136 (2020)
5. Isaza, G., Muñoz, F., Castillo, L., Buitrago, F.: Classifying cybergrooming for child online protection using hybrid machine learning model. Neurocomputing **484**, 250–259 (2022)
6. Pulfrey, J., Hossain, M.S.: Zoom gesture analysis for age-inappropriate internet content filtering. Expert Syst. Appl. **1**(199), 116869 (2022)
7. Stajduhar, A., Ganel, T., Avidan, G., Rosenbaum, R.S., Freud, E.: Face masks disrupt holistic processing and face perception in school-age children. Cognitive Res. Principles Implications **7**(1), 1–10 (2022)
8. Safavi, S., Russell, M., Jančovič, P.: Automatic speaker, age-group and gender identification from children's speech. Comput. Speech Lang. **50**, 141–156 (2018)
9. Roy, S., Sinha, D., Roy, U.: Identifying age group and gender based on activities on touchscreen. Int. J. Biometrics **14**(1), 61–82 (2022)
10. Kamarudin, A.N.A., Ranaivo-Malançon, B.: Simple internet filtering access for kids using Naïve Bayes and blacklisted URLs. In: International Knowledge Conference 2015, Kuching, Sarawak, Malaysia, pp. 1–8 (2015)
11. Alguliyev, R.M., Abdullayeva, F.J., Ojagverdiyeva, S.S.: Protecting children on the internet using deep generative adversarial networks. Int. J. Comput. Syst. Eng. **6**(2), 84–90 (2020)
12. Alguliyev, R.M., Abdullayeva, F.J., Ojagverdiyeva, S.S.: Image-based malicious Internet content filtering method for child protection. J. Inf. Secur. Appl. **65**, 103123 (2022)
13. Barrientos, G.M., Alaiz-Rodríguez, R., González-Castro, V., Parnell, A.C.: Machine learning techniques for the detection of inappropriate erotic content in text. Int. J. Comput. Intell. Syst. **13**(1), 591–603 (2020)
14. Kotenko, I., Saenko, I.., Chechulin, A.., Desnitsky, V.., Vitkova, L.., Pronoza, A..: Monitoring and counteraction to malicious influences in the information space of social networks. In: Staab, S., Koltsova, O., Ignatov, D.I. (eds.) Social Informatics: 10th International Conference, SocInfo 2018, St. Petersburg, Russia, September 25-28, 2018, Proceedings, Part II, pp. 159–167. Springer International Publishing, Cham (2018). https://doi.org/10.1007/978-3-030-01159-8_15
15. Ilyas, M., Fournier, R., Othmani, A., Nait-Ali, A.: BiometricAccessFilter: a web control access system based on human auditory perception for children protection. Electronics **9**(2), 361 (2020)
16. Deva, S.V., Akashe, S.: Implementation of GSM based security system with IOT applications. Int. J. Comput. Network Inf. Secur. **9**(6), 13–20 (2017)
17. Aliyu, M.B., Ahmad, I.S.: Anomaly detection in wearable location trackers for child safety. Microprocess. Microsyst. **91**, 104545 (2022)

18. Akyol, K., Karacı, A., Tiftikçi, M.E.: Care4Student: an embedded warning system for preventing abuse of primary school students. Int. J. Intell. Syst. Appl. **14**(4), 1–10 (2022)

19. Abdullayeva, F.C.: Development of collaborative risk assessment method for dynamic federation of clouds. Problems Inf. Technol. **2**(6), 46–58 (2014)

20. Abdullayeva, F.J., Ojagverdiyeva, S.S.: Multicriteria decision making using analytic hierarchy process for child protection from malicious content on the Internet. Int. J. Comput. Network Inf. Secur. **13**(3), 52–61 (2021)

Identifying the Application of Process Mining Technique to Visualise and Manage in the Healthcare Systems

Arezoo Atighehchian[1] , Tahmineh Alidadi[2] , Reyhaneh Rasekh Mohammadi[3] ,
Farhad Lotfi[4] , and Sima Ajami[5(✉)]

[1] Department of Industrial Engineering and Futures Studies, Faculty of Engineering,
University of Isfahan, Hezarjerib Avenue, Isfahan, Iran
[2] Department of Health Information Technology and Management, School of Medical
Management and Information Sciences, Isfahan University of Medical Sciences,
Hezarjerib Avenue, Isfahan, Iran
[3] Faculty of Medicine, University of Belgrade, Dr. Subotića Street 8, 11000 Belgrade, Serbia
[4] Faculty of Organizational Sciences, University of Belgrade, Jove Ilića 154,
11000 Belgrade, Serbia
[5] Department of Health Information Technology and Management, School of Medical
Management and Information Sciences, Isfahan University of Medical Sciences,
Hezarjerib Avenue, Isfahan, Iran
simaajami@yahoo.com

Abstract. This study aims to identify the application of process mining techniques in health centres for the visualisation of healthcare activities. As a scoping review, this research was used and divided into three phases: literature collection, assessment, and selection. A literature search had done on Google Scholar, Web of Science, PubMed, Elsevier, and ProQuest, along with the impact of inclusion and exclusion criteria. Keywords have been addressed as follows: process mining, visualising, mapping, workflow mining, automated business process, discovery, process discovery, performance mining, healthcare, hospital, emergency department, emergency medical service, and apply. The findings showed that process mining can be used to analyse different activities in the field of healthcare, including workflow in healthcare, clinical and administrative processes, data analysis in information systems, events data in patients' infectious, creation of dashboards, the discovery of unexpected, and hidden relationships. Finally, as the significance of this research, it has been argued that the use of process mining in healthcare allows health professionals to understand the actual implementation of processes.

Keywords: Process Mining · Healthcare · Visualisation · Bottleneck

1 Introduction

The world is faced with resource constraints such as increasing healthcare costs, complexity, and expensive therapies. The use of modern technologies to evaluate and analyse these processes is essential [1, 2]. Healthcare processes are very dynamic, complex, temporary, and increasingly multi-disciplined, so need to analyse and improve continuous

Z. Hu et al. (Eds.): ICCSEEA 2023, LNDECT 181, pp. 299–308, 2023.
https://doi.org/10.1007/978-3-031-36118-0_26

activities of health processes [3]. Process mining is a part of techniques related to the process management and data science that helps to identify the arrangement of treatment activities [4].

Process mining has been able to distinguish characteristics and challenges in the field of healthcare [5]. In addition to these reasons, process mining has been chosen as a research area in this study. Process mining technology has also been able to improve healthcare and reduce the cost of this system by discovering the correlation in data [5–10]. Process management has been considered in many organisations, but healthcare systems had more complex processes than other industries, and it can play as an important role in different fields [11]. Therefore, the use of process mining techniques in healthcare in the form of identifying real processes and methods is essential [1]. In addition, process mining is rapidly becoming an important success for modelling the transmission of the virus, controlling infection, and emergency response analysis of pandemics and epidemic diseases. The existing solutions for process mining include discovering the actual process and presenting it in the form of a model, specifying the type and place of the required change in the process and extracting process information, and calculating KPIs for evaluation by decision-makers. This study's focus was to identify the application of process mining techniques in health centres for visualisation and management in this regard. In this approach, researchers introduced the process mining approach which is a solution to improve productivity, reduce waiting times for receiving health services, reduce costs, improve complex, and time-consuming processes in the health and treatment system and eliminate process bottlenecks.

2 Methods

This study was a scoping review the literature was on the title of visualising and managing the healthcare by process mining techniques. A sub-systematic method was used, which was divided into three phases: literature collection, assessment, and selection. A literature search had done on libraries, databases, and motor engines including Google Scholar, Web of Science, PubMed, Elsevier, and ProQuest, from 2007 to 2021 along with the impact of inclusion and exclusion criteria (Table 1 and Table 2).

Table 1. Inclusion and exclusion criteria.

Criterion			
	Time period	Primary Keywords	Scope
Inclusion	14 recent years (2007–2021)	Process mining OR Workflow mining OR Process Discovery AND health OR Hospital OR Prehospital emergency	Information Technology AND Healthcare
Exclusion	Any study outside this date	Any study without this Title/Abstract/ key word	Any study outside this Scope

The keywords, their synonyms, and their combinations were employed in the searching areas. Combinations in search terms, including MeSH, and Nama thesaurus (science and technology policy information exchange system) approved were used in the searching areas of title, abstract, and keywords with "AND" and "OR" (Table 2).

Table 2. Search terms and strategies.

SPIDER Tool	Search term
S	(Hospital OR Healthcare centre OR Pre Hospital Emergency) AND Process mining
PI	(Barrier OR Challenge) OR (opportunity OR Benefit OR Application) OR (Process mining OR Workflow mining OR Process Discovery) AND ("healthcare centre*" OR "hospital" OR "emergency department")
D/E/R	"Qualitative" OR "Quantitative" OR "Mixed method" OR "Review"

* SPIDER = Sample, Phenomenon of Interest, Design, Evaluation, Research type

By derivation of correlation coefficient and curve alignment of WAMS/SCADA data, When 110 kinds of literature are read, it was found that very limited systematic

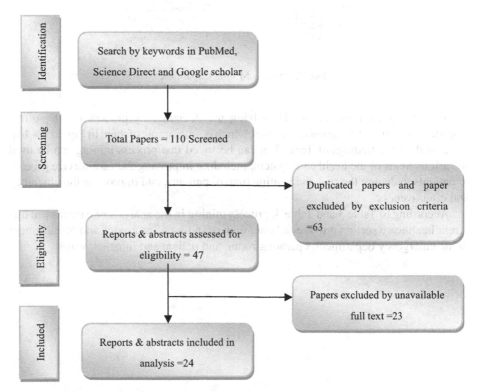

Fig. 1. Flow diagram: The process of PRISMA for data collection and analysis.

information was regarding all of the domains this study wanted to investigate. So, after de-duplication, titles and abstracts of articles and reports were screened, out of which 29 studies were selected for the review based on their relevance and access to full text (Fig. 1).

3 Results

The mortality rate of patients referring to the emergency rose at the time of the COVID-19, and the amount of use ambulances with COVID-19 has roughly was 10% more than other patients. Also, the acceptance rate for COVID-19 disease in hospital was about 49%, while in other groups, 22%. In addition, the acceptance rate for COVID-19 in the ICU was nearly 5% higher than the other patients [12] (Fig. 2).

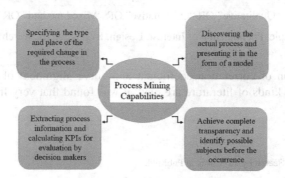

Fig. 2. Process Mining Capabilities.

The global error can be got by solving the deviation by process mining offers countless benefits in the various processes, which is an example of them in Fig. 1 [13–16].

Based on the findings of Table 3, it can be stated that process mining can be used in various fields of the healthcare system, including improving the care service process to patients, reducing the cost and waiting time of patients, and improving the efficiency and effectiveness.

According to Table 3 and Table 4, process mining in the analysis of care flow in different healthcare sectors provides a better understanding of processes and management in the emergency department, operating room, and patient and inpatient wards.

Table 3. Examples of Practical Process Mining Applications in Healthcare.

Outcome	Country	Tool	#
Information, analysis, and production of the model can be obtained by tools and techniques of process mining for using less structured, flexible, and multidisciplinary processes [17]	The Netherlands	ProM	1
Three cluster models were proposed according to the triage level in the emergency department. In the first cluster, green, yellow, and red triage were 34%, 57%, and 8% respectively, which included approximately 50% of patients. In the second cluster, (10% of patients), red triage was much normal, while in the third cluster (almost 40% of cases), yellow triage is exceeded [18]	Greece	ProM Disco Fuzzy, Genetic & Heuristic Mining	2
Better insight has been obtained by analysing the pathway via a process mining approach [19]	Australia	ProM	3
The process of mining showed that the bottleneck emergency process is the "disease diagnosis stage". It took more time in comparison to examining the vital signs and sending sample tests which were 29.3, 11, and 4.5 min respectively [20]	The USA	Disco	4
Combining process mining methods with hospital information systems enabled unskilled managers to support clinical and administrative decisions to fix existing system problems. The first activity period was more than the second activity. The average time of activity (14.38 h) was also lower than the second activity (3.31 days) [21]	Cuba	Disco	5
Using the process mining of different types of data in hospital information system in a spectrum of administrative systems, clinical support systems, logical health systems, and medical equipment information are fully described. There was also a lot of waiting time as follows: the average time before hospitalisation in hospital (12.87 days), outpatient intestinal clinics (average of 4.25 days), and surgery (Average of 6.63 days) [22]	The Netherlands	Questionnaire	6

(*continued*)

Table 3. (*continued*)

Outcome	Country	Tool	#
Results showed that the use of analysis of events, mapping, and process analysis was able to display the paths of the patient and employees to reduce the waiting time, patient re-acceptance, and simplification of clinical processes for better understanding of care-providers [23]	The Netherlands	ProM	7
The combination of the mining and simulation process indicates that the use of prevention techniques significantly reduced the cost of medical care by approximately 45% [24]	France	Various tools in process mining	8
According to the research project, the dashboard allowed discovery and increase the flow of patients based on the spatial data of patients undergoing intervention in the operating room. Therefore, 67% of users wanted to use daily dashboards to organise and monitor operation room. Between 33 and 44% of respondents agreed with quite difficult questions related to simplicity, compatibility, and intuition of the dashboard. Only two questions had 10% severe and 22% strong opposition, and these were related to the ease of use [25]	Spain	Various tools in process mining	9
Process mining was able to redesign and simplify their processes which could be converted into higher-quality processes. It was probably due to underestimating stroke symptoms by patients, general relatives, and physicians. After the onset of the stroke, 27% of patients decided instead of contacting relatives (18%) prefer to contact a general practitioner (22%) [26]	Italy	Various tools in process mining	10
In this research, performance analysis was used by considering the process of cycle time and costs that helped the pre-hospital emergency, improving their processes by identifying bottlenecks. In addition, the ratio of the delay time to the total process time, 0.49 was high [5]	Iran	Disco	13

(*continued*)

Table 3. (*continued*)

Outcome	Country	Tool	#
Waiting times in each of the emergency activities of the treatment were identified. The results showed that the emergency department activities (29% of total waiting time) it was the main emergency bottlenecks. Also, CT scan, triage, and laboratory activities were 21%, 24%, and 25% over standard time respectively [27]	Iran	ProM Disco	14

Table 4. The result of the application of process mining in health.

Country	Application
The Netherlands	A better understanding of the healthcare system [17]
Greece	Studying and analysing processes in the emergency department in-depth [18]
Australia	Patient path analysis from start until the end of care [19]
The USA	Workflow analysis in healthcare [20]
Cuba	Clinical and administrative process analysis of the hospital [21]
The Netherlands	Analysis of various types of data in hospitals [22]
The Netherlands	Analysis of trend care in sepsis patients [23]
France	Analysis of clinical pathways in patients with a hernia [24]
Spain	Create dashboards in the operating room [25]
Italy	Care in strokes [26]
Iran	Improving pre-hospital emergency processes [5]

4 Discussion

Using process mining as an appropriate and flexible tool for improving the performance of treatment processes has been chosen. To perform corrective actions, it can examine the therapeutic processes and identify existing bottlenecks in treatment [12, 21]. Discovering the actual process and identifying possible subjects before the occurrence has been introduced as some of the process mining capabilities so which in previous studies, Zhou et al. the stage of "diagnosis of any diseases", was more than 29.3 min in the emergency department [20]. Also, the results of Mans et al., showed that there was a lot of waiting time before hospitalisation in hospital [22]. The countless capabilities related to the process mining have led to be an important success in modelling for transmitting, controlling, analysing, and responding to the virus. In brief, research findings confirm that using Information and Communications Technology (ICT) can have quite benefits in the field of healthcare. Using traditional methods and not using ICT, as well as Artificial Intelligence (AI) in the medical industry can cause several problems and challenges [28–31]. In particular, process mining techniques could be used to analyse complex data, manage them, and so on. These techniques can also focus on treatment, and the analysis of diagnostics to improve the health systems in real-time responsiveness [5]. Finally, and as mentioned in the result of this study, process mining has improved conditions in hospitals, as well as operating rooms, and emergency departments. Although the previous research has argued that the process redesign stage (5.3%) and a domain expert in each stage (3.8%) were some key findings of process mining of their research [16].

5 Conclusion

The goal of this research was to improve healthcare systems by using process mining. Specifying the type and place of the required change in the process, discovering the actual process and presenting it in the form of a model, extracting process information and calculating KPIs for evaluation by decision-makers, and achieving complete transparency and identifying possible subjects before occurrence were new findings of process mining capabilities in this study. Some main limitations of the process mining are as follows: access to appropriate, correct, and complete process data, also, improved usability for non-experts.

This study found it easy to identify the application of process mining for the visualisation of healthcare activities. In the conclusion of this study, available research related to the use of process mining in healthcare was investigated. The results of these studies showed that the use of process mining in healthcare and treatment can have several benefits. In general, the process of mining by extracting knowledge from current reports, hospital information systems, and existing records can show a clear view of what has happened and reduce costs and the treatment length of stay.

6 Ethical Approval

This research is part of the M.Sc. Thesis that is under the ethics code [IR.MUI.RESEARCH.REC. 1399.497].

Funding. This article resulted from the Master of Sciences thesis in "Health Information Technology" and research project No. 399503 and ethic code IR.MUI.RESEARCH.REC.1399.497that funded by Isfahan University of Medical Sciences, Isfahan, Iran.

Conflict of Interest. None declared.

References

1. Homayounfar, P. (eds.).: Process mining challenges in hospital information systems. 2012 federated conference on computer science and information systems (FedCSIS). IEEE (2012)
2. Wang, L.L., Lo, K.: Text mining approaches for dealing with the rapidly expanding literature on COVID-19. JBiB **22**(2), 781–99 (2021)
3. Alibabaei, A., Badakhshan, P., Alibabaei, H.: Studying BPM success factors differences in various industries. JIJoM**6**(1), 68–74 (2017)

4. Badakhshan, P., Alibabaei A.: Using process mining for process analysis improvement in pre-hospital emergency. ICT for an Inclusive World, 567–580 (2020). https://doi.org/10.1007/978-3-030-34269-2_39
5. Munoz-Gama, J., Martin, N., Fernandez-Llatas, C., Johnson, O.A., Sepúlveda, M., Helm, E., et al.: Process mining for healthcare: characteristics and challenges. J. Biomed. Inform. **127**, 103994 (2022)
6. Qiu, H.-J., Yuan, L.-X., Wu, Q.-W., Zhou, Y.-Q., Zheng, R., Huang, X.-K., et al.: Using the internet search data to investigate symptom characteristics of COVID-19: a big data study **6**(S1), S40–S48 (2020)
7. Guraya, S.Y.: Transforming laparoendoscopic surgical protocols during the COVID-19 pandemic; big data analytics, resource allocation and operational considerations. JIJoS **80**, 21–25 (2020)
8. Ayyoubzadeh, S.M., Ayyoubzadeh, S.M., Zahedi, H., Ahmadi, M., Kalhori, S.R.N.: Predicting COVID-19 incidence through analysis of google trends data in Iran: data mining and deep learning pilot study. JJPH, Surveillance, **6**(2), e18828 (2020)
9. Haleem, A., Javaid, M., Khan, I.H., Vaishya, R.: Significant applications of big data in COVID-19 pandemic. JIJOO **54**(4), 526–528 (2020)
10. Bragazzi, N.L., Dai, H., Damiani, G., Behzadifar, M., Martini, M., Wu, J., et al.: How big data and artificial intelligence can help better manage the COVID-19 pandemic. JIJOER **17**(9), 3176 (2020)
11. Buttigieg, S.C., Prasanta, D., Gauci, D.: Business process management in health care: current challenges and future prospects (2016)
12. Chang, H., Yu, J.Y., Yoon, S.Y., Hwang, S.Y., Yoon, H., Cha, W.C., et al.: Impact of CoViD-19 pandemic on the overall diagnostic and therapeutic process for patients of emergency department and those with acute cerebrovascular disease **9**(12), 3842 (2020)
13. van der Aalst, W.M., Netjes, M., Reijers, H.A.: Supporting the full BPM life-cycle using process mining and intelligent redesign. Contemporary issues in database design and information systems development: Igi Global, 100–132 (2007)
14. van der Aalst, W.: Process mining: overview and opportunities. JATOMIS **3**(2), 1–17 (2012)
15. Van der Aalst, W.: Using process mining to bridge the gap between BI and BPM. MJC **44**(12), 77–80 (2011)
16. de Roock, E., Martin, N.: I. Process mining in healthcare–an updated perspective on the state of the art. JJOBI 103995 (2022)
17. Gupta, S.: Technische Universiteit Eindhoven. Workflow and process mining in healthcare. JMST (2007)
18. Delias, P., Doumpos, M., Grigoroudis, E., Manolitzas, P., Matsatsinis, N.: Supporting healthcare management decisions via robust clustering of event logs. JK-BS **84**, 203–213 (2015)
19. Perimal-Lewis, L., de Vries, D., Thompson, C.H. (eds.).: Health intelligence: discovering the process model using process mining by constructing Start-to-End patient journeys. In: Proceedings of the Seventh Australasian Workshop on Health Informatics and Knowledge Management-Volume 153 (2014)
20. Zhou, Z., Wang, Y., Li, L. (eds.).: Process mining based modeling and analysis of workflows in clinical care-a case study in a Chicago outpatient clinic. In: Proceedings of the 11th IEEE International Conference on Networking, Sensing and Control. IEEE (2014)
21. Orellana García, A., Pérez Alfonso, D., Larrea Armenteros, O.U.: Analysis of hospital processes with process mining techniques. MEDINFO 2015: eHealth-enabled Health. IOS Press, pp. 310–314 (2015)

22. Mans, R.S., van der Aalst, W.M.P., Vanwersch, R.J.B., Moleman, A.J.: Process mining in healthcare: data challenges when answering frequently posed questions. In: Lenz, R., Miksch, S., Peleg, M., Reichert, M., Riaño, D., ten Teije, A. (eds.) KR4HC/ProHealth -2012. LNCS (LNAI), vol. 7738, pp. 140–153. Springer, Heidelberg (2013). https://doi.org/10.1007/978-3-642-36438-9_10

23. Hendricks, R.M.: Process mining of incoming patients with sepsis. JOJOPHI **11**(2) (2019)

24. Phan, R., Augusto, V., Martin, D., Sarazin, M. (eds.).: Clinical pathway analysis using process mining and discrete-event simulation: an application to incisional hernia. In: 2019 Winter Simulation Conference (WSC). IEEE (2019)

25. Martinez-Millana, A., Lizondo, A., Gatta, R., Vera, S., Salcedo, V.T., Fernandez-Llatas, C., et al.: Process mining dashboard in operating rooms: analysis of staff expectations with analytic hierarchy process. JIJOER **16**(2), 199 (2019)

26. Mans, R., Schonenberg, H., Leonardi, G., Panzarasa, S., Cavallini, A., Quaglini, S., et al. (eds.).: Process mining techniques: an application to stroke care. MIE (2008)

27. Taei, M.: Case study: emergency department of Alzahra hospital in Detection and analysis of processes in health systems using mining process techniques Isfahan University of Isfahan, Isfahan, Iran (2017). (Thesis)

28. Pramanik, M.I., Lau, R.Y.K., Demirkan, H., Azad, M.A.K.: Smart health: big data enabled health paradigm within smart cities. Expert Syst. Appl. **87**, 370–383 (2017)

29. Lotfi, F., Fatehi, K., Badie, N.: An analysis of key factors to mobile health adoption using fuzzy AHP. Int. J. Inf. Technol. Comput. Sci. **12**(2), 1–17 (2020)

30. Nayim, A.M.: Comparative analysis of data mining techniques to predict cardiovascular disease, vol. 14, no. 6, pp. 23–32 (2022). https://doi.org/10.5815/ijitcs.2022.06.03

31. Maphosa, V.: E-health implementation by private dental service providers in Bulawayo, Zimbabwe, vol. 15, no. 1, pp. 20–28 (2023). https://doi.org/10.5815/ijieeb.2023.01.02

New About the Principle "Like Begets Like" in Genetics

Sergey V. Petoukhov[✉]

Mechanical Engineering Research Institute of the RAN, Moscow 101990, Russia
spetoukhov@gmail.com

Abstract. The article is devoted to the connection of the ancient principle "like begets like" with the emergent algebraic properties of a complex of a lot of mutually related binary oppositions in the genetic code system (in addition to the known particular connection of this principle with the replication phenomenon in double-stranded DNA). This holistic complex is analyzed in its algebra-matrix representation. Emergence algebraic properties of this complex are firstly studied and revealed on the basis of a matrix-algebraic approach. Appropriate families of algebra-genetic matrices have features for mutual replications of their matrices. It gives a new approach for understanding the important role of the above-named principle in the genetics and genetically inherited physiological systems and also for further development of algebraic biology.

Keywords: DNA alphabets · complementary replication · binary numberings · binary-complementary replication · matrices

1 Introduction

Many sciences turn to biology in connection with the anthropomorphic slogans "man is the measure of all things" and "man is the cosmos" in the hope of finding support for solving their own problems. This point of view can be seen, for example, in the book about the human phenomenon [1], whose author believed the following. To find out how the world was formed and what its further fate is, one should "decipher" the person; because of this, future synthetic science will take man as a basis; this will be a new era in science, in which there will be a complete understanding that man as a "subject of knowledge" is the key to the whole science of nature. The book [2] also supports that the theory of living systems – but not physics – is the most appropriate center for the vision of the world. Accordingly, the knowledge about the genetic code system and inherited properties of physiological ensembles is undoubtedly useful for computer science, engineering and education applications.

The DNA double helix model created by J.D. Watson and F. Crick in 1953 gave a powerful impetus to the development of genetic research. It showed the world a recursive algorithm for the complementary replication of DNA strands, which ensures the replication of the genetic information recorded on these strands. Before the complementary replication, DNA is separated in two complementary strands. Each strand of the original

Z. Hu et al. (Eds.): ICCSEEA 2023, LNDECT 181, pp. 309–317, 2023.
https://doi.org/10.1007/978-3-031-36118-0_27

DNA molecule serves as a template for the production of its new complementary counterpart. This seminal work by Watson and Crick was perceived as the discovery of a key secret of life, corresponding to the ancient notions that "like begets like". Scientists were struck by how simple and beautiful this explanation of the replication and preservation of genetic information based on the mechanism of complementarity turned out to be. It was emphasized that it is this complementarity that provides the most important properties of DNA as a carrier of hereditary information [3]. DNA complementary replication occurs in all living organisms and is required for cell division during growth and tissue repair, as well as ensuring that each new cell receives its own copy of the DNA. Because the cell has the distinct property of division, complementary replication of DNA is required. Complementary replication of DNA strands occurs at breakneck speed. The well-known bacteria E. coli, for example, replicates at a rate of over 1,000 bases per second [4].

The purpose of this article is to present results giving pieces of evidence that the principle "like begets like" is related to the emergent properties of complex of a lot of mutually related binary oppositions in the genetic code system (in addition to the known particular connection of this principle with the replication phenomenon in double-stranded DNA). This relation is revealed on the basis of analysis of an algebra-matrix representation of the complex. Data are described about algebraic binary-complementary replications in the multicomponent binary structures of the genetic coding system, which can be used to model structural features of the system on the language of algebraic matrices and dyadic groups of binary numbers. This result was received by below-described methods of algebra-matrix analysis of emergent properties of the holistic genetic system of binary oppositions. Various algebraic approaches were proposed by different authors for study of genetic structures as it was noted in [5–7] but a limited volume of this article does not allow their review here. The presented algebra-matrix approach in this article is an original and testifies in favor of usefulness of its application in future genetic studies.

2   Binary Oppositions in the Genetic System and Genetic Matrices

The complementary replication of DNA strands is based on the oppositions of complementary nucleobases A-T and C-G, where symbols A, T, C, and G denote correspondingly adenine, thymine, cytosine, and guanine. But these molecular oppositions are just particular cases of a wide set of mutually related binary oppositions in the molecular integral system of genetic coding. Other types of such molecular oppositions are: amino (A and C) vs. keto (G and T); three hydrogen bonds vs. two hydrogen bonds in complementary pairs of nucleotides; purines (A and G) vs. pyrimidines (C and T); n-plets with strong roots vs. n-plets with weak roots. In genetics, each of the types of binary oppositions is usually considered on its own out of their compete ensemble. But the organism is one, and this holistic system of binary oppositions may have important emergent properties that its individual parts do not have and that need to be studied. Below such emergent algebraic properties of the genetic system of binary oppositions are described.

As it was shown in our previous publications [5–8], DNA alphabets of 4 letters, 16 duplets, 64 triplets, 256 tetraplets, etc. can be presented in the form of square tables

(Fig. 1), whose columns are numbered with binary indicators "pyrimidine or purine" (C = T = 1, A = G = 0), and rows are numbered with binary indicators "amino or keto" (C = A = 1, T = G = 0). These tables of DNA alphabets are members of the general tensor family of alphabetic matrices $[C, A; T, G]^{(n)}$, where (n) is the tensor integer power. In such matrices, each of 4 letters, 16 doublets, 64 triplets, ... takes automatically its own individual place.

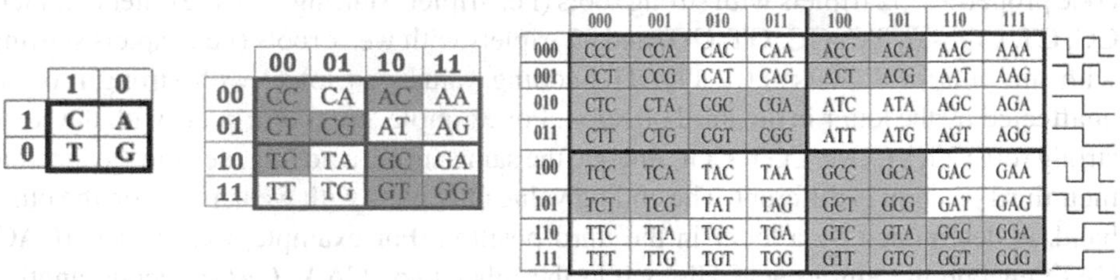

Fig. 1. Matrices of DNA-alphabets of 4 nucleotides, 16 duplets, and 64 triplets. Black cells in the matrices of duplets and triplets contains so-called strong duplets and triplets with strong roots. In each of the rows of the matrices of duplets and triplets, its black-and-white mosaics has a meander-like character and corresponds to meander-like Rademacher functions shown graphically at the right.

Each of the n-plets in such matrices (Fig. 1) can be binary enumerated by the con-catenation of the binary numbers of the row and column at the intersection of which it is located. For example, the triplet CAT becomes numbering 001010 because it belongs to the row 001 and the column 010. In these matrices, all complementary n-plets are located inverse-symmetrically with respect to the center of the appropriate matrix. Cor-respondingly, binary numbering of each n-plet is transformed into numbering of its complementary n-plet (that is, n-plet of the opposite strand of DNA) by the mutual interchanging of digits 0 ↔ 1 in it. For example, by this complementary operation, the numbering 001010 of the triplet CAT becomes numbering 110101 of its complementary triplet GTA. This interchanging 0 ↔ 1 is called the binary-complementary operation and actively used below regarding to an algebraic realization of the ancient principle "like begets like" in matrix genetics.

In the matrices in Fig. 1, which are built on the basis of the noted binary oppositions, the sets of binary numberings of rows and columns correspond to dyadic groups of n-bit binary numbers in which the logic operation of modulo-2 addition serves as the group operation [5–7, 9]. For example, in the matrix of 64 triplets, the dyadic group of binary numberings of its rows and rows contains eight 3-bit binary numbers (1):

$$000, \ 001, \ 010, \ 011, \ 100, \ 101, \ 110, \ 111 \qquad (1)$$

Two binary numbers that are converted into each other under interchange 0 ↔ 1 are called complementary each to another. For example, in the dyadic group (1), the pairs of complementary numbers are the following: 000–111, 001–110, 010–101, 011–100 (in the decimal system, they correspond to pairs of numbers 0–7, 1–6, 2–5, 3–4). In a pair of complementary numbers, one of them is always even and the other is odd, that is any pair of complementary numbers is the pair of even and odd numbers. Accordingly, those columns (rows) that are enumerated by complementary binary numbers are called complementary each to another. In the genetic matrices in Fig. 1, complementary columns are located mirror-symmetrical in the left and right halves of the matrices, and complementary rows are located mirror-symmetrical in the upper and lower halves.

Another type of genetic binary opposition is known. This is the binary-oppositional division of the DNA alphabet of 64 triplets into two equal sub-alphabets based on their code properties: 32 triplets with strong roots (i.e., triplets starting with the other 8 duplets CC, CT, CG, AC, TC, GC, GT, GG) and 32 triplets with weak roots (i.e., triplets starting with the other 8 duplets) [10, 11]. The coding value of triplets with strong roots is unaffected by the letter in the third position. For example, the four triplets with the same strong root CGC, CGA, CGT, CGC encode the same amino acid Arg, despite the fact that their third position is different. The coding value of triplets with weak roots, on the other hand, is determined by a letter in the third position. For example, two triplets (CAC, CAT) encode the amino acid His, while the other two (CAA, CAG) encode another amino acid Gln in the grouping of four triplets with the same weak root CAC, CAT, CAA, and CAG. In Fig. 1, all triplets with strong roots are highlighted in black, while triplets with weak roots are highlighted in white.

Let us concentrate on analysis of the mosaic matrix $[C, A; T, G]^{(3)}$ of 64 triplets in Fig. 1. A black-and-white mosaic of each of its rows has a meander-like character: black fragments and white fragments have identical length. Such meander mosaics of rows correspond to meander-like forms of Rademacher functions $r_n(t) = \text{sign}(\sin 2^n \pi t)$, $n = 1, 2, 3, \ldots$, which are well known in the theories of discrete signals, orthogonal series, and probabilities (here sgn is the sign function of the argument). Rademacher functions take only two values " $+ 1$" and " $- 1$" as it is shown for the considered case in Fig. 1 at right. Black and white cells of the symbolic matrix in Fig. 1 reflect the opposition of two sub-alphabets of n-plets with strong and weak roots and therefore can be represented by elements $+ 1$ and -1 in them. In this representation, the numeric matrix appears, which is conditionally called the genetic Rademacher matrix of the 64 triplets because each of its rows correspond to one of the Rademacher functions (Fig. 2 at the top). This matrix W is a sum of two sparse matrices W_0 and W_1, one of which contains non-zero columns enumerated by even numbers 000, 010, 100, 110 and another contains non-zero columns enumerated by odd numbers 001, 011, 101, 111 (Fig. 2, at the bottom). In this even-odd decomposition of the Rademacher matrix W of 64 triplets, each of the non-zero columns of the first sparse matrix W_0 is a complementary to one of the non-zero columns of the second sparse matrix W_1. By this reason, these two sparse matrices W_0 and W_1 are called a complementary each to another. The sparse matrix W_0 with even-numerated columns is called the even-columns matrix; all its non-zero columns correspond to triplets, which contain pyrimidines C or T at their ends (by this reason, this sparse matrix can be also called the pyrimidine-columns matrix). The sparse matrix W_1 with odd-numerated columns is called the odd-columns matrix; all its non-zero columns correspond to triplets, which contain purines A or G at their ends (by this reason, this matrix can be called the purine-columns matrix).

	000	001	010	011	100	101	110	111
000	+1	+1	-1	-1	+1	+1	-1	-1
001	+1	+1	-1	-1	+1	+1	-1	-1
010	+1	+1	+1	+1	-1	-1	-1	-1
011	+1	+1	+1	+1	-1	-1	-1	-1	=
100	+1	+1	-1	-1	+1	+1	-1	-1
101	+1	+1	-1	-1	+1	+1	-1	-1
110	-1	-1	-1	-1	+1	+1	+1	+1
111	-1	-1	-1	-1	+1	+1	+1	+1

	000	001	010	011	100	101	110	111
000	+1		-1		+1		-1	
001	+1		-1		+1		-1	
010	+1		+1		-1		-1	
011	+1		+1		-1		-1	
100	+1		-1		+1		-1	
101	+1		-1		+1		-1	
110	-1		-1		+1		+1	
111	-1		-1		+1		+1	

+

	000	001	010	011	100	101	110	111
000		+1		-1		+1		-1
001		+1		-1		+1		-1
010		+1		+1		-1		-1
011		+1		+1		-1		-1
100		+1		-1		+1		-1
101		+1		-1		+1		-1
110		-1		-1		+1		+1
111		-1		-1		+1		+1

Fig. 2. The even-odd decomposition of the Rademacher genetic matrix W of 64 triplets as the sum of two sparse complementary matrices W_0 and W_1: at left, the even-columns matrix W_0 containing only non-zero columns having even numberings; at right, the odd-columns matrix W_1 containing only non-zero columns having odd numberings. Empty cells contain zeroes.

3 Algebraic Properties of the Matrices Based on the Genetic Complex of Binary Oppositions

The even-columns $(8 * 8)$-matrix W_0 in Fig. 2 is the sum of 4 sparse $(8 * 8)$-matrices $s_0 + s_1 + s_2 + s_3$ shown in Fig. 3. The set of these 4 matrices s_0, s_1, s_2, s_3 is closed relative to multiplication and corresponds to a certain multiplication table in Fig. 3 at right. This table matches to the multiplication table of the Cockle split-quaternions algebra used in the Poincare conformal disk model of hyperbolic geometry; it is also connected with the theme of the algebraic holography in genetics [12, 13].

Analogically, the odd-columns matrix W_1 (Fig. 2) is the sum of 4 sparse matrices $p_0 + p_1 + p_2 + p_3$ shown in Fig. 4. The set of these 4 matrices p_0, p_1, p_2, p_3 is closed regarding multiplication and defines the multiplication table in Fig. 4 at the right. This multiplication table coincides with the multiplication table of the 4-dimensional algebra, which was received for the even-columns matrix W_0 (Fig. 3). Both the even-columns matrix and the odd-columns matrix present Cockle's split-quaternions with unit coordinates (these split-quaternions have different forms of their matrix representations, with which these even-columns and odd-columns genetic matrices turn out to be associated). Correspondingly, both these genetic matrices are connected with the Poincare conformal disk model of hyperbolic geometry.

The summation of the even-columns matrix W_0 and the odd-columns matrix W_1, which are binary-complementary each other and connected to the same 4-dimensional algebra, gives the summary matrix W (Fig. 2, at the top), which is a new algebraic entity connected already with the 8-dimensional algebra. This summary Rademacher matrix W can be decomposed into the sum of 8 sparse matrices $v_0 + v_1 + v_2 + v_3 + v_4 + v_5 + v_6 + v_7$ shown in Fig. 5. The set of these matrices $v_0, v_1, v_2, v_3, v_4, v_5, v_6, v_7$ is

```
s0 =            s1 =              s2 =               s3 =
10000000        00 -1000 00       00 00 00-10        00 0010 00
01000000        00 -1000 00       00 00 00-10        00 0010 00
00100000        10 0000 00        00 00-10 00        00 0000-10
00010000        10 0000 00        00 00-10 00        00 0000-10
00001000        00 0000-10        00-10 00 00        10 0000 00
00000100        00 0000-10        00-10 00 00        10 0000 00
00000010        00 0010 00        -10 00 00 00       00-1000 00
00000001        00 0010 00        -10 00 0000        00-1000 00
```

*	s0	s1	s2	s3
s0	s0	s1	s2	s3
s1	s1	-s0	s3	-s2
s2	s2	-s3	s0	-s1
s3	s3	s2	s1	s0

Fig. 3. The decomposition of the even-columns matrix W_0 (from Fig. 2) into 4 sparse matrices s_0, s_1, s_2, s_3, whose set is closed relative to multiplication; s_0 plays a role of the identity matrix in this set. The multiplication table for this set is shown at the bottom, which matches with the multiplication table of the 4-dimensional algebra of Cockle split-quaternions used in the Poincare conformal disk model of hyperbolic geometry. A symbol of this model is presented (from https://commons.wikimedia.org/wiki/Category:Poincar%C3%A9_disk_models).

```
P0 =            P1 =              P2 =               P3 =
01000000        000-1000 0        0 00 00 00-1       00 00010 0
01000000        000-1000 0        0 00 00 00-1       00 00010 0
00010000        010 00000         0 00 00-10 0       000 0000-1
00010000        010 00000         0 00 00-10 0       000 0000-1
00000100        000 0000-1        0 00-10 00 0       010 00000
00000100        000 0000-1        0 00-10 00 0       010 00000
00000001        000 00100         0-10 00 00 0       000-1000 0
00000001        000 00100         0-10 00 00 0       000-1000 0
```

*	p0	p1	p2	p3
p0	p0	p1	p2	p3
p1	p1	-p0	p3	-p2
p2	p2	-p3	p0	-p1
p3	p3	p2	p1	p0

Fig. 4. The decomposition of the odd-columns matrix W_1 (from Fig. 2) into 4 sparse matrices p_0, p_1, p_2, p_3, whose set is closed relative to multiplication; p_0 plays a role of the identity matrix inside this set. The shown multiplication table for this set matches the multiplication table of the Cockle split-quaternions algebra used in the Poincare conformal disk model of hyperbolic geometry. The symbol of this model is presented.

closed relative to multiplication and matches already to the multiplication table (Fig. 5, at the right) of a certain 8-dimensional algebra.

This summary matrix W generates algorithmically its complementary-replicated analogue W_R by means of the interchange of numbers $0 \leftrightarrow 1$ in the binary numerating of its columns with the corresponding rearrangement of the columns (that is, rearrangements of columns located mirror symmetrically in the left and right halves

$v_0 =$
```
1 0 0 0 0 0 0 0
1 0 0 0 0 0 0 0
0 0 1 0 0 0 0 0
0 0 1 0 0 0 0 0
0 0 0 0 1 0 0 0
0 0 0 0 1 0 0 0
0 0 0 0 0 0 1 0
0 0 0 0 0 0 1 0
```

$v_1 =$
```
0 1 0 0 0 0 0 0
0 1 0 0 0 0 0 0
0 0 0 1 0 0 0 0
0 0 0 1 0 0 0 0
0 0 0 0 0 1 0 0
0 0 0 0 0 1 0 0
0 0 0 0 0 0 0 1
0 0 0 0 0 0 0 1
```

$v_2 =$
```
0 0 -1 0 0 0 0 0
0 0 -1 0 0 0 0 0
1 0 0 0 0 0 0 0
1 0 0 0 0 0 0 0
0 0 0 0 0 0 -1 0
0 0 0 0 0 0 -1 0
0 0 0 0 1 0 0 0
0 0 0 0 1 0 0 0
```

$v_3 =$
```
0 0 0 -1 0 0 0 0
0 0 0 -1 0 0 0 0
0 1 0 0 0 0 0 0
0 1 0 0 0 0 0 0
0 0 0 0 0 0 0 -1
0 0 0 0 0 0 0 -1
0 0 0 0 0 1 0 0
0 0 0 0 0 1 0 0
```

$v_4 =$
```
0 0 0 0 1 0 0 0
0 0 0 0 1 0 0 0
0 0 0 0 0 0 -1 0
0 0 0 0 0 0 -1 0
1 0 0 0 0 0 0 0
1 0 0 0 0 0 0 0
0 0 -1 0 0 0 0 0
0 0 -1 0 0 0 0 0
```

$v_5 =$
```
0 0 0 0 0 1 0 0
0 0 0 0 0 1 0 0
0 0 0 0 0 0 0 -1
0 0 0 0 0 0 0 -1
0 1 0 0 0 0 0 0
0 1 0 0 0 0 0 0
0 0 0 -1 0 0 0 0
0 0 0 -1 0 0 0 0
```

$v_6 =$
```
0 0 0 0 0 0 -1 0
0 0 0 0 0 0 -1 0
0 0 0 0 -1 0 0 0
0 0 0 0 -1 0 0 0
0 0 -1 0 0 0 0 0
0 0 -1 0 0 0 0 0
-1 0 0 0 0 0 0 0
-1 0 0 0 0 0 0 0
```

$v_7 =$
```
0 0 0 0 0 0 0 -1
0 0 0 0 0 0 0 -1
0 0 0 0 0 -1 0 0
0 0 0 0 0 -1 0 0
0 0 0 -1 0 0 0 0
0 0 0 -1 0 0 0 0
0 -1 0 0 0 0 0 0
0 -1 0 0 0 0 0 0
```

*	v_0	v_1	v_2	v_3	v_4	v_5	v_6	v_7
v_0	v_0	v_1	v_2	v_3	v_4	v_5	v_6	v_7
v_1	v_0	v_1	v_2	v_3	v_4	v_5	v_6	v_7
v_2	v_2	v_3	$-v_0$	$-v_1$	$-v_6$	$-v_7$	v_4	v_5
v_3	v_2	v_3	$-v_0$	$-v_1$	$-v_6$	$-v_7$	v_4	v_5
v_4	v_4	v_5	v_6	v_7	v_0	v_1	v_2	v_3
v_5	v_4	v_5	v_6	v_7	v_0	v_1	v_2	v_3
v_6	v_6	v_7	$-v_4$	$-v_5$	$-v_2$	$-v_3$	v_0	v_1
v_7	v_6	v_7	$-v_4$	$-v_5$	$-v_2$	$-v_3$	v_0	v_1

Fig. 5. The decomposition of Rademacher matrix W (from Fig. 2, at the top), which is the sum of the even-columns matrix W_0 and the odd-columns matrix W_1, into 8 sparse matrices v_0, v_1, v_2, v_3, v_4, v_5, v_6, v_7, whose set is closed relative to multiplication. The multiplication table for this set is shown at the bottom.

of the matrix). This interchanging algorithm $0 \leftrightarrow 1$ in binary numbers provides interchanges in any pair of complementary columns that differ from each other in the content of triplets with purine and pyrimidine endings, in some analogy with the complementarity of purines and pyrimidines in DNA double strands. For example, column with number 110 (which corresponds to the nucleotide order "purine-purine-pyrimidine" in all its triplets) takes the place of column with number 001 (which corresponds to the order "pyrimidine-pyrimidine-purine" in all its triplets). Briefly speaking, molecular complementary-replicated properties of DNA strings exist jointly with algebraic binary-complementary replication properties of the considered alphabetical matrix of the genetic code. Both of these properties are parts of genetics of the whole organisms and so interrelated. These algebraic complementary-replicated properties of genetic matrices allow applying effective algebraic methods for further study of genetics to include it in the field of modern mathematical natural sciences in connection with multi-dimensional algebras, hyperbolic geometry, theory of resonances, etc. The action of complementary-replicated $(8 * 8)$-matrices W and W_R on an arbitrary 8-dimensional vector $\overline{X}$ generates two new vectors that are complementary to each other: the corresponding coordinates of both generated vectors are the same in their absolute values, but have opposite signs. In addition, each of the resulting vectors $\overline{X}*W$ and $\overline{X}*W_R$ is always a complementary palindrome: the sequence of its coordinates, which is read in forward order, coincides with the sequence, which is read in reverse order and having coordinates with the opposite sign. It is interesting because in molecular genetics the problem of complementary palindromes has long been known.

The limited volume of this article allows demonstrating only a fragment of the wide theme of the important role of the principle "like begets like" connected with binary-complemented replications. One can see more detail on this theme in the preprint [14]. The received results can be used in different scientific directions, for example, presented in [15–19].

4   Some Concluding Remarks

The briefly presented algebraic complementary-replicated properties of genetic matrices allow applying effective algebraic methods for further study of genetics to include it in the field of modern mathematical natural sciences.

At various levels of genetically inherited biological organization, different forms of implementation of the fundamental biological principle "like begets like" (or a complementary replication in a broad sense) can be seen. Mirror neurons, for example, are known in the human and animal brains, which have mirror complementary hemispheres. A mirror neuron is a neuron that fires both when an animal acts and when the animal observes another animal performing the same action. As a result, the neuron "mirrors" the behavior of the other, as if the observer were acting. The theme of mirror neurons, whose function is based on one of the forms of the complementary replication principle (in a broad sense), concerns cognitive functions, the origin of language, learning facilitation, automatic imitation, motor mimicry, autism, human emotional capacity such as empathy, and a variety of other issues. [20]. One of the questions that has arisen is: where do mirror neurons come from? [21].

It may be recalled here that mathematical formalisms have a creative power. As Heisenberg noted, *"the Pythagoreans seem to have been the first to realize the creative force inherent in mathematical formulations"* [22]. The proposed matrix formalisms have a creative power in algebraic biology and should be used in the future in theoretical and applied researches. The above-described results of our studies give pieces of evidence that the system of mirror neurons and the system of DNAs complementary replications are not isolated parts of the organism, but they are particular parts of a bio-algebraic complex realizing phenomena "like begets like". It's not that the molecules of two strands of DNA randomly docked, formed a complementary pair and began to repeat the process of complementary replication at breakneck speed. Another point of view is proposed: the DNA filaments replication phenomenon is part of a holistic bio-algebraic genetic complex of complementary replication, parts of which manifest themselves at different levels of organization of the living, up to the functioning of the brain with its mirror neurons and the ability to empathize and imitate external events. This bio-algebraic complex is responsible for the implementation of the ancient principle "like begets like" at different levels of biological organization in the course of biological evolution.

References

1. Teilhard de Chardin, P.: The Human Phenomenon. Brighton: Sussex Academic (1999)
2. Capra, F.: The Tao of Physics: An Exploration of the Parallels Between Modern Physics and Eastern Mysticism. Shambhala Publications Inc, New Jersey (2000)
3. Chapeville, F., Haenni, A.-L.: Biosynthese des proteins. Hermann Collection, Paris Methodes (1974). (in French)
4. Bank, E.: How much time does it take for a DNA molecule to replicate? - sciencing.com.https://sciencing.com/much-time-dna-molecule-replicate-21660.html. Accessed 8 Nov 2022
5. Petoukhov, S.V.: Matrix genetics, algebras of genetic code, noise immunity. Moscow, RCD, p. 316 (2008). (in Russian). ISBN 978-5-93972-643-6
6. Hu, Z.B., Petoukhov, S.V., Petukhova, E.S.: I-Ching, dyadic groups of binary numbers and the geno-logic coding in living bodies. Prog. Biophys. Mol. Biol. **131**, 354–368 (2017)
7. Petoukhov, S.V., He, M. Symmetrical analysis techniques for genetic systems and bioinformatics: advanced patterns and applications. Hershey, USA: IGI Global (2010)
8. Petoukhov, S.V., Svirin, V.I.: Stochastic rules in nucleotide sequences in genomes of higher and lower organisms. IJMSC **7**(2), 1–13 (2021). https://doi.org/10.5815/ijmsc.2021.02.01
9. Harmut, H.F.: Information Theory Applied to Space-Time Physics. The Catholic University of America, DC, Washington (1989)
10. Rumer, Y.: Codon systematization in the genetic code. Dokl. Akad. Nauk SSSR **183**(1), 225–226 (1968)
11. Fimmel, E., Strüngmann, L.: Yury Borisovich Rumer and his 'biological papers' on the genetic code. Phil. Trans. R. Soc. A **374**, 20150228 (2016). https://doi.org/10.1098/rsta.2015.0228
12. Petoukhov, S.V.: Binary oppositions, algebraic holography and stochastic rules in genetic informatics. Biosystems **221**, 104760 (2022). https://doi.org/10.1016/j.biosystems.2022.104760
13. Petoukhov, S.V.: Hyperbolic numbers in modeling genetic phenomena. Preprints **2019**, 2019080284 (2020). https://doi.org/10.20944/preprints201908.0284.v4
14. Petoukhov, S.V.: The principle "like begets like" in molecular and algebraic-matrix genetics. Preprints **2022**, 2022110528 (2022). https://doi.org/10.20944/preprints202211.0528.v2
15. Ganguly, K.K., Asad, M., Sakib, K.: Decentralized self-adaptation in the presence of partial knowledge with reduced coordination overhead. IJITCS **14**(1), 9–19 (2022)
16. Lone, S.A., Mir, A.H.: Smartphone-based biometric authentication scheme for access control management in client-server environment. IJITCS **14**(4), 34–47 (2022)
17. Hamd, M.H., Ahmed, S.K.: Biometric system design for Iris recognition using intelligent algorithms. IJMECS **10**(3), 9 16 (2018)
18. Hissain, A., Muhammad, Y.S., Sajid, M.N.: An efficient genetic algorithm for numerical function optimization with two new crossover operators. IJMSC **4**(4), 42–55 (2018)
19. Almutiri, T., Nadeem, F.: Markov model applications in natural language processing: a survey. IJITCS **14**(2), 1–16 (2022)
20. Ferrari, P.F., Rizzolatti, G.: Mirror neuron research: the past and the future. Philos. Trans. R Soc. Lond. B Biol. Sci. **369**(1644), 20130169 (2014)
21. Heyes, C.M.: Where do mirror neurons come from? Neurosci. Biobehav. Rev. **34**(4), 575–583 (2010)
22. Heisenberg, W.: Physics and Philosophy: The Revolution in Modern Science, Penguin Classics (2000)

A Study of Implementing a Blockchain-Based Forensic Model Integration (BBFMI) for IoT Devices in Digital Forensics

Chintan Singh, Himanshu Khajuria(✉), and Biswa Prakash Nayak

Amity Institute of Forensic Sciences Amity University, Uttar Pradesh, Sector 125, Noida 201301, India
hkhajuria@amity.edu

Abstract. The rapid advancement of Internet of Things (IoT) technology has brought about new challenges in the field of digital forensics. Traditional forensic methods are often inadequate in dealing with the decentralized and distributed nature of IoT devices. IoT has several advantages that have made it appealing to consumers and attackers alike. The technology and resources available to today's cybercriminals allow them to launch millions of sophisticated attacks. In this paper, we propose a blockchain-based forensic model (BBFMI) for investigating IoT devices. The proposed BBFMI model utilizes the immutability and tamper-proof features of blockchain technology to provide a secure and reliable way of collecting, storing, and analyzing forensic evidence from IoT devices. One of the most important benefits of BBFMI is that it provides digital forensics investigators with an immutable chain of evidence that can be used to trace the source of data and its subsequent changes. For instance, when using BBFMI, forensic investigators can easily trace the chain of events that led to the data breach of an IoT device. Our results show that the proposed blockchain-based forensic approach can provide a secure, efficient, and tamper-proof solution for investigating IoT devices in digital crime scenes.

Keywords: Blockchain · Digital forensic · Internet of things · framework · decentralized storage

1 Introduction

The Internet of Things (IoT) has brought about a new era of connectivity, enabling devices to collect and share data in ways that were previously impossible. From smart homes to connected vehicles and medical devices, IoT has the potential to improve our lives in countless ways. However, with this increased connectivity comes new challenges in the field of digital forensics. Traditional forensic methods are often inadequate in dealing with the decentralized and distributed nature of IoT devices [1]. The data generated and stored by these devices is often spread across multiple locations, making it difficult to establish a chain of custody and preserve the integrity of the evidence. One of the main challenges in IoT forensics is the sheer volume of data that is generated by these

© The Author(s), under exclusive license to Springer Nature Switzerland AG 2023
Z. Hu et al. (Eds.): ICCSEEA 2023, LNDECT 181, pp. 318–327, 2023.
https://doi.org/10.1007/978-3-031-36118-0_28

devices. IoT devices are constantly collecting and transmitting data, and this data can be used to track the actions and movements of individuals, making it a valuable tool for forensic investigations. However, due to the decentralized and distributed nature of IoT devices, this data is often spread across multiple locations, making it difficult to collect and forensically analyze. Furthermore, the data stored on IoT devices in incidents are often unstructured and in various formats, making it difficult to extract meaningful information. Another challenge in IoT forensics is the lack of standardization in the way that data is stored and transmitted [2]. Different manufacturers use different protocols and standards for data storage and transmission, making it difficult to extract and analyze data from different devices. This lack of standardization also makes it difficult to establish a chain of custody, as it is difficult to determine who had access to the device and when.

To address these challenges, researchers have proposed the use of blockchain technology for forensic investigations, which provide data security and avoid the contamination of chain of custody. Li et al. (2019) proposed a unique ID method for IoT devices and record them in blockchain to avoid single point failure attack [3]. Zhao et al. (2020) to tackle problems associated with smart homes by developing federated learning systems for IoT devices [4]. In 2021 Gong et al. proposed a framework for IoT devices to make up with computational intensive requirements by introducing blockchain of things gate ways [5]. So, using blockchain technology, it is possible to create a tamper-proof and immutable record of the data collected from IoT devices, making it possible to establish a chain of custody and preserve the integrity of the evidence.

The paper begins by outlining the need for better digital forensics solutions for IoT devices, definitions, guidelines processes and difficulties associated with traditional forensic techniques. It then provides an overview of blockchain technology, discussing its capabilities and potential applications [6]. We aim to demonstrate the effectiveness of our proposed model through a case study and its results. The proposed model will include the use of smart contracts to automate the process of evidence collection and chain of custody, as well as a decentralized storage solution for preserving the integrity of the collected evidence. We also aim to address the lack of standardization in IoT data storage and transmission by proposing a standard for data storage and transmission that can be used by IoT device manufacturers.

In conclusion, the proposed blockchain-based forensic model for IoT is a promising solution for addressing the challenges of IoT forensics. It provides a secure, efficient, and tamper-proof way of collecting, storing, and analyzing forensic evidence from IoT devices. By utilizing the immutability and tamper-proof features of blockchain technology, it is possible to establish a chain of custody and preserve the integrity of the evidence. Furthermore, by proposing a standard for data storage and transmission, we aim to address the lack of standardization in IoT data storage and transmission and make it easier to extract and analyze data from different devices.

2 Background

2.1 Digital Forensic Overview

Digital forensics is the process of identifying, preserving, analyzing, and presenting digital evidence in a manner that is legally admissible. The field of digital forensics has evolved rapidly in recent years, driven by the increasing use of digital technology in all aspects of modern life. Digital forensics is used in a wide range of contexts, including criminal investigations, civil litigation, and incident response [7].

There are several ways to describe digital forensics. Horsman et al., (2022) described computer forensics as "the preservation, identification, extraction, interpretation, and documentation of computer evidence to include the rules of evidence, legal procedure, the integrity of evidence, factual reporting of the information, and providing expert opinion in a court of law or other legal and/or administrative proceedings as to what was found" [8]. The definition of digital forensics that is most often utilized, according to James et al. (2015), is "The use of scientifically derived and proven methods toward the preservation, collection, validation, identification, analysis, interpretation, documentation and presentation of digital evidence derived from digital sources to facilitate or further the reconstruction of events found to be digital in nature" [9].

2.2 IoT in Forensics

IoT forensics, according to Atlam et al., (2017), consists of device, network, and cloud level forensics [10]. Additionally, IoT forensics was defined as a part of digital forensics by Stoyanova et al., 2020 [11]. Forensics, which deals with the identification, collection, organizing, and presentation of evidence, takes place in the context of the IoT ecosystem. IoT devices typically have a range of sensors and other components that can be used to collect data, such as GPS coordinates, audio and video recordings, and even biometric data. This data can be used to reconstruct events, such as a crime or accident, and to identify suspects or victims.

One of the main challenges in IoT digital forensics is the sheer volume of data that can be generated by these devices. For example, a smart home system may generate hundreds of thousands of log entries, each of which must be analyzed to extract relevant information. Additionally, IoT devices can be configured to store data in a variety of formats, which can make it difficult to extract relevant information. Another challenge is the lack of standardization in IoT devices. Many devices use proprietary protocols and data formats, which can make it difficult to extract data. Furthermore, many IoT devices are designed to be disposable, which can make it difficult to recover data from them.

Overall, IoT devices are becoming an increasingly important source of digital evidence, and digital forensics experts must be prepared to deal with the challenges that they present. As IoT continues to evolve, it will be important for digital forensics experts to stay up to date with the latest technologies and techniques in order to effectively analyze the data that these devices generate.

2.3 Blockchain and Digital Forensic

Blockchain technology is becoming increasingly relevant in the field of digital forensics as more and more transactions are conducted on blockchain networks. Blockchain is a distributed ledger technology that records transactions across a network of computers. Each block in the chain contains a number of transactions, and each block is linked to the previous block through a cryptographic hash.

One of the key features of blockchain is its immutability, meaning that once a block is added to the chain, it cannot be altered. This makes blockchain an attractive option for digital forensics as it provides a tamper-proof record of transactions that can be used as evidence in investigations. In a dispersed network of untrusted peers, the blockchain serves as "an immutable ledger for recording transactions" (Gaur et al., 2018). Transactions made by a node are confirmed by other participating nodes in the network. Following validation, the group of transactions is added to the block by a certain class of nodes known as miners, as is the case with bitcoin [11].

Forensic examiners can use blockchain analysis tools to trace the flow of digital assets, track the movement of funds, and identify the parties involved in a transaction. They can also use blockchain data to reconstruct the sequence of events that led to a crime or incident, providing valuable insights for law enforcement and other authorities.

In addition, blockchain can be used to verify the authenticity of digital evidence. For example, hash values can be stored on the blockchain, providing a tamper-proof record of the integrity of the evidence.

3 Previously Related Work

We examine several works that are closely similar to our own in this part. IoT security has made extensive use of blockchain technology. When it comes to IoT forensics, the idea is still in the exploring stage. Hossain et al., 2018 proposed a forensic investigation framework utilizing a public digitalized leadger to gather facts in digital criminal investigations in IoT systems [12]. The framework is set up in three nodes, namely controller to IoT device, controller to cloud, and controller to controller, using a centralised manner. However, it was demonstrated in this work's proof of concept that the suggested system can extract forensic data from IoT devices. However, because of the centralised structure of their strategy, it is challenging to confirm the veracity of the data the framework has gathered. A digital witness strategy was also employed by Li et al., (2019) to enable sharing of IoT device records with guaranteed anonymity [13]. The authors deployed their digital witness model using the Privacy-aware IoT Forensics (PRoFIT) methodology that they had previously developed. The approach suggested in their study aims to support the collection of digital evidence in IoT contexts while preserving the privacy of the data amassed. The suggested approach supports the PRoFIT methodology's 11 privacy criteria as documented by Nieto et al. (2018) [14]. A distributed logging approach for IoT forensics was suggested by Nieto et al. in their work from 2017 [15]. A modified information dispersal algorithm (MIDA) was employed by the authors of this study to guarantee the accessibility of logs produced in IoT contexts. The logs are distributed, authenticated, encrypted, and consolidated. The distributed methodology utilised in this study solely addresses the storage of logs, not their verification. An IoT forensics framework using a permissioned-based blockchain was presented by Noura et al., (2020) by using the immutability virtue of blockchain technology. The integrity, legitimacy, and non-repudiation of evidence are improved by this paradigm [16]. Kumar et al., (2021) provided a forensic framework for distributed computing, transparency and decentralization in digital forensic investigations [17]. The provided Ethereum virtual machine framework results in clear view of process performed under digital forensic investigations that include all the stakeholders such as cloud service providers, internet service providers and heterogeneous devices. Thus, this framework could be utilized in security operations centers also. Furthermore, in after that Jacob et al., (2022) also designed framework model for forensic IoT using blockchain to help in preventing duplication of data in investigation and also developed secure level of confidence between evidence entities. More recently Gao et al., (2022) provided a botnet framework which is enhanced by blockchain in order to provide strong resilience against Distributed Denial of Service Attack (DDoS), Sybils, and digital forensic investigations [18]. Although the proposed method of botnets lacks in terms of censorship, resilience against single point of failure, high latency, and cost.

4   Results and Discussions

4.1   Proposed Model for IoT Incidence in Digital Forensics

In this section, we demonstrate our blockchain-based forensic model for the Internet of Things (BBFMI). IoT setups are divided into three layers. These are the network, device, and cloud layers. Our design allows leverage of the decentralized nature of blockchain to make sure that the logs created in IoT settings are preserved on the network and are accessible for verification by any of the participating nodes in the network. Numerous artefacts should be considered during a forensic investigation. However, our technique just considers system and event logs. The entities used in BBFMI are tabulated in Table 1, with their assigned abbreviations. Also Fig. 1. Shows the suggested BBFMI model that can enhance the IoT evidence in digital forensic investigations.

Table 1. Entities used in BBFMI framework model

S No	Entities Used	Abbreviations Used
1	Cloud Service Providers	(CPs)
2	Blockchain Centre	(BCs)
3	Log Processing Centre	(LPs)
4	User Centre	(UCs)

4.1.1   Blockchain Centre Entity (BCs)

A blockchain is a distributed ledger technology that is used to record transactions across a network of computers. In a blockchain model, each block on the chain contains a group of transactions, and once a block is added to the chain, the information in that block cannot be altered [19]. This allows for a secure and transparent way to store and share information. The blockchain is maintained by a network of nodes, each of which has a copy of the entire ledger. The nodes work together to validate new transactions and add them to the chain, ensuring that the ledger is accurate and up to date. The distributed nature of the blockchain means that no single entity controls the data, making it more resistant to tampering and fraud. Before being written onto a block on the distributed ledger that makes up the BCs, each log is processed. The distributed ledger consists of all the blocks that have been committed to the network. Each block contains a transactional value of hashed values derived from logs. The logs are collected from the various entities, hashed, and added to the blockchain network as transactions. The nodes in the BCs are made up of forensic investigators, CPs, and IoT devices. The blocks are proposed once the nodes have reached a consensus.

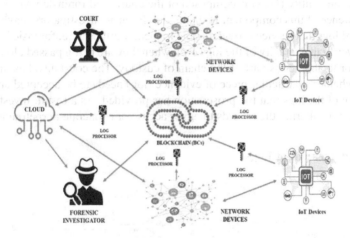

Fig. 1. Proposed BBFMI Model for IoT using blockchain

4.1.2 Log Processing Centre Entity (LPs)

A log processing center in a blockchain model refers to the network of nodes that are responsible for processing and validating transactions before they are added to the blockchain. In this model, each node has a copy of the entire blockchain ledger and works together to validate new transactions and add them to the chain. This is done by following a set of consensus rules, which are designed to ensure the integrity and security of the blockchain. When a new transaction is submitted to the network, it is broadcast to all the nodes in the network. Each node then verifies the transaction, checking that it is valid and that the sender has sufficient funds to complete the transaction [20]. Once a transaction is validated, it is grouped with other transactions into a block and added to the blockchain. The block is then broadcast to all the nodes in the network, who update their copies of the ledger to reflect the new block.

LPs handled the processing of logs in this BBFMI model. An Application Programming Interface (API) called the LCs serves as a bridge between different entities and the blockchain network. The IoT, network, and cloud layers all produce logs that are extracted by the LCs. The hashed data from the retrieved logs are written onto the block as a transaction after being hashed with the SHA-256 hash algorithm demonstrated by formula (1). Since the logs may include sensitive information, we decided to hash the logs. So it's not a good idea to preserve the logs in plaintext. Second, by reducing the size of the logs by hashing, processing the logs takes less time overall.

$$\text{Transaction} = \text{Hash} \times (\text{log}) \tag{1}$$

4.1.3   User Centre Entity (UCs)

The user centre entity (UCs) is composed of the courts and forensic investigators. Due to the existence of this component in the model, forensic investigators may validate the accuracy of logs that service providers send to them. Furthermore, forensic investigators may still verify the veracity of the logs even when they are being passed along from one investigator to another as part of the chain of custody. The court also has the power to evaluate whether a particular piece of evidence (log) needs to be accepted and to check the validity of the logs that the prosecution has provided. As a result, investigators are prevented from altering the records to either accuse or exonerate perpetrators.

4.1.4   Endorsement of Logs

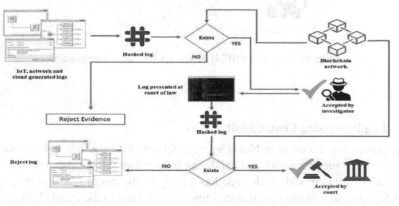

Fig. 2. Endorsement process of proposed BBFMI framework

Investigators completely rely on the CPs, networks, and IoT devices for evidence when it comes to IoT forensics, in contrast to traditional digital forensics. This dependence on CPs might result in unreliable evidence. Our suggested approach uses a decentralised ledger to ensure log integrity while also allowing several forensic process parties to examine logs. In our proposed approach, the forensic investigator still depends on

CPs and IoT devices for the evidence. However, the evidence—in this case, the various logs generated by the cloud instances, network, and IoT devices—is encrypted when a forensic investigator acquires such data from a CPs. The hashed result Fig. 2. Endorsement process of proposed BBFMI framework is then checked against hashes that are used as transaction values on the public blockchain. The prosecutor then searches for the hashed value on the BCs system. If the hash value is present on the blockchain, the prosecutor validates the log and delivers it to the court as trustworthy evidence. On the other hand, if the hash value does not exist on the blockchain network, the log is rejected. When a court acquires a log from a forensics expert, it may evaluate the credibility of the log by hashing it precisely and comparing the result to the hash values on the blockchain network. The court further than decide the acceptance and rejection on the basis credibility of the evidences. Figure 2. Demonstrate attestation of the proposed BBFMI framework.

Our proposed approach is based on blockchain, making it totally decentralised, in contrast to the centralised manner described in previous works [13, 15]. Because our architecture is decentralised, it is feasible to audit logs to see if they are authentic or not. Additionally, it prevents forensic analysts and service providers from surreptitiously altering logs. Our approach makes use of the blockchain's immutability property to ensure the accuracy of logs collected in an IoT environment during digital forensic investigations. It also has the benefit of being verifiable. This benefit offers the forensic organisations the opportunity to confirm the veracity of the logs generated in IoT devices, a benefit that is either not present or is not critically examined in the previous related work [14, 15].

5 Conclusion

The Internet of Things (IoT) has brought about a proliferation of connected devices in our homes, workplaces, and cities. These devices, which include everything from smartphones and smart home devices to industrial control systems and medical equipment, generate vast amounts of data that can be used for forensic investigations. IoT devices are often vulnerable to tampering, both physically and remotely, which can make it difficult to trust the authenticity of the data they store while investigation. This makes it difficult for forensic investigators to use the data as evidence in criminal and civil proceedings. Blockchain technology, with its decentralized and tamper-proof nature, can be used to ensure the integrity and immutability of digital evidence from IoT devices. Implementing and endorsement of log processes entities modification, user centre entity workflow of a BBFMI for IoT devices in digital forensic investigation has the potential to revolutionize the way of incident response and better reliability of IoT devices. By leveraging the tamper-proof and decentralized nature of blockchain technology, BBFMI can ensure the integrity and immutability of digital evidence, making it a valuable tool for both forensic investigators and organizations looking to improve their incident response capabilities. However, it is important to note that further research and development is needed to fully realize the potential of BBFMI and overcome any technical challenges that may arise due to BBFMI.

References

1. Okoh, S.A., Dibal, P.Y.: Performance analysis of IoT cloud-based platforms using quality of service metrics. Int. J. Wirel. Microw. Technol. **13**, 1–4 (2020). https://doi.org/10.5815/IJWMT.2023.01.05
2. Jain, R., Tata, S.: Cloud to edge: distributed deployment of process-aware IoT applications. In: Proceedings - 2017 IEEE 1st International Conference on Edge Computing, EDGE 2017, pp. 182–189 (2017). https://doi.org/10.1109/IEEE.EDGE.2017.32
3. Li, D., Peng, W., Deng, W., Gai, F.: A blockchain-based authentication and security mechanism for IoT. In: Proceedings - International Conference on Computer Communications and Networks, ICCCN. Institute of Electrical and Electronics Engineers Inc (2018). https://doi.org/10.1109/ICCCN.2018.8487449
4. Zhao, Y., et al.: Privacy-preserving blockchain-based federated learning for IoT Devices. IEEE Internet Things J. **8**, 1817–1829 (2021). https://doi.org/10.1109/JIOT.2020.3017377
5. Gong, L., Alghazzawi, D.M., Cheng, L.: Bcot sentry: a blockchain-based identity authentication framework for IoT devices. Information **12**, 203 (2021). https://doi.org/10.3390/info12050203
6. Agrawal, S., Kumar, S.: MLSMBQS: design of a machine learning based split & merge blockchain model for QoS-aware secure IoT deployments. Int. J. Image Graph. Signal Process. **14**, 58–71 (2022). https://doi.org/10.5815/IJIGSP.2022.05.05
7. Mishra, A., Singh, C., Dwivedi, A., Singh, D., Biswal, A.K.: Network forensics: an approach towards detecting cyber crime. In: 2021 International Conference in Advances in Power, Signal, and Information Technology, APSIT 2021. Institute of Electrical and Electronics Engineers Inc (2021). https://doi.org/10.1109/APSIT52773.2021.9641399
8. Motha, J., Maduranga, M., Jayatilaka, N.: Design of an IoT-enabled solar tracking system for smart farms. Int. J. Wirel. Microw. Technol. **12**, 1–13 (2022). https://doi.org/10.5815/IJWMT.2022.06.01
9. James, J.I., Shosha, A.F., Gladyshev, P.: Digital forensic investigation and cloud computing. In: Cloud Technology: Concepts, Methodologies, Tools, and Applications, pp. 1231–1271. IGI Global (2014). https://doi.org/10.4018/978-1-4666-6539-2.ch057
10. Atlam, H.F., Alenezi, A., Walters, R.J., Wills, G.B., Daniel, J.: Developing an adaptive risk-based access control model for the internet of things. In: Proceedings - 2017 IEEE International Conference on Internet of Things, IEEE Green Computing and Communications, IEEE Cyber, Physical and Social Computing, IEEE Smart Data, iThings-GreenCom-CPSCom-SmartData 2017, pp. 655–661 (2018). https://doi.org/10.1109/iThings-GreenCom-CPSCom-SmartData.2017.103
11. Stoyanova, M., Nikoloudakis, Y., Panagiotakis, S., Pallis, E., Markakis, E.K.: A survey on the internet of things (IoT) forensics: challenges, approaches, and open issues (2020) https://ieeexplore.ieee.org/abstract/document/8950109/. https://doi.org/10.1109/COMST.2019.2962586
12. Rauf, A., Shaikh, R.A., Shah, A.: Trust modelling and management for IoT healthcare. Int. J. Wirel. Microw. Technol. **12**, 21–35 (2022). https://doi.org/10.5815/IJWMT.2022.05.03
13. Li, S., Choo, K.K.R., Sun, Q., Buchanan, W.J., Cao, J.: IoT forensics: amazon echo as a use case. IEEE Internet Things J. **6**, 6487–6497 (2019). https://doi.org/10.1109/JIOT.2019.2906946
14. Nieto, A., Rios, R., Lopez, J.: IoT-forensics meets privacy: towards cooperative digital investigations. Sens. (Switz.) **18**, 492 (2018). https://doi.org/10.3390/s18020492
15. Nieto, A., Rios, R., Lopez, J.: A methodology for privacy-aware IoT-forensics. In: Proceedings - 16th IEEE International Conference on Trust, Security and Privacy in Computing and Communications, 11th IEEE International Conference on Big Data Science and Engineering

and 14th IEEE International Conference on Embedded Software and Systems, pp. 626–633 (2017). https://doi.org/10.1109/Trustcom/BigDataSE/ICESS.2017.293

16. Noura, H.N., Salman, O., Chehab, A., Couturier, R.: DistLog: a distributed logging scheme for IoT forensics. Ad Hoc Netw. **98**, 102061 (2020). https://doi.org/10.1016/j.adhoc.2019.102061

17. Kumar, G., Saha, R., Lal, C., Conti, M.: Internet-of-Forensic (IoF): a blockchain based digital forensics framework for IoT applications. Futur. Gener. Comput. Syst. **120**, 13–25 (2021). https://doi.org/10.1016/j.future.2021.02.016

18. Gao, H., et al.: BlockchainBot: a novel botnet infrastructure enhanced by blockchain technology and IoT. Electron. **11**, 1065 (2022). https://doi.org/10.3390/electronics11071065

19. Novo, O.: Blockchain meets IoT: an architecture for scalable access management in IoT. IEEE Internet Things J. **5**, 1184–1195 (2018). https://doi.org/10.1109/JIOT.2018.2812239

20. Atlam, H.F., Alenezi, A., Alassafi, M.O., Wills, G.B.: Blockchain with Internet of Things: benefits, challenges, and future directions. Int. J. Intell. Syst. Appl. **10**, 40–48 (2018). https://doi.org/10.5815/ijisa.2018.06.05

The Method of Determining the Adequacy of Virtual Community's Themes

Synko Anna[(⊠)]

Lviv Polytechnic National University, Stepana Bandery Street 12, Lviv 79013, Ukraine
anna.i.synko@lpnu.ua

Abstract. Due to the large increase in information content, there is a need for selection, verification data and sources which contain this data. The article presents the method for determining the adequacy of a source that may contain useful information. It is known that not all open sources of information on the Internet contain the necessary data for the user searching. Themes posted in virtual communities were chosen as the data source. The proposed method consistently checks the selected theme according to the established requirements (criteria): the measure of relevance of the information content and the measures of actuality of the data. The relevance score reflects the degree of compliance of the theme of the source, which may contain valuable information, to the user's request. To calculate the relevance, statistical measure the TF-IDF and stemming methods (lemmatization and a lookup table) were applied. The actuality score is represented through an activity of creation or placement of posts in the theme. If over certain periods of time there is a significant increase in posts (information content) in the theme, then this theme and the data in it are urgent and needed for users. Indicators that determine the threshold value for each requirement are also provided. If each requirement is successfully met, the theme as a data source can be considered adequate. This method quickly finds relevant and "active" data sources that may contain the necessary information based on the experience of community users. To display the operation of the method, the industry of software is chosen. Among the selected 29 themes of different communities, only five are adequate to the user's request, which is 17%. This means that, with a high probability, 83% of the information posted in other themes is not necessary for the user.

Keywords: Virtual communities · Theme · Adequate source · Data relevance · social networks · Actuality · TF-IDF

1 Instruction

Social networks are the most popular and most visited resources of the Internet. In 2021, over 4.26 billion people were using social media worldwide, a number projected to increase to almost six billion in 2027 (Fig. 1) [1]. Social networks are one of the platforms where users can share knowledge. On average, internet users spend 144 min per day on social media and messaging apps, an increase of more than half an hour since 2015 [1].

© The Author(s), under exclusive license to Springer Nature Switzerland AG 2023
Z. Hu et al. (Eds.): ICCSEEA 2023, LNDECT 181, pp. 328–340, 2023.
https://doi.org/10.1007/978-3-031-36118-0_29

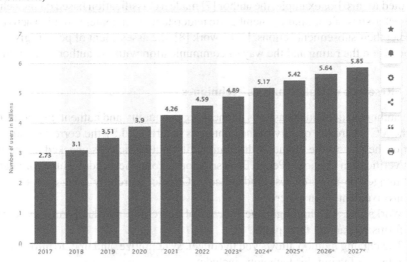

Fig. 1. A chart to represent number of social media users worldwide

It is known that social networks contain virtual communities (VC) on various themes. There are thematic communities and communities containing many themes. Virtual communities become a good source of data because it contains data not only published by product manufacturers, but also consumer feedback [2, 3]. Which gives an opportunity to learn about a certain product from different points of view [4].

The amount of open information is growing rapidly every day. Data on one specific question can be on thousands of different sites. Searching for information is one of the most demanded tasks in practice, which any Internet user has to solve [5]. Therefore, there is a need to develop a method for checking the relevance and actuality of data sources that may contain useful information.

2 Related Work

VC research is a long-term experience among various sciences (including communication and social sciences, mathematical analysis and graph theory, marketing, artificial intelligence and software). These industries reflect the behavior of different segments of the population in a certain period of time on the Internet. The development of the source adequacy method is based on the research of the following aspects: virtual communities and data analysis techniques.

2.1 Virtual Communities Research

In previous studies, the authors [6] conducted an analysis of types of communities and their features. As a result, a community model was built, which shows the general structure characteristic of all types of virtual communities. The main components of the VC structure are participants and information content. Therefore, many studies are devoted to this. For example, the author [7] made a classification based on the behavioral characteristics of community members to understand their content creation activities and predict their subsequent actions. In the work [8], the assessment of posts is given, which depends on the rating and the way of communication with the author who published it.

2.2 Research of Data Analysis Techniques

The author [9] defined the aspects of checking the quality and authenticity of the content in the VC where the quality of the content is determined by the correspondence of the text of the post to the theme of the community; authenticity is checked on the basis of text verification services (PeopleBrowsr, Snopes.com, Geofeedia, HuriSearch and etc.) and images (Foto Forensics, Findexif.com, Google Search by Image and etc.) or with the involvement of an expert.

Works [10, 11] highlight the process of searching and analyzing the content of platforms by parsing their pages.

The parsing method allows you to collect, systematize, and update information. The advantages of using this technique include:

1) Fast data collection in any mode;
2) Prevention of errors;
3) Carrying out a regular inspection at a given time interval;
4) Presentation of data in any format;
5) Ensuring an even load on the website where parsing takes place (so as not to create the effect of a DDoS attack).

This parsing technique works only for one VC. To analyze and collect data from other websites, it is necessary to perform similar actions with their HTML pages.

Linguistic methods are used to analyze the content of the text. One of them is stemming, which has several variants of algorithms that differ in accuracy and performance.

The study [12] proposes to compare document retrieval precision performances based on language modeling techniques – particularly stemming and lemmatization. As a result, it was established that lemmatization gives a better result in terms of accuracy in contrast to the baseline algorithm. But lemmatization has an important drawback – dependence on the correct recognition of parts of speech.

As opposed to previous studies, the author proposed to consider the theme itself as a source of useful information, because communities can have many themes and, accordingly, many thematic situations [13]. This will reduce the volume of data and provide the user with the data corresponding to his request. The quality of the selected content is suggested to be determined by the end user.

The innovativeness of the work lies in the fact that in previous studies there was no verification of communities – their themes as a source of data. In works [13–15], the correspondence of the content is determined by one of the stemming methods – lemmatization. In this work, two approaches are used (lemmatization and a lookup table), the sequential application of which eliminates each other's shortcomings and increases the accuracy of the algorithm. Also was used a dictionary containing auxiliary terms that expands and supplements the user's information request and theme names [16].

Fast gains content in VC effect on data volume and time consumption for processing them as well. Viewing the theme as a source of data will allow:

1) To ignore themes, which don't semantically meet the user's query but are in communities.
2) To reduce the risk of block due to excessive activity to download data.
3) To speed up the search for information by reducing the amount of data.

The purpose of the works is to develop the method for determining the adequacy of the theme for downloading discussions from them.

The software industry was selected to display the operation of this method.

3   The Method for Determining of Adequacy of a Theme as Source of Valuable Data

Before uploading discussions, it is necessary to check themes as sources of data for their adequacy. The adequacy is a broad term covering all science (philosophy, psychology, jurisprudence, mathematics, etc.). In probability theory the adequacy is to comply with the actual results (for example, results of the research), with the results obtained by the calculation. In the technique the adequacy is compliance of the basic characteristics of the model to specifications of the object, which simulated [17].

The adequacy source of information is the themes of virtual communities that meets requirements:

1) To be relevant.
2) To be actual – the growth of the count of posts over a certain period of time.

The relevance score ($\mu_{relevance}$) specifies if themes of VC have information about a particular subject area. This estimate we'll be considering using the relevance score of the user's query to set off all keywords characterizing the theme. Restrictions on score $\mu_{relevance}$ are range [0, 1]. The source can be considered relevant if performed the next the following criterion:

$$\mu_{relevance} \geq \alpha \tag{1}$$

α – the indicator, which threshold for estimate of relevance score of sources.

The actuality score (μ_{actual}) determines the rate of content data growth for a certain period of time. Restrictions on score μ_{actual} are range [0,1]. The source can be considered relevant if performed the next the following criterion:

$$\mu_{actual} \geq \beta \tag{2}$$

β – the indicator, which threshold for estimate of the actuality score of sources.

These requirements define the adequacy of the theme are performed sequentially. If the condition of relevance doesn't fulfill, then the next rate won't be calculating and the source will not be able to fulfill criterion the adequacy estimate. If the condition of relevance had been fulfilled but the actuality score doesn't satisfy the given criterion the theme is not adequate. The UML activity diagram displays the method for determining the adequacy of a theme (Fig. 2). Activity diagrams describe the dynamic aspects and behavior of the system. Activity diagrams represent the flow from one activity to another activity.

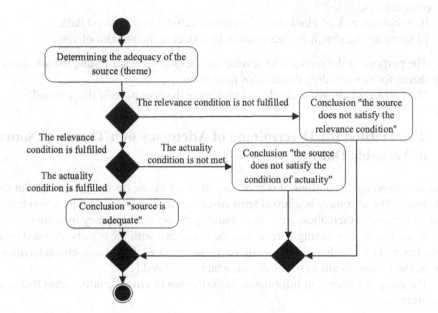

Fig. 2. The UML activity diagram to determine of adequacy of a theme

4 Determination of Relevance of the Theme

The relevance is measure of correspondence of obtained result to desired or expected. Search relevance is the measure of how closely and accurately the generated search results relate to the user's search query. There are many search engines that will offer his own search priority in terms of indices. Therefore, the same query in different systems displays different results.

After analyzing the criteria of increase of relevance of the search content [18] from VC was allocated the following: the title of VC, theme and post (message) should contain keywords and logical structure about certain object or subject (for example title or function the software); absence of errors in the text (for correct indexing of the text parts of the site); presence of completed meta tags that may contain useful data (for example, Title, Description, Keywords, Heading ect.), keywords in the URL-address and in internal links to the theme page, etc.; the content of post should be unique; loading time of community pages; visual representation of the content of the VC (text size).

Usually, the relevance is determined by statistical research methods. The main functions are to collect, construction, group, synthesis, analysis and evaluation of data.

So, the relevance theme is a theme which is in VC and with a certain degree of relevance contains content in the form of a set of discussions to the user's request.

For selection and further evaluation of themes of virtual communities it is necessary to perform and fill the database according to the following attributes: URL address, title of the VC, titles of themes. All this data needed for the automated data collection (parsing). After that, it is necessary to select those themes that can contain information about a software.

Communities that contain software data often list this information in the community title (if this source is entirely focused on the specified software) or in the theme(s). To compare the titles, the author introduced the concept of the measure relevance of title software to themes titles. If the title of the software is not mentioned in the community themes, we will assume that this source does not contain information about it. And therefore, based on that, the author developed the method of selecting VC that may contain valuable content for a specific software (Fig. 3).

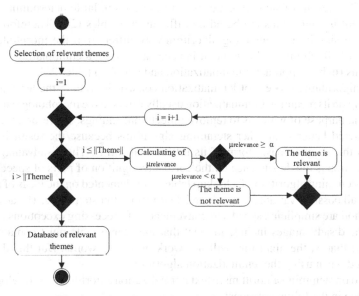

Fig. 3. A flowchart of the selection of VC themes

The method of selection VC themes is to perform the next steps:

1) For each VC themes calculated the relevance measure according to the user's request – $\mu_{relevance}(SW(Keyword), VC(Theme_i))$.
2) If the relevance value of the title of the i-th theme to the keywords of the software is greater than or equal to the specified value, then we will assume that the source (theme) is relevant.
3) Otherwise, we will consider that the source (theme) is not relevant.

4) The method is made until all themes are mapped to the keywords of the software.

The result of the method has selected communities and rated themes that contain information about a certain software. Therefore, the relevance indicator is fulfilled.

Later, the selected themes should be investigated according to the following assessment – an actuality of the content. After all, it is this evaluation that influences the subsequent decision regarding acceptance or rejection of the information content of a certain theme of the VC.

4.1 Calculation of Relevance of the Theme

To determine the estimate of relevance of the theme to the user's request, it is necessary to perform text normalization [19]. Normalization is a method in which a set of words in a sentence is turned into a sequence to reduce search time. Words that have the same meaning but have some differences depending on the context or sentence are normalized.

A stemming operation is usually used to normalize the text. The result of stemming is finding the base of the word (stem) by removing auxiliary parts such as endings and suffixes. The results of stemming are sometimes very similar to determining the root of a word, but its algorithms are based on different principles [20]. Therefore, the word after processing by the stemming algorithm may differ from the morphological root of the word. To eliminate this issue, it is necessary to use a combination of stemming algorithms (hybrid approach): lemmatization and a lookup table [21].

The algorithmic process of lemmatization consists in finding the lemma of a word depending on its meaning. Lemmatization usually refers to the morphological analysis of words, the purpose of which is to remove inflectional endings. Therefore, this approach is considered better than other stemming algorithms because the result is a lemma, which is the original, basic form of its inflectional forms. The disadvantage of using lemmatization is the dependence on the correct recognition of parts of speech.

The stemming algorithm of a lookup table is implemented on the basis of a table that contains all possible variants of words and their forms after stemming. The advantages of this method are simplicity, speed and convenience of processing exceptions to language rules. The disadvantages include the fact that the search table must contain all forms of words: that is, the algorithm will not work with new words. But this drawback is eliminated when using the lemmatization algorithm.

Based on stemming algorithms, a text normalization algorithm was developed, which takes place in the following steps:

1) Divide the text into lexemes (words).
2) Clean words from apostrophes, punctuation marks.
3) Remove words that are not taken into account during the linguistic analysis because they do not carry important information for the text (pronouns, articles, prepositions, conjunctions) [22].
4) Apply the lemmatization algorithm, which will allow you to obtain lemmas (the basic form of words). And if a lemma for a word is not found, it is handed over to an expert for consideration.
5) Apply the stemming algorithm – lookup table.

6) The words that were not processed as a result of passing the search algorithm according to a lookup table are transferred to the expert, and the algorithm for cutting off endings and suffixes is performed.

It should be taken into account that the keywords specified by the user may not be sufficient for the selection of themes. Therefore, in order to expand the query of the selection of sources, it is necessary to develop a dictionary of terms containing important, useful words for the selected subject area. As a result, it is possible to combine the user query and the dictionary as follows into an expression ($\|Vyraz\|$) containing a set of meaningful (unique, non-repeated) words:

$$\|Vyraz\| = \|norm(Query)\| \cup \|Dicrionary\| \tag{3}$$

where $\|norm(Query)\|$ – set of normalized words that are in the user's request;
$\|Dicrionary\|$ – a dictionary of terms for the selected subject area.

We will calculate the degree of correspondence of the theme of the VC of the chosen subject area based on the word frequency formula (TF-IDF measures):

$$\mu_{relevance} = \frac{\|Vyraz\| \cap \|norm(Theme)\|}{\|norm(Theme)\|} \tag{4}$$

where $\|Vyraz\|$ – set of normalized words that are in the user's request;
$\|norm(Theme)\|$ – a dictionary of terms for the selected subject area.

The text normalization process according to a user request is depicted using the UML sequence diagram in Fig. 4.

The result of passing this algorithm is the normalization of the words contained in the text using stemming methods. The consistent application of several methods allowed to eliminate the shortcomings of each and make the normalization method universal for any words.

4.2 Determining the Actuality Requirement of Virtual Communities' Themes

The actuality is the subjective value that people attribute to information if it meets their needs and interests. The actuality of data is a property that reflects their significance in today's reality [23].

Accordingly, an actuality post within the community discourse is a post containing relevant data for user queries. The actuality of the discourse as an element of the theme is determined by the presence of relevant posts in it. So, the actuality of the theme is characterized by the number of published relevant discourse.

To measure the value of relevance, the author introduced the concept of the level of actuality, which is determined by the number of relevant posts in the theme, as a source of data. It allows the following statement: the more posts that have been created in a certain period of time (T) in a certain theme, the greater the importance of the actuality level. The natural limitations of the actuality level are [0, 1].

Therefore, to calculate the level of actuality, it is necessary to apply the concept of activity (A), which determines the rate of increase in information content in a certain theme.

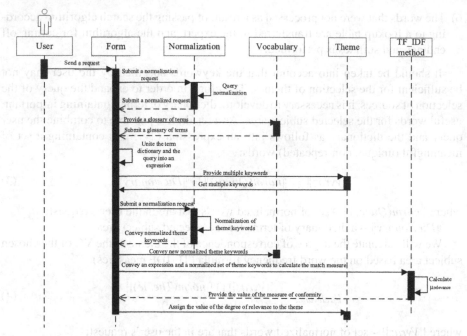

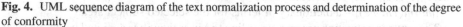

Fig. 4. UML sequence diagram of the text normalization process and determination of the degree of conformity

To calculate the actuality of the theme, author applied a logarithmic function, which has the following advantages:

1) The set of definitions of the logarithmic function $D = (0, \infty)$, and the range of obtained results for the theme will be within the natural limits $[0,1]$.
2) The logarithmic function is monotonic and increasing (to calculate the activity of the theme). When $b > 1$ (the increase of messages in the theme must be more than one message):
   a) if $b = 1$ – an increase in the theme by one post, which is not a good indicator for confirming activity. Accordingly, we will consider the theme not actual;
   b) if $b = 0$ – there was no information growth during the comparison period or the theme is new and has no post yet. Accordingly, the logarithm will be undefined (which does not meet the condition for calculating the logarithm). Such a theme will not be considered;
3) Logarithmic functions are proportional to different bases.
4) Simplicity when working with a large amount of data (posts), since the logarithm function grows more slowly with increasing numbers [24].

Taking into account the above conditions for the use of the logarithm, the author developed a formula for calculating the activity of the theme using the binary logarithm:

$$A = \frac{\log_2(y_i)}{\log_2(y_n)} \tag{5}$$

where y_i – logarithmic value of the number of posts of the current level (compared period);

y_n – the logarithmic value of the total number of posts.

Therefore, using this method will allow you to eliminate insignificant themes that have a small increase in the number of posts over a certain period of time, as well as get rid of sources that have no posts (new theme) or have only one post, which cannot be considered a discussion.

5 Results

An experimental study was conducted to determine the adequacy of themes as sources of valuable information. A parsing process was applied to automatically collect data from the VC.

User request: «Development of web applications in Java».

Accordingly, it was necessary to find exactly those themes that will contain actual and relevant information for this request.

Five communities were selected for information selection – Dou.ua, replace.org, senior.ua, proger.com, cyberforum.ru (Fig. 5). These communities contain themes on various subjects. Each theme has a score calculated according to the following estimate relevance. For indicator α, which threshold for these estimated the following value is applied $\alpha \geq 0, 75$.

Therefore, the theme is adequate if all indicators are higher than $0, 75$.

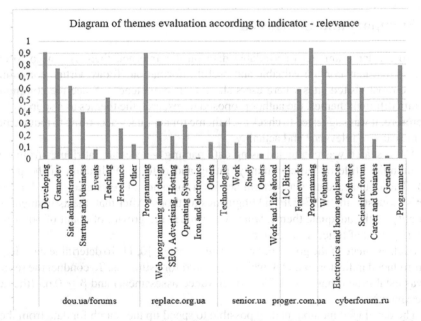

Fig. 5. A diagram to represent a selection of communities and themes by relevance of data

As can be seen from the research, among 29 themes, only 6 turned out to be relevant sources of data (Fig. 5). The next step is to check relevant themes about the activity of creating posts for certain periods – the last six months (Table 1).

Table 1. Results regarding the growing of count of post in adequate themes over the past six months

Community title	Theme title	Post growth rate (monthly)						Total score
		Aug	Sept	Oct	Nov	Dec	Jan	
dou.ua/forums	Developing	0.61	0.69	0.79	0.9	0.82	0.81	0.77
	Gamedev	0.93	0.94	0.92	0.96	0.99	0.96	0.95
proger.com.ua	Programming	0.72	0.8	0.84	0.79	0.85	0.88	0.81
cyberforum.ru	Webmaster	0.56	0.59	0.52	0.66	0.69	0.59	0.6
	Software	0.72	0.68	0.71	0.75	0.81	0.8	0.75
	Programmers	0.88	0.89	0.86	0.94	0.92	0.91	0.9

Total score in Table 1 represents the arithmetic mean of the activity of filling messages in the theme for the last six months. For indicator β, which threshold for these estimates the following value is applied $\beta \geq 0,75$. As can be seen from the research, among six adequate themes, only 5 satisfy the condition of relevance.

6 Summary and Conclusion

Due to the large amount of available data on the Internet, there is a need to check sources that may contain valuable and useful information. Today, virtual communities are the most visited sites where users share their experiences. Since the communities contain different themes, the author proposed to research the themes as a source of data. Therefore, it was suggested to check each theme for adequacy according to requirements: the measure of relevance and actuality.

The research was conducted using data parsing – necessary attributes from virtual communities (title of the community, rating and date of publication of the post and ect). The obtained data was determined as a measure of relevance according to the user's request using a TF-IDF and linguistics methods. After that, the assessment of the actuality of the adequate theme, which determines the growth of count of posts over a certain period of time, is calculated.

Each estimate of the given measures is in the range [0, 1]. To determine the adequacy is introduced indicators α and β, which threshold for estimates. To conduct the research, it was established that $\alpha \geq 0,75$ (the relevance assessment) and $\beta \geq 0,65$ (the actual assessment).

The developed method made it possible to speed up the search for data from the VC and to focus the end user's attention on those themes that are relevant and actual. And therefore, with a high probability, adequate themes contain the necessary data for the end user. Selection and verification of data posted in adequate themes should be carried out by the user searching for information.

As a result of the research, it was established that only 17% of themes are adequate for the user. These results are compatible with the Pareto principle, which states that roughly 80% of consequences come from 20% of causes. Based on this, it can be argued that 20% of themes contain data that provide 80% of the necessary data to the user.

7 Future Research

Today, it is important not only to research data sources which can have useful information, but also the content of this source. Posted information can be analyzed according to the following components: content and rating of post. In VC, users rate a post, thereby forming its rating. The rating reflects how relevant the post was for other users. Published data can be analyzed for unwanted content, links, malicious files, etc., as well as establish authenticity and uniqueness based on text verification services.

In the next study, the author plans to develop a software complex that collects authority posts from adequate themes.

References

1. Statista. Number of social media users worldwide from 2017 to 2027. Social Media & User-Generated Content. https://www.statista.com/statistics/278414/number-of-worldwide-social-network-users/
2. Korzh, R., Peleschyshyn, A., Syerov, Y., Fedushko, S.: The cataloging of virtual communities of educational thematic. Webology 11(1), 117 (2014)
3. Trach, O., Peleshchyshyn, A., Korzh, R.: Methods for creating a team for managing a virtual community. CEUR Workshop Proceedings. In: Proceedings of the 1st International Workshop IT Project Management (ITPM 2020), vol 2565, pp. 83–92 (2020)
4. Winanti, F., Goestjahjanti, S.: Component learning community for informal education to support culinary community at era new normal covid-19: A systematic literature review. Int. J. Inf. Technol. Comput. Sci. 14(1), 20–31 (2022). https://doi.org/10.5815/ijitcs.2022.01.03
5. Faryad, S., Batool, H., Asif, M., Yasin, A.: Impact of internet of things (IoT) as persuasive technology. Int. J. Inf. Technol. Comput. Sci. (IJITCS) 13(6), 16–28 (2021). https://doi.org/10.5815/ijitcs.2021.06.02
6. Peleshchyshyn, A., Kravets, R., Sierov, Y.: Analysis of existing types of virtual communities on the internet and building a virtual community model based on a web forum. Inf. Syst. Netw. J. Lviv Polytech. Nat. Univ. 699, 212–221 (2011)
7. Fedushko, S.: Development of a software for computer-linguistic verification of socio-demographic profile of web-community member. Webology, 11(2), 126 (2014). http://www.webology.org/2014/v11n2/a126.pdf
8. Fedushko, S., Mastykash, O., Syerov, Y., Shilinh, A.: Model of search and analysis of heterogeneous user data to improve the web projects functioning. In: Zhengbing, H., Petoukhov, S., Dychka, I., He, M. (eds.) Advances in Computer Science for Engineering and Education IV, pp. 56–74. Springer International Publishing, Cham (2021). https://doi.org/10.1007/978-3-030-80472-5_6

9. Korzh, R., Peleshchyshyn, A., Trach, O., Tsiutsiura, M.: Increasing the efficiency of the processes of formation of the informational image of the HEI. In: Shakhovska, N., Medykovskyy, M.O. (eds.) CSIT 2019. AISC, vol. 1080, pp. 661–679. Springer, Cham (2020). https://doi.org/10.1007/978-3-030-33695-0_45

10. Zhou, B., Hang, Z., Xavier, P., Sanja, F., Adela, B., Antonio, T.: Scene parsing through ADE20K dataset. In: Proceedings of the IEEE conference on computer vision and pattern recognition, pp. 633–641 (2017). https://doi.org/10.1109/CVPR.2017.544

11. Zhao, H., Hengshuang, Z., Jianping, S., Xiaojuan, Q., Xiaogang, W., Jiaya, J.: Pyramid scene parsing network. In: Proceedings of the IEEE Conference on Computer Vision and Pattern Recognition, pp. 2881–2890 (2017)

12. Balakrishnan, V., Lloyd-Yemoh, E.: Stemming and lemmatization: A comparison of retrieval performances. In: Proceedings of SCEI Seoul Conferences, Seoul, Korea, pp.174–179 (2014)

13. Synko, A.: The method of trust level of publications hosted in virtual communities. Sci. J. TNTU (Tern.) **105**(1), 68–79 (2022). https://doi.org/10.33108/visnyk_tntu2022.01.068

14. Lamberth, C.: The power of cross-indication testing: agrochemicals originally stemming from a different indication. Pest Manage. Sci. **78**(11), 4438–4445 (2022). https://doi.org/10.1002/ps.7100

15. Alhaj, Y.A., Xiang, J., Zhao, D., Al-Qaness, M.A., Dahou, M.A.: A study of the effects of stemming strategies on Arabic document classification. IEEE Access **7**, 32664–32671 (2019). https://doi.org/10.1109/ACCESS.2019.2903331

16. Babich, O., Vyshnyvskiy, V., Mukhin, V., Zamaruyeva, I., Sheleg, M., Kornaga, Y.: The technique of key text characteristics analysis for mass media text nature assessment. Int. J. Modern Educ. Comput. Sci. (IJMECS) **14**(1), 1–16 (2022). https://doi.org/10.5815/ijmecs.2022.01.01

17. Tedeschi, L.O.: Assessment of the adequacy of mathematical models. Agric. Syst. **89**(2–3), 225–247 (2006)

18. Dogan, A., Birant, D.: Machine learning and data mining in manufacturing. Expert Syst. Appl. **166**, 114060 (2021)

19. Singh, D., Singh, B.: Investigating the impact of data normalization on classification performance. Appl. Soft Comput. **97**, 105524 (2020)

20. Ballabio, D., Grisoni, F., Todeschini, R.: Multivariate comparison of classification performance measures. Chemom. Intell. Lab. Syst. **174**, 33–44 (2018)

21. Jivani, A.G.: A comparative study of stemming algorithms. J. Comput. Sci. Technol. **2**(6), 1930–1938 (2011)

22. Manalu, S.R.: Stop words in review summarization using TextRank. In: 14th International Conference on Electrical Engineering/Electronics, Computer, Telecommunications and Information Technology (ECTI-CON) (2017)

23. Ngiam, K.Y., Khor, I.W.: Big data and machine learning algorithms for health-care delivery. Oncology **20**(5), 262–273 (2019)

24. Reynolds, R., Stauffer, A.: A method for evaluating definite integrals in terms of special functions with examples. Int. Math. Forum **15**(5), 235–244 (2020)

A Feature Selection and Regression Refinement Network for Power Facility Detection in Remote Sensing Images

Yu Zhou[1], Wenhao Mo[2], Changyu Cai[2], Yuntao Yao[2(✉)], and Yanli Zhi[1]

[1] State Grid Jiangxi Electric Power Supply Co. Ltd., Nanchang 330096, China
[2] China Electric Power Research Institute, Beijing 100192, China
js_xiao@qq.com

Abstract. Intelligent identification and detection of power facility is crucial to power grid security, and remote sensing equipment is usually used to collect images in power routing inspection. However, the characteristics of remote sensing images, such as complex background, large scale change and arbitrary orientation, make it challenging to detect the objects in them. To improve the detection accuracy of power facility in remote sensing images, a feature selection and regression refinement network is proposed in this paper. Specifically, a bidirectional feature fusion module (BFFM) is devised which fuses deep and shallow features to obtain a multi-scale feature map with stronger representation ability. A feature selection module (FSM) with hybrid attention is devised which utilizes the fusion of spatial and channel attention to obtain attention response maps and generates the features specific to classification task and regression task through different polarization functions. Furthermore, a regression refinement module (RRM) is designed to adjust the anchors based on regression features to improve the regression accuracy. Experiments on DOTA and a power facility dataset achieve excellent detection results, the mAP on DOTA reaches 73.1 and the mAP of high voltage tower reaches 71.2.

Keywords: Object detection · Remote sensing · Power facility

1 Instruction

Detection of power facility plays an important role in maintaining the security of the grid system. Large power facilities such as high-voltage towers and thermal power chimneys have high spatial complexity, and require remote sensing acquisition equipment with wide coverage to obtain their source images, so it is of great significance to research the object detection of power facility in remote sensing images. However, the background of remote sensing images is complex, which cause irrelevant background features to seriously disturb the feature information of objects. Moreover, some objects are densely arranged and have small sizes, making it impossible to effectively capture the feature distribution of objects. In addition, the orientation of objects is arbitrary, which makes the object detection of remote sensing image more difficult.

Z. Hu et al. (Eds.): ICCSEEA 2023, LNDECT 181, pp. 341–348, 2023.
https://doi.org/10.1007/978-3-031-36118-0_30

With the development of deep learning, many object detection algorithms [1–3] have achieved relatively accurate detection results. RetinaNet [4] is a typical one-stage algorithms, it uses Focal Loss to balance positive and negative samples, while using feature pyramids to make predictions on multiple layers of feature maps, which achieves high accuracy and speed. CAD-Net [5] and method [6] alleviate the problem that the object scale varies greatly in remote sensing images by constructing more powerful feature representations. R3Det [7] develops a feature refinement module to increase detection performance by obtaining more accurate features. However, there are few researches on object detection for power facilities in remote sensing images. DeepWind [8] is a weakly supervised localization model which automatically localize wind turbines in satellite imagery. The model takes a satellite image as input and output a predicted response map which indicates the location of the turbines in the image. In the response map, a peak detection module finds local maxima and combines them based on according to proximity. Using deep convolutional neural networks, Sal-MFN [9] is an end-to-end detection framework for Thermal Power Plants. It employs a multi-scale feature network for the adaptation of the thermal power plants of various sizes and a saliency enhanced module to reduce background distractions for enhanced feature maps.

In this paper, a feature selection and regression refinement network (FSRR-Net) for power facility detection in remote sensing images is proposed. Specifically, to enhance the network's feature representation ability, a bidirectional feature fusion module (BFFM) is designed. To generate task-specific features, a feature selection module (FSM) with hybrid attention is designed. To improve the regression accuracy, a regression refinement module (RRF) is designed to adjust the anchors for better spatial alignment with the object feature.

2 Proposed Method

Figure 1 displays the overall framework of the proposed FSRR-Net. ResNet [10] is used as the backbone. Firstly, the bidirectional feature fusion module (BFFM) fuses shallow features and deep features to generate a multi-scale feature pyramid. Then, the features specific to classification task and regression task are extracted separately through the feature selection module (FSM) with hybrid attention. Finally, the regression refinement module (RRM) adjusts anchors in multiple levels based on regression features. The BFFM can improve the ability of network to represent features, the FDM can generates the features specific to different tasks, the RRM can improve spatial misalignment between anchor and object feature. In this way, the detection performance can be significantly improved.

2.1 Bidirectional Feature Fusion Module

In order to detect objects of various scales, we add a bottom-up fusion path on the basis of the top-down fusion path of FPN [11]. Shallow features no longer need to pass through as many convolution layers to reach the top layer due to the architecture, so as to reduce the feature loss during the transfer process. We also introduce P_6 and P_7 layers in the feature pyramid to predict larger scale objects. The new feature extraction module is

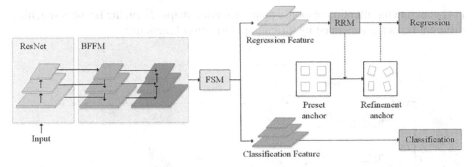

Fig. 1. The overall framework of FSRR-Net

called Bidirectional Feature Fusion Module (BFFM). Details of the module are shown in Fig. 2, where 1 × 1Conv denotes the convolution process with the 1 × 1 convolution kernel, which serves to change the feature map's channel number; 3 × 3Conv denotes the convolution process with the step size of 1 and the 3 × 3 convolution kernel; 2 × Down denotes the two-fold down-sampling operation, which is implemented with the step size of 2 and the 3 × 3 convolution kernel; 2 × Up represents the two-fold up-sampling operation, which is implemented with the bilinear interpolation.

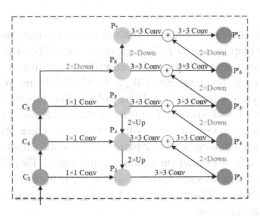

Fig. 2. Bidirectional Feature Fusion Module

2.2 Feature Selection Module with Hybrid Attention

The majority of object detection frameworks use the same feature maps for classification and regression tasks. However, due to the inconsistency between the two tasks, the shared features reduce the detection performance [12]. We design Feature Selection Module (FSM) with hybrid attention to avoid this problem, which can extract the features specific to the classification and the regression separately. Figure 3 shows the structure of FSM. Firstly, a hybrid attention mechanism with channel attention [13] and spatial attention

[14] generate the attention response maps of feature maps. Then, the features specific to different tasks are generated by different polarization functions.

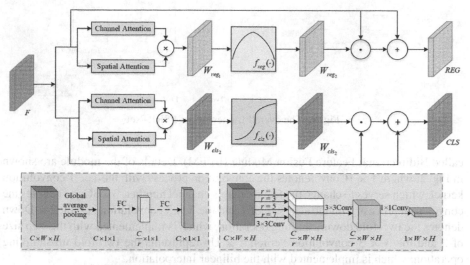

Fig. 3. Feature Selection Module with hybrid attention

The channel attention model achieves enhancement or suppression of certain channels of the feature map by learning the importance of different channels. The spatial attention model selects feature regions that contribute significantly to the task for processing. After attention response maps $W_{reg} \in R^{C \times W \times H}$ and $W_{cls} \in R^{C \times W \times H}$ are generated by the hybrid attention model, the polarization functions for different tasks are used to obtain the task-specific features. In the regression task branch, network focuses more on the boundary features of the object. Therefore, the polarization function needs to enhance the boundary information of the feature map. In the classification task branch, network focuses more on semantic information. Therefore, the polarization function needs to enhance the useful information of the feature map and ignore the background noise information. The polarization functions for regression and classification tasks respectively are as follows:

$$\begin{cases} f_{reg}(x) = 4x(1-x) \\ f_{cls}(x) = \dfrac{1}{1 + e^{-(x-0.5)}} \end{cases} \quad (1)$$

The extracted features for the regression task are spread uniformly over the object, which helps to recognize the boundaries of the object and locate it accurately. The extracted features for the classification task are mainly clustered in the most easily recognized object's parts to prevent disruptions from its other parts, which leads to more accurate classification results.

3 Regression Refinement Module

Anchor-based object detection algorithms need to pre-set anchor of a certain scale and a certain aspect ratio, and during the training, the class which the anchor belongs to and the offset from the ground-truth are learned. In remote sensing images, because of the large scale-variation and arbitrary orientation of objects, it is difficult to align the fixed anchor with the object. Based on the above analysis, we design Regression Refinement Module (RRM), Based on the above analysis, this paper designs an anchor frame refinement network based on image features, which apply multi-level refinement to pre-set anchors to alleviate spatial unalignment between anchors and objects, enabling anchors to capture more object features. Figure 4 displays the structure of RRM.

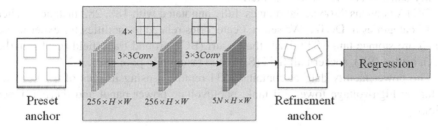

Fig. 4. Feature Selection Module with hybrid attention (r in spatial attention – dilation rate)

The proposed algorithm uses the five-parameter approach to describe the bounding box. The anchor after the nth refinement is noted as $(x^n, y^n, w^n, h^n, \theta^n)$, which represents the pre-set anchor when $n = 0$. The offset learned in the $n + 1$ th refinement is $(t_x^{n+1}, t_y^{n+1}, t_w^{n+1}, t_h^{n+1}, t_\theta^{n+1})$, defined as follows:

$$
\begin{cases}
t_x^{n+1} = \dfrac{x_{n+1} - x_n}{w_n}, \; t_y^{n+1} = \dfrac{y_{n+1} - y_n}{h_n}, \\[2mm]
t_w^{n+1} = \log(\dfrac{w_{n+1}}{w_n}), \; t_h^{n+1} = \log(\dfrac{h_{n+1}}{h_n}), \\[2mm]
t_\theta^{n+1} = \tan(\theta_{n+1} - \theta_n)
\end{cases}
\tag{2}
$$

where $(x_i, y_i)w_i, h_i, \theta_i$ represent the center point coordinates, width, height and angle of the anchor in the i th refinement, respectively.

Because of the serious boundary problems with the five-parameter approach, we use IoU-Smooth L1 loss to address this problem [15]. IoU-Smooth L1 is defined as follows:

$$
IoU - \text{Smooth}_{L_1} = -\log(IoU) \sum_{j \in \{x,y,w,h,\theta\}} \frac{L_{reg}\left(v'_{nj}, v_{nj}\right)}{\left|L_{reg}\left(v'_{nj}, v_{nj}\right)\right|}
\tag{3}
$$

where $L_{reg}(\cdot)$ represents the Smooth L1 loss [1], v'_{nj} represents the prediction's offset, v_{nj} represents the ground-truth's offset, and IoU represents the intersection ratio between the predicted and true bounding boxs. The loss function is approximately equal to zero in the boundary situation, which eliminates the abrupt increase in loss.

The proposed detector contains two tasks. The loss of the regression task consists of refinement regression loss and detection regression loss, both calculated by IoU-Smooth L1. The loss of the classification task is the anchor category prediction loss, calculated by Focal Loss [16]. The weighted sum of the classification and regression losses constitutes the overall loss function of the proposed detector.

4 Experiments

Experiments are carried out on a public remote sensing dataset DOTA [16] and a power facility dataset collected from the Internet.

DOTA contains 2806 aerial images, fully annotated with 188, 282 instances. There are 15 categories in DOTA, We select 7 categories related to facilities for experiments, which are storage tank (ST), basketball court (BC), ground track field (GTF), harbor (HA), bridge (BR), soccer ball field (SBF) and swimming pool (SP).

The power facility dataset contains 1534 remote sensing images of 4 categories, including high-voltage tower, oil tank, photovoltaic power panel and thermal power chimney.

4.1 Results on DOTA

To evaluate the performance of the proposed FSRR-Net, we compare it to RetinaNet [4], CAD-Net [5], R3Det [7] on the DOTA dataset. In the experiments, FSRR-Net uses ResNet152 as the backbone and uses rotation and random flip for image enhancement. Table 1 displays the experimental results.

Table 1. Detection results on DOTA (%) (The best results of each category are bolded.)

Methods	BR	GTF	BC	ST	SBF	HA	SP	mAP
RetinaNet	43.7	66.7	80.2	75.5	53.9	63.9	65.9	68.7
CAD-Net	49.4	66.5	79.2	73.3	48.4	62.0	67.0	69.9
R^3Det	48.5	62.5	81.4	83.5	62.0	65.4	67.5	71.7
FSRR-Net	**52.1**	**68.6**	**86.7**	**86.3**	**64.8**	**66.9**	**68.7**	**73.1**

From Table 1, the mAP of the proposed FSRR-Net reaches 73.1% for fifteen categories of objects, which is the best among the four algorithms, and achieves best detection results in all seven categories objects related to facilities. Results from the experiments demonstrate the proposed detector's superior performance.

Table 2. Detection results on the power facility dataset (%)

Category	mAP
High-voltage tower	71.2
Oil tank	87.7
Thermal power chimney	76.1
Photovoltaic power panel	21.5

4.2 Results on the Power Facility Dataset

Table 2 displays the FSRR-Net's detection mAPs of the high-voltage tower, oil tank, photovoltaic power panel and thermal power chimney, and Fig. 5 shows the corresponding visualization results. It can be seen that the proposed FSRR-Net obtains high mAPs on the high-voltage tower, oil tank, and thermal power chimney, indicating that FSRR-Net is able to accurately detect objects of different scale and densely arranged within the images. However, the mAP is low on the photovoltaic power panel, which is probably due to the fact that the size of a photovoltaic panel unit in remote sensing images is too small and the arrangement of photovoltaic panels is too dense, making it difficult for the network to extract useful features.

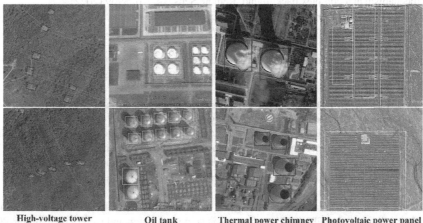

Fig. 5. The visualization detection results on the power facility dataset

5 Summary and Conclusion

To increase the detection accuracy of power objects in remote sensing images, this paper proposes a feature selection and regression refinement network (FSRR-Net). Specifically, the bidirectional feature fusion module (BFFM) is designed to detect objects of various scales. The feature selection module (FSM) is designed to generate the features required for classification task and regression task correspondingly, which can avoid the interaction between the two tasks. Furthermore, the regression refinement module

(RRM) is designed to enable the anchor to capture more effective object features. Experiments on DOTA dataset and a power facility dataset achieves excellent detection results, indicating that our method has superior detection performance for typical power facilities and can be well used for intelligent monitoring of power facilities.

Funding Information. This work is funded by the Science and Technology Project of State Grid Corporation(Grant No. 5700-202225178A-1-1-ZN).

References

1. Girshick, R.: Fast R-CNN. In: Proceedings of the IEEE International Conference on Computer Vision, pp. 1440–1448 (2015)
2. Li, L., Zhou, Z., Wang, B., et al.: A novel CNN-based method for accurate ship detection in HR optical remote sensing images via rotated bounding box. IEEE Trans. Geosci. Remote Sens. **59**(1), 686–699 (2020)
3. Xiao, J., Guo, H., Zhou, J., et al.: Tiny object detection with context enhancement and feature purification Expert Syst. Appl. 118665 (2023)
4. Lin, T.Y., Goyal, P., Girshick, R., et al.: Focal loss for dense object detection. In: Proceedings of the IEEE International Conference on Computer Vision, pp. 2980–2988 (2017)
5. Zhang, G., Lu, S., Zhang, W.: CAD-Net: A context-aware detection network for objects in remote sensing imagery. IEEE Trans. Geosci. Remote Sens. **57**(12), 10015–10024 (2019)
6. Xiao, J., Guo, H., Yao, Y., et al.: Multi-scale object detection with the pixel attention mechanism in a complex background. Remote Sens. **14**(16), 3969 (2022)
7. Yang, X., Yan, J., Feng, Z., et al.: R3det: Refined single-stage detector with feature refinement for rotating object. In: Proceedings of the AAAI Conference on Artificial Intelligence, vol. 35, no. 4, pp. 3163–3171 (2021)
8. Zhou, S., Irvin, J., Wang, Z., et al.: Deepwind: weakly supervised localization of wind turbines in satellite imagery. In: 33rd Conference on Neural Information Processing Systems (NeurIPS 2019) (2019)
9. Yin, W., Sun, X., Diao, W., et al.: Thermal power plant detection in remote sensing images with saliency enhanced feature representation. IEEE Access **9**, 8249–8260 (2021)
10. He, K., Zhang, X., Ren, S., et al.: Deep residual learning for image recognition. Proceedings of the IEEE Conference on Computer Vision and Pattern Recognition, pp. 770–778 (2016)
11. Lin, T.Y., Dollár, P., Girshick, R., et al.: Feature pyramid networks for object detection. In: Proceedings of the IEEE Conference on Computer Vision and Pattern Recognition, pp. 2117–2125 (2017)
12. Wu Y, Chen Y, Yuan L, et al. Rethinking classification and localization for object detection[C]. Proceedings of the IEEE/CVF conference on computer vision and pattern recognition. 2020: 10186–10195
13. Hu, J., Shen, L., Sun, G.: Squeeze-and-excitation networks. In: Proceedings of the IEEE Conference on Computer Vision and Pattern Recognition, pp. 7132–7141 (2018)
14. Woo, S., Park, J., Lee, J.-Y., Kweon, I.S.: CBAM: convolutional block attention module. In: Ferrari, V., Hebert, M., Sminchisescu, C., Weiss, Y. (eds.) ECCV 2018. LNCS, vol. 11211, pp. 3–19. Springer, Cham (2018). https://doi.org/10.1007/978-3-030-01234-2_1
15. Yang, X., Yang, J., Yan, J., et al.: Scrdet: towards more robust detection for small, cluttered and rotated objects. In: Proceedings of the IEEE/CVF International Conference on Computer Vision, pp. 8232–8241 (2019)
16. Xia, G.S., Bai, X., Ding, J., et al.: DOTA: a large-scale dataset for object detection in aerial images. In: Proceedings of the IEEE Conference on Computer Vision and Pattern Recognition, pp. 3974–3983 (2018)

Development of Arithmetic/Logical Computational Procedures: Computer Aided Digital Filter Design with Bilinear Transformation

Abdul Rasak Zubair[1][(⊠)] and Adeolu Johnson Olawale[1,2]

[1] Electrical and Electronic Engineering Department, University of Ibadan, Ibadan, Nigeria
ar.zubair@ui.edu.ng
[2] Electrical & Electronics Engineering Department, Lead City University, Ibadan, Nigeria

Abstract. The development of a software system for computer aided digital filter design using bilinear transformation is presented. Starting from first principles, patterns in the manual determination of filter coefficients are studied and converted to arithmetic/logical computational procedures or operations which are coded into computer subprograms. These subprograms are packaged as a software system which is user friendly, and which can be regarded as a filter coefficients' calculator. The software system is subjected to rigorous tests and is found to be accurate. The software system also plots pole-zero diagram and the frequency response. Mathematics coupled with the Arithmetic/Logical computational capabilities of the computer have been used to assist digital filter design in Electrical and Electronic Engineering.

Keywords: Filter · Approximation functions · Filter coefficients · Bilinear transformation · Patterns · Arithmetic and logical operations · Frequency response

1 Introduction

A filter is said to be a frequency selective electrical network. It passes some frequencies and blocks some frequencies. Filtering is used to extract a desired signal from the available mixture of signals. Analog filter processes analog signals while digital filter processes discrete time signals. Analog signal or continuous-time signal can be converted to discrete-time signal before being processed by a digital filter and later converted back to analog signal [1–3].

Designing a digital filter basically entails determination of a set of coefficients of the filter. Infinite Impulse Response (IIR) Filter design begins with normalized analog low pass filter specifications of Fig. 1 [3, 4]. ε is the passband ripple parameter, δ_p is the passband deviation, δ_s is the stopband deviation, w_{pass} is the passband edge frequency, and w_{stop} is the stopband edge frequency. A_{pass} and A_{stop} are the passband Gain and the stopband Gain respectively as described in Eqs. (1) and (2) [3, 4].

Z. Hu et al. (Eds.): ICCSEEA 2023, LNDECT 181, pp. 349–360, 2023.
https://doi.org/10.1007/978-3-031-36118-0_31

An approximation function provides a realizable frequency response which is approximate to the ideal frequency response. Five approximation functions are studied in details in [4]. These are the Butterworth, Chebyshev, Inverse Chebyshev, Elliptic, and Bessel approximation functions [4–10].

A narrow transition band and a high negative stopband Gain are desirable for effective frequency discrimination between the passband and stopband frequencies. The narrower the transition band (lower w_{stop}/w_{pass} ratio), the higher the order number n. Higher negative stopband Gain A_{stop} also requires higher order number n. The higher the order number n, the more complex are the design computations and the filter network [4].

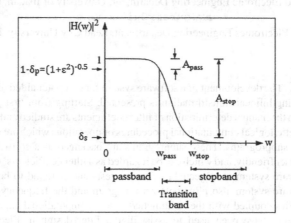

Fig. 1. Low pass filter specifications [3, 4]

$$A_{pass} = 10log_{10}(1 + \varepsilon^2) = -20log_{10}(1 - \delta_p) \tag{1}$$

$$A_{stop} = -20log_{10}(\delta_s) \tag{2}$$

Digital Filter design is an active area of research in Electrical and Electronic Engineering. A number of complex approaches are used to arrive at the filter coefficients using optimization techniques or tedious approaches using extensive tables coupled with manual expansion and computations [11–15]. A simple approach is developed in this work.

There are basic equations in the literature for designing normalized analog low pass filter with passband edge frequency of 1 rad/sec for the different approximation functions. There are substitutions in the form of additional equations to be made to convert a normalized analog low pass filter to analog low pass or high pass or band pass or band stop filter with given edge frequencies. There is another equation which is known as bilinear transformation to convert analog filter to digital filter. The collection of these equations are referred to as first principles or theory of filter design [1–10]. Filter design requires complex manual mathematical expansions of expressions and complex computations of filter's coefficients. The manual expansions and computations from first principles are studied, observed patterns in these tasks are converted to arithmetic/logical computational procedures or operations which are coded into computer programs. Mathematics coupled with the Arithmetic/Logical computational capabilities of the computer are employed to assist digital filter design in Electrical and Electronic Engineering.

2 Methodology

2.1 Normalized Analog Low Pass Filter Design from First Principles

w_{pass} of a normalized analog low pass filter is 1 rad/sec. w_{stop}, A_{pass}, and A_{stop} must be specified. There are equations for finding the filter order n for four of the five approximation functions except the Bessel approximation function [4–10]. The filter function is the ratio of two polynomials as given by Eqs. (3), (4), and (5). s is equal to jw, where w is the angular frequency. There are equations for finding the filter function for all the five approximation functions. The polynomials PolyNF and PolyDF are matrices with filter coefficients as elements. The Gain of the filter is given by Eq. (6). K is a constant.

$$H(s) = \frac{KPolyNF}{PolyDF} = \frac{K(a_1s^n + a_2s^{n-1} + a_3s^{n-2} + \dots + a_{n-1}s^2 + a_ns + a_{n+1})}{(b_1s^n + b_2s^{n-1} + b_3s^{n-2} + \dots + b_{n-1}s^2 + b_ns + b_{n+1})} \quad (3)$$

$$PolyNF = \left[a_1, a_2, a_3, a_4, \dots, a_{n-2}, a_{n-1}, a_n, a_{n+1} \right] \quad (4)$$

$$PolyDF = \left[b_1, b_2, b_3, b_4, \dots, b_{n-2}, b_{n-1}, b_n, b_{n+1} \right] \quad (5)$$

$$Gain = 20log_{10}|H(jw)| \quad (6)$$

2.2 Analog Filter Design

Analog low pass filter, Analog high pass filter, Analog band pass filter, and Analog band stop filter are obtainable from the normalized analog low pass filter function. New polynomials PolyNNF and PolyNDF are formed for the proposed analog filter from the polynomials PolyNF and PolyDF of the normalized analog low pass filter function as shown in Eq. (7).

$$H(s) = \frac{KPolyNNF}{PolyNDF} = \frac{K(A_1s^n + A_2s^{n-1} + A_3s^{n-2} + \dots + A_{n-1}s^2 + A_ns + A_{n+1})}{(B_1s^n + B_2s^{n-1} + B_3s^{n-2} + \dots + B_{n-1}s^2 + B_ns + B_{n+1})}$$
$$(7)$$

2.2.1 Analog Low Pass Filter Design

To obtain an Analog low pass filter with bandpass edge frequency of w_c from the normalized analog low pass filter with bandpass edge frequency of 1 rad/sec, Eq. (8) is substituted into Eq. (3). Steps 1, 2 and 3 are carried out.

$$s = \frac{s}{w_c} \tag{8}$$

Step 1: To obtain PolyNNF, substitution of Eq. (8) in the polynomial PolyNF $a_1 s^n + a_2 s^{n-1} + a_3 s^{n-2} + \ldots + a_{n-1} s^2 + a_n s + a_{n+1}$ leads to $a_1 \frac{s^n}{w_c^n} + a_2 \frac{s^{n-1}}{w_c^{n-1}} + a_3 \frac{s^{n-2}}{w_c^{n-2}} + \ldots + a_{n-1} \frac{s^2}{w_c^2} + a_n \frac{s^1}{w_c^1} + a_{n+1} \frac{s^0}{w_c^0}$

Step 2: Multiplication of all terms by w_c^n gives
$a_1 s^n + a_2 w_c^1 s^{n-1} + a_3 w_c^2 s^{n-2} + \ldots + a_{n-1} w_c^{n-2} s^2 + a_n w_c^{n-1} s^1 + a_{n+1} w_c^n s^0$
Compare with $A_1 s^n + A_2 s^{n-1} + A_3 s^{n-2} + \ldots + A_{n-1} s^2 + A_n s^1 + A_{n+1} s^0$
$\therefore A_m = a_m w_c^{m-1}$

Step 3: Repeat steps 1 and 2 for the polynomial PolyDF to obtain PolyNDF. Similarly, $B_m = b_m w_c^{m-1}$

The observed patterns in steps 2 and 3 are converted to arithmetic/logical computational procedures or operations which are coded into a computer program. The filter order remains as n.

2.2.2 Analog High Pass Filter Design

To obtain an Analog high pass filter with bandpass edge frequency of w_c from the normalized analog low pass filter with bandpass edge frequency of 1 rad/sec, Eq. (9) is substituted into Eq. (3). Steps 1, 2 and 3 are carried out.

$$s = \frac{w_c}{s} \tag{9}$$

Step 1: To obtain PolyNNF, substitution of Eq. (9) in the polynomial PolyNF. $a_1 s^n + a_2 s^{n-1} + a_3 s^{n-2} + \ldots + a_{n-1} s^2 + a_n s + a_{n+1}$ leads to

$$a_1 \frac{w_c^n}{s^n} + a_2 \frac{w_c^{n-1}}{s^{n-1}} + a_3 \frac{w_c^{n-2}}{s^{n-2}} + \ldots + a_{n-1} \frac{w_c^2}{s^2} + a_n \frac{w_c^1}{s^1} + a_{n+1} \frac{w_c^0}{s^0}$$

Step 2: Multiplication of all terms by s^n gives

$a_1 w_c^n s^0 + a_2 w_c^{n-1} s^1 + a_3 w_c^{n-2} s^2 + \ldots + a_{n-1} w_c^2 s^{n-2} + a_n w_c^1 s^{n-1} + a_{n+1} w_c^0 s^n$

Flip from left to right

$a_{n+1} w_c^0 s^n + a_n w_c^1 s^{n-1} + a_{n-1} w_c^2 s^{n-2} + \ldots + a_3 w_c^{n-2} s^2 + a_2 w_c^{n-1} s^1 + a_1 w_c^n s^0$

Compare with $A_1 s^n + A_2 s^{n-1} + A_3 s^{n-2} + \ldots + A_{n-1} s^2 + A_n s^1 + A_{n+1} s^0$
$\therefore A_m = a_{n+2-m} w_c^{m-1}$

Step 3: Repeat steps 1 and 2 for the polynomial PolyDF to obtain PolyNDF. Similarly, $B_m = b_{n+2-m} w_c^{m-1}$

The observed patterns in steps 2 and 3 are converted to arithmetic/logical computational procedures or operations which are coded into a computer program. The filter order remains as n.

2.2.3   Analog Band Pass Filter Design

To obtain an Analog band pass filter with bandpass edge frequencies of w_{c1} and w_{c2} from the normalized analog low pass filter with bandpass edge frequency of 1 rad/sec, Eq. (10) is substituted into Eq. (3). Steps 1, 2, 3, and 4 are carried out.

$$s = \frac{s^2 + w_{c1}w_{c2}}{s(w_{c2} - w_{c1})} = \frac{s^2 + p}{sg} = \frac{s}{g} + \frac{p}{sg} \tag{10}$$

where p is the product $(w_{c2}w_{c1})$ and g is difference $(w_{c2} - w_{c1})$.

Step 1: To obtain PolyNNF, substitution of Eq. (10) in the polynomial PolyNF.
$a_1 s^n + a_2 s^{n-1} + a_3 s^{n-2} + \dots + a_{n-1}s^2 + a_n s + a_{n+1}$ leads to

$$a_1 \left(\frac{s}{g} + \frac{p}{sg}\right)^n + a_2 \left(\frac{s}{g} + \frac{p}{sg}\right)^{n-1} + \dots + a_n \left(\frac{s}{g} + \frac{p}{sg}\right)^1 + a_{n+1} \left(\frac{s}{g} + \frac{p}{sg}\right)^0$$

Step 2: The expansion of $\left(\frac{s}{g} + \frac{p}{sg}\right)^m$ in Table 1 is studied for obvious patterns. The matrices in Table 1 can be built logically and sequentially from m = 0, 1, 2 up to a desired value of m by simply starting with 1, adding two consecutive elements in the matrix for (m-1) to get new element in the matrix for m and ending with a 1. Negative powers of s are introduced in addition to the positive powers of s. The number of coefficients will therefore increase from (n + 1) to (2n + 1). The order of the filter is doubled.

Table 1. Expansion of $\left(\frac{s}{g} + \frac{p}{sg}\right)^m$.

	Expansion	Matrix	m
$\left(\frac{s}{g} + \frac{p}{sg}\right)^0$	1	[1]	0
$\left(\frac{s}{g} + \frac{p}{sg}\right)^1$	$\frac{sp^0}{g} + \frac{p^1}{sg}$	[1, 1]	1
$\left(\frac{s}{g} + \frac{p}{sg}\right)^2$	$\frac{s^2 p^0}{g^2} + 2\frac{p^1}{g^2} + \frac{p^2}{s^2 g^2}$	[1, 2, 1]	2
$\left(\frac{s}{g} + \frac{p}{sg}\right)^3$	$\frac{s^3 p^0}{g^3} + 3\frac{s^1 p^1}{g^3} + 3\frac{p^2}{sg^3} + \frac{p^3}{s^3 g^3}$	[1, 3, 3, 1]	3
$\left(\frac{s}{g} + \frac{p}{sg}\right)^4$	$\frac{s^4 p^0}{g^4} + 4\frac{s^2 p^1}{g^4} + 6\frac{p^2}{g^4} + 4\frac{p^3}{s^2 g^4} + \frac{p^4}{s^4 g^4}$	[1, 4, 6, 4, 1]	4

$$A_1 s^n + A_2 s^{n-1} + \dots + A_n s^1 + A_{n+1} s^0 + A_{n+2} s^{-1} + \dots + A_{2n} s^{-(n-1)} + A_{2n+1} s^{-n}$$

Each coefficient a_{n-m+1} in PolyNF will contribute to the terms s^m, s^{m-2}, s^{m-4}, ... up to s^{-m} in PolyNNF.

Step 3: Multiplication of all terms by s^n gives

$$A_1 s^{2n} + A_2 s^{2n-1} + ... + A_n s^{n+1} + A_{n+1} s^n + A_{n+2} s^{n-1} + ... + A_{2n} s^1 + A_{2n+1} s^0$$

Step 4: Repeat steps 1, 2 and 3 for the polynomial PolyDF to obtain PolyNDF.

The observed patterns in steps 2, 3, and 4 are converted to arithmetic/logical computational procedures or operations which are coded into a computer program. The filter order becomes 2n.

2.2.4 Analog Band Stop Filter Design

To obtain an Analog band stop filter with bandstop edge frequencies of w_{c1} and w_{c2} from the normalized analog low pass filter with bandpass edge frequency of 1 rad/sec, Eq. (11) is substituted into Eq. (3). Steps 1, 2, 3, and 4 are carried out.

$$s = \frac{s(w_{c2} - w_{c1})}{s^2 + w_{c1} w_{c2}} = \left(\frac{s^2 + p}{sg}\right)^{-1} = \left(\frac{s}{g} + \frac{p}{sg}\right)^{-1} \tag{11}$$

where p is the product $(w_{c2} w_{c1})$ and g is difference $(w_{c2} - w_{c1})$.

Step 1: To obtain PolyNNF, substitution of Eq. (11) in the polynomial PolyNF. $a_1 s^n + a_2 s^{n-1} + a_3 s^{n-2} + ... + a_{n-1} s^2 + a_n s + a_{n+1}$ leads to

$$a_1 \left(\frac{s}{g} + \frac{p}{sg}\right)^{-n} + a_2 \left(\frac{s}{g} + \frac{p}{sg}\right)^{-(n-1)} + ... + a_n \left(\frac{s}{g} + \frac{p}{sg}\right)^{-1} + a_{n+1} \left(\frac{s}{g} + \frac{p}{sg}\right)^0$$

Multiplication of all terms by $\left(\frac{s}{g} + \frac{p}{sg}\right)^n$ leads to

$$a_1 \left(\frac{s}{g} + \frac{p}{sg}\right)^0 + a_2 \left(\frac{s}{g} + \frac{p}{sg}\right)^1 + ... + a_n \left(\frac{s}{g} + \frac{p}{sg}\right)^{n-1} + a_{n+1} \left(\frac{s}{g} + \frac{p}{sg}\right)^n$$

It is noted that the difference between band pass filter and band stop filter is that m decreases from n to 0 in the step 1 for band pass filter (Sect. 2.2.3) but m increases from 0 to n in the step 1 for band stop filter (Sect. 2.2.4). The coefficient in front of $\left(\frac{s}{g} + \frac{p}{sg}\right)^m$ is a_{n-m+1} for band pass filter but a_{m+1} for band stop filter.

Steps 2 to 4 are the same for band pass filter in Sect. 2.2.3.

The computer program for the band pass filter is copied and adjusted slightly to form the program for band stop filter. The filter order is also 2n.

2.3 Digital Filter Design with Bilinear Transformation

An analog filter can be converted to a digital filter of the same "type" with the substitution of Eq. (12) in Eq. (7). New polynomials PolyZNF and PolyZDF are formed for the proposed digital filter from the polynomials PolyNNF and PolyNDF of the corresponding analog filter function as shown in Eq. (13).

$$s = \frac{2}{T_s} \frac{[z-1]}{[z+1]} \tag{12}$$

$$H(z) = \frac{KPolyZNF}{PolyZDF} = \frac{K(C_1 z^n + C_2 z^{n-1} + C_3 z^{n-2} + \dots + C_{n-1} z^2 + C_n z + C_{n+1})}{(D_1 z^n + D_2 z^{n-1} + D_3 z^{n-2} + \dots + D_{n-1} z^2 + D_n z + D_{n+1})} \tag{13}$$

where $z = e^{jw}$.

"type" refers to low pass, high pass, band pass and band stop. T_s is the sampling interval which is the reciprocal of the sampling frequency f_s. The order n of the Digital filter is the same as the order n of the corresponding analog filter. Equation (12) is known as Bilinear Transformation. Analog frequency (w_c) and digital frequency (w_{dc}) are related by Eqs. (14) and (15). The frequency response of a digital filter is plotted over the range of 0 to π radian. Beyond this range, the frequency response repeats itself. Steps 1, 2, 3, and 4 are carried out.

$$w_c = \frac{2}{T_s} \tan\left(\frac{w_{dc}}{2}\right) \tag{14}$$

$$w_{dc} = 2\tan^{-1}\left(\frac{T_s w_c}{2}\right) \tag{15}$$

Step 1: To obtain PolyZNF, substitution of Eq. (12) in the polynomial PolyNNF $A_1 s^n + A_2 s^{n-1} + A_3 s^{n-2} + \dots + A_{n-1} s^2 + A_n s + A_{n+1}$ leads to

$$A_1 \left(\frac{2}{T_s} \frac{[z-1]}{[z+1]}\right)^n + A_2 \left(\frac{2}{T_s} \frac{[z-1]}{[z+1]}\right)^{n-1} .. + A_n \left(\frac{2}{T_s} \frac{[z-1]}{[z+1]}\right)^1 + A_{n+1} \left(\frac{2}{T_s} \frac{[z-1]}{[z+1]}\right)^0$$

Multiplication of all terms by $[z+1]^n$ leads to

$$A_1 \left(\frac{2}{T_s}\right)^n [z-1]^n [z+1]^0 + A_2 \left(\frac{2}{T_s}\right)^{n-1} [z-1]^{n-1} [z+1]^1 + .$$

$$.. + A_n \left(\frac{2}{T_s}\right)^1 [z-1]^1 [z+1]^{n-1} + A_{n+1} \left(\frac{2}{T_s}\right)^0 [z-1]^0 [z+1]^n$$

Each term is like $A_m \left(\frac{2}{T_s}\right)^{m1} [z-1]^{m1} [z+1]^{m2}$
where $m1 = n+1-m$, $m2 = m-1$, and $n = m1 + m2$.
m varies from 1 to $n+1$, m1 varies from n to 0, and m2 varies from 0 to n.

Step 2: The expansions of $[z-1]^{m1}$ and $[z+1]^{m2}$ in Table 2 are studied for obvious patterns. The matrices similar to that in Table 1 are involved in the expansions in Table 2 and can be handled in the same way. The product $[z-1]^{m1}[z+1]^{m2}$ is obtainable by multiplying the expansions of $[z-1]^{m1}$ and $[z+1]^{m2}$. Each of the coefficients C_m in

PolyZNF will receive contributions from the coefficients $A_1, A_2, A_3, \ldots$, and A_{n+1} in PolyNNF.

$$C_1 z^n + C_2 z^{n-1} + C_3 z^{n-2} + \ldots + C_{n-1} z^2 + C_n z + C_{n+1}$$

Step 3: Multiplication of all terms by z^{-n} gives

$$C_1 z^0 + C_2 z^{-1} + C_3 z^{-2} + \ldots + C_{n-1} z^{-(n-2)} + C_n z^{-(n-1)} + C_{n+1} z^{-n}$$

Table 2. Expansions of $[z - 1]^{m1}$ and $[z + 1]^{m2}$.

m1	Expansion of $[z - 1]^{m1}$ (m1 + 1) terms	m2	Expansion of $[z + 1]^{m2}$ (m2 + 1) terms
0	1	0	1
1	$z - 1$	1	$z + 1$
2	$z^2 - 2z + 1$	2	$z^2 + 2z + 1$
3	$z^3 - 3z^2 + 3z - 1$	3	$z^3 + 3z^2 + 3z + 1$
4	$z^4 - 4z^3 + 6z^2 - 4z + 1$	4	$z^4 + 4z^3 + 6z^2 + 4z + 1$
5	$z^5 - 5z^4 + 10z^3 - 10z^2 + 5z - 1$	5	$z^5 + 5z^4 + 10z^3 + 10z^2 + 5z + 1$

Step 4: Repeat steps 1, 2, and 3 for the polynomial PolyNDF to obtain PolyZDF.

The observed patterns in steps 2, 3, and 4 are converted to arithmetic/logical computational procedures or operations which are coded into a computer program. The filter order of the digital filter is the same as the order of the corresponding analog filter.

2.4 Digital Filter Design Software

The computer programs developed for the normalized analog low pass filter, the four types of analog filter and the digital filter are packaged as a software system for digital filter design. The software system developed in MATLAB working environment can plot the pole-zero diagram for the normalized analog low pass filter and frequency response for the various filters. User friendliness is incorporated in the software system which can be regarded as a filter coefficients' calculator. The software and its user manual are available on the research gate pages of the authors.

3   Tests and Results

The software system is subjected to rigorous tests. Sample results are presented in this section. The user is expected to supply some inputs based on the desired filter type and specifications.

3.1 Digital Band Pass Filter

3.1.1 Inputs

$w_{pass} = 1$ rad/sec, $w_{stop} = 2$ rad/sec, $A_{pass} = -3$ dB, and $A_{stop} = -100$ dB for the normalized analog low pass filter. Elliptic approximation function was selected. Bandpass edge frequencies are $w_{c1} = 1.8850 \times 10^4$ rad/sec (3 kHz) and $w_{c2} = 6.2832 \times 10^4$ rad/sec (10 kHz) for the analog band pass filter. Sampling period T_s was selected as 1/50000 s (Sampling Frequency is 50 kHz which is greater than twice the maximum frequency of 10 kHz).

3.1.2 Outputs

n was found to be 7 for the normalized analog low pass filter and 14 for the analog band pass filter and the digital band pass filter. Equivalent bandpass edge digital frequencies were found to be $w_{dc1} = 0.3726$ rad/sec and $w_{dc2} = 1.122$ rad/sec. Filter functions of Eqs. (16), (17), and (18) are obtained for the normalized analog low pass filter, analog band pass filter and digital band pass filter respectively. The frequency response curves for the normalized analog low pass filter, the analog band pass filter, and the digital band pass filter are presented in Fig. 2 (a), 2(b), and 3(a) respectively.

$$H(s) = \frac{3.533 \times 10^{-5}(s^6 + 29.3437s^4 + 222.78s^2 + 491.5639)}{s^7 + 0.5053s^6 + 1.9405s^5 + 0.7659s^4 + 1.1025s^3 + 0.2939s^2 + 0.1640s + 0.174} \tag{16}$$

$$H(s) = \frac{\begin{array}{c}3.533 \times 10^{-5}(1.3814 \times 10^{-28}s^{13} + 8.8232 \times 10^{-18}s^{11} + 1.5522 \times 10^{-7}s^9 + 834.9448s^7 \\ +2.1773 \times 10^{11}s^5 + 1.7360 \times 10^{19}s^3 + 3.8126 \times 10^{26}s)\end{array}}{\begin{array}{c}3.1409 \times 10^{-33}s^{14} + 6.9800 \times 10^{-29}s^{13} + 3.7830 \times 10^{-23}s^{12} + 7.0068 \times 10^{-19}s^{11} \\ +1.7530 \times 10^{-13}s^{10} + 2.5901 \times 10^{-9}s^9 + 3.9778 \times 10^{-4}s^8 + 4.4189s^7 + 4.7111 \times 10^5 s^6 \\ +3.6331 \times 10^9 s^5 + 2.9122 \times 10^{14}s^4 + 1.3786 \times 10^{18}s^3 + 8.8153 \times 10^{22}s^2 + 1.9264 \times 10^{26}s^1 \\ +1.0266 \times 10^{31}\end{array}} \tag{17}$$

$$H(z) = \frac{\begin{array}{c}3.533 \times 10^{-5}(0.3430z^0 - 1.4837z^{-1} + 2.6771z^{-2} - 3.2476z^{-3} + 3.7185z^{-4} - 3.1998z^{-5} \\ +1.3503z^{-6} + 1.1912 \times 10^{-15}z^{-7} - 1.3503z^{-8} + 3.1998z^{-9} - 3.7185z^{-10} + 3.2476z^{-11} \\ -2.6771z^{-12} + 1.4837z^{-13} - 0.3430z^{-14})\end{array}}{\begin{array}{c}0.108z^0 - 1.08z^{-1} + 5.284z^{-2} - 16.739z^{-3} + 38.250z^{-4} - 66.580z^{-5} \\ +90.935z^{-6} - 98.929z^{-7} + 86.128z^{-8} - 59.726z^{-9} + 32.496z^{-10} - 13.469z^{-11} \\ +4.027z^{-12} - 0.780z^{-13} + 0.074z^{-14}\end{array}} \tag{18}$$

3.2 Digital High Pass Filter

3.2.1 Inputs

$w_{pass} = 1$ rad/sec, $w_{stop} = 2$ rad/sec, $A_{pass} = -3$ dB, and $A_{stop} = -50$ dB for the normalized low pass filter. Chebyshev approximation function was selected. Bandpass edge frequency $w_{dc} = 1.2$ rad/sec for the digital high pass filter.

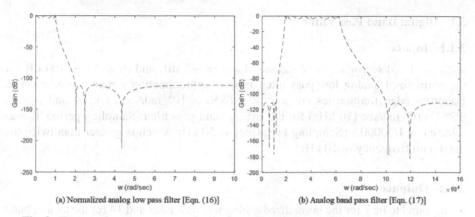

(a) Normalized analog low pass filter [Eqn. (16)]   (b) Analog band pass filter [Eqn. (17)]

Fig. 2. Frequency response curves for the normalized low pass filter and analog pass filter

3.2.2 Outputs

n was found to be 5 for the normalized analog low pass filter, the analog high pass filter and the digital high pass filter. Equivalent bandpass edge analog frequency was found to be $w_c = 5.654x10^4$ rad/sec (9 kHz). Sampling period T_s was selected as 1/50000 s (Sampling Frequency is 50 kHz which is greater than twice the maximum frequency of 9 kHz). Filter functions of Eqs. (19), (20), and (21) are obtained for the normalized analog low pass filter, the analog high pass filter and the digital high pass filter respectively. The frequency response of the digital high pass filter is presented in Fig. 3(b).

$$H(s) = \frac{0.0626}{s^5 + 0.5745s^4 + 1.415s^3 + 0.5489s^2 + 0.4080s^1 + 0.0626} \tag{19}$$

$$H(s) = \frac{0.0626(5.6549x10^4 s^5)}{3.5427x10^3 s^5 + 1.3046x10^9 s^4 + 9.9264x10^{13} s^3 + 1.447x10^{19} s^2 + 3.322x10^{23} s + 3.2699x10^{28}} \tag{20}$$

$$H(z) = \frac{0.626(0.5655z^0 - 2.8274z^{-1} + 5.6549z^{-2} - 5.6549z^{-3} + 2.8274z^{-4} - 0.5655z^{-5})}{4.7576z^0 - 2.5992z^{-1} + 5.207z^{-2} + 0.7633z^{-3} + 0.9356z^{-4} + 1.3994z^{-5}} \tag{21}$$

3.3 Discussion

Manual calculation is easily handled for order 1 and 2. Manual calculation for order 3 and above is not convenient and is prone to human error. With this software, calculation for higher order is made easy and error free.

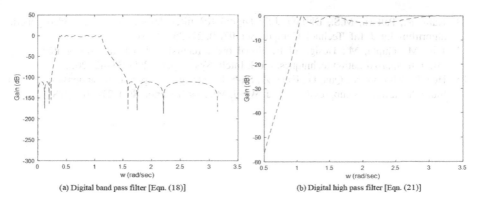

(a) Digital band pass filter [Eqn. (18)]   (b) Digital high pass filter [Eqn. (21)]

Fig. 3. Frequency response curves for the digital band pass filter and digital high pass filter

4 Conclusion

The development of a software system for the computation of filter coefficients has been presented and tested. Analog and Digital filter design has been made simple. Test results and frequency response plots confirm the accuracy of the arithmetic/logical computational procedures deduced from patterns in manual computations. The software is recommended for use in Electrical and Electronic Engineering practice.

References

1. Oppenheim, V.A.: Discrete Time Signal Processing. Prentice-Hall, Englewood Cliffs (1999)
2. Antoniou, A.: Digital Signal Processing: Signals, Systems and Filters. McGraw Hill, New York (2006)
3. Orfanidis, S.J.: Introduction to Signal Processing. Prentice Hall, Rutgers (2010)
4. Zubair, A.R., Olawale, A.J.: Active learning strategy: computer aided numerical class project on pole-zero plot and transfer function of five low pass filter approximation functions. Global J. Eng. Technol. Adv. **12**(1), 038–063 (2022)
5. Thede, L.: Practical Analog and Digital Filter Design. Artech House, Inc., Ohio (2004)
6. Rice, J.R., Usow, K.H.: The Lawson algorithm and extensions. Math. Comput. **22**, 118–127 (1968)
7. Smith, S.W.: The Scientist and Engineer's guide to Digital Signal Processing. California Technical Publishing, California (1999)
8. Ingle, V.K., Proakis, J.G.: Digital Signal Processing Using MATLAB, 3rd edn. Global Engineering, Stamford (2010)
9. Rader, B.G.: Digital Processing of Signals. McGrawHill, New York (1969)
10. Abramowitz, M., Stegun, I.A. (eds.): Handbook of Mathematical Functions with Formulas. Dover Publications, New York (1965)
11. Ferdous, H., Jahan, S., Tabassum, F., Islam, M.: The performance analysis of digital filters and ANN in de-noising of speech and biomedical signal. Int. J. Image Graph. Signal Process. **15**(1), 63–78 (2023)
12. Adamu, Z.M., Dada, E.G., Joseph, S.B.: Moth flame optimization algorithm for optimal FIR filter design. Int. J. Intell. Syst. Appl. **13**(5), 24–34 (2021)

13. Kaur, R., Patterh, M.S., Dhillon, J.S.: Digital IIR filter design using real coded genetic algorithm. Int. J. Inf. Technol. Comput. Sci. 07(7), 27–35 (2013)
14. Jain, M., Gupta, M.: Design of fractional order recursive digital differintegrators using different approximation techniques. Int. J. Intell. Syst. Appl. 12(1), 33–42 (2020)
15. Hea, L., Zhonga, X., Qua, D., Fana, Z., Shib, K.: Design on open-loop bandpass filter with multiple harmonics suppression. Int. J. Wirel. Microw. Technol. 8(5), 27–36 (2018)

Multidimensional Ranking and Taxonomic Analysis of the Regional Socio-Economic Development in Ukraine

Mariana Vdovyn and Larysa Zomchak[✉]

Ivan Franko National University of Lviv, Lviv 79000, Ukraine
lzomchak@gmail.com

Abstract. The research paper deals with the problems of evaluation, analysis and modeling of socio-economic development of the regions of Ukraine. The main focus is on the main indicators that affect the maintenance of the physical, moral, economic and intellectual state, such as the income of the population and the level of wages. The population's expenses and the unemployment rate were also considered. Taxonomic analysis and multidimensional ranking approach was used for modeling of socio-economic development. Reasons for using multidimensional statistical methods for assessing regional socio-economic development are substantiated in the paper. A brief description of the method of taxonomic analysis and multidimensional ranking was made. A comparative analysis of the obtained results was also carried out. The correlation coefficient was calculated for the evaluation of ranking results using different methods, such as the method of taxonomic analysis and multidimensional ranking. The place of each region among other regions of Ukraine was determined and recommendations were made for outsider regions. The value and importance of the research is to develop these recommendations for improving the socio-economic indicators for the outsider regions, according to the determination of the place of the region in the aggregate of the rest of the regions. It is the methods of taxonomic analysis and multidimensional ranking that make it possible to identify leaders and outsiders.

Keywords: Socio-economic development · Socio-economic indicators · Taxonomic analysis · Multidimensional ranking

1 Introduction

Socio-economic development reflects the dependence between the level of economic development and the solution of social problems of the country as a whole, its regions, business entities, etc. Government institutions need the development of anti-crisis programs and strategies of socio-economic development at the state and regional levels. So, the study of socio-economic development is permanently relevant. Clarification of regional socio-economic development in Ukraine is important for their further recovery and development stimulation. There are many indicators of socio-economic development, but it is important to focus on those that are closely related to the standard

Z. Hu et al. (Eds.): ICCSEEA 2023, LNDECT 181, pp. 361–370, 2023.
https://doi.org/10.1007/978-3-031-36118-0_32

of living of the population. Such indicators are: unemployment rate, average nominal wage, income and expenditure of the population. Obviously, there are regional socio-economic development programs that offer mechanisms for improving socio-economic development indicators. However, it is important to understand what positions and places a specific region has in the rating. Multidimensional classification methods provide an opportunity to evaluate and analyze indicators of socio-economic development. Taxonomic analysis and multidimensional ranking are precisely those methods that help to identify the place of a specific element among the rest, in our case, the region.

2 Literature Review

The study of socio-economic development of the country is one of the topical issues. In particular, socio-economic development in EU countries is considered in [1], the impact of socio-economic factors on population poverty in Pakistan is investigated in [2] and the correlation between socio-economic development and the level of tourism development in EU countries is offered in [3]. Rodríguez Martin and others in [4] offer an indicator of social and economic development for southern European countries based on the DP2 distance method. The socio-economic development in the EU countries, especially those that became members of the integration association after 2004, is described in detail in [5].

Unemployment as one of the indicators of socio-economic development is described in many researches. In particular, unemployment during COVID 19 D.L. Blustein, P.A. and others research in [6] and [7]. Chi-Wei Su, Ke Dai, Sana Ullah & Zubaria Andlib also investigate the impact of COVID 19 on the unemployment rate in European countries in [8]. Some scholars focus on the costs of health insurance as a form of social protection [9], others consider COVID 19 modeling [10] and regions development modeling [11].

A quantitative approach for analysis of the socio-economic state of the country is offered in many papers. Pareto, A. in [12] suggests the correlation and regression analysis of socio-economic indicators, Barrington-Leigh, C. and Escande, A. in [13] offer comparative analysis of the indicators of welfare. The evaluation and modeling of socio-economic processes are considered in [14]. Cluster analysis of data is suggested in [15] ang forecasting of economic processes in [16]. Logistic regression analysis of economic development was made in [17] and modeling the impact of the instability of the external environment on the enterprise in [18].

Methods of taxonomic analysis and multidimensional ranking were used in a lot of researches too. Mlodak, A. in [19] offers the indices of feature variability in a multidimensional space-time model and individual feature variability. Malina, A. in [20] considers the taxonomic analysis of economic structures in Poland. Bogliacino, F., & Pianta, M. in [21] investigate manufacturing and services using taxonomy. A statistical study of economic regional development was carried out in [22]. The authors use the taxonomic method for the analysis of economic processes. Hristov Metodi uses the methods of assessing the socio-economic indicators of EU countries in [23] and rating analysis in [24]. Different methods of data standardization are considered in [25–27]. So, the literature review showed that, despite the large number of scientific works on the problems of studying socio-economic development and various methods of modeling economic processes, there is still a need to study regional socio-economic development and find out the place of each Ukrainian region in the ranking.

3 Methodology. Methods of Multidimensional Statistics

3.1 Taxonomic Analysis in Modeling of the Regional Socio-Economic Development

Various methods are used to estimate the state of socio-economic development, in particular, methods of descriptive statistics, game-theoretic modeling, optimization methods, heuristics, etc. The research considers the methods of multidimensional statistics, such as taxonomic analysis and multidimensional ranking. The paper deals with the problems of determining the place of each region of Ukraine among the rest, in order to understand the socio-economic development of which regions should be paid attention to by state administration bodies. That is why the methods chosen for the study are rating methods.

Taxonomic analysis involves the formation of a matrix of observations, standardization of data, formation of a reference vector and determination of the distance between observations and this vector, as well as the calculation of a taxonomic indicator.

The distance between each element and the reference vector is often defined as follows

$$C_{j0} = \left(\sum_{i=1}^{m} (z_{ij} - z_0)^2 \right)^{1/2} \tag{1}$$

where z_{ij} − standardized indicators and z_0− the reference value. This is the Euclidean distance.

The taxonomic index is defined as the ratio of the Euclidean distance of the distance to the distance between the upper and lower points of the range.

Taxonomic indicators are often ranked to clearly visualize the place of each element.

3.2 Multidimensional Ranking in Modeling of the Regional Socio-Economic Development

Multidimensional ranking is the construct of an appropriate sequence of placement of units of a statistical population according to priorities, which makes it possible to determine the superiority of one unit over another at once based on several characteristics. The method involves the calculation of an integral indicator. The integral estimate (Gj) is geometrically interpreted as a point in multidimensional space whose coordinates indicate the scale or position of the j-th unit. The integral estimate Gj is often defined as the mean of the standardized values of features Z_{ij}. All indicators are standardized and analyzed, which indicators are stimulators, and which indicators are dissimulators. The results of the integrated assessment are also ranked.

The following stages of calculating the integral indicator can be distinguished:

1. formation of a feature set (matrix of input information);

2. implementation of data standardization, since, as a rule, indicators have different units of measurement;
3. choice of methods for calculating weighting factors, or understanding that all indicators are equally important;
4. determination of the procedure for aggregation of indicators;
5. ranking of the obtained results.

The methods of multidimensional ranking and taxonomic analysis are precisely those methods that will help to identify outsiders, so that it is possible to propose specific steps for improving certain processes for them. In our case, to identify outsider regions for which it is necessary to develop a mechanism for improving socio-economic indicators.

4 Results. Model the Socio-Economic Development of Ukraine

4.1 General Overview of the Socio-Economic Development of Ukraine

In 1991, with the restoration of independence, Ukraine was seen as a country with great prospects for potential high socio-economic development. However, the great integration into the unified national economic complex of the USSR did not give the opportunity to fully use this potential. Obstacles to high socio-economic development, in our opinion, were also a distorted system of political power influence, a high level of corruption and, as a consequence, shadow processes in the economy (black economy).

In 2000–2020, there was an increase in indicators of socio-economic development, except, of course, the global crisis in 2008. The impact of COVID-19 cannot be ignored either. However, we believe that the full-scale invasion of the aggressor country into Ukraine in 2022 will have the greatest impact. Given the possibility of using official statistical data, we will conduct our analysis for 2021.

Indicators of socio-economic development may change rapidly in 2022 due to the war, but our main task is to examine the situation in peacetime. It is clear that the need to develop a mechanism for improving indicators of socio-economic development is a priority for the leadership in the regions of Ukraine. It is also important to forecast the state of socio-economic development. Forecasting methods are carried out in [28, 29].

4.2 Indicators of Socio-Economic Development in Different Regions of Ukraine

The following main indicators of socio-economic development were chosen for the study: income of the population (million UAH), unemployment rate in the age group of 15–70 years (%), average nominal wage (thousand UAH), population expenses (million UAH). All indicators are considered in terms of 24 regions of Ukraine. Crimea is not taken into account, since it is annexed by Russia and the State Statistics Service does not publish official statistical data. Luhansk and Donetsk regions were also considered without taking into account the temporarily occupied territories.

It is worth noting that the highest level of unemployment in 2021 was in Luhansk region (16%), and the lowest in Kharkiv region (6.7%). Despite the Covid-19 pandemic, Kharkiv region managed to create 9.5 thousand new jobs only for the first quarter of 2021 at the expense of all sources of funding.

Regarding the average nominal salary in 2021: the highest in the Kyiv region (17,409 UAH), and the lowest - in the Chernihiv region (13,537 UAH).

The income of the population was the highest in the Dnipropetrovsk region (434,791 million UAH), and the lowest in the Chernivtsi region (72,792 million UAH).

Household expenses are the highest in Kyiv region (328,874 million UAH), and the lowest in Luhansk region (55,870 million UAH).

Obviously, according to different indicators, different regions are leaders and outsiders. Therefore, it is advisable to use the methods of multivariate statistics. In our opinion, it is appropriate to consider ranking methods, since it is important for us to determine the place of each region. We chose two methods: taxonomic analysis and multidimensional ranking.

4.3 Taxonomic and Integral Indicators of Regional Socio-Economic Development

In the taxonomic analysis, Euclidean distance was used and in the multidimensional ranking, standardization was performed on the basis of root mean square deviation taking into account that personal income and average nominal wages are stimulators, while spending and unemployment are dissimulators.

The results are presented in Table 1 and visualized in Fig. 1.

Table 1. Assessment of the socio-economic development in Ukrainian regions

Region	Taxonomic indicators	Rank	Integral indicators	Rank
Vinnytsia	0,795	6	−0,001	11
Volyn	0,744	21	−0,426	21
Dnipropetrovsk	0,729	22	0,302	4
Donetsk	0,755	18	0,007	10
Zhytomyr	0,769	17	−0,232	18
Zakarpattia	0,788	8	−0,024	12
Zaporizhzhia	0,807	4	0,218	8
Ivano-Frankivsk	0,776	12	−0,078	13
Kyiv	0,803	5	0,522	1
Kirovohrad	0,709	24	−0,719	24
Luhansk	0,716	23	−0,478	22
Lviv	0,797	7	0,293	5
Mykolayiv	0,808	3	0,482	2
Odesa	0,786	9	0,272	6
Poltava	0,783	10	−0,072	14

(continued)

Table 1. (*continued*)

Region	Taxonomic indicators	Rank	Integral indicators	Rank
Rivne	0,825	1	0,436	3
Sumy	0,781	11	−0,092	15
Ternopil	0,751	19	−0,348	19
Kharkiv	0,769	16	0,266	7
Kherson	0,749	20	−0,393	20
Khmelnytsk	0,810	2	0,175	9
Cherkassy	0,775	13	−0,158	17
Chernivtsi	0,770	14	−0,115	16
Chernihiv	0,725	15	−0,588	23

Source: authors' construct based on the data of the State Statistics Service[30, 31]

The table shows that Rivne and Mykolaiv regions are the clear leaders in both methods, and Luhansk and Kirovohrad regions are outsiders. The current standard of living of the population of the Kirovohrad region creates significant limitations for regional development due to low wages. In general, in the Kirovohrad region there are conditions for regional development, in particular large areas of land resources, significant potential in engineering and food industry, a developed agro-industrial complex. There are also deposits of unique minerals in the region, sufficient human resources. There are opportunities for expansion in the region's transit potential, the growth of tourism. All these characteristics, of course, affect regional development, but they are not decisive. Thus, it decreases labor productivity, the outflow of qualified engineers and technicians increases personnel, employees of labor professions outside the region and Ukraine. The regional leadership should take into account all factors and make decisions on improving social and economic development. Certain recommendations for decision-making are offered in [32]. As for the Luhansk region, the main reason for the low socio-economic development is the armed aggression of Russia in the east of Ukraine and the temporary occupation of a part territories of Luhansk region. In 2022, the situation became worse, as the majority of the territory of Luhansk region remains occupied as a result of a full-scale invasion.

Figure 1. Clearly shows that according to the results of the taxonomic analysis, the difference in the distribution of different regions is insignificant, but according to the integrated assessment, the picture is slightly different.

It is also noticeable (Table 1) that certain regions are leaders according to the first method, and outsiders according to the other. First of all, this concerns the Dnipropetrovsk region.

It is worth illustrating the results also by the ranks according to both methods (Fig. 2).

Figure 2 shows that only a few regions received different rating results. The correlation coefficient between the results of multidimensional ranking and taxonomic analysis CORREL = 0.683, which indicates a noticeable correlation according to the Chaddock scale.

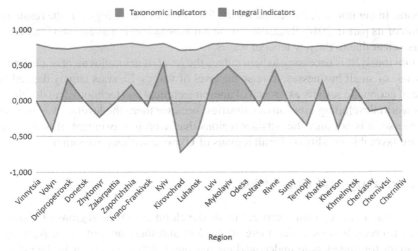

Fig. 1. Taxonomic and integral indicators of socio-economic development in the regions of Ukraine, Source: authors' construct based on the data of the State Statistics Service [30, 31]

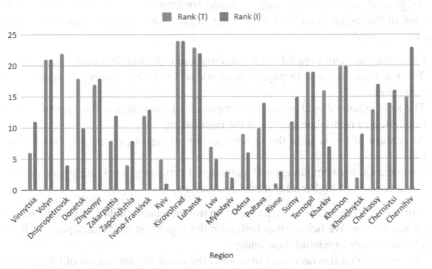

Fig. 2. Chart Comparison of rating results by taxonomic analysis and integral assessment, Source: authors' construct based on the data of the State Statistics Service [30, 31]

5 Discussion

So, by examining the place of each region, we were able to understand who is a clear outsider and what conclusions should be made by the healers of these regions regarding the future improvement of socio-economic development. Taxonomic analysis and multidimensional ranking show that the outsider regions are the Luhansk and Kirovohrad regions. In our opinion, the reason for the fall of the Luhansk region is the russian occupation of its part in 2014. Regarding to the Kirovohrad region as an outsider, everything is also clear due to the low level of wages.

Obviously, it is necessary to help reduce the level of unemployment, including at the expense of small businesses, to raise the level of wages. Salaries largely depend on the type of economic activity as a result of uneven technical and technological development. It is worth developing innovative industries, because there the level of wages is higher. Of course, it is not only the outsider regions that need improvement. It is important to ensure favorable conditions for all regions of Ukraine without exception.

6   Summary and Conclusion

The indicators of regional socio-economic development of the regions of Ukraine are quite different. It is clear that there are leaders and there are outsiders. Actually, it is important for outsiders to understand the reasons for their positions in the rating. The methods of economic analysis and multidimensional ranking make it possible to identify outsiders, and in dynamics to assess the change in positions.

In general, the following main proposals for improving the socio-economic development of the region in terms of well-being and the standard of living of the population can be identified:

1. Contribute to increasing labor productivity and reducing the gender pay gap.
2. To ensure an increase in payments of social funds for vulnerable sections of the population.
3. The fight against illegal business, corruption and assistance in the detection and legal regulation of non-labor income of the population.
4. Allocation of funds from the regions for retraining of persons who have lost their jobs.
5. Cooperation with employers and financing of programs involving young professionals, graduates of higher educational institutions.

However, all these conclusions based on the indicators of 2021 would be relevant if there was no war. We believe that first, after the liberation of the occupied territories, it will be necessary to rebuild the country.

We believe that it is necessary to monitor the place of each region of Ukraine among the rest of the regions every year. This is necessary for the further development of mechanisms for improving indicators of socio-economic development. The methods of multidimensional statistics will help in this case. It is important, in the future, to take into account various indicators of socio-economic development for rating.

References

1. Fura, B., Wang, Q.: The level of socioeconomic development of EU countries and state of ISO 14001 certification. Qual. Quant. **51**(1), 103–119 (2017). https://doi.org/10.1007/s11135-015-0297-7

2. Ayyoub, M., Gillani, D.Q.: Role of socioeconomic factors in poverty alleviation: an assessment of urban informal sector. Kashmir Econ. Rev. **26**(1), 28–40 (2017)
3. Stec, M., Grzebyk, M.: Socio-economic development and the level of tourism function development in European Union countries – a comparative approach. European Rev. **30**(2), 172–193 (2022). https://doi.org/10.1017/S106279872000099X
4. Rodríguez Martin, J.A., Holgado Molina, M.D., Salinas Fernández, J.A.: An index of social and economic development in the community's objective-1 regions of countries in southern Europe. Eur. Plan. Stud. **20**(6), 1059–1074 (2012)
5. Stec, M., Filip, P., Grzebyk, M., Pierścieniak, A.: Socio-economic development in the EU Member States – concept and classification. Eng. Econ. **25**(5), 504–512 (2014)
6. Blustein, D.L., Duffy, R., Ferreira, J.A., Cohen-Scali, V., Cinamon, R.G., Allan, B.A.: Unemployment in the time of COVID-19: a research agenda. J. Vocat. Behav. **119**, 103436 (2020)
7. Blustein, D.L., Guarino, P.A.: Work and unemployment in the time of COVID-19: the existential experience of loss and fear. J. Humanist Psychol. **60**(5), 702–709 (2020)
8. Chi-Wei, S., Dai, K., Ullah, S., Andlib, Z.: COVID-19 pandemic and unemployment dynamics in European economies. Econ. Res.-Ekonomska Istraživanja **35**(1), 1752–1764 (2022)
9. Tkachenko, R., et al.: Piecewise-linear approach for medical insurance costs prediction using SGTM neural-like structure. CEUR Workshop Proc. **2255**, 170–179 (2018)
10. Abriham, A., Dejene, D., Abera, T., Elias, A.: Mathematical modeling for COVID-19 transmission dynamics and the impact of prevention strategies: a case of Ethiopia. Int. J. Math. Sci. Comput. (IJMSC) **7**(4), 43–59 (2021). https://doi.org/10.5815/ijmsc.2021.04.05
11. Koroliuk, Y., Hryhorenko, V.: ANN model of border regions development: approach of closed systems. Int. J. Intell. Syst. Appl. (IJISA) **11**(9), 1–8 (2019). https://doi.org/10.5815/ijisa.2019.09.01
12. Pareto, A.: A robust method for regression and correlation analysis of socio-economic indicators. Qual. Quant.2022). https://doi.org/10.1007/s11135-022-01599-z
13. Barrington-Leigh, C., Escande, A.: Measuring progress and well-being: a comparative review of indicators. Soc. Indic. Res. **135**, 893–925 (2018). https://doi.org/10.1007/s11205-016-1505-0
14. Podgorna, I., Babenko, V., Honcharenko, N., Sáez-Fernández, F.J., Fernández, J.A.S., Yakubovskiy, S.: Modelling and analysis of socio-economic development of the European Union countries through DP2 method. Wseas Trans. Bus. Econ. **17**, 454–466 (2020). https://doi.org/10.37394/23207.2020.17.44
15. Lytvynenko, V., Kryvoruchko, O., Lurie, I., Savina, N., Naumov, O., Voronenko, M.: Comparative studies of self-organizing algorithms for forecasting economic parameters. Int. J. Modern Educ. Comput. Sci. (IJMECS) **12**(6), 1–15 (2020). https://doi.org/10.5815/ijmecs.2020.06.01
16. Chittineni, S., Bhogapathi, R.B.: Determining contribution of features in clustering multidimensional data using neural network. Int. J. Inf. Technol. Comput. Sci. (IJITCS) **4**(10), 29–36 (2012). https://doi.org/10.5815/ijitcs.2012.10.03
17. Zomchak, L., Starchevska, I.: Macroeconomic determinants of economic development and growth in Ukraine: logistic regression analysis. In: Hu, Z., Wang, Y., He, M., eds Advances in Intelligent Systems, Computer Science and Digital Economics IV. CSDEIS 2022. Lecture Notes on Data Engineering and Communications Technologies. Springer Cham, p. 158 (2023). https://doi.org/10.1007/978-3-031-24475-9_31
18. Poplavska, Z., Komarynets, S.: Modelling the External Economic Environment Instability Impact on the Organizational Flexibility of the Enterprise. In: Kryvinska, N., Greguš, M. (eds.) Developments in Information & Knowledge Management for Business Applications. SSDC, vol. 330, pp. 1–20. Springer, Cham (2021). https://doi.org/10.1007/978-3-030-62151-3_1

19. Mlodak, A.: Evaluation of the variability of statistical features in a taxonomic model. Wiadomości Statystyczne Polish Stat. **9**, 5–18 (2005). (in Polish)

20. Malina, A.: Multi-criteria taxonomy in comparative analysis of economic structures in Poland [M], [w:] A. Zeliaś (red.) Przestrzenno-czasowe modelowanie i prognozowanie zjawisk gospodarczych, Wydawnictwo Akademii Ekonomicznej w Krakowie, pp. 305–312 (2002) (in Polish)

21. Bogliacino, F., Pianta, M.: The Pavitt taxonomy, revisited. Patterns of innovation in manufacturing and services. Documento Escuela de Economía, p. 57 (2015)

22. Andrusiv, U., et al.: Analysis of economic development of Ukraine regions based on taxonomy method. Manag. Sci. Lett. **10**(3), 515–522 (2020)

23. Metodi, H.: Opportunities for assessing the socio-economic Condition of the countries of the European Union, March 30 (2022). https://doi.org/10.2139/ssrn.4070368.SSRN: https://ssrn.com/abstract=4070368

24. Metodi, H.: Rating analysis of cohesion policy of the European Union and the member countries 29 June 2020. SSRN https://ssrn.com/abstract=3637950

25. Zelias, A.: Some notes on the selection of normalization of diagnostic variables. Stat. Transit. **5**(5), 787–802 (2002)

26. Singh, D., Singh, D.: Feature wise normalization: an effective way of normalizing data. Pattern Recogn. **122**, 108307 (2022). https://doi.org/10.1016/j.patcog.2021.108307. ISSN 0031–3203

27. Jain, S., Shukla, S., Wadhvani, R.: Dynamic selection of normalization techniques using data complexity measures. Expert Syst. Appl. **106,** 252-262 (2018). https://doi.org/10.1016/j.eswa.2018.04.008. ISSN 0957-4174

28. Izonin, I., et al.: Stacking-based GRNN-SGTM ensemble model for prediction tasks. In: Proceedings of the DASA, pp. 60–66 (2020)

29. Hassan, M.M., Mirza, T.: Using time series forecasting for analysis of GDP growth in India. Int. J. Educ. Manag. Eng. **11**(3), 40–49 (2021). https://doi.org/10.5815/ijeme.2021.03.05

30. Official site: State Statistics Service of Ukraine. http://www.ukrstat.gov.ua. Accessed 20 Jan 2023

31. Official site: State Employment Center. https://www.dcz.gov.ua. Accessed 20 Jan 2023

32. Sahai, A., Sankat, C.K., Khan, K.: Decision-making using efficient confidence-intervals with Meta-analysis of spatial panel data for socioeconomic development project-managers. Int. J. Intell. Syst. Appl. (IJISA) **4**(9), 92–103 (2012). https://doi.org/10.5815/ijisa.2012.09.12

Advanced Method for Compressing Digital Images as a Part of Video Stream to Pre-processing of UAV Data Before Encryption

Zhengbing Hu[1], Myroslav Ryabyy[2], Pylyp Prystavka[2], Sergiy Gnatyuk[2(✉)], Karina Janisz[3], and Dmytro Proskurin[2]

[1] Hubei University of Technology, Wuhan 430068, Hongshan, China
[2] National Aviation University, 1 Liubomyra Huzara Avenue, Kyiv 03058, Ukraine
{m.riabyi,s.gnatyuk}@nau.edu.ua
[3] University of Applied Sciences in Nowy Sącz, 1 Staszica Street, 33-300 Nowy Sącz, Poland

Abstract. Reducing the volume of video and photo information while maintaining high-quality visual perception remains an urgent task that specialists in specific subject areas are constantly working on. However, without questioning the success of the developers of compression methods (both lossy and lossless), reducing the size during streaming of video information carries a deterioration in the quality of information that is quite noticeable to the human eye or distorts the information in such a way that its further processing becomes impossible. In the work, a basic method to increase the level (percentage) of compression of a digital image (as a part of the video stream) with losses was proposed, and it is based on both transformations that ensure the invariance of that digital image and background extraction methods. The main value of result is that a high level of compression was demonstrated. Future study will be related to the universal dataset of crypto-algorithms development as well as AI/ML subsystem formation.

Keywords: Data compression · digital images · video stream · image processing · digital image compression · UAV

1 Introduction

The comprehensive development of information technologies, aimed at improving life, also brings some problems that need to be solved. One of the global problems is related to data storage, including video information, the volume of which is constantly and relentlessly growing. The other side of this problem is the communication channels required to transmit this information. In the absence of broadband access to the network, compression of source information without loss or with loss, that is not visible to the human eye, remains an urgent task.

Reducing the volume of video and photo information, while maintaining high-quality visual perception, is a problem on which specialists from specific subject areas are

Z. Hu et al. (Eds.): ICCSEEA 2023, LNDECT 181, pp. 371–381, 2023.
https://doi.org/10.1007/978-3-031-36118-0_33

constantly working. However, without questioning the successes of the developers of compression methods (both with loss and without), it is worth noting that usually, today, reducing the size of streaming video information brings a deterioration in the quality of information that is quite noticeable to the human eye or distort information so that its further processing becomes impossible.

2 Literature Review

Up-to-date image processing methods [1–3] analysis allowed defining compression approaches and technologies [4–6], including AI/ML-based techniques use to improve efficiency of the image processing [7–10]. Also there are some cases of image processing in the various spheres of UAV implementation from civil needs to special military tasks [11–15]. All mentioned approaches have some advantages as well as disadvantages that can be object for research and improvement.

3 Problem Statement

Mentioned approaches and existing solutions have many disadvantages but can be realized in practical cases after improvements. The main objective of the work is to propose an improved method of digital image compression for data pre-processing before encryption in the absence of broadband network access.

4 Methodology

Compression methods are divided into two groups, lossy compression and lossless compression. Lossless compression methods give a lower compression ratio, but retain the exact value of the original pixel. Lossy compression methods provide high compression ratios, but do not allow reproduction of the original data with accuracy to the intensity of the color components of a pixel. However, when viewing photos and video information, the human eye does not perceive all shades of colors, therefore, some details can be ignored without significantly distorting the image. Therefore, when compressing photo and video data with losses, a part of the information that is not visible to the observer, or that which does not have a significant impact on the perception of the information, will be discarded. Consider a digital image raster (a video data frame can act as a digital image (DI)) (Fig. 1 a) 729 * 601 pixels with a color depth of 24 bits/pixel, which is a two-dimensional array of 729 elements in width and 601 elements in length, and each element carries 24 bits of information about the color of a given image point.

The size of this image is 1.25 MB, the corresponding image size in PNG format is 585 KB and in JPEG format is 286 KB (in this work, compression using known compression methods will be used to retain the best quality). The image shown in the figure (Fig. 1b) is also 729 * 601 pixels with a color depth of 24 bits/pixel, its size in BMP format is also 1.25 MB, but in PNG format it is reduced to 2.59 Kb and in JPEG format - to 7.44 Kb. This difference is explained by the different detailing of the data of

a)   b)

Fig. 1. Color DI

the DI, and more precisely, in the second case, a straight line will be presented on the graph.

After considering the three-dimensional graph (Fig. 2) of a part of the image (Fig. 1a) and the two-dimensional graphs 1st and 25th lines of the DI for one color component (Fig. 3), it can be seen that the image function is oscillating, which affects the compression level of the DI [18–20], since smoother functions obviously compress better.

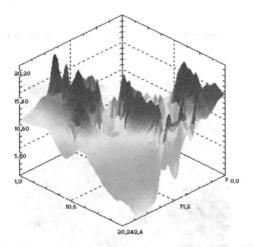

Fig. 2. A three-dimensional plot of a part of the image (Fig. 1)

To reduce the oscillation of the function, we can apply a "smoothing" (low frequency) filter to the image (Fig. 1a) [17, 18], which will make the function smoother, which in turn will reduce oscillation and possibly increase the compression level.

Applying the filter to the image (Fig. 1a) immediately shows that the distortions are not visually identifiable (Fig. 4). After looking at the graphs shown in Figs. 5 and 6, we can say that the function has become smoother. New image size in PNG format is 443 KB and in JPEG format - 258 KB. More precisely, for the PNG format the image size decreased by 11.36%, and for the JPEG format by almost 4% [21, 22].

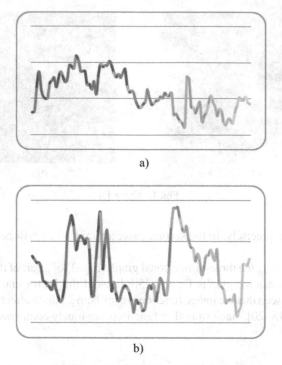

a)

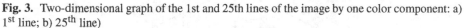

b)

Fig. 3. Two-dimensional graph of the 1st and 25th lines of the image by one color component: a) 1st line; b) 25th line)

Fig. 4. Filtered DI (Fig. 1a)

However, if a video frame is a raster image in RGB format, and a stream of video frames is a sequence of video frames received from a digital video recording device (digital video camera), then the movement in the frame (frame change) is the translational movement of a certain group of pixels, which is observed in the adjacent frames of the video stream. A necessary condition for finding motion in a frame is the possibility of locating the same group of pixels in adjacent frames.

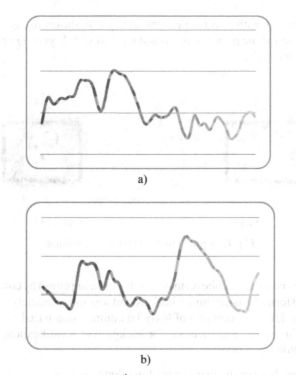

Fig. 5. Two-dimensional plot of 1st and 25th lines of the filtered image using one color component: a) 1st line; b) 25th line)

A moving object is a group of pixels that keeps its shape on adjacent frames, and the rectangle surrounding it, which keeps its dimensions, and with its center point slightly displaced between those frames (within a few pixels).

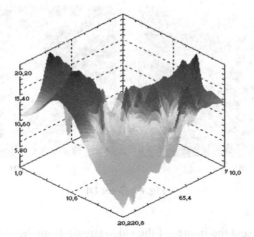

Fig. 6. A 3D plot of a portion of the filtered image (Fig. 5)

A group of pixels with a rectangle surrounding it is shown in Fig. 7. The center of the group is located at the point Ci(xi, yi) and Ci + 1(xi + 1, yi + 1) on the i-th and i + 1-th frames, respectively.

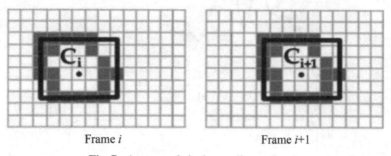

Frame *i*   Frame *i*+1

Fig. 7. A group of pixels on adjacent frames

Taking into account all the above, these result can be improved by combining methods of invariant transformations and image background separation, namely, a general method of increasing the level (percentage) of lossy DI compression based on transformations that ensure DI invariance and methods for background identification. This method is based on the following actions:

- preliminary analysis of the first frame of the flow;
- removing the background parts of the image by the selected method;
- image filtering using a low-pass filter of the required power;
- DI compression by any method or algorithm with or without losses;
- restoration of the image using a CF filter, which is the pseudo-inverse of the LF filter, that was applied to smooth the DI.

a)   b)

Fig. 8. Color DI

Figures 8, 9 present the frames of the video stream from the video surveillance and the UAV cameras with the results of their processing. After putting a filter on the image where the moving part was separated from the background (Fig. 8), it is immediately clear that the distortions are not visually identifiable (Fig. 9, 10). The size of the original and processed images is given in the table (Table 1) for images a and b under numbers 1 and 2, respectively.

a)   b)

Fig. 9. Color DI

a)   b)

Fig. 10. Color DI

5 Results and Discussion

Based on the data in the table (Table 1), we can talk about the effectiveness of the proposed method, which poses the following tasks:

- research and analysis of existing and promising background extraction methods;
- research of possible options for improving the use of invariant transformations;
- verification of the proposed method in "real-time" operation mode;
- statistical substantiation of the results of the study of the PSNR indicator and the relative error.

After the compression (pre-processing stage) UAV data will be encrypted in AI-based security system (Fig. 11).

To achieve this target, it is necessary to create the universal set (dataset) of crypto algorithms that could solve various types of problems in different conditions. The encryption algorithm will be selected by AI component (Fig. 12).

Table 1. Picture's size

№	Output size, MB	The size of the processed image, MB	Compression percentage, %
1	2,64	0,34	87,12
2	2,63	0,24	90,87

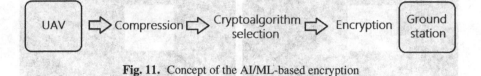

Fig. 11. Concept of the AI/ML-based encryption

There are many cryptographic algorithms that can be used in UAVs, it will be necessary to conduct analysis based on the selected criteria [23–26], for example: size of the encrypted data block; encryption key sizes; resistance against crypto-analytical attacks known at this time; structure of the algorithm; no weak and equivalent keys; data encryption speed; requirements for operational and non-volatile memory; restrictions on the use of the algorithm.

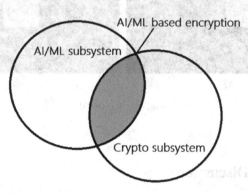

Fig. 12. Concept of the AI/ML-based encryption

Also other criterion and requirements should be added for more detail analysis based on USA, EU and domestic crypto contests.

6   Conclusions and Future Work

In the work, a new method is proposed to increase the level (percentage) of compression of the DI (as a part of the video stream) with losses. It is based on transformations that ensure the invariance of the DI and methods for extracting the background. The main value of result is that a high level of compression was demonstrated and it requires further refinement and experimental research.

It is promising to focus further research in the direction of choosing or refining an existing background identification method or creating an own one, which can be used on different types of devices for receiving a video stream.

Also the universal dataset of crypto-algorithms will be developed as well as AI/ML subsystem [27] will be proposed in the next research studies.

Acknowledgement. This work is carried out within the framework of research project № 0122U002361 "Intellectualized system of secure packet data transmission based on intelligence and search UAV" (2022–2024), funded by the Ministry of Education and Science of Ukraine.

References

1. Pang, S., et al.: SpineParseNet: spine parsing for volumetric MR image by a two-stage segmentation framework with semantic image representation. IEEE Trans. Med. Imaging **40**(1), 262–273 (2021). https://doi.org/10.1109/TMI.2020.3025087
2. Perdios, D., Vonlanthen, M., Martinez, F., Arditi, M., Thiran, J.-P.: CNN-based image reconstruction method for ultrafast ultrasound imaging. IEEE Trans. Ultrason. Ferroelectr. Freq. Control **69**(4), 1154–1168 (2022). https://doi.org/10.1109/TUFFC.2021.3131383
3. Küstner, T., et al.: LAPNet: non-rigid registration derived in k-space for magnetic resonance imaging. IEEE Trans. Med. Imaging **40**(12), 3686–3697 (2021). https://doi.org/10.1109/TMI.2021.3096131
4. Tellez, D., Litjens, G., van der Laak, J., Ciompi, F.: Neural image compression for gigapixel histopathology image analysis. IEEE Trans. Pattern Anal. Mach. Intell. **43**(2), 567–578 (2021). https://doi.org/10.1109/TPAMI.2019.2936841
5. Xu, Y., Zhang, J.: Invertible resampling-based layered image compression. In: 2021 Data Compression Conference (DCC), Snowbird, UT, USA, pp. 380–380 (2021). https://doi.org/10.1109/DCC50243.2021.00064
6. Mali, A., Ororbia, A.G., Kifer, D., Giles, C.L.: Neural JPEG: end-to-end image compression leveraging a standard JPEG encoder-decoder. In: 2022 Data Compression Conference (DCC), Snowbird, UT, USA, pp. 471–471 (2022). https://doi.org/10.1109/DCC52660.2022.00082
7. Baba Fakruddin Ali, B.H.,Prakash, R.: Overview on machine learning in image compression techniques. In: 2021 Innovations in Power and Advanced Computing Technologies (i-PACT), Kuala Lumpur, Malaysia, pp. 1–8 (2021). https://doi.org/10.1109/i-PACT52855.2021.9696987
8. Pilikos, G., Horchens, L., Batenburg, K.J., van Leeuwen, T., Lucka, F.: Deep data compression for approximate ultrasonic image formation. In: 2020 IEEE International Ultrasonics Symposium (IUS), Las Vegas, NV, USA, pp. 1–4 (2020). https://doi.org/10.1109/IUS46767.2020.9251753
9. Khan Gul, M.S., Suleman, H., Bätz, M., Keinert, J.: RNNSC: recurrent neural network-based stereo compression using image and state warping. In: 2022 Data Compression Conference (DCC), Snowbird, UT, USA, pp. 455–455 (2022). https://doi.org/10.1109/DCC52660.2022.00066
10. Patwa, N., Ahuja, N., Somayazulu, S., Tickoo, O., Varadarajan, S., Koolagudi, S.: Semantic-preserving image compression. In: 2020 IEEE International Conference on Image Processing (ICIP), Abu Dhabi, United Arab Emirates, pp. 1281–1285 (2020). https://doi.org/10.1109/ICIP40778.2020.9191247
11. Zhou, J., Xi, W., Li, D., Huang, H.: Image enhancement and image matching of UAV based on histogram equalization. In: 2021 28th International Conference on Geoinformatics, Nanchang, China, pp. 1–5 (2021). https://doi.org/10.1109/IEEECONF54055.2021.9687650

12. Liu, B., Zhang, J., Li, Z.: An improved APAP algorithm via line segment correction for UAV multispectral image stitching. In: IGARSS 2022 - 2022 IEEE International Geoscience and Remote Sensing Symposium, Kuala Lumpur, Malaysia, pp. 6057–6060 (2022). https://doi.org/10.1109/IGARSS46834.2022.9884605

13. Gademer, A., et al.: Faucon noir UAV project development of a set of tools for managing, visualizing and mosaicing centimetric UAV images. In: 2009 IEEE International Geoscience and Remote Sensing Symposium, Cape Town, South Africa, pp. III-228–III-231 (2009). https://doi.org/10.1109/IGARSS.2009.5417833

14. Yang, X.H., Xin, S.X., Jie, Z.Q., Jing, X., Dan, Z.D.: The UAV image mosaic method based on phase correlation. In: 11th IEEE International Conference on Control & Automation (ICCA), Taichung, Taiwan, pp. 1387–1392 (2014). https://doi.org/10.1109/ICCA.2014.6871126

15. Zhang, R., Xu, F., Yu, H., Yang, W., Li, H.-C.: Edge-driven object matching for UAV images and satellite SAR images. In: IGARSS 2020 - 2020 IEEE International Geoscience and Remote Sensing Symposium, Waikoloa, HI, USA, pp. 1663–1666 (2020). https://doi.org/10.1109/IGARSS39084.2020.9324021

16. Naveen Kumar, R., Jagadale, B.N., Bhat, J.S.: An improved image compression algorithm using wavelet and fractional cosine transforms. Int. J. Image Graph. Signal Process. (IJIGSP) 10(11), 19–27 (2018). https://doi.org/10.5815/ijigsp.2018.11.03

17. Prystavka, P., Cholyshkina, O.: Pyramid image and resize based on spline model. Int. J. Image Graph. Signal Process. (IJIGSP) 14(1), 1–14 (2022). https://doi.org/10.5815/ijigsp.2022.01.01

18. Prystavka, P.O., Matsuga, O.M.: Data Analysis. Education. Manual, Publishing House of Dnipro (2008). 92 p.

19. Kumar, M., Vaish, A.: Lossy compression of color images using lifting scheme and prediction errors. Int. J. Mod. Educ. Comput. Sci. (IJMECS) 8(4), 1–8 (2016). https://doi.org/10.5815/ijmecs.2016.04.01

20. Sharath Kumar, P.N., Devanand, P., Alexander, G., Saritha, Sujathan, K., Deepak, R.U.: Automated matching of pixel of interest between two digital images from two different microscope imaging devices. In: 2020 IEEE Recent Advances in Intelligent Computational Systems (RAICS), Thiruvananthapuram, India, pp. 96–100 (2020). https://doi.org/10.1109/RAICS51191.2020.9332522

21. Wang, Z., Liu, Z.: Image quality improvement algorithm in the process of film shooting under digital imaging technology. In: 2022 International Conference on Applied Artificial Intelligence and Computing (ICAAIC), Salem, India, pp. 957–960 (2022). https://doi.org/10.1109/ICAAIC53929.2022.9792797

22. Tang, M.: Image segmentation technology and its application in digital image processing. In: 2020 International Conference on Advance in Ambient Computing and Intelligence (ICAACI), Ottawa, ON, Canada, pp. 158–160 (2020)

23. Gnatyuk, S., Akhmetov, B., Kozlovskyi, V., Kinzeryavyy, V., Aleksander, M., Prysiazhnyi, D.: New secure block cipher for critical applications: design, implementation, speed and security analysis. In: Hu, Z., Petoukhov, S., He, M. (eds.) AIMEE 2019. AISC, vol. 1126, pp. 93–104. Springer, Cham (2020). https://doi.org/10.1007/978-3-030-39162-1_9

24. Kalimoldayev, M., Tynymbayev, S., Gnatyuk, S., Ibraimov, M., Magzom, M.: The device for multiplying polynomials modulo an irreducible polynomial. News Natl. Acad. Sci. Repub. Kazakhstan Ser. Geol. Techn. Sci. 2(434), 199–205 (2019)

25. Hu, Z., Gnatyuk, S., Okhrimenko, T., Tynymbayev, S., Iavich, M.: High-speed and secure PRNG for cryptographic applications. Int. J. Comput. Netw. Inf. Secur. 12(3), 1–10 (2020)

26. Saxena, A.K., Sinha, S., Shukla, P.: Design and development of image security technique by using cryptography and steganography: a combine approach. Int. J. Image Graph. Signal Process. (IJIGSP) 10(4), 13–21 (2018). https://doi.org/10.5815/ijigsp.2018.04.02

27. Lanubile, F., Calefato, F., Quaranta, L., Amoruso, M., Fumarola, F., Filannino, M.: Towards productizing AI/ML models: an industry perspective from data scientists. In: 2021 IEEE/ACM 1st Workshop on AI Engineering - Software Engineering for AI (WAIN), Madrid, Spain, pp. 129–132 (2021). https://doi.org/10.1109/WAIN52551.2021.00027

Ternary Description Language and Categorical Systems Theory

G. K. Tolokonnikov[(✉)]

VIM RAS, Moscow, Russia
admcit@mail.ru

Abstract. Categorical systems theory is an integral part of algebraic biology that predicts the properties of organisms based on the genome, including intellectual properties, which leads to artificial intelligence models. A wide range of system approaches is formalized in the language of ternary description, so it is very important to find relationships between the systems of the ternary description language (TDL) and category systems theory. This paper proposes a version of the TDL, in which it is possible to present the definition of the system in the TDL in the required form, in which it is possible to establish the relationship of this type of systems with the system block of algebraic biology. The irreducibility of categorical systems to systems based on TDL is shown. The proposed new version of the TDL is also useful for solving existing problems in the TDL formalism.

Keywords: Categorical systems · Algebraic biology · Logic · Ternary description language · Artificial intelligence

1 Introduction

There are dozens of definitions of the concept of a system, the most profound of which is the functional system of P.K. Anokhin. These approaches largely fall under the language of ternary description (TDL) developed for systems [5–9], where the definition of functional systems was also considered. However, this consideration turns out to be incomplete, since the role of the system-forming factor, which also underlies categorical systems, remains unclear. This paper proposes a version of the TDL, in which it is possible to present the definition of the system in the TDL in the required form, in which it is possible to establish the relationship of this type of systems with the system block of algebraic biology. The irreducibility of categorical systems to systems based on TDL is shown.

2 Comparison of the Concept of a System in TDL and Categorical Systems Theory

The development of various logics is usually based on the analysis of natural language. The modern predicate calculus was built in the works of G. Frege, the well-known contradiction discovered in them by B. Russell led to the creation of type theory, axiomatics

© The Author(s), under exclusive license to Springer Nature Switzerland AG 2023
Z. Hu et al. (Eds.): ICCSEEA 2023, LNDECT 181, pp. 382–390, 2023.
https://doi.org/10.1007/978-3-031-36118-0_34

of set theory, where this contradiction was eliminated, and no other contradictions were found. The justification of G. Frege's logic is based on his fundamental works "On the meaning and meaning" and "On the concept and subject" and his deep analysis of natural language, adopted in modern classical logic [10]. A rigorous construction of constructive logic in the form of a Markov tower, which has many applications (for example, [11]), on which the theory of Markov algorithms, the theory of computation and programming languages is based, was carried out in [12] and in the works of A.A. Markov cited in [12]. At the same time, the construction of the Markov tower also relies on the analysis of natural language. The TDL logic of A.I. Uyemov, which interests us, is also based on the analysis of natural language, however, as we will see, it has a number of flaws and is actually not completed in a number of essential sections (well-formed formulas, inference rules, interpretation of symbols, etc.), which is easy is seen when compared with the construction of the Markov tower [12]. Nevertheless, A.I. Uyemov's approach contains original ideas that, apparently, can be saved. First of all, it is desirable to do this in relation to the definition of systems that interests us, for which we have proposed in this paper a suitable interpretation of TDL.

The fundamental pillar of logic according to G. Frege is the assertion that for science the key logical concept is the concept of truth (falsity) of natural language statements, which are narrative sentences. The goal of science is to find true statements about the objects under study. G. Frege cleared the content of natural language statements from everything that is not related to truth (falsity) and received classical logic in a modern way. The Markov tower was built on the same support, its construction was made possible thanks to a very narrow class of studied real-life objects in the form of words in alphabets, in contrast to which G. Frege considers arbitrary objectively existing things and actual infinity.

As can be seen from the texts of A.I. Uyemov, it is not based on the specified requirement to discard everything in scientific statements that does not relate to questions of the truth of statements. As a result, the logic he develops is not strictly scientific; here, in order to compare his systems approach with the scientific approach of the categorical systems theory, we propose an adjustment to the TDL. We will demonstrate a couple of places where A.I. Uyemov deviates from scientific justification.

TDL is built on the basis of two triads of philosophical categories "thing - property - relation" and "definite-indefinite-arbitrary". In contrast to the position accepted in mathematical logic (in the formulation of G. Frege: "What is simple cannot be decomposed, and what is logically simple cannot be determined … if we find something that is simple, … then it must be given a clear name … A definition intended to give a name to something that is logically simple is impossible. Therefore, there is nothing left but to let the reader or listener know by hints how to understand what is meant by this word" ([13], p. 253)) the following is accepted in the TDL, which, as is well known, leads to contradictions due to logical circles, the position: "Between the definitions of these categories (things, properties and relations) there is the following dependence (Fig. 1):

All these categories are defined through each other, and the central main category among them is the category of a thing. Through it, the category of property and relation is directly determined, while the category of property is determined with the help of

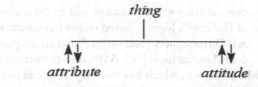

Fig. 1. Things, properties, and relations

the category of relation, and vice versa indirectly, through the category of thing" [7]. In mathematical logic, it is unacceptable to lay such circles as the basis of the theory.

The second of the unacceptable features of the TDL is the violation of A.I. Uyemov's requirement in mathematical logic, also discovered by G. Frege, of the semantic triangle "sign - meaning - meaning" in relation to the sign: "It must be demanded from a logically perfect language that … it is not provided with a value" ([13] p. 240). Note that just as G. Frege does not explicitly separate syntax from semantics (unlike the formal theories of mathematical logic), so there is no such separation in TDL. But in TDL there is no specified requirement that signs have an interpretation in the form of a meaning. A.I. Uyemov introduces three symbols in TDL t, a, A. At the same time, he associates the symbols with the words of the natural language: t - "definite" ("this"), a - "indefinite" ("something"), A - "arbitrary" ("any") ([8] p. 70). So, t stands for a certain thing, a stand for an indefinite thing. But "… a certain object t, … an indefinite object a" contradicts the main further statement of A.I. Uyomov that "thing (object)-property-relationship" makes sense only in the context. Let's see what t, a, A are from their description in the second book [9]: "… if we replace some words or groups of <natural language> words with symbols, it will be much easier for us to understand what is written. … denote the defined by t. Indefinite, let's denote some as a. Arbitrary, we will denote as A. Thus, the symbols t, a, A Uyemov A.I. compared some words of natural language. A sign (a name, a proper name, a word, and so on, something that points to an object) has a single meaning (an object, including an imaginary one) and a single meaning. The utterance is written as a declarative sentence in natural language. The value of a proposition is the truth value "false" or "true". The sign differs from the concept, which is predicative, for example, the word "definite" defines the concept ("___ is certain"), since there are certain objects, that is, just the meanings of the signs from the Frege triangle, and there are many certain objects.

We emphasize that we do not give definitions of the listed: sign, meaning, meaning (subject), concept. We refer to Frege's thought cited above, only "hint" at these entities, which we will further treat as the basis for rigorous definitions.

Let's return again to attempts to understand what connects with the signs t, a, A in TDL. Of the two contradictory interpretations, the second, published 20 years after the first, is more understandable, as well as what A.I. Uyemov does. He tries to write out some part of the natural language in the form of signs for groups of words in the natural language, using the two triads "thing-property-relation" and "definite-indefinite-arbitrary". So, having written out a set of characters, he tries to match it with a set of words of a natural language. It has two kinds of sets of words, declarative sentences (judgments) and, apparently, one can say concepts ("a phrase denoting a concept" [14]).

Further, he determines well-formed formulas from symbols, providing such formulas with an interpretation in the natural language, builds axioms and inference rules, again relying on the interpretation of symbols and formulas from them in the form of natural language words. Unfortunately, he did not give a complete construction of the ternary description language, he refers to the fact that this is not bad, since it is possible to introduce, in addition to the existing ones, new operations for constructing formulas and new inference rules. Of course, here one cannot bypass the questions of consistency and other questions common to logical calculi. A.I. Uyemov provided some fragment of his language in [9].

So, A.I. Uyemov does not explain what the strict unambiguous interpretation of the symbols he introduced consists in, that is, he does not give a clear definition based on his understanding of the words "definite", "indefinite", "arbitrary".

In his opinion, A.I. Uyemov believes that one of the motivations for constructing TDL is the existing difficulties in mathematical logic, in particular, in relation to many-place predicates. However, the examples he cites do not stand up to scrutiny. The main example, also cited by his followers [15], is as follows.

Properties and relations in classical logic are specified by one-place and many-place predicates, respectively (see p.81 in [13]), while the number of places and the order of arguments are fixed. "... Mathematical logic requires fixing both the number of objects entering into a relationship and their order. But from the point of view of the ordinary understanding of the relationship, this is not required. For example, let's take the statement "the merchants quarreled". It is clear to everyone that a certain relationship between merchants is expressed here, and for understanding this relationship, both the number of these merchants and their order are completely indifferent. Therefore, it is quite possible to speak of relationships with an indefinite number of related objects. In mathematical logic, such relations are considered incorrect" (p. 82 in [8]). The example is, in fact, incomplete. Obviously, "it is clear to everyone" not only that there is a ratio of quarreling merchants among the merchants, but also how many merchants quarreled in that clear situation. The transition to an indefinite number of merchants is not an easy abstraction, it is clear to A.I. Uyemov, but far from "everyone", to whom he referred. But, indeed, it is clear to everyone when any person directly observes the fight and sees how many of these merchants are there. The second point of incomprehensibility is that the predicate $P(x_1, \ldots, x_n)$ that n merchants fought contains variables that run through the set of K merchants. But the candidate for the relation "merchants fought" must have an interpretation, otherwise it will remain indefinite: how is the term "merchants" interpreted in this candidate? A.I. Uyemov does not give an answer to the question on what (on what set and how it is connected with K) the candidate for the relationship is defined. In fact, there is no particular difficulty in defining the predicate "the merchants fought" in the classical predicate calculus. This is a simple task, there is, for example, the following solution. Let M denote the collection of people. Denote by $Q(x) = "\_\_x\_\_$ is a merchant", - the predicate "$P_n(x_1, \ldots, x_n)n$ people $x_1, \ldots, x_n$ fought". Then the formula $P_n(x_1, \ldots, x_n)$ & $Q(x_1)$ & $\ldots$ & $Q(x_n)$ has the interpretation "the merchants fought." This is a second-order weak logic formula.

As you can see, there is no such difficulty with multi-place predicates in the traditional predicate calculus, and the example cited by A.I. Uyemov is a simple problem of mathematical logic that has the above solution.

In the section on systems, we will discuss other deviations from the scientific method present in A.I. Uyemov's systems theory based on TDL.

3 Ternary Description Language Fragment with Interpretation

Unlike A.I. Uyemov, we will consider the roots of words and from them (from the roots) we will consider the formation of things, properties and relationships: "brother" - the root coincides with the thing (Uyemov's triads "thing-property-relationship"), further, "fraternal" - property, "fraternal" - attitude. We will also present in our interpretation the indicated fragment of the ternary description language.

We propose to consider the set of word roots as the basis of explicit interpretation for the language of ternary description. On this set, we introduce further some operations that will generate new elements in such a way that the roots of words will be generators for the complete set of interpretation of the language. The analogue here is the free algebra with generators and n-ary operations.

Let us introduce variables $x, y, z, \ldots$

The symbols t, a, A will be defined as partial functional symbols applicable to all word roots, that is, we have expressions tx, ay, Az interpreted as follows, but possibly having limitations in application both to these expressions and to some other words of the language.

A.I. Uyemov, in addition to the symbols t, a, A, introduces into the language brackets (), [], an asterisk *, the symbol i (more precisely, the Greek letter "iota" ι, but it is more convenient for us to use the Latin i). These symbols and their use, he interprets as follows.

For the time being, let B denote a thing, O a relation, C a property.

Then:

$(B)C$ is interpreted as "thing B has property C".

$O(B)$ - "thing B has the property O".

$(B*)C$ - "property C is attributed to thing B".

$O(*B)$ - "relation O is established in thing B".

Brackets [] translate the listed judgments into things, properties and relations:

$[(B)C]$ - "a thing with property C".

$[(B*)C]$ - "property C attributed to thing B" and so on.

A.I. Uyemov postulates that t, a, A become things, properties and relations (respectively, defined, indefinite and arbitrary) depending on the "context", as in the language - what is in the formula to the right of the parentheses - it's a property, what's on the left is a relation, what's in parentheses is a thing. Taking into account the interpretation of the symbols t, a, A, which should be substituted for B, C, O, we have, for example, the expression for the last formula $[(t*)a]$, interpreted by A.I. Uyemov as "an indefinite property attributed to a certain thing".

With such a formalization, the problem arises of the occurrence of the same symbol in the formula twice.

Let us immediately give the definition of the system (denoted S), which is given by A.I. Uyemov

$$(iA)s = ([a(*iA)])t \text{ or } t([iA * a]).$$

The letter i plays the role of indicating that the symbol A on the right and on the left means the same thing.

A.I. Uyemov interprets this formula as follows.

"Any object is a system by definition, if some relation that has a predetermined property is realized in this object, or if some properties that are in a predetermined relation are realized in this object" (p. 35 and p. 40 in [nine]).

We interpret this definition using the proposed set of word roots.

We see two variants of the formulas of the language of ternary description, the proposition (for example, (a)t) and the proposition placed in square brackets.

Note that a "thing" (property or relation) makes sense only in a judgment corresponding to the place it occupies, therefore it is impossible to call some judgments in square brackets a thing. In our interpretation, this roughness is removed: we postulate that the judgment placed in square brackets is a root (perhaps a new root, in addition to the roots of individual words). In our version of the language of ternary description - LTO - the above formulas have the form:

$$(B)C \sim (\xi)\eta \sim P\xi x\eta y, \xi, \eta = t, a, A\big(P^* for(\xi^*)\eta\big).$$

For example, when choosing $\xi = a, \eta = t$, we have the judgment *Paxty* with the interpretation "some thing ax has a particular property ty".

$$O(B) \sim \eta(\xi) \sim Q\xi x\eta y, \xi, \eta = t, a, A\big(Q^* for\eta(\xi^*)\big).$$

When choosing $\xi = a, \eta = t$, we have the judgment Qaxty with the interpretation "some thing ax has a particular relation ty", and so on.

Let us introduce the symbol μ into the language corresponding to the comparison of the judgments of the roots, so, for example, $\mu Paxty$ is the root.

This is a new root because it corresponds to the two existing x and y word roots. To this root are applicable, as well as to other roots t, a, A.

Since, for different x and y, the expressions Ax and Ay are graphically different, and for $x = y$ they coincide and, thus, have the same meaning, there is no need to introduce the operator i in our version of TDL.

Let us introduce a binary operation δ on the judgments U, V, which corresponds in interpretation to the union of the disjunction "or" ("either one or the other"): δUV.

Now we can write out the definition of the system according to A.I. Uyemov in our version of the language of ternary description.

A system is any root of the following proposition.

$$\delta P\mu Q^* AUaVtWQ\mu P^* AUaVtW.$$

U, V, W - arbitrary roots.

4   Comparison of the Concept of a System in TDL and Categorical Systems Theory

In this section, we will begin a comparison of systems theory according to A.I. Uyemov and categorical systems theory. In a separate paper, we will continue the analysis of TDL and its logic, in particular, the section on implications and rules of inference. Here, as indicated in [15], there are unfinished and contradictory aspects of A.I. Uyemov's approach. Having explained these points, we will be able to advance in understanding the attributive theory of systems developed by A.I. Uyemov on the basis of TDL.

Above, we developed the concept of a system according to A.I. Uyemov, correcting the fragment of the TDL to a variant that can be understood, while obtaining a definition of the system, the rationale for which A.I. Uyemov is discussed below.

In the categorical theory of systems, which generalized the theory of functional systems by P.K. Anokhin, a system is understood as a convolution of categorical splices corresponding to subsystems of a given system [16]. The concept of a system is functional in nature, the system is formed on the basis of a system-forming factor, which is formalized by convolution. In one of the main interpretations, a polyarrow as a system translates a set of objects that have a certain property, expressed by the corresponding predicate, into a new property of these (or transformed) objects, which is compared toresult, according to the terminology of P.K. Anokhin. The categorical splices themselves can be objects for which it is possible to construct categorical splices of the next order, similarly to how higher-order categories arise in category theory. In our opinion, one of the main differences between the definitions of systems is the static definition of the system according to A.I. Uyemov, it states the state of a set of objects with some properties that implements a certain order on a set of objects, described by the corresponding predicate. The second difference, which shows the impossibility of covering the concept of a categorical system by the definition of a system according to A.I. Uyemov (relation, and a relation having a predetermined property) are capable of realizing only the first two levels of the hierarchy. Systems according to A.I. Uyemov do not satisfy the important requirement for systems put forward in [22], which is fulfilled for functional systems and their generalizing categorical systems and consists in the need to include sources for its occurrence and functioning in the concept of a system. We proceed from the fact that the system is an objective phenomenon, it can be revealed by other researchers, while the system properties are as objective as the properties of the electron that physicists study. It is generally accepted in science that the real object under study can have various mathematical models, and in the process of studying these models, although they more and more accurately describe the properties of the object, it is impossible to achieve the "final" model. When one model replaces another, the previous model is said to be partially wrong, and the next model is said to be more true. A classic example here is a set of models of the atom: the Thomson atom, the Bohr atom, the atom in nonrelativistic quantum mechanics and in quantum field theory. The method applied by A.I. Uyemov, in this case, is reduced to the analysis of various definitions of the atom in these theories and the identification of what is common in these definitions. Obviously, this method has nothing to do with science. With a scientific approach to such a phenomenon as a system, its definition in one theory should also turn out to be more accurate, closer to the essence of the phenomenon. Thus, the substantiation of the definition of the system given by A.I. Uyemov is not scientific. Therefore, the definition of a system given by A.I. Uyemov can only be viewed as another one separate from other definitions. It is necessary, which is part of our task, to find out the relationship between the concept of a system in TDL and the concept of a system in categorical systems theory.

5 Conclusions

There are numerous systems theories. In particular, the general theory of systems of A.I. Uyemov (GTS), based on TDL and covering several dozens of system approaches. Categorical systems theory is an integral part of algebraic biology that predicts the properties of organisms based on the genome, including intellectual properties, which leads to models of artificial intelligence (AI) related to strong AI, in contrast to approaches based on neural networks [17–21]. In this paper, we propose a version of the TDO fragment, in which it was possible to present the definition of the system according to A.I. Uyemov in the form necessary for comparison with the system block of algebraic biology. It is shown that categorical systems are not reducible to GTS at TDO. The proposed new version of the ternary description language is also useful for solving existing problems in the formalism of the ternary description language.

References

1. Anokhin, P.K.: Fundamental questions of the general theory of functional systems. In: Principles of Systemic Organization of Functions, pp. 5–61. Nauka (1973)
2. Tolokonnikov, G.K.: Informal category systems theory. Biomachsystems **2**(4), 7–58 (2018)
3. Tolokonnikov, G.K., Petoukhov, S.V.: From algebraic biology to artificial intelligence. In: Hu, Z., Petoukhov, S., He, M. (eds.) CSDEIS 2019. AISC, vol. 1127, pp. 86–95. Springer, Cham (2020). https://doi.org/10.1007/978-3-030-39216-1_9
4. Tolokonnikov, G.K.: Convolution polycategories and categorical splices for modeling neural networks. In: Hu, Z., Petoukhov, S., Dychka, I., He, M. (eds.) ICCSEEA 2019. AISC, vol. 938, pp. 259–267. Springer, Cham (2020). https://doi.org/10.1007/978-3-030-16621-2_24
5. Uyemov, A. The language of ternary description as a deviant logic: Part 1, 2, 3. Boletim da Sociedade Paranaense de Matematica **15**(1–2), 25–35 (1995), **17**(1–2), 71–81 (1997), **18**(1–2), 173–190 (1998)
6. Uyemov A.I. Some issues of the development of modern logic. In: Scientific Notes of the Tauride National University and Vernadsky, vol. 21, no. 1 (2008)
7. Uyemov, A.I.: Things, properties and relations. Publishing House of the Academy of Sciences of the USSR (1963). 184 p.
8. Uyemov A.I.: System approach and general theory of systems. Thought (1978). 272 p.
9. Uyemov, A.I., Saraeva, I.N., Tsofnas, A.Yu.: General systems theory for the humanities. W., Wydawnictwo "Uniwersitas Rediviva" (2001). 276 pp.
10. Church, A.: Introduction to mathematical logic, Moscow, IL (1960). 486 pp.
11. Rajan, E.G.: Theory and Application of Symbolic Computing with Artificial Intelligence Perspective. London Journals Press (2023). 1074 p.
12. Markov, A.A., Nagorny, N.M.: Theory of Algorithms. Nauka (1984). 432 p.
13. Frege, G.: Logic and logical semantics. Aspkt Press (2000). 512 p.
14. Uyemov, A.I. Some questions of the development of modern logic. In: Scientific notes of the Taurida National University and Vernadsky. Philosophy, Sociology, vol. 21(60), no. 1 (2008)

15. Leonenko, L.L.: Definitions in the language of ternary description. In: Scientific Notes of the Taurida National University and Vernadsky. Philosophy, Culturology, Political Science, Sociology, vol. 24(63), no. 3–4. pp. 397–404 (2011)
16. Tolokonnikov, G.K.: Categorical gluings, categorical systems and their applications in algebraic biology. Biomachsystems 5(1), 148–235 (2021)
17. Karande, A.M., Kalbande, D.R.: Weight assignment algorithms for designing fully connected neural network. IJISA (6), 68–76 (2018)
18. Dharmajee Rao, D.T.V., Ramana, K.V.: Winograd's inequality: effectiveness for efficient training of deep neural networks. IJISA (6), 49–58 (2018)
19. Hu, Z., Tereykovskiy, I.A., Tereykovska, L.O., Pogorelov, V.V.: Determination of structural parameters of multilayer perceptron designed to estimate parameters of technical systems. IJISA (10), 57–62 (2017)
20. Awadalla, H.A.: Spiking neural network and bull genetic algorithm for active vibration control. IJISA 10(2), 17–26 (2018)
21. Ojo, J.S., Ijomah, C.K., Akinpelu, S.B.: Artificial neural networks for earth-space link applications: a prediction approach and inter-comparison of rain-influenced attenuation models. IJISA 14(5), 47–58 (2022)
22. Lektorsky, V.A., Sadovsky, V.N.: On the principles of systems research. Quest. Philos. (8), 67–79 (1960)

Radial Mean LBP (RM-LBP) for Face Recognition

Shekhar Karanwal[✉]

CSE Department, Graphic Era University (Deemed), Dehradun, UK, India
shekhar.karanwal@gmail.com

Abstract. In literature various local descriptors has been invented for Face Recognition (FR) in unconstrained conditions. Most of these local descriptors are inspired from the LBP concept. With the result numerous LBP variants are introduced. Some achieved good results, and some are not. The reason behind unsatisfactory performance is the lack of building the discriminant concept in forming the descriptor. The major significance of any research is to develop the novel concept/method in form of descriptor, and it should be discriminant and robust, which should justify it by defeating the various benchmark methods. In proposed work such novel concept is introduced in the form of novel local descriptor called as Radial Mean LBP (RM-LBP), in different unconstrained conditions. In RM-LBP, initially mean value is obtained from 4 radial pixels located at 4 different radii, in eight directions of 9×9 patch. This forms the 3×3 patch, in which neighbors are filled with mean values and center place is filled with median of those. Ultimately for making the RM-LBP code, the neighbors are compared with the center pixel. After obtaining RM-LBP map image (by computing codes in each location) the size derived is 256. To compress and match reduced feature, PCA and SVMs are used. For experiments evaluation ORL face dataset is used. Results suggests that RM-LBP is better and efficient descriptor than many others. The 6 other evaluated descriptors and many literature approaches are easily outclassed by RM-LBP. RM-LBP achieves the ACC of [72.22% 84.37% 92.85%], which proves higher than many others. The significance of RM-LBP is justified when it defeats the numerous benchmark methods. The matlab version used for evaluation is MATLAB R2021a.

Keywords: Face recognition · local feature · global feature · compression · matching · gray images

1 Introduction

In Computer Vision and Pattern Recognition fields, the one class of descriptors which has significantly marks its influence is the local descriptors. The local descriptors achieve astonishing outcomes in such fields as a result invention of local descriptors has been progressing day by day. The local descriptors build its size by working on different image portions by using the small and larger size patches. Some descriptor used small

© The Author(s), under exclusive license to Springer Nature Switzerland AG 2023
Z. Hu et al. (Eds.): ICCSEEA 2023, LNDECT 181, pp. 391–400, 2023.
https://doi.org/10.1007/978-3-031-36118-0_35

scale patches and some uses the large scale patches for feature extraction. The descriptors which build the feature size by using integration of small and large scale patches produces finer results than either of them. In local there are two ways of feature extraction, (i) Either local features are extracted globally or (ii) the regional manner feature extraction (with integration) is done to achieve the entire feature size. Each phase has its advantages. In first phase, the time complexity is on the lower side and in second phase the time complexity is on the higher side. The accuracy in second phase is much superior in contrast to the first phase. The various image transformations which lower the performance are occlusion, corruption, blur, light, age, scale, emotion and pose. One of the major patches based local descriptor developed in literature is LBP [1]. LBP is the first descriptor in which extraction of features are done by using 3×3 patch. In LBP, the neighbors are replaced with binary value 1 or 0 as per the thresholding function. For higher or equal value to center, 1 is given as the label else 0 is given. The 8 bit pattern is then transformed to LBP code by weights allocation. By continuation of calculating codes (in each position) constructs LBP image, from which the histogram size produced is 256. LBP has the two basic advantages, (i) invariance monotonic property of gray scale and (ii) less complex concept. There are numerous disadvantages are also noticed in LBP and these are, (i) noisy function, which is used for thresholding, (ii) limited spatial patch ability, (iii) large size of the feature and (iv) un-affective in harsh lightning variations. This makes the researchers to propose the various variants of LBP. Some achieves good results, and some are not.

In recent times, other class of methods which attracted significant attention is deep learning methods. The success achieved by deep learning methods is relied on the complex structure built up as per the unconstrained conditions. As a result, their performances are much superior to local and global methods. CNN, LetNet, AlexNet and VGG are some of the versions of deep methods. In CNN methods, the completely connected final layers of pre-trained CNN models are used for making the size. The other activation functions (like softmax and classification layers) can also be used for developing the feature size. Most of these deep methods possesses complex structure, which results in various demerits, and these are, (i) high computational cost, (ii) training data requirement in huge amount and, (iii) adaption as per the parameter settings. These demerits put a question in the mind for developing such method. Literature portrays that there are some works based on the local descriptors, which achieves better performance than these deep methods. So, to develop such descriptor is more beneficial in contrast to methods which are complex and whose performances are not adequate.

In literature various local descriptors has been invented for Face Recognition (FR) in unconstrained conditions. Most of these local descriptors are inspired from the LBP concept. With the result numerous LBP variants are introduced. Some achieved good results and the performance of others are not satisfactory. The main limitation behind unsatisfactory performance is the lack of building the discriminant concept in forming the descriptor size. The developed concept in form of descriptor should be discriminant and robust which should prove it by defeating the various benchmark methods. This should be prime objective for any research. The proposed descriptor fulfills this objective (the prime) comprehensively. In proposed work such novel concept is introduced in form of novel local descriptor called as Radial Mean LBP (RM-LBP), in different unconstrained conditions. In RM-LBP, initially mean value is obtained from 4 radial pixels located at 4 different radii, in eight directions of 9×9 patch. This forms the 3×3 patch, in which neighbors are filled with mean values and center place is filled with median of those. Ultimately for making RM-LBP code, the neighbors are compared with center pixel. After obtaining RM-LBP map image (by computing codes in each location) the size derived is 256. To compress and match the reduced feature PCA [2] and SVMs [3] are used. For experiments evaluation ORL [4] face dataset is used. Results suggests that RM-LBP is better and efficient descriptor than many others. The 6 other evaluated descriptors i.e. LBP [1], HELBP [5], NI-LBP [6], LPQ [7], RD-LBP [6] and 6x6 MB-LBP [8], and many literature approaches are easily outclassed by RM-LBP. RM-LBP achieves the ACC of [72.22% 84.37% 92.85%], which proves higher than many others. The matlab version used for evaluation is MATLAB R2021a.

Road map: In Sect. 2 the work related to local descriptors are discussed, Sect. 3 provides proposed descriptor description, results and discussions are communicated in Sect. 4 and 5, Finally conclusion and future prospect are given in Sect. 6.

2   Related Works

Karanwal et al. [9] presented two LBP variants in Face Recognition (FR) called as ND-LBP and NM-LBP. In ND-LBP, clockwise directioned pixels are collated to each other, for making its feature size. In NM-LBP, the feature size is formed after the neighbors are compared with mean of those. Eventually the most robust feature is formed by merging ND-LBP and NM-LBP features, called as ND-LBP + NM-LBP. PCA and SVMs are consumed for the compaction and classification. For evaluating results ORL and GT face datasets are used. Result shows that ND-LBP + NM-LBP performance is far better than individual descriptors and also from various literature methods. The performance of ND-LBP, NM-LBP and ND-LBP + NM-LBP is not as effective as it should be in the unconstrained conditions. The methodology used for making their feature size is not discriminant and robust, which reflects in their results. If the developed methodology is not effective, then the results are not encouraging. In [10] Khanna et al. discovered the Emotion Recognition (FR) method, by utilizing the two best performing local methods i.e. LBP and STFT. LBP is used as local feature for extracting features and STFT is used as the frequency feature for extracting the features. The feature amalgamation of both is further shortened by FDR, variance threshold and chi-square test. SVMs is the classification method used for matching. Results on different datasets confirms the capability of invented feature. In this, those two methods are used which are already exists in literature. No new methodology is evaluated in this research, as a result the outcome is not as encouraging as it should be in the unconstrained conditions. Karanwal et al. [11] discovered three LBP versions in FR i.e. MLBP, MnLBP and CLP. MLBP is mean based method in which desired code is developed by neighbor's comparison to whole patch mean. MnLBP is the median based method in which desired code is developed by neighbor's comparison to whole patch median. In contrast to the center pixel comparison (as LBP does), the mean and median comparison proves much better and discriminant. To build more informative and influential face descriptor the LBP, MLBP and MnLBP features are merged totally. This feature is termed as CLP. For compression and matching

PCA and SVMs are used. On ORL and GT, the CLP conquers performance of alone and various literature techniques. Due to the usage of 3×3 patch, the existing methodology is not as impressive as it should be in the unconstrained conditions. This sacrifices the robustness to huge extent. Under noisy conditions, the higher patch must be used for attaining the better results. In proposed work the higher patch is used for feature extraction and it achieves better results than the other patches.

Bedi et al. [12] presented MLBP for classification of Liver images, ultrasound in nature (LUI). MLBP transforms mutual relationship among neighboring pixels, to the pattern (binary) based on the Standard deviation and Euclidean distance templates from center place. The GLCM and the color features are also used for making the size. On two distinct applications i.e. natural and facial images, MLBP proves its ability, by defeating various methods. The one thing which is lacking in this work is that the feature extraction is performed from small scale patch. It must be performed from the higher scale patch for achieving better results. Karanwal et al. [13] invented novel FR descriptor MB-ZZBP. By taking 6x6 patch, initially mean is generated in different regions (of size 2x2). Then the zigzag ordered pixels are compared to each other. The higher ordered pixel is differentiated from the lower ordered pixel to form MB-ZZLBP code. On several datasets MB-ZZLBP defeats the accuracy of numerous methods. In MB-ZZLBP, the feature extraction is performed through the 6x6 patch. Although 6x6 patch achieve better accuracy but not effective as proposed descriptor achieve in the unconstrained conditions. The RM-LBP extracts features from higher patch and achieves encouraging results. Kola et al. [14] imposed noise discriminant descriptor for ER. LBP feature is generated first, by computing the diagonal and four neighbors individually. For effective description of feature, adaptive window process and radial orientation mean are also launched. For matching SVMs is used. On distinct datasets, result confirms invented method ability. Despite using various things the discriminativity achieved by this method is not as impressive as required in the unconstrained conditions. Some better methodology could be created in unconstrained conditions to achieve the better outcome.

3 The Invented Descriptor RM-LBP

Literature presented various local descriptors in unconstrained conditions. Most of these local descriptors are inspired from the methodology of LBP. LBP encodes the neighbors based on gray pixel value and value of center pixel. After the development of LBP, various LBP variants were introduced. Most of these descriptors (compared descriptors, implemented in proposed work) lacks somewhere in their methodology as a result, the desired results are not generated when it is required. The main objective of this research is to develop novel methodology, and which beats accuracy of several existing methods. In proposed RM-LBP, the research objective i.e. discriminativity and robustness is achieved by introducing the novel descriptor Radial Mean LBP (RM-LBP) for FR. The RM-LBP beats several descriptors in terms of accuracy. The RM-LBP is implemented by using 9 $\times$ 9 patch for the FR application. Results suggests that RM-LBP is far better descriptor than the compared descriptors. The advancing features of RM-LBP is the extraction of features from 9×9 patch. The detailed RM-LBP illustration is given as.

In RM-LBP, 9 × 9 patch is used for extracting the feature size. Initially, the mean value is derived from 4 radial pixels (located at different radii) in eight directions of the 9 × 9 patch. After mean value generation there is the evolution of 3 × 3 patch. In 3 × 3 patch, the eight neighbor locations are filled with the mean values (derived from the respective directional radial pixels (located at different radii). The center location is replaced with the median of those eight neighborhoods. Further, the neighbors are thresholded to 1 for being larger or same mean value to median else 0 is given. This construct the pattern of size 8 bit, which is transfigured to RM-LBP code by deployment of the weights (binomial) and values summation. The RM-LBP code computation process is done for every pixel location and it results in the map image of RM-LBP. The RM-LBP map forms 256 size. Equation 1 and Eq. 2 shows RM-LBP code computation procedure for single location. In Eq. 1, the mean is obtained from each radial directions. In Eq. 1, P, (R1-R4) and $U_{(R1-R4),p}$ signifies the size of the neighborhood, radial pixels (at R1-R4 i.e. 4) and individual pixels located at different radii. In Eq. 2, P, R, $\mu_{(R1-R4),p}$ and $U_{c(median)}$ signifies the size of the neighborhood, radius, individual mean values and median value. Figure 1 shows RM-LBP illustration. The proposed FR framework is delivered in Fig. 2.

$$\mu_{(R1-R4),P} = \frac{1}{P/2} \sum_{p=0}^{P/2-1} U_{(R1-R4),p}(\text{for each direction}) \qquad (1)$$

$$RM - LBP_{P,R}(x_c) = \sum_{p=0}^{P-1} f\left(\mu_{(R1-R4),p} - U_{c(median)}\right)2^p, \ f(x) = \begin{pmatrix} 1 \ x \geq 0 \\ 0 \ x < 0 \end{pmatrix} \qquad (2)$$

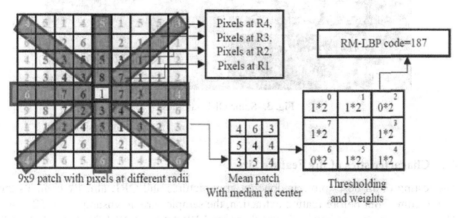

9x9 patch with pixels at different radii   Mean patch With median at center   Thresholding and weights

Fig. 1. RM-LBP example

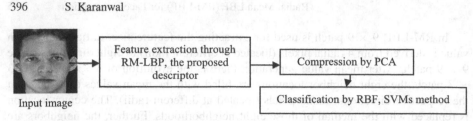

Input image

Fig. 2. Block diagram of the invented FR framework

4 Experiments

4.1 Specifications of the Dataset

The dataset which has been taken for the testing and evaluation is ORL. ORL is also familiar as AT&T dataset, is one of the oldest and the prime datasets used for many times in the literature. This dataset holds 400 samples in gray format for 40 humans. Every human consists of 10 images which shows different challenges. The different challenges of ORL dataset are light, emotion and pose variations. The pose challenge is the major challenge encountered mostly and the other two are on lower side. The resolution of these images are 112×92. Some ORL samples are delivered in Fig. 3.

Fig. 3. Some ORL samples

4.2 Characteristics of the Feature Size

The testing of all descriptors is done on gray pictures and ORL already contains the gray pictures. But before feature extraction, the samples are downsampled to 52×48. Then 7 descriptors are imposed, and these are LBP, HELBP, NI-LBP, LPQ, RD-LBP, 6x6 MB-LBP and RM-LBP. Former 6 are compared ones and last one is developed one. Size of all are 256. This is the histogram size, as all are local neighborhood-based techniques. To further compress the feature and select the relevant features for classification, PCA is used. The size taken by the RBF for evaluating all descriptors is 25, after PCA deployment. The feature extraction size and compression size are same for all the descriptors (to make the comparison fair). The matlab environment used for evaluation is R2021a.

4.3 Accuracy Generation on Distinct Subsets

The evaluation metric utilized for evaluation is accuracy. In this work accuracy is recorded on test data size by using the SVMs (RBF) classifier. The division of correct matched samples to all in the test dataset is the protocol used for measuring the accuracy. The formula is shows in Eq. 3. The elements of the Eq. 3 are ACC, Tt and ICMS. These elements states accuracy, test size and incorrect matched samples. The correct matched samples are obtained by differentiating the unmatched samples from the test size. The remaining element Tn signifies the training size.

$$ACC = \frac{\text{Correct Matched Samples}(Tt - ICMS)}{\text{Test dataset}(Tt)} * 100 \qquad (3)$$

On ORL, the Tn = 1:3 and Tt = 9:7. The Tn/Tt formed subsets are 1/9, 2/8 and 3/7. This mean for Tn = 1(i.e. 40 samples) the Tt = 9(i.e. 360 samples). For Tn = 2(i.e. 80 samples) the Tt = 8(i.e. 320 samples) and for Tn = 3(i.e. 120 samples) the Tt = 7(i.e. 280 samples). After running 30 times RBF classifier, the best/finest ACC is recorded. Table 1 communicates ACC formation. It is clearly evident from Table 1 that best/finest ACC is accomplished by the RM-LBP. RM-LBP beats completely all the compared descriptors. RM-LBP attains the astonishing ACC of [72.22% 84.37% 92.85%] on Tn = 1:3. On same Tn values other descriptors procures the ACC of [66.66% 79.68% 88.57%], [66.94% 80.00% 88.92%], [66.11% 79.68% 88.21%], [65.27% 81.25% 87.85%], [68.88% 82.18% 89.28%] and [69.44% 83.12% 90.35%]. The ACC analysis in terms of graph is displayed in Fig. 4.

Table 1. The investigation of ACC on ORL

	SVM (RBF) classifier		
	Tn/Tt details		
	1/9	2/8	3/7
	40/360	80/320	120/280
Descriptors	ACC in %		
LBP	66.66	79.68	88.57
HELBP	66.94	80.00	88.92
NI-LBP	66.11	79.68	88.21
LPQ	65.27	81.25	87.85
RD-LBP	68.88	82.18	89.28
6 × 6 MB-LBP	69.44	83.12	90.35
RM-LBP	**72.22**	**84.37**	**92.85**

To ensure reliability and accuracy of the results the justification is done by using following factors (i) the implementation is performed in MATLAB R2021a, (ii) a novel local descriptor RM-LBP is introduced which is programmed in matlab, (iii) For feature compaction and classification, the steps of PCA and SVMs are used, which are also programmed in matlab and finally in step (iv) the dataset used for the evaluation is ORL.

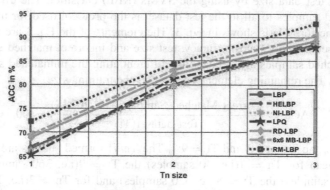

Fig. 4. Graph analysis

4.4 Comparison of ACC Against the Literature Approaches

The approaches which are used for comparison against the RM-LBP are in total 9. These 9 approaches are evaluated on same settings in which RM-LBP is evaluated. Their ACC illustration is as defined: IGFC [15], CLBP [15], MLDP [16], SRRS [16] and RSLDA [16] attains the ACC of 88.57%, 73.93%, 89.29%, 85.36% and 88.93% on Tn = 3. ILBP

Table 2. ACC Comparison on ORL

Approaches	Tn details		
	Tn = 1	Tn = 2	Tn = 3
	ACC in %		
IGFC [15]	A	A	88.57
CLBP [15]	A	A	73.93
MLDP [16]	A	A	89.29
SRRS [16]	A	A	85.36
RSLDA [16]	A	A	88.93
ILBP [17]	61.38	80.00	88.21
DLBP [17]	**72.22**	**87.18**	**93.57**
CNN-LCDRC [18]	A	83.65	90.27
LCDRC [18]	A	74.47	80.18
RM-LBP	**72.22**	84.37	**92.85**
A-Absent			

[17] and DLBP [17] succeeds in obtaining the ACC of [61.38% 80.00% 88.21%] and [72.22% 87.18% 93.57%] on Tn = 1:3. CNN-LCDRC [18] and LCDRC [18] secures the ACC of [83.65% 90.27%] and [74.47% 80.18%] when Tn = 2:3. RM-LBP conquer the ACC of 8 approaches on the compared Tn size. One approach which secures better ACC than RM-LBP is DLBP [17] on Tn = 2:3. On Tn = 1, both achieves similar ACC. Table 2 delivers the ACC comparison on ORL.

5 Discussions

This work develops the novel local descriptor in FR called as RM-LBP. RM-LBP has been proposed to overcome the previous methodologies, which lacks in their concept as a result their performances are not adequate as it should be in the unconstrained conditions. Some of these descriptors has been evaluated for comparison with respect to the RM-LBP and RM-LBP beats completely the performance of these compared descriptors. The compared descriptors are LBP, HELBP, NI-LBP, LPQ, RD-LBP and 6x6 MB-LBP. RM-LBP also outperforms various literature approaches. In RM-LBP, initially mean value is obtained from 4 radial pixels located at 4 different radii, in eight directions of 9 × 9 patch. This forms the 3 × 3 patch, in which neighbors are filled with mean values and center place is filled with median of those. Ultimately for making the RM-LBP code, the neighbors are compared with the center pixel. After obtaining RM-LBP map image (by computing codes in each location) the size derived is 256. To compress and match the reduced feature PCA and SVMs are used. For experiments evaluation ORL face dataset is used. RM-LBP achieves the astonishing ACC of [72.22% 84.37% 92.85%] on Tn = 1:3. The ACC of other compared ones are not as effective as RM-LBP. The other descriptors achieves the ACC of [66.66% 79.68% 88.57%], [66.94% 80.00% 88.92%], [66.11% 79.68% 88.21%], [65.27% 81.25% 87.85%], [68.88% 82.18% 89.28%] and [69.44% 83.12% 90.35%], in order as defined earlier. RM-LBP beats eight literature approaches also.

6 Conclusions with Future Prospect

Most of local methods are inspired from the LBP, which lacks somewhere in their methodology, as a result the desired performance is not achieved. Therefore to cope up with these methodologies, the proposed work invented the novel local descriptor RM-LBP, in unconstrained conditions. For RM-LBP, totally new methodology is introduced in the form of the local descriptor. In RM-LBP, initially mean value is obtained from 4 radial pixels located at 4 different radii, in eight directions of 9 × 9 patch. This forms the 3 × 3 patch, in which neighbors are filled with mean values and center place is filled with median of those. Ultimately for making the RM-LBP code, the neighbors are compared with the center pixel. After obtaining RM-LBP map image (by computing codes in each location) the size derived is 256. To compress and match the reduced feature PCA and SVMs are used. For experiments evaluation ORL face dataset is used. Results suggests that RM-LBP is better and efficient descriptor than many others. The 6 other evaluated descriptors and many literature approaches are easily outclassed by RM-LBP. The 6 descriptors are evaluated on same applicable parameters on which RM-LBP is evaluated. RM-LBP achieves the ACC of [72.22% 84.37% 92.85%], which proves higher than many others. The matlab version used for evaluation is MATLAB R2021a. Due to the development of RM-LBP descriptor in various unconstrained conditions, by using 9×9 patch, it can be used for real world applications. This enhances the present state of the FR application by gifting the novel local descriptor RM-LBP. The proposed work is justified by comparing and defeating the numerous benchmark methods.

In variety of ways, the future work can be extended: (i) the extraction of features regionally, (ii) performance evaluation on large scale datasets, and (iii) proposing the more discriminant method in unconstrained conditions. These three points will be covered up in the forthcoming article.

References

1. Ojala, T., Pietikainen, M., Harwood, D.: A comparative study of texture measures with classification based on featured distributions. Pattern Recogn. **29**(1), 51–59 (1996)
2. Kravchik, M., Shabtai, A.: Efficient cyber attack detection in IC systems using LWNN and PCA. IEEE Trans. Dependable Secur. Comput. (2021)
3. Hazarika, B.B., Gupta, D.: Density weighted-SVMs for binary class imbalance learning. Neural Comput. Appl. **33**, 4243–4261 (2021)
4. http://www.cl.cam.ac.uk/research/dtg/attarchive/facedatabase.html
5. Nguyen, H.-T., Caplier, A.: Elliptical local binary patterns for face recognition. In: Park, J.-I., Kim, J. (eds.) ACCV 2012. LNCS, vol. 7728, pp. 85–96. Springer, Heidelberg (2013). https://doi.org/10.1007/978-3-642-37410-4_8
6. Liu, L., Zhao, L., Long, Y., Kuang, G., Fieguth, P.: Extended LBP for texture classification. Image Vis. Comput. **30**(2), 86–99 (2012)
7. Pradeep, M., Rao, S.H.: Identification of facial expressions using LPQ. In: IC3P (2022)
8. Liao, S., Zhu, X., Lei, Z., Zhang, L., Li, S.Z.: Learning multi-scale block local binary patterns for face recognition. In: Lee, S.-W., Li, S.Z. (eds.) ICB 2007. LNCS, vol. 4642, pp. 828–837. Springer, Heidelberg (2007). https://doi.org/10.1007/978-3-540-74549-5_87
9. Karanwal, S.: Fusion of two novel local descriptors for face recognition in distinct challenges. In: ICSTSN (2022)
10. Khanna, K., Gambhir, S., Gambhir, M.: A novel technique for image classification using STFT and LBP. Multimedia Tools Appl. **81**, 20705–20718 (2022)
11. Karanwal, S.: Combined local pattern (CLP): a novel descriptor for face recognition. In: ICIRCA (2022)
12. Bedi, A.K., Sunkaria, R.K.: Mean distance LBP: a novel technique for color and texture IR for LUI. Multimedia Tools Appl. **80**, 20773–20802 (2021)
13. Karanwal, S., Diwakar, M.: MB-ZZLBP for face recognition. In: MARC, pp. 613–622 (2021)
14. Kola, D.G.R., Samayamantula, S.K.: A novel approach for facial expression recognition using LBP with adaptive window. Multimedia Tools Appl. **80**, 2243–2262 (2021)
15. Zhang, Y., Yan, L.: A fast face recognition based on image gradient compensation for feature description. Multimedia Tools Appl. **81**(18), 26015–26034 (2022). https://doi.org/10.1007/s11042-022-12804-4
16. Meng, M., Liu, Y., Wu, J.: RDP via joint margin and LSP. Neural Process. Lett. **53**, 959–982 (2021)
17. Karanwal, S.: Improved local binary pattern and discriminative local binary pattern: two novel local descriptors for face recognition. In: ICDSIS (2022)
18. Hosgurmath, S., Mallappa, V.V., Patil, N.B., Petli, V.: A face recognition system using CFE with LCDRC. Int. J. Electr. Comput. Eng. **12**(2), 1468–1476 (2022)

Waste Classification Using Random Forest Classifier with DenseNet201 Deep Features

Kemal Akyol[1(✉)] and Abdulkadir Karacı[2]

[1] Faculty of Engineering and Architecture, Computer Engineering, Kastamonu University, Kastamonu, Turkey
kakyol@kastamonu.edu.tr
[2] Faculty of Engineering, Software Engineering, Samsun University, Samsun, Turkey

Abstract. The successful management of solid waste, which has become a major issue in urban life, reduces environmental pollution. To address this issue, numerous studies using various deep learning models have been conducted. This study focuses on traditional convolutional neural networks and transformer-based deep learning architectures and provides a detailed examination summary. In this context, many experimental studies dealing with the PoolFormer and ResNet transformers and the DenseNet201 and ResNet50 pre-trained CNN models were run on the TrashNet dataset containing waste images. The Random Forest (RF) classifier was trained on different deep features extracted from 80% of this dataset, and the remaining 20% was reserved as the test dataset. Test dataset was used to validate the performance of the RF. According to the results, RF with DenseNet201 deep features presents 96.4% classification accuracy. Furthermore, the RF-based hybrid model fed with the trained DenseNet201 outperformed other studies using the same dataset in the literature in terms of classification accuracy. The hybrid method employed used in this study may serve as a model for future research.

Keywords: Waste classification · Recycling · DenseNet201 deep features · Random Forest

1 Introduction

Environmental pollution due to poor management of solid waste is a global problem [1]. The World Bank has reported that there will be 3.40 billion tons of waste in 2050 [2]. Combating solid waste, which has become the most serious problem in urban life, will play an important role in the future [3]. Mismanagement of massive solid waste generation can have a huge impact on the natural environment. Establishing a sustainable waste management system based on a "Zero Waste" approach is a necessary step in the development strategy [4]. Controlling a sharp increase in the amount of urban household waste produced has become a serious social problem as urbanization has accelerated, the urban population has grown, and living standards have improved [5]. Domestic waste management and recycling can help reduce waste and environmental damage [6]. Waste classification can help with waste disposal and recycling, and so contributing to better resource use and conservation [7]. Waste that is not treated properly

Z. Hu et al. (Eds.): ICCSEEA 2023, LNDECT 181, pp. 401–409, 2023.
https://doi.org/10.1007/978-3-031-36118-0_36

causes increasing environmental problems in the long run and moreover becomes the biggest obstacle to sustainable development [8]. Smart garbage classification solutions will play an important role in the future for the growing population due to population migration, urbanization, and modernization [3]. With the purity and accuracy of the recycled materials, the desired level of recycling is achieved [10]. Recycling is the best way to reduce waste, mitigate environmental impacts, and contribute to the country's economy [9]. Seike et al. reported that recycling requires too much cost [11].

Convolutional neural networks (CNN) have proven their performance in computer vision with their highly automated feature extraction processes and excellent accuracy [12]. Computer vision and deep learning approaches are used to perform automatic detection and classification of waste types [13]. CNN models have been used to improve recognition performance in computer vision [5].

Many studies based on CNN models have been conducted on various datasets to effectively determine waste classes in literature. For example, Zhang et al. (a) proposed a ResNet18-based self-monitoring module for recyclable waste images. The authors also provided the integration of their proposed model into mobile terminals [7]. Zhang et al. (b) reported that the classification accuracy of their proposed DenseNet169 was 82.8% [8]. Yang and Tung used scale-invariant feature transform features, support vector machine (SVM), and also CNN models on the TrashNet dataset, and they showed that SVM achieved better results than the CNN [14]. Mao et al. trained the Yolo-v3 model on the TrashNet dataset and detected domestic waste in Taiwan [15]. Qin et al. presented a study that included image enhancement and deep learning-based image classification stages. In comparison to other methods, the authors' proposed framework improved accuracy by 0.50% to 15.79% on test sets with complex backgrounds [16]. Mao et al. proposed a data augmentation method, as well as a genetic algorithm, optimized DenseNet121 to improve classification accuracy on the TrashNet dataset [17]. Wang et al. proposed a deep learning-based model and cloud computing technique to perform high-accuracy waste classification in their municipal waste management system [18].

The main aim of this study is to classify solid wastes with high performance. For this, hybrid models with the Random Forest (RF) classifier trained on features extracted with transformers and conventional CNN models are used. In this context, numerous experimental studies concerning the performance of pre-trained deep learning models and transformers for classifying waste images are conducted. Pretrained models offer higher accuracy compared to transformers.

The contributions of this paper can be summarized as follows:

1. To classify solid wastes, a hybrid model called DenseNet201 + RF, which has not been used in previous studies, is proposed.
2. DenseNet201 + RF is a two-stage model that uses DenseNet201 to extract feature maps of solid waste and RF to classify.
3. When compared to pre-trained models, the proposed model performs significantly better classification. It is also more successful than previous studies.

The remainder of this paper is structured as follows. Section 2 introduces the material and methodology used in this study. Section 3 presents model training and testing experiments. Section 4 contains the experimental results and finally, Sect. 5 concludes with the final remarks.

2   Material and Methodology

This section provides a brief overview of the dataset used in the experiments, as well as CNN and transformer architectures.

2.1   Dataset

In this study, the TrashNet waste image dataset, which contains a total of 2527 images, was used. This dataset is a public benchmark dataset published by Yang and Thung [14]. This dataset which can be downloaded from the Kaggle website considered a general criterion for waste classification and includes 6 classes consisting of cardboard, glass, metal, paper, plastic, and trash. Sample images for each class are presented in Fig. 1.

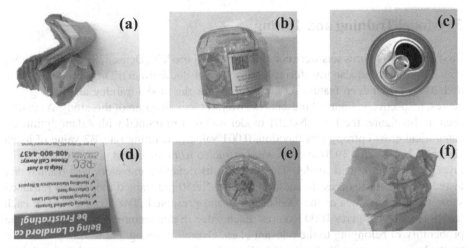

Fig. 1. Sample images in the dataset: (a) cardboard, (b) glass, (c) metal, (d) paper, (e) plastic, (f) trash

2.2   Convolutional Neural Networks and Transformers

Convolutional Neural Networks have made tremendous advances in computer vision [19]. CNN models are built with multiple building blocks, including convolution, pooling, and fully-connected layers [20, 21]. CNN employs a convolution layer to distinguish important parts of the image [22]. The convolution layer is quite critical because it is responsible for feature extraction. This layer, which is the foundation of CNN, employs the convolution operation rather than general matrix multiplication. The convolution layer's parameters are made up of a collection of learnable filters, also known as kernels

[23]. The convolution method is based on wandering a selected filter on the input image. The filter's dimensions can be 3 x 3, 5 x 5, or 7 x 7.

A filter (m1) is present in each convolution layer. The output of Layer l consists of $m_1^{(l)}$ feature maps in $Y_i^{(l)}$, $m_2^{(l)} x m_3^{(l)}$. Here, i. feature map shown with $Y_i^{(l)}$ is calculated according to Eq. 1. Here, $B_i^{(l)}$ represents the deviation matrix and $K_{i,j}^{(l)}$ represents the filter dimension [24].

$$Y_i^{(l)} = f\left(B_i^{(l)} + \sum_{j=1}^{m_i^{l-1}} K_{i,j}^{(l)} x Y_j^{(l-1)}\right) \tag{1}$$

The transformer architecture which is a current deep learning model, and was first proposed to solve natural language processing tasks has high computational efficiency and is scalable. This method can capture local and global long-range dependencies using a self-attention mechanism, which goes far beyond the receptive field of traditional convolutional filters [25].

3 Model Training and Testing

First, the deep features sets were composed of CNN models (Densen201 and ResNet50) which retrained and also transformers (ResNet and PoolFormer). Then, randomly 80% and 20% of each deep feature set are reserved for the model training and testing processes, respectively. Figure 2 presents the overall block diagram of this study. As can be seen in this figure, the DenseNet201 model was first re-trained with Adam optimizer, categorical cross-entropy loss function, 0.001 value of learning rate, 32 value of batch size, and 300 epochs. 1920 features were extracted from the global_average_pooling2d layer of this pre-trained model. The RF with its default parameters was trained and tested using these features. In addition, the ResNet50 pre-trained model was retrained same parameter values as the DenseNet201 and presented 2048 features. Also, each transformer model gives 1000 features. Each value in this feature vector represents the probability of belonging to the relevant class. Overall, training and testing procedures were also carried out with the RF classifier using deep features extracted from all deep learning architectures.

The accuracy and sensitivity metrics given in Eqs. 2 and 3 were utilized to validate the performances of models used in experiments. The True Positive (TP), True Negative (TN), False Positive (FP), and False Negative (FN) basic criteria in these equations were used to measure the correct classification success of the models. These metrics are based on the confusion matrix. Here, TP and TN indicate the correctly classified waste class and the number of other waste samples for each class, respectively. FP and FN represent the misclassified sample numbers. FP represents the false number of instances assigned to the relevant class from other relevant classes, and FN represents the false number of instances assigned to other classes from the relevant class.

$$\text{Sensitivity} = \frac{TP}{TP + FN} \tag{2}$$

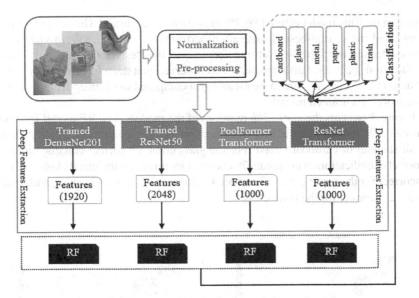

Fig. 2. General block diagram of this study

$$Accuracy = \frac{TN + TP}{TN + FP + TP + FN} \tag{3}$$

4 Experimental Results and Discussion

Transformer and CNN models were run using Tensorflow and Keras libraries in Python programming language. It is a very difficult and long process to train these models without using a graphics processor unit (GPU). For this reason, the training and testing of each model were carried out on Google Colab, a product of Google Research, using Tesla P100-PCIE-16 GB GPU, Intel(R) Xeon(R) 2.30 GHz CPU, and 25 GB RAM. Furthermore, the RF classifier, which is fed with deep features, is imported from the *sklearn* library.

Table 1 presents the accuracy and sensitivity metrics for testing dataset. According to these metrics, the RF fed with DenseNet201 features yielded the best classification

Table 1. Experimental results on the test dataset

Classifier	Deep features	Avg Acc (%)
RF	PoolFormer	81.8
	ResNet	79.4
	DenseNet 201	**96.4**
	ResNet50	79.05

Note: The best value is indicated in bold font

performance. This model offered 96.4% classification accuracy. The RF fed with Pool-Former had the second-best classification performance with 81.8%. Models fed with features derived from the ResNet transformer and the ResNet50 model gave very low classification accuracy. Furthermore, it was observed that deep features extracted using re-trained CNN models were more meaningful compared to the deep features extracted using frozen pre-trained models.

Figure 3 presents the confusion matrices of the models. The RF model fed with the DenseNet201, which has the best classification performance, has correctly classified the all samples of cardboard class. For the glass class, 95 correct classifications and 5 incorrect classifications were made. 79 correct 3 incorrect for the metal class, 115 correct 4 incorrect for the paper class, 93 correct 4 incorrect for the plastic class, and finally 25 correct 2 incorrect for the trash class.

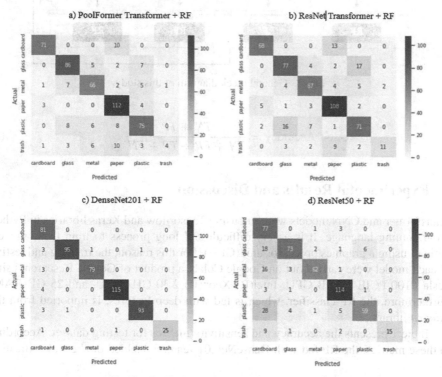

Fig. 3. Confusion matrices of each experiment

Table 2 compares this study with previous studies that used the same dataset. According to this table, the classification performance obtained in this study is superior to that of other studies in the literature. Yang and Thung [14], who composed the dataset used in this study, achieved a very low classification accuracy of 63%. The authors used SVM and CNN in their study and stated that the SVM had a better classification performance than CNN. Aral et al. [26] obtained 95% classification accuracy with DenseNet169 and DenseNet121 fine-tuned models. The DenseNet201 and RF based hybrid model

(DenseNet201 + RF) proposed in this study gave very high classification performance. Other studies that classify solid waste images are also presented in the introduction. However, they were not discussed in this section for a fairer comparison since they were not validated on the dataset used in this study.

Table 2. Comparison of the performances of the studies performed on the TrashNet dataset

Study	Method	Accuracy (%)
Aral et al. [26]	DenseNet169	95
Aral et al. [26]	DenseNet121	95
Yang and Thung [14]	SIFT + SVM	63
Proposed Study	**DenseNet201 + RF**	**96.4**

5 Conclusion

In this study, six different waste types were classified well with the proposed model. The RF classifier was used for the classification task. Various experiments were carried out with this classifier fed using features extracted with transformers and pre-trained CNN models. The RF model fed with the DenseNet201 deep features yielded the highest accuracy ratio of 96.4%. Moreover, deep features extracted from retrained CNN models are more meaningful compared to transformers. The features extracted from the PoolFormer transformer are more meaningful than the ResNet transformer for the RF classifier. Another important point to note here is that the CNN model retrained on the TrashNet dataset is more meaningful compared to the CNN model frozen. Furthermore, DenseNet201 + RF hybrid model is higher than previous studies using the same dataset in the literature. In the future, it is planned to perform real-time classification on a larger dataset and use more different deep learning models.

Acknowledgment. Authors would like to thank Yang and Thung [14] for the publicly available waste dataset.

References

1. Majchrowska, S., et al.: Deep learning-based waste detection in natural and urban environments. Waste Manage. **138**, 274–284 (2022). https://doi.org/10.1016/J.WASMAN.2021. 12.001
2. Worldbank: What a Waste. https://datatopics.worldbank.org/what-a-waste/
3. Soni, G., Kandasamy, S.: Smart garbage bin systems – a comprehensive survey. Commun. Comput. Inf. Sci. **808**, 194–206 (2018). https://doi.org/10.1007/978-981-10-7635-0_15/ COVER

4. Wu, T.W., Zhang, H., Peng, W., Lü, F., He, P.J.: Applications of convolutional neural networks for intelligent waste identification and recycling: a review. Resour. Conserv. Recycl. **190**, 106813 (2023). https://doi.org/10.1016/J.RESCONREC.2022.106813
5. Chen, S., et al.: Carbon emissions under different domestic waste treatment modes induced by garbage classification: case study in pilot communities in Shanghai, China. Sci. Total Environ. **717**, 137193 (2020). https://doi.org/10.1016/J.SCITOTENV.2020.137193
6. Sarc, R., Curtis, A., Kandlbauer, L., Khodier, K., Lorber, K.E., Pomberger, R.: Digitalisation and intelligent robotics in value chain of circular economy oriented waste management – a review. Waste Manage. **95**, 476–492 (2019). https://doi.org/10.1016/J.WASMAN.2019.06.035
7. Zhang, Q., et al.: Recyclable waste image recognition based on deep learning. Resour. Conserv. Recycl. **171**, 105636 (2021). https://doi.org/10.1016/J.RESCONREC.2021.105636
8. Zhang, Q., Yang, Q., Zhang, X., Bao, Q., Su, J., Liu, X.: Waste image classification based on transfer learning and convolutional neural network. Waste Manage. **135**, 150–157 (2021). https://doi.org/10.1016/J.WASMAN.2021.08.038
9. Wang, Z., Peng, B., Huang, Y., Sun, G.: Classification for plastic bottles recycling based on image recognition. Waste Manage. **88**, 170–181 (2019). https://doi.org/10.1016/J.WASMAN.2019.03.032
10. Tachwali, Y., Al-Assaf, Y., Al-Ali, A.R.: Automatic multistage classification system for plastic bottles recycling. Resour. Conserv. Recycl. **52**, 266–285 (2007). https://doi.org/10.1016/J.RESCONREC.2007.03.008
11. Seike, T., Isobe, T., Harada, Y., Kim, Y., Shimura, M.: Analysis of the efficacy and feasibility of recycling PVC sashes in Japan. Resour. Conserv. Recycl. **131**, 41–53 (2018). https://doi.org/10.1016/J.RESCONREC.2017.12.003
12. Dhillon, A., Verma, G.K.: Convolutional neural network: a review of models, methodologies and applications to object detection. Prog. Artif. Intell. **9**, 85–112 (2020). https://doi.org/10.1007/S13748-019-00203-0/FIGURES/15
13. Vo, A.H., Hoang Son, L., Vo, M.T., Le, T.: A novel framework for trash classification using deep transfer learning. IEEE Access **7**, 178631–178639 (2019). https://doi.org/10.1109/ACCESS.2019.2959033
14. Yang, M., Thung, G.: Classification of Trash for Recyclability Status. Project (2016)
15. Mao, W.L., Chen, W.C., Fathurrahman, H.I.K., Lin, Y.H.: Deep learning networks for real-time regional domestic waste detection. J. Clean. Prod. **344**, 131096 (2022). https://doi.org/10.1016/J.JCLEPRO.2022.131096
16. Qin, J., Wang, C., Ran, X., Yang, S., Chen, B.: A robust framework combined saliency detection and image recognition for garbage classification. Waste Manage. **140**, 193–203 (2022). https://doi.org/10.1016/J.WASMAN.2021.11.027
17. Mao, W.L., Chen, W.C., Wang, C.T., Lin, Y.H.: Recycling waste classification using optimized convolutional neural network. Resour. Conserv, Recycl. **164**, 105132 (2021). https://doi.org/10.1016/J.RESCONREC.2020.105132
18. Wang, C., Qin, J., Qu, C., Ran, X., Liu, C., Chen, B.: A smart municipal waste management system based on deep-learning and Internet of Things. Waste Manage. **135**, 20–29 (2021). https://doi.org/10.1016/J.WASMAN.2021.08.028
19. Maduranga, M., Nandasena, D.: Mobile-based skin disease diagnosis system using convolutional neural networks (CNN). Image, Graphics and Signal Process. **2022**, 47–57 (2022). https://doi.org/10.5815/ijigsp.2022.03.05
20. Yamashita, R., Nishio, M., Do, R.K.G., Togashi, K.: Convolutional neural networks: an overview and application in radiology. Insights Imaging **9**, 611–629 (2018). https://doi.org/10.1007/S13244-018-0639-9/FIGURES/15

21. Mali Patil, B., Rani Raigonda, M., Anakal, S., Bhadrashetty, A.: Early detection of dementia using deep learning and image processing. Int. J. Eng. Manuf. **1**, 14–22 (2023). https://doi.org/10.5815/ijem.2023.01.02

22. Shashidhar, R., Arunakumari, B.N., Manjunath, A.S., Roopa, M.: Indian sign language recognition using 2-D convolution neural network and graphical user interface A S Manjunath. Image, Graphics and Signal Process. **2022**, 61–73 (2022). https://doi.org/10.5815/ijigsp.2022.02.06

23. Singhal, T.: A review of coronavirus disease-2019 (COVID-19). Indian J. Pediatr. **87**, 281–286 (2020). https://doi.org/10.1007/S12098-020-03263-6/METRICS

24. Toğaçar, M., Ergen, B., Cömert, Z., Özyurt, F.: A Deep feature learning model for pneumonia detection applying a combination of mRMR feature selection and machine learning models. IRBM **41**, 212–222 (2020). https://doi.org/10.1016/J.IRBM.2019.10.006

25. Yang, M., et al.: Transformer-based deep learning model and video dataset for unsafe action identification in construction projects. Autom. Constr. **146**, 104703 (2023). https://doi.org/10.1016/J.AUTCON.2022.104703

26. Aral, R.A., Keskin, S.R., Kaya, M., Haciömeroğlu, M.: Classification of TrashNet dataset based on deep learning models. In: Proceedings - 2018 IEEE International Conference on Big Data, Big Data 2018, pp. 2058–2062 (2019). https://doi.org/10.1109/BIGDATA.2018.8622212

An Increased Implementation of Generalized Sidelobe Canceller Based on the Expectation - Maximization Algorithm

Quan Trong The[1]([✉]) [iD] and Sergey Perelygin[2] [iD]

[1] Digital Agriculture Cooperative, Ha Noi, Vietnam
quantrongthe1984@gmail.com
[2] Faculty of Media Technologies, State University of Film and Television St. Petersburg, Saint - Petersburg, Russian Federation

Abstract. An effective way of improving the speech quality and increasing the evaluation of speech system, communication in complex noise scenario, where exist several types of noise, interference, third-party speaker is studied for many decades and still a challenging problem. The single-channel algorithm usually uses spectral subtraction that attenuates speech component and decreases the quality of system. Microphone array beamforming is the most popular research topic for overcoming this drawback. However, the performance of microphone array beamforming often deteriorated for many reasons: microphone mismatches, the different microphone sensitivities, phase error or imprecise geometry of microphone array. In this paper, the author introduced the utilizing of EM algorithm for enhancing performance of GSC beamformer in real recording situation. Experimental results confirmed the effectiveness of this approach in terms of the signal-to-noise ratio from 3.8 (dB) to 8.0 (dB) and better noise suppression. From the obtained numerical result, the EM algorithm can be used in many speech applications with arbitrary microphone array geometry.

Keywords: The expectation maximization · generalized sidelobe canceller · speech enhancement · noise suppression · adaptive filter

1 Instruction

Speech enhancement and separation are the importance tasks in audio signal processing for obtaining the desired target speaker. Realistic audio signal usually involve: noise, interference, third-party talker, transport vehicle, annoying speaker, reverberation or imprecise sampling frequency. In this scenario, source extraction refers to separate the useful signal, suppress background noise or cancel different other speech source. Therefore, in general speech enhancement is referred to the complicated saving speech component while alleviating interfering noise. Speech enhancement and noise reduction are also crucial pre-processing steps for robust understanding in many available speech applications, such as hearing aid, medical devices, teleconference, video game consoles, surveillance systems.

© The Author(s), under exclusive license to Springer Nature Switzerland AG 2023
Z. Hu et al. (Eds.): ICCSEEA 2023, LNDECT 181, pp. 410–418, 2023.
https://doi.org/10.1007/978-3-031-36118-0_37

The important task for robust implementation in adverse complex and annoying surrounding environments has led to increase performance of speech enhancement system, which use an appropriate method for suppressing noise and saving the target speech. Single-channel algorithm often uses the subtraction spectral, which leads cancelling the desired useful signal. And in almost real-life scenarios, the multiple speech sources or noise requires an appropriate solution to separate each directional point source without speech distortion of other speakers. The number of microphones per device has significantly installed in the last few decades, most acoustic instruments are equipped with two or more special microphones. All these devices require real-time processing, extracting and successful achieving useful information.

Recently, the using of multiple-microphone systems [1–5] are introduced for dealing the above challenge problem by utilizing the spatial information for achieving the necessary speech component. Many works have been evaluated for extracting the desired speaker while eliminating interference and noise. Among them, due to the simplicity and low computation, beamforming technique has become one of the most widely used for suppressing the non-target directional noise and extract the only target directional signals.

Using the spatial diversity to achieve an advantage of noise reduction and extract the target directional useful signal. The spatial beampattern, which based on the information of DOA, the characteristic of surrounding environment, the distribution of microphone array. This technique plays a major role in microphone array signal processing. Many scholars have proposed several types of beamformer, which utilizes a constrained criteria to obtain an optimum solution.

A generalized sidelobe canceller (GSC), that has a simple scheme for implementation, is one of the most popular beamformer is used widely. A GSC contains a fixed beamformer (FBF), a blocking matrix (BM), and an adaptive noise canceller (ANC). The FBF ensures a beampattern toward the directional sound source while attenuating the other environmental signals, which from different non-target directions. The BM has the task of suppressing target component signal, so the only noise can pass through the block. The ANC uses adaptive method to enhance the obtained target signal from FBF and BM.

The FBF often uses the delay-and-sum (DAS) algorithm to form beampattern toward the desired target speaker. Otherwise, the BM plays an important role in speech enhancement of system. For example, unsuitable phase differences between the microphone array signals are unavoidable. Such phase errors are incoherent due to the imprecise the sound propagation in real recording environment. The microphone array position, the different sensitivities, the mismatch between the true direction-of-arrival (DOA) and the assumed DOA or the inexactly sampling lead to speech leakage in the output of the BM block. In results of GSC beamformer, the attenuation of target speech.

As usually, the blocking matrix (BM) is assumed to pass through the only noise and reject the desired speech signal. This issue is that the traditional BM need retaining noise well and suppressing signal leakage, which corrupts the performance of GSC beamformer. Therefore, the adjustment of coefficients for BM is essential problem, which studied by many researchers. In [6], Hoshuyama suggested a new BM, which use constrained weights vector adaptive filter and a norm-constrained ANC. A similar in

GSC beamformer was illustrated in the frequency domain [7]. The authors [8] utilized the sound-source presence probability, which calculated from the instantaneous DOA of useful target directional talker and voice activity detection into BM block to mitigate the speech leakage to retain only the noise components. In [9], the author proposed using Linear Constrained Minimum Variance (LCMV) beamformer to replace BM block to decrease the leakage of desired speech signal and improve the efficient noise reduction. Li [10] used the estimation of DOA to enhance the BM block and alleviate the speech leakage.

Under the condition of highly non-stationary noise and low SNR, the accurate control of coefficients of BM or ANC block is still challenging despite the effectiveness of several methods. Recently, a deep neural network was used to increase the speech enhancement [11].

For mitigating the problem of leakage signal, the author proposed using the EM algorithm for suppressing the speech leakage at the output of BM to enhance the overall of the quality output signal. The EM algorithm is applied to alleviate the influence of unwanted phase errors due to the microphone mismatch, the different microphone array sensitivities. The BM may not incompletely block the target directional signal, so with the using the resulting EM algorithm, a priori speech presence probability allows blocking speech component from the output of BM block. In this contribution, the EM algorithm is used for estimating the value of speech presence probability, which is calculated for the observed dual - microphone system (DMA2) signals. The illustrated experiments were shown that the effectiveness of suggested method while increasing the signal-to-noise ratio (SNR) while saving the desired speech. The advantages of DMA2 is low cost, compact, low power consumption and has been widely implemented in wearable devices such as smart glassed, hearing aids, earphones.

The remainder of this contribution is organized as follows. The next section, the structure of GSC beamformer is introduced to eliminate noise while keeping target directional signal. Section 3 explains the principle of EM algorithm and the obtained results. The proposed method of using EM algorithm is illustrated for alleviating the speech leakage component. In Sect. 4, a dual-microphone array (DMA2) system is used for demonstrating the effectiveness of suggested method in comparison the quality output signal between conventional GSC beamformer (con-GSC) and the proposed method (pro-EM-GSC). Conclusion and future research from this work in Sect. 5.

2   GSC Beamformer

In this section, the principle working of GSC beamformer is illustrated in frequency domain. The received microphone array signals are modeled as the following equation:

$$X_1(t, f) = S(t, f)e^{j\Phi_s} + V_1(t, f) \tag{1}$$

$$X_2(t, f) = S(t, f)e^{-j\Phi_s} + V_2(t, f) \tag{2}$$

where t, f is the index of time and frequency, two noisy signals $X_1(t, f), X_2(t, f)$, the additive noise $V_1(t, f), V_2(t, f)$ on each microphones, $\Phi_s = \pi f \tau_0 \cos\theta_s$, $\tau_0 = d/c$, $c = 343(m/s)$ is the speed of sound in the air, d is the inter-microphone distance, θ_s is the direction-of-arrival of interest useful signal (Fig. 1).

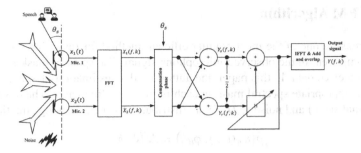

Fig. 1. The scheme of GSC beamformer

The most widely used for FBF beamformer is algorithm delay-and-sum algorithm, and subtraction is the method for BM block after compensation phase. Therefore, the main and the reference signal $Y_s(t,f)$, $Y_r(t,f)$ can be expressed as:

$$Y_s(t,f) = \frac{X_1(t,f)e^{-j\Phi_s} + X_2(t,f)e^{j\Phi_s}}{2} \tag{3}$$

$$Y_r(t,f) = \frac{X_1(t,f)e^{-j\Phi_s} - X_2(t,f)e^{j\Phi_s}}{2} \tag{4}$$

The auto and cross power spectral density of $Y_s(t,f)$, $Y_r(t,f)$ are very importance for further signal processing:

$$P_{Y_s Y_r}(t,f) = (1-\alpha)P_{Y_s Y_r}(t-1,f) + \alpha Y_s(t,f)Y_r^*(t,f) \tag{5}$$

$$P_{Y_r Y_r}(t,f) = (1-\alpha)P_{Y_r Y_r}(t-1,f) + \alpha Y_r(t,f)Y_r^*(t,f) \tag{6}$$

where α is the smoothing parameter, in range of $\{0..1\}$

The coefficients of adaptive filtering are calculated as:

$$H(t,f) = \frac{P_{Y_s Y_r}(t,f)}{P_{Y_r Y_r}(t,f)} \tag{7}$$

And the final desired output signal according to the model of GSC beamformer can be represented as:

$$Y(t,f) = Y_s(t,f) - Y_r(t,f) \times H(t,f) \tag{8}$$

The adaptive algorithm, which used in GSC, helps suppressing the noise component from output of FBF beamformer with the reference signal, which obtained from BM block. GSC beamformer allows achieving high spatial gain, noise reduction, diversity beamformer, speech quality and intelligibility. The degree of noise suppression depends on correctness of direction-of-arrival of the interest useful signal, microphone mismatches, phase error, microphone sensitivity or position of MA. In real-life scenario, these factors effect on performance of GSC beamformer and reduce speech quality. To overcome this drawback, in the next section, the author introduced EM algorithm, which will be applied to improve the evaluation of GSC beamformer.

3 The EM Algorithm

Spectral mask is one of the most popular efficient method for removing the remaining speech component in the observed microphone signal. The spectral mask is calculated from the received data. In this paper, the author used an estimation of speech presence to form an appropriate spectral mask in each frame, each frequency. The desired target useful signal $s(t, f)$ and noise $n_m(t, f)$ are distributed by the following equation:

$$p\big(n_m(t,f); \phi_{tfm}\big) = \mathcal{N}\big(0, \phi_{tfm}\big) \tag{9}$$

A Gaussian distribution with variance $|n_m(t,f)|^2 = \phi_{tfm}$ and zero mean. The received signals are modeled by a Gaussian mixture as:

$$p(x(t,f); \lambda) = \sum\nolimits_{m=1}^{N+1} \alpha_{fm} p(y(t,f)|C(t,f) = l; \lambda) \tag{10}$$

$$p(x(t,f)|C(t,f) = m; \lambda) = \mathcal{N}_c\big(0, \phi_{tfm} \boldsymbol{B}_{fm}\big) \tag{11}$$

where α_{fm} stand by a mixture weight $\left(\sum_{m=1}^{M+1} \alpha_{fm} = 1\right)$,

$\boldsymbol{B}_{fm} = \hat{\boldsymbol{h}}_m(f)\hat{\boldsymbol{h}}_m^H(f)$ is spatial correlation matrix of noise source m. $C(t,f) = m$, $m = 1, \ldots, M$ according to source classes, and $C(t,f) = M + 1$ denotes to noise class. The function log likelihood can be derived by:

$$\mathcal{L}(\lambda) = \sum\nolimits_t \sum\nolimits_f log p(x(t,f); \lambda) \tag{12}$$

$$= \sum\nolimits_t \sum\nolimits_f log \sum\nolimits_m \alpha_{fm} \mathcal{N}_c\big(0, \phi_{ftm} \boldsymbol{B}_{fm}\big) \tag{13}$$

where $\lambda = \{\lambda_m\} = \{\alpha_{fm}, \phi_{tfm}, \boldsymbol{B}_{fm}\}$ is the setting parameters. EM algorithm helps maximum the value of log likelihood. The posteriori probability is defined as $M_m(t,f) = p(C(t,f) = m|x(t,f), \lambda)$.

At E-step: $M_m(t,f)$ is determined in each frame:

$$M_m(t,f) = p(C(t,f) = m|x(t,f), \lambda) \frac{\alpha_{fm} p(x(t,f)|\lambda_m)}{\sum_m \alpha_{fm} p(x(t,f)|\lambda_m)} \tag{14}$$

At M-step: The parameter λ is updated as:

$$\phi_{tfm} = \frac{1}{M} x^H(t,f) \boldsymbol{B}_{fm}^{-1} x(t,f) \tag{15}$$

$$\boldsymbol{B}_{fm} = \frac{\sum_t^T \frac{M_m(t,f)}{\beta_{tfm}} x(t,f) x^H(t,f)}{\sum_t^T M_m(t,f)} \tag{16}$$

$$\alpha_{fm} = \frac{1}{T} \sum\nolimits_t^T M_m(t,f) \tag{17}$$

The spectral mask, which is utilized for suppressing the speech leakage component at the reference signal, is derived by:

$$M_n(t,f) = \sum_{m=1}^{M} M_m(t,f) \tag{18}$$

Finally, $Y_r(t,f)$ is pre-processing as:

$$\widehat{Y}_r(t,f) = Y_r(t,f) \times M_n(t,f) \tag{19}$$

With the working principles, the perfect evaluation of GSC beamformer is depends on the amount noise in the reference signal. The EM algorithm, which based on the a priori information of speech presence, helps removing the remaining speech component. The obtained signal $\widehat{Y}_r(t,f)$ allows achieving a better result of speech quality at the output of GSC beamformer. The next section will illustrate the promising result of EM algorithms.

4 Experiments and Discussion

In this section, the author performed an experiment with DMA2 in realistic environment in presence of noise, third-party speaker, the sound of fan, telephone. The inter - microphone distance $d = 5(cm)$, the observed input signals with sampling frequency $Fs = 16kHz$. These data were transformed to the STFT domain, $NFFT = 512$, Hamming window, overlap 50%, smoothing parameter $\alpha = 0.1$. The DOA of interest signal is $\theta_s = 90^0$. The experiment was performed to verify the effectiveness of the proposed method (proEMGSC) in comparison with the conventional GSC beamformer (conGSC). An objective method for measuring the signal-to-noise ratio (SNR) is used [12]. The purpose of experiment is verifying the effectiveness of EM algorithm in enhancing speech quality of final output signal.

The captured microphone array signal is shown in Fig. 2.

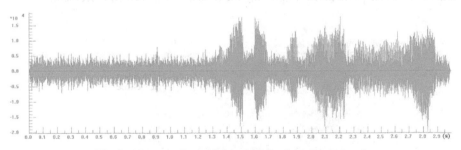

Fig. 2. The waveform of the original microphone signal

In Fig. 3(a), (b) the final output signal processed by conGSC and proEMGSC are presented. From the obtaining results, proEMGSC reduce speech distortion of the outputs signal while suppressing surrounding noise. Due to the utilizing the EM algorithm, the speech leakage in the reference signal is mitigated. Herein, the speech quality of the output signal is increased. The degree of speech distortion is to 4.0 (dB) from Fig. 4.

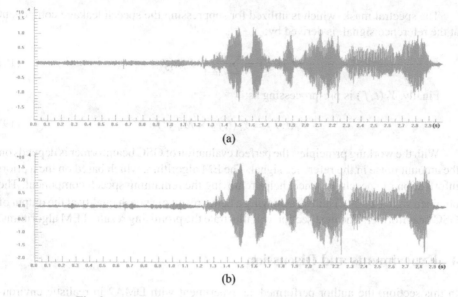

(a)

(b)

Fig. 3. The processed signal by conGSC (a) and proEMGSC (b)

A comparison of the speech quality of two signals in term of the signal-to-noise ratio (SNR) is shown in Table 1.

The advantage of the proposed method is the increasing of SNR from 3.8 to 8.0 (dB). From the obtained numerical results, as we can see that, EM algorithm based - the proposed method allows reducing the component speech leakage, which often deteriorates speech enhancement of GSC Beamformer. In the case, if only noise in the reference signal, GSC Beamformer perfectly extracts the desired signal with speech distortion. EM algorithm can be integrated into multi-microphone system for dealing other complicate problem for improving the robustness speech enhancement.

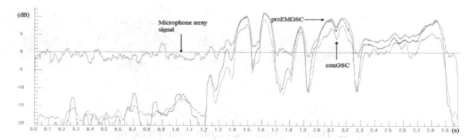

Fig. 4. The energy of microphone array signal, the processed by conGSC, proEMGSC

The lack of almost microphone array processing signal is the speech leakage, due to the error of interested signal's DOA or microphone mismatches. This problem leads to the degraded evaluation of beamformer. EM algorithm blocks the existing speech component in the reference signal and outperforms an improvement of GSC beamformer. The EM

Table 1. The signal-to-noise ratio (dB)

Method Estimation	Microphone array signal	conGSC	proEMGSC
NIST STNR	9.2	26.0	28.8
WADA SNR	6.3	21.8	29.8

algorithm uses the observed microphone array signal to take into account the probability of speech presence, therefore this approach is an adaptive method. The numerical results have shown the capability of EM algorithm in overcoming the conventional drawback of GSC beamformer. The promising simulated signals has satisfied the aim of evaluated experiment.

5 Conclusion

MA technology is the most common popular method applied into numerous speech applications and has been consistently an attractive research field. MA have been an intensive realistic application for sound source localization, separation and recognition, teleconference systems, hearing aid. GSC beamformer is an effective method for extracting desired target directional signal while eliminating surrounding noise. In this paper, the author addressed the speech leakage, which often occurs in realistic situation in presence of many objective reasons. The simulation results show that the EM algorithm can help block speech component from the output of BM block and achieve better speech enhancement in term the signal-to-noise ratio and noise suppression. In future, the author will combine the property of considered environment to suppress more speech leakage from BM block and enhance performance of GSC beamformer.

References

1. Brandstein, M., Ward, D.: Microphone Arrays, vol. 18, p. 398. Springer-Verlag, Heidelberg, Germany (2001). https://doi.org/10.1007/978-3-66204619-7
2. Benesty, J., Chen, J., Huang, Y.: Microphone array signal processing, p. 240. Springer-Verlag, Berlin, Germany (2008). https://doi.org/10.1007/978-3-540-78612-2
3. Benesty, J., Chen, J., Pan, C.: Fundamentals of Differential Beamforming. Springer Singapore, Singapore (2016). https://doi.org/10.1007/978-981-10-1046-0
4. Benesty, J., Cohen, I., Chen, J.: Fundamentals of Signal Enhancement and Array Signal Processing. Wiley (2017). https://doi.org/10.1002/9781119293132
5. Griffiths, L., Jim, C.: An alternative approach to linearly constrained adaptive beamforming. IEEE Trans. Antennas Propag. **30**(1), 27–34 (1982). https://doi.org/10.1109/TAP.1982.1142739
6. Hoshuyama, O., Sugiyama, A., Hirano, A.: A robust adaptive beamformer for microphone arrays with a blocking matrix using constrained adaptive filters. IEEE Trans. Signal Process. **47**, 2677–2684 (1999)

7. Herbordt, W., Kellermann, W.: Computationally efficient frequency-domain robust generalized sidelobe canceller. In: Proceedings of the 7th International Workshop on Acoustic Echo and Noise Control (IWAENC), Darmstadt, Germany, 10–13 September 2001 (2001)

8. Yoon, B.J., Tashev, I., Malvar, H.: Robust adaptive beamforming algorithm using instantaneous direction of arrival with enhanced noise suppression capability. In: Proceedings of the 2007 IEEE International Conference on Acoustics, Speech and Signal Processing, Honolulu, HI, USA, 15–20 April 2007, pp. 133–136 (2007)

9. Khayeri, P., Abutalebi, H.R., Abootalebi, V.: A nested superdirective generalized sidelobe canceller for speech enhancement. In: Proceedings of the 2011 8th International Conference on Information, Communications & Signal Processing, Singapore, 13–16 December 2011; pp. 1–5 (2011)

10. Li, B., Zhang, L.H.: An improved speech enhancement algorithm based on generalized sidelobe canceller. In: Proceedings of the 2016 International Conference on Audio, Language and Image Processing (ICALIP), Shanghai, China, 11–12 July 2016; pp. 463–468 (2016)

11. Liu, H.P., Yu, T., Fuh, C.S.: Bone-conducted speech enhancement using deep denoising autoencoder. Speech Commun. **104**, 106–112 (2018)

12. https://labrosa.ee.columbia.edu/projects/snreval/

13. Liu, J., Guo, J., Xu, G., Li, K.: Model-based synthesis method of multiple patterns linear arrays with the minimum number of antenna elements: a state space approach. Int. J. Image, Graph. Sign. Process. **9**(6), 23–28 (2017). https://doi.org/10.5815/ijigsp.2017.06.03

14. Rind, Y.M.: Iterative shrinkage operator for direction of arrival estimation. Int. J. Inf. Eng. Electron. Bus. **8**(5), 26–35 (2016). https://doi.org/10.5815/ijieeb.2016.05.04

15. Errifi, H., Baghdad, A., Badri, A., Sahel, A.: Radiation characteristics enhancement of microstrip triangular patch antenna using several array structures. Int. J. Wireless Microwave Technol. **5**(3), 1–17 (2015). https://doi.org/10.5815/ijwmt.2015.03.01

Pedestrian Re-identification Combined with Triplet Metric Learning and Dual-Domain Filtering in Noisy Scene

Wenyu Song[1], Liyuan Wang[2(✉)], Qi Huang[3], Shaohuai Yu[2], and Shurui Wang[4]

[1] China Electric Power Equipment and Technology Co., Ltd., Beijing 100052, China
[2] CCCC Second Highway Consultants Co., Ltd., Wuhan 430056, China
19671869@qq.com
[3] China Ship Development and Design Center, Wuhan 430068, China
[4] School of Electronic Information, Wuhan University, Wuhan 430072, China

Abstract. In actual power grid surveillance system, it is difficult for many pedestrian re-identification methods to extract pedestrian features at night, rainy or foggy days because of the insufficient illumination or the bad weather. Focusing on the pedestrian re–identification in noisy situations, this paper proposes to use the dual-domain filter decomposition to construct triples, which are used to train metric learning model to learn the robust features. Our proposed method mainly consists of two parts. Firstly, we analyze the distribution characteristic of image noise in surveillance videos and propose an image denoising method based on dual-domain filtering. Secondly, based on the separation effect of dual-domain filtering on image noise, this paper draws on the clustering ability of the triple loss and proposes a new way to construct the training triplet. In the training stage, different from the traditional triplet matrix learning, the original images with the low-frequency component and the high-frequency component generated by the dual-domain filtering are used as the input triplet. So the network can further suppress the noise components and improve the robustness of the feature. The experiments on Market-1501 and CUHK03 data sets show that our Rank-1 only drops 1.9% and 7.8% in noisy scene, which means the proposed method can make the accuracy of noisy pedestrian re-identification performance be close to that of the original non-noise pedestrian re-identification performance.

Keywords: Pedestrian Re-identification · Dual-domain Filter · Matrix Learning · Triplet Loss

1 Instruction

Deep learning has achieved great success in many fields [1]. Pedestrian re-identification (referred as ReID) is an important one among them. It is a technology that uses computer vision to determine whether a specific pedestrian exists in an image or a video sequence. However, due to power systems are usually located in remote and complex terrain areas, complex situations such as light change, complex background transformation, low-resolution images, occlusion often occur, pedestrian re-identification is still

Z. Hu et al. (Eds.): ICCSEEA 2023, LNDECT 181, pp. 419–428, 2023.
https://doi.org/10.1007/978-3-031-36118-0_38

very challenging in actual power grid surveillance system. At present, in the field of pedestrian re-identification, a new challenge is that in some situations with insufficient illumination and harsh weather conditions, such as night, rain and fog, most models are difficult to obtain pedestrian characteristics with sufficient expressive ability. Pedestrian re-identification can be roughly divided into two types: feature expression-based and metric learning-based.

The method based on feature expression is a commonly used method for pedestrian re-identification. Early extracted human features usually come from apparent shallow features, such as Haar features, SIFT features and LBP features. In [2] the HSV color histogram is combined with the SILTP feature and proposes the LOMO feature. The advantage of LOMO is that it has good adaptability to changes in viewing angle, but it is less robust to changes in background. Deep learning technology can adaptively learn the salient features of images, extract more abstract and deep-level attributes, such as semantic information, etc., and effectively improve the ability to express pedestrian features [3]. Luo Hao and Jiang Wei et al. propose a dynamic matching minimum path algorithm based on the local area correlation of pedestrians to achieve low-cost pedestrian image alignment [4]. In addition, many studies also obtain the pose information by extracting the bone joint points of the human body, which can be used not only to segment the pedestrian image, but also to achieve the alignment of the pedestrian image [5]. Liang Zheng and Yujia Huang et al. propose a pose-invariant descriptor PIE [6]. Haiyu Zhao and Maoqing Tian propose a SpindleNet to solve the problem of pedestrian mismatch caused by posture changes or occlusion by fusing the features of the seven local regions of interests of the human body [7].

Initially, in the calculation of pedestrian similarity, simple distance measurement methods such as cosine distance, Euclidean distance, etc. are usually used, without considering the primary and secondary relationships and importance between various dimensions. Therefore, many metric learning algorithms have been proposed by researchers to replace the way of distance measurement. Köstinger et al. propose a simple direct measurement (KISSME) algorithm [8]. Subsequently, Shengcai Liao and Yang Hu continue to improve on the basis of KISSME, and propose a cross-view quadratic discriminant analysis (XQDA) algorithm [2]. In metric learning based on convolutional neural networks, many models use the loss function as a sample similarity metric function, which reduces the intra-class distance and increases the inter-class distance through supervised learning, thereby improving discrimination. The triplet loss function is widely used in discriminant networks. On the basis of triplet loss, Hermans et al. [9] introduce hard case mining strategy and propose a hard sample mining triplet loss (TriHard Loss). This method is used in each batch of training samples. Chen et al. [10] add a negative sample in the triple, and propose a quadruple loss function, which considers the absolute distance between positive and negative samples, making the network have a better ability to express features.

For the re-identification of pedestrians in noisy environments, a more intuitive solution is to use image enhancement technology for noise reduction. However, there are two problems with this solution: one is that there is a problem of loss of details in image enhancement, which makes it difficult to extract features; the other is that the steps are cumbersome and it is impossible to achieve pedestrian feature extraction end-to-end. In addition, the generalization ability of the conventional triplet metric learning network, shown in Fig. 1, will be significantly reduced under noisy conditions even if it has a better clustering effect on different samples. The main reason is that the training and test samples are noisy images, and the difference between the samples is small, which makes the model unable to obtain sufficient discrimination.

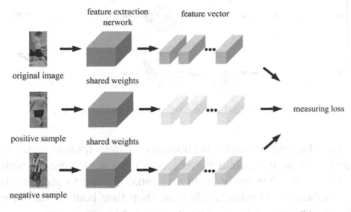

Fig. 1. Structure diagram of triplet measurement model

Therefore, in this paper, the characteristics of image noise is analyzed in surveillance videos and find that the decomposition characteristics based on dual-domain filtering can well reflect the characteristics of image noise. The dual-domain filter decomposition is used to generate two decomposition graphs, together with the original graph to form a triplet. The triplet is used to let the network learn the noise characteristics of the image and suppress the noise components in the output feature map to improve the robustness of pedestrian features to noise.

2   Triplet Measurement Model Based on Dual-Domain Filter Decomposition

Noise belongs to the high frequency part of the image. The denoising method based on dual-domain filtering has a good enhancement effect on the image. So the triplet network structure proposed in this paper is shown in Fig. 2.

2.1   Improved Triplet Network Architecture

Unlike the conventional triplet measurement model, this paper combines dual-domain filter decomposition and triple measurement learning to propose a new triplet construction method, using the original pedestrian image as a fixed triple, and the low-frequency image obtained by image decomposition as a positive sample image, and the high-frequency image as a negative sample image, which enables the network to learn the distribution of image noise and improves the ability of pedestrian features to generalize to noise [11].

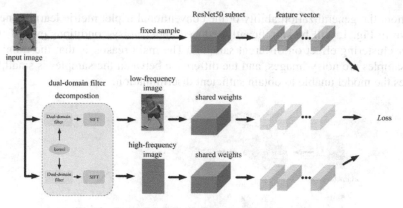

Fig. 2. Overall structure diagram of model training

Dual-domain filtering belongs to the transform-domain filtering methods. The basic idea is to transform the image into the frequency domain, denoise the high-frequency part of the image, and then restore it to the spatial domain to generate the denoised image. The combination of bilateral filter and short-time Fourier transform can better preserve the edge, texture and detail information of the image.

First, the input image I_x is subjected to joint bilateral filtering [18] as formula (1), to obtain the background layer image $\tilde{I}_x = F(I_x)$ and the guide layer $\tilde{I}_g = F(I_g)$.

$$F(I) = \frac{\sum\limits_{q \in S_p} k(p, q) I(p)}{\sum\limits_{q \in S_p} k(p, q)} \tag{1}$$

where S_p is the neighborhood window centered on the pixel point p and the radius r, $k(p, q)$ is the kernel function of the bilateral filter. At the first iteration, $I_g = I_x$ is consistent with the input image, and the filtering result will serve as the guide layer for the next iteration.

Then perform the short-time Fourier transform [12] as formula (2), to obtain $\tilde{I}_X = SF(I_x)$ and $\tilde{I}_G = SF(I_x)$, calculate the neighborhood residuals $\nabla I_x(p, q)$ and $\nabla I_g(p, q)$ of the input image I_x and the guide layer I_g at point p, and use the kernel function $k(p, q)$ to perform windowing to perform coefficient contraction.

$$SF(I) = \sum\limits_{q \in S_p} \exp\left(\frac{-i \cdot 2\pi (q - p) \cdot f}{2 \cdot r + 1}\right) \cdot k(p, q) \cdot \nabla I(p, q) \tag{2}$$

Use $\tilde{I}_G$ to construct the Gaussian kernel function $\tilde{k}(p, f)$, combined with $\tilde{I}_X$ to calculate the detail layer image $\tilde{I}_Z$, as shown below:

$$\tilde{I}_Z = \frac{1}{|F_p|} \sum\limits_{f \in F_p} \tilde{k}(p, f) \tilde{I}_X(p, f) \tag{3}$$

Among them, F_p is the frequency distribution matrix corresponding to $\tilde{I}_X$, $|F_p|$ is the number of elements of F_p.

The background layer $\tilde{I}_x$ and the detail layer $\tilde{I}_Z$ obtained above correspond to the required low-frequency image and high-frequency image, respectively.

2.2 Improved Loss Function

The loss function of the model in this paper mainly includes two parts: contrast loss and triplet loss. This paper has made corresponding improvements based on their basic forms. Because the image information contained in the low-frequency image and high-frequency image obtained by the dual-domain filter decomposition is not completely equal, when calculating the loss value with the original image, there will be a certain gap in the value range. Therefore, a weighting factor is added to adjust their relative size. Specifically, after feature calculation by the model in this paper, the feature vector f_A of the original image is the fixed image, the feature vector f_P of the positive sample image is the low-frequency image, and the feature vector f_N of the negative sample image is the high-frequency image.

The contrast loss can be expressed as:

$$L_C - \sum y \|f_A - f_P\|_2^2 + \lambda(1-y)\max\left(w - \|f_A - f_N\|_2, 0\right)^2 \tag{4}$$

Among them, the low-frequency graph corresponds to $y = 1$, the high-frequency graph corresponds to $y = 0$, and w refers to the minimum interval.

The triplet loss can make the positive and negative samples form an aggregation effect. It can be expressed as:

$$L_T = \sum \max\left(\|f_A - f_P\|_2^2 - \mu\|f_A - f_N\|_2^2 + w, 0\right) \tag{5}$$

Among them, w refers to the smallest interval, and the weighting factor u has the same effect as λ in the contrast loss, and is used to adjust the weight of the two losses in the low-frequency graph and the high-frequency graph.

Therefore, the overall loss function of the network can be expressed as:

$$L = L_C + L_T \tag{6}$$

2.3 Feature Extraction and Reordering

Figure 3 shows the flow chart of the triplet measurement model in the testing and application phases. In model testing, dual-domain filtering is not required for denoising. Pedestrian images containing noise are directly input into a branch network after training, and feature vectors of pedestrian samples are obtained end-to-end. This is due to the fact that the training triples of the model come from the dual-domain filter decomposition, which shows the distribution characteristics of noise well. It is based on these decomposition graphs that the simple feature extraction network can also learn the noise characteristics of the image and suppress the noise feature expression.

Considering the effects of lighting, posture, angle of view and occlusion, the order of matching results may not be accurate, so this paper uses the k-mutual neighbor reordering method [11] to expand the ranking table and improves the accuracy of re-identification.

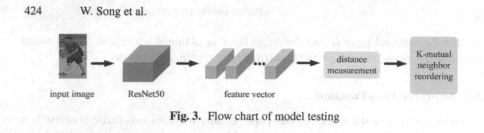

input image   ResNet50   feature vector   distance measurement   K-mutual neighbor reordering

Fig. 3. Flow chart of model testing

3 Experiment and Analysis

3.1 Experimental Platform and Parameter Setting

Market-1501 and CUHK03 datasets are used as training samples to optimize the model and to evaluate the performance. Two evaluation indicators are used to measure the performance of the algorithm, namely Rank-N and mAP.

The Caffe framework is used to implement our model. The hardware of the experimental platform is Intel Core i5-8300H CPU 2.30GHz, NVIDIA Tesla K40 GPU 12GB, 32GB RAM. The feature extraction backbone network in this paper is based on the ResNet50 network structure. The input pedestrian image is uniformly scaled to a resolution of 224 × 224. Each sub-network finally obtains 1024-dimensional feature vectors after maximum pooling and fully connected layers, which are used to calculate the loss. The size of the Mini-batch is set to 64 for each iteration period. The basic learning rate is 0.0002, and it decays exponentially every 100 cycles.

3.2 Experimental Results and Analysis

This paper mainly compares three types of representative pedestrian re-identification methods [13–17], which are traditional feature extraction method, deep learning feature fusion method and reordering method. The comparison algorithm specifically includes LOMO+XQDA [2], SpindleNet [7] and IDE_ResNet_50 [11]. In addition, this paper further compares with the recent deep learning-based [19] pedestrian re-identification algorithms SVDNet [20], APR [21], PIE [6].

Experimental Results on Market-1501. Table 1 lists the results of pedestrian re-identification experiments on the Market-1501 dataset. It can be observed that the mAP of the triplet measurement model designed in this paper is superior to most methods in different noise scenarios [22]. Rank-1 is slightly inferior to SVDNet and APR, and the main reason is that although SVDNet and APR are based on ResNet50 similar to this paper, this paper mainly adjusts the output structure of the network to adapt to the loss function calculation and it does not specifically improve the accuracy of the model, but focuses more on improving the robustness of the model to noisy scenes.

Figure 4 shows how the CMC curves of the three representative methods vary with the sample scenes. It can be observed from the figure that the triplet measurement model proposed in this paper is highly robust in different noise situations and can be closest to the performance of pedestrian re-identification in ideal scenes.

Experimental Results on CUHK03. Similar to the experiments on the Market-1501 dataset, Table 2 lists the results of the pedestrian re-identification and comparison experiments in different noise scenarios on the CUHK03 dataset. Compared with some recent

Table 1. Rank-1 and mAP of each method of Market-1501 dataset

Method	Original image		Gaussian noise		Salt and pepper noise		Rain noise	
	Rank-1	mAP	Rank-1	mAP	Rank-1	mAP	Rank-1	mAP
LOMO+XQDA [2]	43.8	22.7	37.5	19.5	39	21.6	35.7	20.3
SpindleNet [7]	76.9	–	72.4	–	74.4	–	73.7	–
IDE_ResNet_50 [11]	78.9	55	74.2	49.4	75.5	51.9	75.1	50.6
SVDNet [20]	82.3	62.1	78.6	58.4	80.3	59.2	78.1	57.5
APR [21]	**84.3**	64.7	**79.2**	60.2	**81.4**	61.9	**80.8**	60.8
PIE [6]	79.3	56.0	74.8	50.5	76.1	53.0	77.1	54.2
Ours	79.8	**68.3**	77.5	**65.8**	78.6	**67.1**	78.7	**67.8**

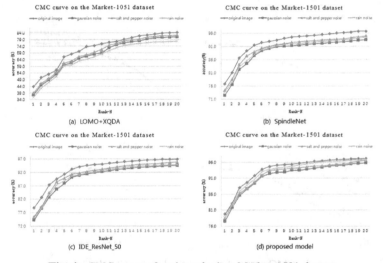

Fig. 4. CMC curve of each method on Market-1501 dataset

deep learning methods, the performance of this method on the CUHK03 dataset still has a certain gap, and we will further improve the network structure in subsequent studies.

Figure 5 also shows how the CMC curve obtained by three representative methods changes with the sample scenes on the CUHK03 dataset. The experimental results show that the CMC curve and mAP index of the triplet measurement model designed in this paper are less attenuated, and are more robust to different noises, and can be closest to the pedestrian re-identification index in the ideal scene.

Table 2. Rank-1 and mAP of each method of CUHK03 dataset

Method	Original image		Gaussian noise		Salt and pepper noise		Rain noise	
	Rank-1	mAP	Rank-1	mAP	Rank-1	mAP	Rank-1	mAP
LOMO+XQDA [2]	14.8	13.6	8.2	9.5	12.7	10.8	10.5	10.1
SpindleNet [7]	33.8	–	31.7	–	32.2	–	31.5	–
IDE_ResNet_50 [11]	22.2	21.0	18.4	17.6	19.5	18.8	18.9	17.8
SVDNet [20]	40.9	37.8	37.2	28.8	38.4	31.6	36.4	27.1
APR [21]	45.7	46.8	42.4	41.6	43.3	42.7	42.8	42.1
PIE [6]	34.2	31.1	31.4	25.8	33.1	29.6	29.4	22.6
Ours	23.5	22.7	21.1	19.8	21.4	20.2	22.5	21.5

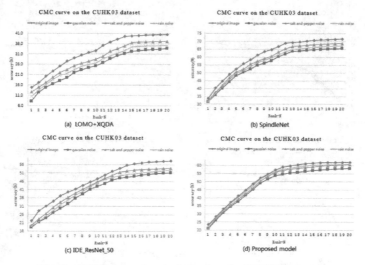

Fig. 5. CMC curve of each method on the CUHK03 dataset

4 Conclusion

This paper focuses on the problem of pedestrian re-identification of low-quality images in actual power grid surveillance system. By analyzing the characteristics of image noise in surveillance videos, it is proposed to use dual-domain filter decomposition to construct triples for training triple metric models, which makes it possible to extract pedestrian features that are robust to noise, and realize the re-identification of pedestrians in the surveillance video of the noise situations. Through the comparative analysis of experiments on public datasets, the effectiveness of this method for re-identification of noisy image pedestrians is verified, and the feature extraction of noise-recognized pedestrian re-identification is realized.

References

1. Xiao, J., Guo, H., Zhou, J., Zhao, T., Yu, Q., Chen, Y.: Tiny object detection with context enhancement and feature purification. Expert Syst. Appl. **211**, 118665 (2023)
2. Liao, S., Hu, Y., Zhu, X., Li, S.Z.: Person re-identification by local maximal occurrence representation and metric learning. In: Proceedings of the IEEE Conference on Computer Vision and Pattern Recognition, pp. 2197–2206 (2015)
3. Bai, Z., Wang, Z., Wang, J., Hu, D., Ding, E.: Unsupervised multi-source domain adaptation for person re-identification. In: Proceedings of the IEEE/CVF Conference on Computer Vision and Pattern Recognition, pp. 12914–12923 (2021)
4. Luo, H., Jiang, W., Zhang, X., Fan, X., Qian, J., Zhang, C.: AlignedReID++: dynamically matching local information for person re-identification. Pattern Recogn. **94**, 53–61 (2019)
5. Wang, T., Liu, H., Song, P., Guo, T., Shi, W.: Pose-guided feature disentangling for occluded person re-identification based on transformer. In: Proceedings of the AAAI Conference on Artificial Intelligence, vol. 36, pp. 2540–2549 (2022)
6. Zheng, L., Huang, Y., Lu, H., Yang, Y.: Pose-invariant embedding for deep person re-identification. IEEE Trans. Image Process. **28**(9), 4500–4509 (2019)
7. Zhao, H., et al.: Spindle Net: person re-identification with human body region guided feature decomposition and fusion. In: Proceedings of the IEEE Conference on Computer Vision and Pattern Recognition, pp. 1077–1085 (2017)
8. Koestinger, M., Hirzer, M., Wohlhart, P., Roth, P.M., Bischof, H.: Large scale metric learning from equivalence constraints. In: 2012 IEEE Conference on Computer Vision and Pattern Recognition, pp. 2288–2295. IEEE (2012)
9. Hermans, A., Beyer, L., Leibe, B.: In defense of the triplet loss for person re-identification. arXiv preprint arXiv:1703.07737 (2017)
10. Chen, W., Chen, X., Zhang, J., Huang, K.: Beyond triplet loss: a deep quadruplet network for person re-identification. In: Proceedings of the IEEE Conference on Computer Vision and Pattern Recognition, pp. 403–412 (2017)
11. Zhong, Z., Zheng, L., Cao, D., Li, S.: Re-ranking person re-identification with k-reciprocal encoding. In: Proceedings of the IEEE Conference on Computer Vision and Pattern Recognition, pp. 1318–1327 (2017)
12. Xiao, J., Zou, W., Zhang, S., Lei, J., Wang, W., Wang, Y.F.: Video denoising algorithm based on improved dual-domain filtering and 3D block matching. IET Image Proc. **12**(12), 2250–2257 (2018)
13. Aharon, M., Elad, M., Bruckstein, A.: K-SVD: an algorithm for designing overcomplete dictionaries for sparse representation. IEEE Trans. Signal Process. **54**(11), 4311–4322 (2006)
14. Huang, D.A., Kang, L.W., Yang, M.C., Lin, C.W., Wang, Y.C.F.: Context-aware single image rain removal. In: 2012 IEEE International Conference on Multimedia and Expo, pp. 164–169. IEEE (2012)
15. Kang, L.W., Lin, C.W., Fu, Y.H.: Automatic single-image-based rain streaks removal via image decomposition. IEEE Trans. Image Process. **21**(4), 1742–1755 (2011)
16. Luo, Y., Xu, Y., Ji, H.: Removing rain from a single image via discriminative sparse coding. In: Proceedings of the IEEE International Conference on Computer Vision, pp. 3397–3405 (2015)
17. Son, C.H., Zhang, X.P.: Rain removal via shrinkage-based sparse coding and learned rain dictionary. arXiv preprint arXiv:1610.00386 (2016)
18. Xiao, J., Shen, M., Lei, J., Zhou, J., Klette, R., Sui, H.: Single image dehazing based on learning of haze layers. Neurocomputing **389**, 108–122 (2020)
19. Zeng, K., Ning, M., Wang, Y., Guo, Y.: Hierarchical clustering with hard-batch triplet loss for person re-identification. In: Proceedings of the IEEE/CVF Conference on Computer Vision and Pattern Recognition, pp. 13657–13665 (2020)

20. Sun, Y., Zheng, L., Deng, W., Wang, S.: SVDNet for pedestrian retrieval. In: Proceedings of the IEEE International Conference on Computer Vision, pp. 3800–3808 (2017)
21. Lin, Y., et al.: Improving person re-identification by attribute and identity learning. Pattern Recogn. **95**, 151–161 (2019)
22. Zheng, L., Shen, L., Tian, L., Wang, S., Wang, J., Tian, Q.: Scalable person re-identification: a benchmark. In: Proceedings of the IEEE International Conference on Computer Vision, pp. 1116–1124 (2015)

Performance Analysis of Free Text Keystroke Authentication Using XGBoost

Ievgeniia Kuzminykh[1,2]($\boxtimes$), Saransh Mathur[1], and Bogdan Ghita[3]

[1] Department of Informatics, King's College London, 30 Aldwych, London WC2B 4BG, UK
ievgeniia.kuzminykh@kcl.ac.uk
[2] Department of Infocommunication Engineering, Kharkov National University of Radio Electronics, Nauki Av. 14, Kharkiv 61000, Ukraine
[3] School of Engineering, Computing and Mathematics, University of Plymouth, Drake Circus, Plymouth PL4 8AA, UK

Abstract. Authentication based on keystroke dynamics is a form of behavioral biometric authentication that uses the user typing patterns and keyboard interaction as a discriminatory input. This type of authentication can be coupled with a fixed text password in a traditional login system to contribute to a multifactor authentication or provide continuous user authentication in a usable security system, where the typing patterns are continuously analysed to validate the user at run time. This paper investigates the effectiveness of free text keystroke for continuous authentication in real-world systems. Evaluation is performed using XGBoost multiclass classification, applied to an unbalanced free-text keystroke dataset. The introduction of additional activity-based features and removal of inaccuracies in the timing between keys allowed a reduction of the EER for the Clarkson II dataset from 14–24%, as achieved by previous studies, to 8% when employing the proposed method.

Keywords: Usable security · Keystroke dynamics · Continuous authentication · XGBoost

1 Instruction

Recent years witnessed an increasing interest towards unobtrusive and usable security systems as the surface area of the Internet continuously expands, particularly to accommodate transparent authentication systems. In this context, keystroke dynamics provides a solid solution and has become the de-facto solution during the past decade. Nevertheless, it is yet to see a promising mass-scale deployment for keystroke dynamics-based authentication systems due to a number of inherent challenges, including the need for an extensive analysis of the keystroke dynamics-based authentication systems to be put in use and, amongst those, to establish if the systems are accurate enough to deploy. A significant number of studies, such as [1–4], investigated the feasibility of keystroke dynamics as a unique or additional means for user continuous authentication. Amongst their conclusions, the authors agreed that performance of the free-text dynamics for user

© The Author(s), under exclusive license to Springer Nature Switzerland AG 2023
Z. Hu et al. (Eds.): ICCSEEA 2023, LNDECT 181, pp. 429–439, 2023.
https://doi.org/10.1007/978-3-031-36118-0_39

authentication is always worse when compared to specific text, such as username and password, as the variability of the input, has a significant impact on accuracy. In this study we also investigate whether the performance of free-text keystroke authentication can be improved by implementing a novel set of features for free-text keystroke dynamics.

The aim of this study is to investigate the applicability of free-text for biometric authentication with applying additional activity-based characteristics. One of the challenges encountered was that the initial sanitising of the dataset led to inaccuracies in the timing between successive keys; additional filtering was necessary to remove these artifacts and correct timing. This allowed us to reduce the EER for Clarkson II dataset from 14–24% that was achieved by the authors in [5] to 8%.

2   Related Works

The summary of the related studies is presented in Table 1 and shows the machine learning algorithms that researchers were used for classification, the dataset used and what type of keystroke, free text or fixed text, was under experiment.

A recent study by Daribay et al. [6] focused on the performance evaluation of a fixed-text keystroke dynamics showed promising results with XGBoost Classifier giving an accuracy of 90.91%. XGBoost is a powerful machine learning algorithm, but it requires parameter tuning to leverage the full potential, which was missing from this analysis. While the overall result is encouraging, it does not address the continuous authentication aspect due to the fixed text employed.

Baynath et al. [7] further tested the large-scale applicability of keystroke dynamics, a dataset size for this study was also way larger than the previous ones, as they worked on a combination of the Killourhy Database (CMU database) [3] and their own inbuilt database consisting of fixed text of four different strong passwords. One of the most important conclusions of the study was that the cost of implementation for such system remains low even for large datasets, both computationally and financially.

There have also been some efforts to increase the usability component of continuous authentication, such as the unsupervised approach by Ananya and Singh [8] which did not require any preregistration or the method proposed by Sim et al. [9], where the mouse could be mounted with a fingerprint sensor for initial login and then periodic re-verification in conjunction with continuous authentication via keystroke dynamics. Since eliminating the onboarding process does not seem feasible, one way to increase the usability can be by reducing the number of keystrokes or inputs required from the user itself, particularly appealing for free-text keystroke systems, but prone to error as keystroke sample size is directly correlated with accuracy. To solve this issue, one can use the strategy adopted by Ayotte et al. in [10], aiming for frequent and cumulative authentication, using flexible thresholds. The authors proposed an instance-based tail area density (ITAD) metric to help reduce the number of keystrokes required to perform authentication.

Several other works [2, 3, 11–13] studied using keystroke dynamics for user continuous authentication. In works [11, 12] the authors considered continuous authentication for users who remotely access the desktop machine via RDP protocol using distance-based algorithms to identify differences in the keystroke patterns. The authors of study

[13] collected and used a mixed dataset, consisting of fixed-text and free text keystrokes, which was used by many subsequent studies [10, 14, 15].

Table 1. Related studies and their results (in the chronological order)

Study	Dataset, subjects	Type	Classifier	Metrics	Accuracy, %
A. Lo et al. 2020 [16]	133	Fixed	RF, SVM, Manhattan, Euclidian	Accuracy	74.4–95.6
S. Singh et al. (CMU) 2020 [17]	51	Fixed	KNN, SVC (RBF), RF, XGBoost	Accuracy	70.4–93.6
A. Daribay et al. [6] 2019	51	Fixed	XGBoost	Accuracy	90.91
K. Elliot et al. 2019 [18]	23	Fixed	RF, NN, DT, SVM	Accuracy	71–100
C. Murphy, et al. (Clarkson II) 2017	103	Free	Degree of disorder, n-graph reject ratio	EER	88.64
A. Bansal 2016 [19]	5	Free	GMM	Accuracy	78.4
Y. Sun et al. (Buffalo) 2016 [13]	148	Free	GMM	EER	96.6
A. Darabseh et al. 2015 [20]	28	Fixed	KNN, SVM	Accuracy	81–84
E. Vural et al. (Clarkson I) 2014 [21]	39	Mixed	Degree of disorder, n-graph reject ratio	FAR, FRR, EER	96.9
J. Roth (MSU) 2014 [22]	51 / 30	Fixed / Free	Distance, n-graph reject ratio	EER	94.5
A. Messerman et al. 2011 [23]	55	Free	Degree of disorder	FAR, FRR	N/A
K. Killourhy and Maxion (CMU) 2009 [3]	51	Fixed	Manhattan, Mahalanobis and 12 more	EER, FAR	63–90

(continued)

Table 1. (*continued*)

Study	Dataset, subjects	Type	Classifier	Metrics	Accuracy, %
C. Loy, et al. 2007 [24]	100	Fixed	ARTMAP- FD	EER	88
D. Gunetti and C. Picardi 2005 [25]	40	Free	Distance, n-graph reject ratio	FAR, IPR	N/A
D.T. Lin 1997 [26]	125	Fixed	BPNN	FAR, IPR	N/A
R. Joyce and G. Gupta 1990 [27]	27	Fixed	Manhattan (filtered)	IPR, FAR	86.7

To summarise, it is apparent that the area of free-text based keystroke dynamics, although extremely capable, has presented less interest, particularly due to its potentially higher computational demands and error rates. Most of the studies test the practicality of a keystroke dynamics-based authentication system using the password-based fixed-text, which cannot be used for continuous, transparent authentication. Within the free text keystroke dynamics field, studies focused on a similar feature-set, including digraphs and n-graphs. Although, these methods have generally resulted in an accuracy approaching almost 90% most of the time, the potential of other types of features remains untapped.

3   Methodology

This study aims to investigate and improve the accuracy of free text keystroke dynamics as a method for user authentication by introducing an additional, activity-based set of features, which were not used by prior research relating to free text patterns. The proposed features are derived from the physical characteristics of the keyboard; the user profile aggregates the hold times of specific keys in combination with their location. The process of finding out how good or bad such system would perform is done through XGBoost classification algorithm. The methodology, outlined in Fig. 1, consists of data pre-processing to filter inaccurate data, followed by feature extraction, classification and evaluation steps. Based on the conclusions of prior research, combined with the emergency of better performing algorithms, the process is using XGBoost for classification.

3.1   Dataset

Given the requirements of the study and the available datasets, the Clarkson II dataset [19] was used as input. This dataset includes keystroke timing information for 103 subjects in a completely uncontrolled environment collected over a period of 2.5 years using a keylogger tool installed on each computer to record user interaction. The subjects used different hardware and OS platforms, different keyboards, different browsers, different

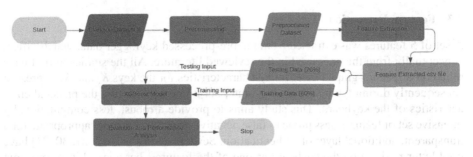

Fig. 1. Flowchart for implementation

software, and even different tasks. For each key interaction, the dataset contains the user ID, the key event (0 if pressed and 1 if released), the timestamp, logged in .NET ticks, and the key name, as shown below in Table 2.

Table 2. Sample of dataset

User ID	Time Stamp (ticks)	Action Type	Key Name
4302075	636172286538589004 (2016-12-13 12:24:13)	'KeyDown'	'A'
4302075	636172286539669002 (2016-12-13 12:24:13)	'KeyDown'	'Space'
4302075	636172286541684820 (2016-12-13 12:24:14)	'KeyUp'	'A'

3.2 Data Sanitation and Pre-processing

As discussed in [10] and [16], the performance of algorithms on the Clarkson II dataset compared to other more controlled free text datasets is always worse. The root cause of the difference is the fact that users have a specific pattern when typing text sequences that they are familiar with, such as usernames and passwords, but use less discriminative typing for free text. The Clarkson database includes an additional element of error, as the authors sanitised the text to remove any sensitive sequences; in the process, timing of adjacent key presses was therefore also affected. To alleviate these artifacts, the dataset was filtered to remove incomplete patterns or accidentally pressed keys.

Based on the ability to filter and the removal of incomplete data (such as key events where only the key press was registered, with no key release), 24 users were removed from the dataset, leaving a total of 79 users with data suitable for analysis. The pre-processing also converted the .NET ticks to a *YYYY-MM-DD T HH:MM:SS.zzzz* format as well as replaced the key names with their respective ASCII values.

3.3 Feature Extraction

A set of 5 features was extracted from the pre-processed keylogger data, which differs conceptually from the ones used in the reviewed literature. All the studies in past have relied on digraph/n-graph based timing characteristics for the keys K_i and K_{i+1} pressed subsequently during typing. This is accurate, but it does not capture the physical characteristics of the keyboard. This study aims to provide a robust, less computationally intensive set of features, less prone to false negative errors, hence more appropriate for a transparent, additional layer of authentication. Several studies [11, 12, 21, 30, 31] have used letter position on the keyboard as one of the features, but applied it to fixed text keystroke dynamics. Besides features based on how a user's hand interacts with letters on the keyboard, the dataset also captures the use of shift and CAPSLock, as well as backspace and space keys are also assigned their separate features. All the features are based on average hold times and calculated per 10 min of keylogging. The labelled features are shown in the Table 3.

Table 3. Features extracted from keylogger data

Name	Label	Description
l	f2	Hold time (ms) of keys at left part without shift
r	f3	Hold time (ms) of keys at right part without shift
L	f0	Hold time (ms) of keys at left part with shift
R	f1	Hold time (ms) of keys at right part with shift
SPACE	f4	Hold time (ms) of Space key
BACKSPACE	f5	Hold time (ms) of Backspace key

The keyboard was divided into two parts- left and right - as per the traditional placement of a hand on the keyboard, as shown in Fig. 2.

Fig. 2. Finger positions on a keyboard

3.4 Machine Learning Algorithms

The dataset, after filtering, includes 79 users and the process is aiming for authentication, the classification problem is an imbalanced multiclass classification. Based on the existing studies, as summarised in Sect. 2, XGBoost appears to be one of the most promising and successful in resolving such problems [31, 32] and is a public domain classifier, hence, it will be used as part of this study. Gradient boosting (GB) methods are usually very powerful classifiers because of the ensemble training techniques that typically perform very well on unbalanced data. Amongst the available solutions, the sklearn library includes an excellent Python implementation of this algorithm, which was also used for this study.

For creating an XGBoost classifier model, the dataset must be split into input and target arrays. The input array contains all the feature rows, while the target array includes the corresponding entries for usernames. The input array is assigned to a variable 'x' and target array is assigned to a variable 'y'. We also split the data into training and testing datasets in proportion 80% and 20%, respectively.

4 Results and Analysis

4.1 Feature Importance

It is essential to evaluate the relative contribution and efficiency of the features as discriminants for the classes of outcomes of the model. Conducting the evaluation on the model gave the following results presented on Fig. 3.

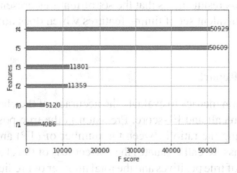

Fig. 3. Feature importance

The labels from f0 to f5 correspond to the features from Table 3. The F scores are the relevance indicators of the respective features when applied to the decision tree; the higher the F score of a feature, the higher its relative importance.

According to Fig. 3, the combination of *space* and *backspace* average hold-time features have the highest importance in predicting the users. This is followed by the *l* and *r* features, with the lowest importance being for that of *L* and *R*. The low F Scores for *L* and *R* features are justifiable, as the frequency of uppercase characters is significantly lower than their lowercase counterparts. One possible explanation for the *backspace* feature is that the typing mistakes a user makes while typing are a good discriminator for the behaviour of different users. As, for *space*, having the highest importance could be a result of the frequency of use of space bar in general. Given their limited impact, the *L* and *R* features may be removed from the list of inputs with a minimal impact on the model accuracy; to be traded for a significant reduction in computational complexity.

4.2 Classification Accuracy

The accuracy of model is the primary concern in classification model that gives us a fraction for the samples that were predicted correctly. The accuracy can be calculated using Eq. (1):

$$Accuracy = \frac{TP + TN}{TP + TN + FP + FN} \tag{1}$$

where True Positive (TP) is the rate of correct positive predictions, True Negative (TP) is the rate of correct negative predictions, False Positive (FP) is the rate of incorrect positive prediction, and False Negative (FN) is the rate incorrect negative predictions.

In order to get an accurate indication of the ability of the model to generalise, the testing subset accuracy is also determined. After training the model with the training subset, test subset to the model, the model was tested against the testing subset. The results indicated an accuracy of is 91.72% for the testing subset. This accuracy is among highest among the related studies from Table 1 in area of free-text keystroke dynamics, and highest for such a large, unbalanced dataset. The original paper that used the Clarkson II dataset reported an accuracy of 88.64% and a 10.36% EER, while our study showed only 8.28% EER. This result shows that the set of features presented in Table 3 is more effective than the equivalent set of timing features which does not take into account the key location.

4.3 Classification Report

We used three common metrics to evaluate the accuracy of the classifiers for each of the 79 users: precision, recall and F1-score. Precision is the true positive predictive value of a class, representing the ratio between the number of (TP) and the total number of predicted positive class. Recall evaluates the correctness of the class, defined as the ratio between the number of true positives and the total number of predictions of the respective class. The F1-score is the harmonic mean of recall and precision.

The results presented in Table 4 showed that the precision, recall, and F1 score for most of the users are high (0.85–1.0), except for those where support, which represents number of actual occurrences of the class in our dataset, has low values, as there are not enough training samples for these users to form a unique enough signature. As the dataset used is imbalanced, support in the training data will differ for each class. Support doesn't change between models but instead influences the evaluation process.

Table 4. A snippet of the results of classification for 6 users

User	Precision	Recall	F1-Score	Support
0	0.98	0.98	0.98	165
1	0.95	0.87	0.91	112
2	0.88	0.51	0.64	55
3	0.88	0.90	0.89	1110
16	0	0	0	1
17	0	0	0	5

From the report, it is apparent that classes with a lower number of samples are generally being predicted incorrectly and perform poorly. From a modelling perspective, there is not enough training data for these users to form a unique enough signature, hence they are classified as belonging to other classes. The users with few samples are the ones contributing the most to the False Positives for the other users, and False Negative for their own classes.

5 Conclusion

This study analyses the efficiency of deploying free text keystroke dynamics-based authentication as a continuous authentication method. The results showed an overall accuracy of 91.72%, and up to 98% for unique users, and the XGBoost based classifier can be implemented as a continuous authentication system, which ensures that the user does not change after an initial sign-on.

The paper proposes a user classification approach using a novel features set for free text keystroke dynamics, focused on the positioning of keys on the keyboard. The additional features, with the average hold-times of frequently pressed keys (backspace, space bar, and shift), lead to a higher accuracy in comparison with prior studies.

The results also indicated that accuracy for specific users is highly dependent on the amount of training data available, therefore users with a limited amount of data are likely to be incorrectly classified by the system.

References

1. Embroker Team: 2022 Must-Know Cyber Attack Statistics and Trends, Embroker Business and Advice Research, 31 January 2022. https://www.embroker.com/blog/cyber-attack-statistics. Accessed 15 Mar 2022
2. Kang, P., Cho, S.: Keystroke dynamics-based user authentication using long and free text strings from various input devices. Inf. Sci. **308**, 72–93 (2015)
3. Killourhy, K.S., Maxion, R.A.: Comparing anomaly-detection algorithms for keystroke dynamics. In: Proceedings of the IEEE/IFIP International Conference on Dependable Systems and Networks, pp. 125–134 (2009)

4. Leggett, J., Williams, G., Usnick, M., Longnecker, M.: Dynamic identity verification via keystroke characteristics. Int. J. Man Mach. Stud. **35**(6), 859–870 (1991)
5. Lu, X., Zhang, S., Hui, P., Li, P.: Continuous authentication by free-text keystroke based on CNN and RNN. Comput. Secur. **96**, 101861 (2020)
6. Daribay, A., Obaidat, M.S., Krishna, P.V.: Analysis of authentication system based on keystroke dynamics. In: Proceedings of the International Conference on Conference on Computer, Information, and Telecommunication Systems (CITS), pp. 1–6 (2019)
7. Baynath, P., Soyjaudah, K.M.S., Khan, M.H.-M.: Machine learning algorithm on keystroke dynamics pattern. In: Proceedings of the IEEE Conference on Systems, Process and Control (ICSPC), pp. 11–16 (2018)
8. Ananya, Singh, S.: Keystroke dynamics for continuous authentication. In: Proceedings of the 8th International Conference on Cloud Computing, Data Science and Engineering (Confluence), pp. 205–208 (2018)
9. Sim, T., Zhang, S., Janakiraman, R., Kumar, S.: Continuous verification using multimodal biometrics. IEEE Trans. Pattern Anal. Mach. Intell. **29**(4), 687–700 (2007)
10. Ayotte, B., Banavar, M., Hou, D., Schuckers, S.: Fast free-text authentication via instance-based keystroke dynamics. IEEE Trans. Biom. Behav. Ident. Sci. **2**(4), 377–387 (2020)
11. Kuzminykh, I., Ghita, B., Silonosov, A.: On keystroke pattern variability in virtual desktop infrastructure. In: Proceedings of the International workshop on Computer Modeling and. Intelligent Systems, pp. 238–248 (2021)
12. Kuzminykh, I., Ghita, B., Silonosov, A.: Impact of network and host characteristics on the keystroke pattern in remote desktop sessions. arXiv:2012.03577 [cs] (2020)
13. Sun, Y., Ceker, H., Upadhyaya, S.: Shared keystroke dataset for continuous authentication. In: Proceedings of the IEEE International Workshop on Information Forensics and Security (WIFS), pp. 1–6 (2016)
14. Kiyani, A.T., Lasebae, A., Ali, K., Rehman, M., Haq, B.: Continuous user authentication featuring keystroke dynamics based on robust recurrent confidence model and ensemble learning approach. IEEE Access **8**, 156177–156189 (2020)
15. Huang, J., Hou, D., Schuckers, S., Law, T., Sherwin, A.: Benchmarking keystroke authentication algorithms. In: IEEE Workshop on Information Forensics and Security (WIFS), pp. 1–6 (2017)
16. Lo, A., Ayma, V.H., Gutierrez-Cardenas, J.: A comparison of authentication methods via keystroke dynamics. In: Proceedings of the IEEE Engineering International Research Conference (EIRCON), pp. 1–4 (2020)
17. Singh, S., Inamdar, A., Kore, A., Pawar, A.: Analysis of algorithms for user authentication using keystroke dynamics. In: Proceedings of the International Conference on Communication and Signal Processing (ICCSP), pp. 0337–0341 (2020)
18. Elliot, K., Graham, J., Yassin, Y., et al.: A comparison of machine learning algorithms in keystroke dynamics. In: Proceedings of the International Conference on Computational Science and Computational Intelligence (CSCI), pp. 127–132 (2019)
19. Murphy C, Huang J, Hou D, Schuckers S. Shared dataset on natural human-computer interaction to support continuous authentication research. Proceedings of the IEEE International Joint Conference on Biometrics (IJCB), pp. 525–530 (2017)
20. Bansal, A.: keystrokeDynamics/Readme, March 2017. https://github.com/ankiteciitkgp/key strokeDynamics/blob/master/Readme.md. Accessed on 15 Mar 2022
21. Darabseh, A., Namin, A.S.: On accuracy of classification-based keystroke dynamics for continuous user authentication. In: Proceedings of the International Conference on Cyberworlds, pp. 321–324 (2015)
22. Vural, E., Huang, J., Hou, D., Schuckers, S.: Shared research dataset to support development of keystroke authentication. In: Proceedings of the IEEE International Joint Conference on Biometrics, pp. 1–8 (2014)

23. Roth, J., Liu, X., Metaxas, D.: On continuous user authentication via typing behavior. IEEE Trans. Image Process. **23**(10), 4611–4624 (2014)
24. Messerman, A., Mustafifa, T., Camtepe, S.A., Albayrak, S.: Continuous and non-intrusive identity verification in real-time environments based on free-text keystroke dynamics. In: Proceedings of the IEEE International Joint Conference on Biometrics, pp. 1–8 (2011)
25. Loy, C.C., Lai, W.K., Lim, C.P.: Keystroke patterns classification using the ARTMAP-FD neural network. In: Proceedings of the International Conference on Intelligent Information Hiding and Multimedia Signal Processing, vol. 1, pp. 61–64 (2007)
26. Gunetti, D., Picardi, C.: Keystroke analysis of free text. ACM Trans. Inf. Syst. Secur. (TISSEC) **8**(3), 312–347 (2005)
27. Lin, D.-T.: Computer-access authentication with neural network based keystroke identity verification. In: Proceedings of the International Conference on Neural Networks, vol. 1, pp. 174–178 (1997)
28. Joyce, R., Gupta, G.: Identity authentication based on keystroke latencies. Commun. of the ACM **33**(2), 168–176 (1990)
29. Alsultan, A., Warwick, K., Wei, H.: Improving the performance of free-text keystroke dynamics authentication by fusion. Appl. Soft Comput. **70**, 1024–1033 (2018)
30. Singh, P.I.: Robust security system for critical computers. Int. J. Inf. Technol. Comput. Sci. (IJITCS) **4**(6), 24–29 (2012)
31. Dhar, P., Guha, S.: Fish image classification by XgBoost based on Gist and GLCM Features. Int. J. Inf. Technol. Comput. Sci. (IJITCS) **4**, 17–23 (2021)
32. Manju, N., Harish, B.S., Prajwal, V.: Ensemble feature selection and classification of internet traffic using XGBoost classifier. Int. J. Inf. Technol. Comput. Sci. (IJITCS) **7**, 37–44 (2019)

Cyberbullying Detection and Classification in Social Media Texts Using Machine Learning Techniques

Joseph D. Akinyemi[1]([⊠]), Ayodeji O. J. Ibitoye[2], Christianah T. Oyewale[1], and Olufade F. W. Onifade[1]

[1] Department of Computer Science, University of Ibadan, Ibadan, Nigeria
jd.akinyemi@ui.edu.ng, jd.akinyemi@mail1.ui.edu.ng
[2] Department of Mathematical and Computing Science, Kola Daisi University, Ibadan, Nigeria

Abstract. Cyberbullying (CB) is both a public health concern as well as an emotional problem. There have been efforts to mitigate this problem from different discipline dimensions. Artificial Intelligence (AI) has recently emerged as a solution to CB and outshined many earlier solutions. In this work, we have investigated the problem of CB on social media platforms using Machine/Deep Learning and Natural Language Processing (NLP) techniques. Using a dataset containing 47,692 tweets, we investigated the task of detecting CB from social media posts and classifying them as either age-based, religious, ethnic, and political CB or neutral (non-CB). We spot-checked 5 Machine Learning (ML) algorithms (Gradient Boosting, Logistic Regression, Naïve Bayes, Random Forest, and Support Vector Machine) and one Deep Learning algorithm (a sequence model). The algorithms were evaluated based on accuracy, precision, recall, and F1 score. Random Forest reported the best accuracy of 93% while Naive Bayes reported the worst accuracy of 84%, while the DL model had a classification accuracy of 91%. The developed models can help detect and classify CB sentiments in social media posts, thus reducing the harm caused by CB in the social media space.

Keywords: Cyberbullying · Deep Learning · Machine Learning · NLP · Social media

1 Introduction

In recent times, irrespective of one's color, ethnicity, age, or location, online information sharing has continued to be more popular, especially, among "generation Z" who spend their time on different social networking sites like Snapchat, TikTok, Twitter, etc. for different purposes. While people act and react to perspectives differently, nearly anyone can abuse and hurt other users through the publishing of defamatory remarks, which could spread as rumors. Bullying is defined as the use of threat, force, or compulsion to maltreat, intimidate, or overpower others [1]. It is common to engage in such behavior, which is characterized by an imbalance of power and may involve physical or verbal abuse as well as threats or intimidation. Among other types of bullying, the use of electronic devices to

© The Author(s), under exclusive license to Springer Nature Switzerland AG 2023
Z. Hu et al. (Eds.): ICCSEEA 2023, LNDECT 181, pp. 440–449, 2023.
https://doi.org/10.1007/978-3-031-36118-0_40

send threatening and insulting communications with the goal to harass or injure others is known as cyberbullying [1]. Additionally, CB is defined as "willful and repetitive violence committed through the use of computers, cell phones, and other electronic devices" by [2]. These wick actions could include but are not limited to sending, posting, or sharing negative, harmful, false, or mean content about someone else using text, audio, videos, and images respectively. It can also include sharing personal or private information about someone else to cause embarrassment or humiliation. These actions no doubt could have negative effects like anxiety, depression, fear, and in the long run, suicide on the victims. Due to its effects on adolescents' emotional and developmental health, CB among adolescents is really regarded as a public health issue [3]. Surprisingly, the likelihood of getting rid of these contents is oftentimes seemingly impossible.

Due to unrestrained communication at any time, and with a large number of individuals at once, social lifestyles now exist in multifaceted contexts in the digital space. While Internet-based sharing, learning, and social interactions are generally believed to help increase awareness of its potential advantages, its negative effects like CB may outweigh its positive effects [4]. With people relating freely and uncontrollably with others in the digital space, a continuous need to identify and analyse potential bullying is inevitable because CB has become one of the dark threats of the Internet [5]. Consequently, it is relatively important to develop pattern-driven approaches to identify and prevent CB. The essence of the proposed system will be to minimise CB attacks and its effect on innocent young individuals. Hence, the goal of the research is to detect and classify social media texts using machine learning techniques into age-based, religious, ethnic, and gender-based CB or non-CB contents. This can then become useful for determining the presence of psychologically harmful contexts within social media posts and possibly flagging their author(s). The rest of this work is organised as follows; in Sect. 2, related works on CB are presented, the developed models are described in Sect. 3, experiments, results, and discussions are presented in Sect. 4, and the conclusion and future works are presented in Sect. 5.

2 Related Previous Works

Before the advent of technology and the Internet, investigations on traditional (physical) bullying and its impacts have been ongoing for about 40 years [6]. However, the focus of this study switched from nominal bullying to CB with the introduction of personal computers, which were widely accessible and reasonably priced in the 1990s, it may be claimed that traditional bullying has made its way into the technological world. Conversely, it was not until it started to cause grave consequences like despair and even death that CB started to garner genuine attention as a problem. By this time, there have been several research directions and models built for CB detection and prevention in the digital space. In the works of [7–9], studies were conducted on how frequently children use the internet to send instant messages, emails, and social networking sites, as well as the various experiences with harassment received through the internet. This could include, for example, sending the victim the same messages repeatedly in a way that can have an emotional impact on the victim, harassing and creating fear through textual expressions, and sending scary images and videos. These data are difficult to collect due to data privacy regulations [10]. With the freedom of expression through text, by using text mining techniques, [11] created a model for automatically identifying instances of CB on social networks. [12] also used a variety of classifiers, including AdaBoost, Light Gradient Boosting Machine (LGBM), Support Vector Machine (SVM), Random Forest, Stochastic Gradient Descent, Logistic Regression (LR), and Multinomial Naïve Bayes for the purpose of identifying CB incidents in tweets. [13] employed a pre-trained BERT model to identify CB on social media networks. In [14], Term Frequency-Inverse Document Frequency (TF-IDF) vectorizer was used to extract features and two classifiers, SVM and Naïve Bayes (NB) were used to detect CB on a Twitter dataset with accuracies of 71.25% and 52.75% respectively. By integrating the social media and textual content features respectively using information gained for feature ranking, [15] developed an approach for identifying CB. In [16], a transformer network-based word embedding approach was used to solve the limitations of conventional word embedding techniques by using RoBERTa to generate word embeddings and LGBM to classify CB tweets. A deep learning-based models based on 1D convolutions was used in [17] to characterise CB in Twitter and YouTube posts.

With the goal to classify hate speech, which could be a product of CB, [18] developed a deep convolutional neural network on Twitter dataset methodology to forecast tweets that contain hate speech. They experimented with a number of Machine Learning algorithms and adopted SVM to predict hate speech in tweets with an accuracy of 53%. An ML model that recognises CB across many social media sites was suggested by [19] using Long Short-term Memmory (LSTM) layers on data from Formspring, Twitter, and Wikipedia. Similarly, [20] performed an analysis to compare classification algorithms like NB, Decision Tree, LR and SVM for detecting CB crime using the Indonesian Twitter social media. As in previous studies, SVM had the best performance with an accuracy of 76% and F-score of 82%. Other methods which have been used include text mining techniques [21], Multinomial NB [22] and LSTM [23].

In this work, we investigated the performance of 6 ML models (including a DL model) on a fairly large (47,692) dataset of Twitter posts and compared their performances on the dataset.

3   The Cyberbullying Detection and Classification Models

In this work, the CB detection and classification models built were based on five Machine Learning (ML) models and one Deep Learning (DL) model. The ML models were selected to cut across most of the popular types of supervised learning algorithms (i.e., the probabilistic, the tree-based, and the kernel-based models). The DL model was primarily a sequence model.

Apart from the learning algorithms, we also used different feature representation algorithms to vectorize the text and create word embeddings. From empirical experiments, we found the Term Frequency-Inverse Document Frequency (TF-IDF) vectorizer [24] to be more efficient for the ML models and the pre-trained Glove word embeddings [25] to be more efficient for the sequence DL model.

The generic process followed in building our CB detection and classification models is described as follows. First, input texts from the dataset were preprocessed by cleaning, tokenizing, and padding the text, and then an appropriate word embedding (TF-IDF or Glove) is generated based on the specific class of learning algorithms (ML or DL) to be used. The word embedding, which is now a vector of numeric values is passed into the learning algorithm which detects whether the input contains CB sentiments and classifies them accordingly.

3.1 Dataset

The dataset used in this work was authored by [26] and obtained from Kaggle (kaggle.com). It contains 47,692 tweets which covered 4 specific categories of CB (age, gender, ethnicity, religion), one unspecified category of CB, and non-CB posts. The dataset was stored in a Comma-Separated Value (CSV) file containing two columns - one for the post and the other for the corresponding CB label of the post.

The dataset was loaded from the CSV file into a Pandas DataFrame and thereafter preprocessed. The dataset was preprocessed using python libraries such as Demoji (to remove the emojis), re (used for regular expression matching in order to remove web links, mentions, hashtags, and punctuations), and Natural Language Toolkit (NLTK) which was used to perform word stemming (converting derived words to their base words e.g., 'seeing' becomes 'see') and stop words removal (removing frequently occurring words that do not give special semantics to the statement e.g., is, a, of, etc.). Also, 2887 duplicate entries were removed and all the posts labeled as unspecified CB type "others", were removed. Eventually, the cleaned dataset contained a total of 39,869 tweets in the religion (7998), age (7992), gender (7973), ethnicity (7961), and non-CB (7945) classes. As seen from the number of tweets in each class, the dataset is nearly balanced, but to completely avoid bias due to class imbalance, the dataset was split into training (80%) and test (20%) sets, and the training set was balanced so that each class in the training set had exactly the same number of tweets.

3.2 Features Representation

Learning algorithms often work with real-valued features. Therefore, there is a need to convert the text in our dataset samples to numeric values. We investigated 2 methods for this purpose, the Term Frequency-Inverse Document Frequency (TF-IDF) algorithm [24] and the pre-trained Glove embeddings [25].

TF-IDF consists of two parts, the TF part, which measures the frequency of a word across a dataset, and the IDF part, which measures the degree of scarcity or frequency of a word in a dataset. Thus, while the TF part may score a word high because of its frequency in a dataset, IDF, being a log-normalised value, may score it low if it appears across too many samples in the dataset. TF is simply the number of times a word occurs in a dataset, while IDF divides the frequency of the word by the number of samples in which the word occurs in the document. Glove uses an unsupervised learning algorithm to obtain vector representations for words by training on aggregated word-to-word co-occurrence statistics from a large dataset and the obtained representations are made available for creating word embeddings for training learning algorithms.

Based on our findings from empirical observations, we used TF-IDF for all 5 ML algorithms and used Glove embeddings for the DL model.

3.3 Learning Algorithms

Five ML algorithms were trained and tested on the dataset of study. These earning algorithms were chosen for their reported performance in previous studies and in a bid to compare their performances on this dataset. The 6 learning algorithms studied are Gradient Boosting (GB), Logistic Regression (LR), Naive Bayes (NB), Random Forest (RF), Support Vector Machine (SVM), and Bi-directional Long Short-Term Memory (BiLSTM). The first five algorithms are classical ML algorithms while the last one is a deep neural network that is known for its ability to capture the forward and backward relationship between words in a sentence, thus giving an efficient representation of the semantics of a sentence by capturing word contexts. As the problem of CB detection and classification is a classification problem, all the algorithms used are supervised algorithms – we have the target classes of the dataset samples.

4 Experiments, Results, and Discussions

All experiments were carried out on the Kaggle jupyter notebook environment using Kaggle's Graphics Processing Units (GPU). The programming language used was Python 3.7 with libraries such as Matplotlib (for visualization), NLTK (a library for several NLP tasks), Numpy (for array handling), Pandas (for data analysis), Scikit-learn: (for ML), and Tensorflow (for DL). As seen in Fig. 1, the cleaned/preprocessed dataset contains texts that are all in lowercase and are void of stopwords, punctuations, web links, hashtags, emojis, and mentions.

	tweet_text	cyberbullying_type	cleaned_text
0	In other words #katandandre, your food was cra...	not_cyberbullying	word food crapilicious
1	Why is #aussietv so white? #MKR #theblock #ImA...	not_cyberbullying	whi white
2	@XochitlSuckkks a classy whore? Or more red ve...	not_cyberbullying	classi whore red velvet cupcakes
3	@Jason_Gio meh. :P thanks for the heads up, b...	not_cyberbullying	gio meh thank head concern anoth angri dude tw...
4	@RudhoeEnglish This is an ISIS account pretend...	not_cyberbullying	isi account pretend kurdish account islam lies

Fig. 1. Some dataset samples before and after preprocessing

After preprocessing the dataset, the text of the dataset was converted into numeric values with TF-IDF using a maximum of 5000 features, while 100-dimensional Glove embeddings were also used to create feature vectors from texts. Without specifying the maximum features, TF-IDF would use the entire vocabulary of the dataset and this would result in a very huge feature dimension (based on the entire vocabulary of the dataset) that will make learning very difficult. The TF-IDF vectors are created as a sparse matrix to save space since many of the feature values are very close to 0. Because of the huge computational demand of deep neural networks, training BiLSTM with huge 5000-dimensional features actually slowed down learning and limited the algorithm's chances of convergence. Thus, we used 100-dimensional Glove embeddings for BiLSTM.

As mentioned earlier, we used 80% of the cleaned dataset for training and 20% for testing. The splitting was done using sci-kit-learn's train_test_split function which ensures that roughly equal amounts of samples per class are present in both splits. The training dataset was further balanced using a random oversampling method which uses random samples of an under-sampled class to increase the class size until it has exactly the same number of samples as the largest class. All the ML models used were from the scikit-learn library while the DL model used was from the Tensorflow library. A randomization seed of 42 was used to ensure reproducibility. LR used the 'saga' solver, NB used a multinomial classifier and SVM used a linear classifier with no kernel. All the models were trained with grid-search for the best set of hyperparameters. The DL model is primarily made up of BiLSTM and dense layers; it also included drop-out layers to combat overfitting. Figure 2 shows a summary of the DL architecture. The DL model was compiled with an "Adam" optimizer with a learning rate of 5E-4 and categorical cross-entropy loss function and trained for 15 epochs with a 10% validation set to obtain final training loss and accuracy values of 0.2528 and 0.9185, respectively. The training was stopped at 15 epochs to prevent overfitting using the EarlyStopping callback of Tensorflow to stop the training if the validation accuracy drops three consecutive times.

```
Model: "sequential"
_____
Layer (type)                 Output Shape              Param #
=================================================================
embedding (Embedding)        (None, 200, 100)          3772900

bidirectional (Bidirectional (None, 200, 128)          84480

dropout (Dropout)            (None, 200, 128)          0

bidirectional_1 (Bidirection (None, 200, 64)           41216

dropout_1 (Dropout)          (None, 200, 64)           0

global_average_pooling1d (Gl (None, 64)                0

dense (Dense)                (None, 16)                1040

dropout_2 (Dropout)          (None, 16)                0

dense_1 (Dense)              (None, 5)                 85
=================================================================
Total params: 3,899,721
Trainable params: 126,821
Non-trainable params: 3,772,900
_____
```

Fig. 2. Architecture of the deep learning model

The test set contained 7788 samples with 1582, 1567, 1558, 1552, and 1529 samples in the age, ethnicity, gender, religion, and non-CB classes respectively. The next subsections present the results of these models in terms of accuracy, precision, recall, and f1-score on the test set.

Table 1 shows the performances of the 6 models for CB detection and classification. The algorithms have been listed in ascending order of their accuracies. As seen from the table, Random Forest achieved the best performance with an accuracy of 93% while Naïve Bayes had the worst performance with an accuracy of 84%. Coming close in performance to RF are SVM, GB, and LR which all had 92% accuracy and BiLSTM had an accuracy of 91%. Generally speaking, all the models achieved their best predictions on the ethnicity class and the worst predictions on the non-CB class. This is likely because the most frequent words in the ethnicity class were almost very specific to that class while the ones in the non-CB class were also common to the other classes so, they were often misclassified.

The outstanding performance of RF on this dataset is supposedly due to its ensemble of trees which enables it to make better-informed decisions on the classes of the samples. Also, the result of BiLSTM, when compared to other classical ML models shows the DL model does not perform well on this particular dataset, most likely because of the small size of the dataset compared to the number of classes to consider. Also, when compared with other classical ML algorithms, it uses a feature dimension that is 50 times less than what the other ML algorithms used. When we used a feature dimension of 100 features for TF-IDF, RF's accuracy stood at 90%. Figure 3 shows the plots of training and validation loss and accuracy of the DL model and the smooth curves show good training progression as well as a good fit of the BiLSTM model.

Table 1. Precision, recall, accuracy, and F1 score of the 6 cyberbullying classification models

Learning Algorithm	Class	Precision	Recall	F1 score
BiLSTM	Age	0.96	0.98	0.97
	Ethnicity	0.95	0.98	0.96
	Gender	0.78	0.92	0.84
	Not-CB	0.88	0.74	0.81
	Religion	0.95	0.94	0.95
	Accuracy	–	–	0.91
GB	Age	0.98	0.97	0.98
	Ethnicity	0.99	0.98	0.99
	Gender	0.96	0.79	0.86
	Not-CB	0.75	0.91	0.82
	Religion	0.96	0.95	0.95
	Accuracy	–	–	0.92
LR	Age	0.96	0.97	0.96
	Ethnicity	0.98	0.98	0.98
	Gender	0.92	0.84	0.88
	Not-CB	0.80	0.85	0.82
	Religion	0.94	0.96	0.95
	Accuracy	–	–	0.92

(continued)

Table 1. (*continued*)

Learning Algorithm	Class	Precision	Recall	F1 score
NB	Age	0.81	0.97	0.88
	Ethnicity	0.88	0.91	0.90
	Gender	0.86	0.80	0.83
	Not-CB	0.83	0.55	0.66
	Religion	0.81	0.96	0.88
	Accuracy	–	–	0.84
RF	Age	0.98	0.98	0.98
	Ethnicity	0.98	0.98	0.98
	Gender	0.92	0.86	0.89
	Not-CB	0.81	0.86	0.84
	Religion	0.95	0.96	0.96
	Accuracy	–	–	0.93
SVM	Age	0.94	0.98	0.96
	Ethnicity	0.97	0.98	0.98
	Gender	0.94	0.81	0.87
	Not-CB	0.79	0.85	0.82
	Religion	0.95	0.96	0.96
	Accuracy	–	–	0.92

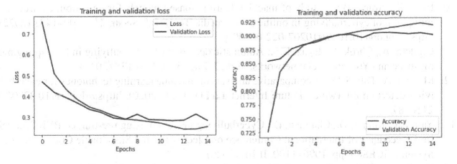

Fig. 3. Training and validation loss and accuracy of the DL model

5 Summary and Conclusion

Cyberbullying has become a real concern across the globe and its easiest means of spreading is through social interaction platforms. Therefore, every effort to combat CB on these platforms is praiseworthy. In this work, we investigated the performance of 6

Machine Learning models (BiLSTM, GB, LR, NB, RF, SVM) using two feature representation schemes (TF-IDF and Glove). With 5000-dimensional TF-IDF feature vectors, RF recorded the best performance while NB had the worst performance. Future works should employ more features representation schemes and more hyperparameter tuning schemes to obtain optimal settings for improved performance on larger cyberbullying datasets.

References

1. Mahanta, D., Khatoniyar, S. Cyberbullying and its impact on mental health of adolescents. IRA Int. J. Manag. Soc. Sci. **14**(2) (2019). https://doi.org/10.21013/jmss.v14.n2sp.p1
2. Hinduja, S., Patchin, J.W.: Bullying Beyond the Schoolyard: Preventing and Responding to Cyberbullying, 2nd edn. Sage Publications, Thousand Oaks (2015)
3. Raskauskas, J., Huynh, A.: The process of coping with cyberbullying: a systematic review. Aggress. Viol. Behav. **23**, 118–125 (2015). https://doi.org/10.1016/j.avb.2015.05.019
4. Kaluarachchi, C., Warren, M., Jiang, F.: Responsible use of technology to combat cyberbullying among young people **24**, 1–17 (2020)
5. Balakrishnan, V., Khan, S., Arabnia, H.R.: Improving cyberbullying detection using Twitter users' psychological features and machine learning. Comput. Secur. **90**, 101710 (2020)
6. Espelage, D.L., Hong, J.S.: Cyberbullying prevention and intervention efforts: current knowledge and future directions. Can. J. Psychiatr. **62**(6), 374–380 (2017). https://doi.org/10.1177/0706743716684793
7. Finkelhor, D., Mitchell, K.J., Wolak, J.: Online Victimization: A Report of the Nation's Youth, Washington, DC (2000)
8. Finn, J.: A survey of online harassment at a university campus. Inter-Pers. Viol. **19**, 468–483 (2004)
9. Lenhart, A., Maddeen, M., Hitlin, P.: Pew, internet & American life project: teens and technology: youth are leading the transition to a fully S30 (2005)
10. Raju, K., Aruna, B.: A study of machine learning-based models for detection, control, and mitigation of cyberbullying in online social media. Int. J. Inf. Secur. **21**, 1409–1431 (2022). https://doi.org/10.1007/s10207-022-00600-y
11. Çiğdem, A., Çürük, E., Eşsiz, E.S.: Automatic detection of cyberbullying in formspring.me, myspace and Youtube social networks. Turk. J. Eng. **3**(4), 168–178 (2019)
12. Muneer, A., Fati, S.M.: A comparative analysis of machine learning techniques for cyberbullying detection on Twitter. Future Internet **12**(11), 1–21 (2020). https://doi.org/10.3390/fi12110187
13. Yadav, J., Kumar, D., Chauhan, D.: Cyberbullying detection using pre-trained BERT model. In: Proceedings of International Conference on Electronics and Sustainable Communication Systems (ICESC), pp. 1096–1100. IEEE (2020)
14. Dalvi, R.R., Baliram Chavan, S., Halbe, A.: Detecting a twitter cyberbullying using machine learning. In: Proceedings of the 4th International Conference on Intelligent Computing and Control Systems (ICICCS), Madurai, India, 13–15 May 2020. https://ieeexplore.ieee.org/document/9120893
15. Huang, Q., Singh, V.K., Atrey, P.K.: Cyber bullying detection using social and textual analysis. In: Proceedings of the 3rd International Workshop Socially-Aware Multimedia (SAM), pp. 3–6 (2014). https://doi.org/10.1145/2661126.2661133
16. Pericherla, S., Ilavarasan, E.: Transformer network-based word embeddings approach for autonomous cyberbullying detection. Int. J. Intell. Unmanned Syst. (2021). https://doi.org/10.1108/IJIUS-02-2021-0011

17. Kargutkar, S.M., Chitre, V.: A study of cyberbullying detection using machine learning techniques. In: ICCMC, pp. 734–739 (2020). https://doi.org/10.1109/ICCMC48092.2020.ICCMC-00013
18. Roy, P.K., Tripathy, A.K., Das, T.K., Gao, X.-Z.: A framework for hate speech detection using deep convolutional neural network. IEEE Access **8**, 204951–204962 (2020). https://doi.org/10.1109/ACCESS.2020.3037073
19. Mahat, M.: Detecting cyberbullying across multiple social media platforms using deep learning. In: Proceedings of International Conference on Advance Computing and Innovative Technologies in Engineering (ICACITE), pp. 299–301. IEEE (2021)
20. Ari, M., Hadi, S., Febriyanti, P.: A comparative analysis of classification algorithm for cyberbullying crime detection: an experimental study of Twitter social media in Indonesia. Sci. J. Inform. **9**(2), 133–138 (2022). https://doi.org/10.15294/sji.v9i2.35149
21. Patacsil, F.: Analysis of cyberbullying incidence among Filipina victims: a pattern recognition using association rule extraction. Int. J. Intell. Syst. Appl. **11**(11), 48–57 (2019). https://doi.org/10.5815/ijisa.2019.11.05
22. Akhter, A., Uzzal, K.A., Polash, M.M.A.: Cyber bullying detection and classification using multinomial Naïve Bayes and fuzzy logic. Int. J. Math. Sci. Comput. **5**(4), 1–12 (2019). https://doi.org/10.5815/ijmsc.2019.04.01
23. Bashir, E., Bouguessa, M.: Data mining for cyberbullying and harassment detection in Arabic texts. Int. J. Inf. Technol. Comput. Sci. **13**(5), 41–50 (2021). https://doi.org/10.5815/ijitcs.2021.05.04
24. Luhn, H.P.: A statistical approach to mechanized encoding and searching of literary information. IBM J. Res. Dev. **1**(4), 309–317 (1957). https://doi.org/10.1147/rd.14.0309
25. Pennington, J., Socher, R., Manning, C.D.: GloVe: global vectors for word representation. In: Empirical Methods in Natural Language Processing (EMNLP), pp. 1532–1543 (2014). http://www.aclweb.org/anthology/D14-1162
26. Wang, J., Fu, K., Lu, C.-T.: SOSNet: a graph convolutional network approach to fine-grained cyberbullying detection. In: Proceedings of IEEE International Conference on Big Data (Big Data), pp. 1699–1708 (2020). https://doi.org/10.1109/BigData50022.2020.9378065

Detection of Potential Mosquito Breeding Sites Using CNN and Fewshot Learning

Gabaalini Ananthajothy, Rudsika Navaratnam, Niluksha Thevarasa, and Maheshi B. Dissanayake[✉]

Department of Electrical and Electronic Engineering, Faculty of Engineering, University of Peradeniya, Peradeniya, Sri Lanka
{e16028,e16249,e16256,maheshid}@eng.pdn.ac.lk

Abstract. As of World Health Organization (WHO), effective mosquito-vector control is critical in mitigating and controlling vector-borne diseases such as Dengue Fever. In this research, we proposed an automated technological solution to aid the public health inspectors (PHIs) task by automating the detection of potential mosquito-vector breeding sites using drone images. With this scope, we compare two strategies for detecting potential mosquito breeding sites (water pools) in aerial images: convolutional neural network (CNN) and few-shot learning. In this research, we adopt a supervised learning approach, and a labeled custom dataset with aerial images is generated to aid the model training process. Out of the proposed two methods, the CNN achieved the highest testing accuracy of 84.29% after training for 100 epochs, whereas FSL showed only 79% of accuracy after training the model for 2000 epochs. In conclusion, the proposed CNN-based system has the potential to support PHIs in locating potential mosquito breeding sites, especially those that are inaccessible to humans.

Keywords: Convolutional Neural Networks · Fewshot Learning · Image classification

1 Introduction

Dengue fever is a viral infection, affecting 0.4 to 1.3 million people yearly in tropical and subtropical countries. It is transmitted by female Aedesaegypti and Aedes Albopictus mosquitoes, which thrive in clean water, such as puddles, tanks, plastic containers, and old tires. Poor sanitation and garbage collection practices contribute to the increase in mosquito density. Public health inspectors (PHIs) in Sri Lanka are tasked with physically checking every potential site for the presence of mosquito vectors during the rainy season. However, it is physically impossible to check above ground level areas such as rooftops, roof gutters, and overhead water tanks for stagnant water.

Unmanned aerial vehicles (UAV) based misquote vector surveillance have emerged as a potential technological solution for the above problem [1]. It is possible to remotely identify potential mosquito breeding sites with stagnant water that are inaccessible to the human observer, using drone cameras and visual inspection [2–4]. In our work, we present a supervised deep-learning approach to automatically detect objects and locations with high potential to become mosquito breeding sites using aerial images captured by a simple commercial UAV with a regular camera without any filtering. Most of the learning algorithms in the literature, especially deep learning, require an extra-large dataset representing the task in hand. However, it is impractical in specific scenarios to collect such task-specific larger datasets. This particular research is one such scenario. Hence, in our solution, we propose to utilize learning algorithms which has the potential to perform highly under the constraint of a smaller dataset. We specifically utilized Fewshot Learning (FSL) [5, 6] and shallow Convolutional Neural Network (CNN) [7, 8] for the task at hand. Also, during the training process, the proposed models were trained to adapt to different geographical settings with different camera specifications. Further, the images captured through moving UAVs contain dynamic and spatial information about the environment. In our proposed solutions to detect mosquito breeding sites, i.e. stagnant water, these spatial and temporal characteristics present in the aerial images are utilized to detect water areas.

2 Literature Review

In recent years, many computer vision-based solutions coupled with intelligent machines/models with self-learning capabilities have been proposed to find solutions to everyday problems. Unmanned aerial vehicles (UAV) based vector surveillance has emerged as a potential technological intervention to the above bottleneck [1]. Potential malaria vector breeding site detection using UAVs in contrasting environments such as paddy fields and ponds is presented in [2]. Here images captured by a drone flying over a potential site are used to generate orthophotos. Then these orthophotos are visually inspected by the PHIs to detect potential mosquito breeding areas. In [3] modified drone camera with multispectral imaging and a high magnification factor is used to locate potential mosquito breeding sites and visualize vectors at different stages of mosquito lifestyle. An in-depth survey of modern practices and state-of-the-art research on UAV-based vector surveillance is presented in [1]. Most literature on this topic mainly utilizes visual inspection or image processing techniques to detect water, and recent literature attempts to automate this process using machine learning and deep learning approaches. For instance, in [4] deep learning based image segmentation methods are compared for the task of automated segmentation of water regions in aerial images. In our study, we use FSL and Shallow CNN to provide an intelligent solution to address the shortcomings of earlier works.

2.1 Fewshot Learning (FSL)

FSL is a popular deep-learning technique used when training data are either hard to find or costly to generate. In such cases, we need to train the model with lesser data. Furthermore, FSL is a concept that can be used for training the models with a lower amount of data. N-Shot Learning, Few-Shot Learning, One-Shot Learning, and Zero-Shot Learning are some variations of FSL algorithms [5]. The main difference between FSL from the standard supervised learning-based classification is that it does not aim to explicitly train the model to recognize the input image class at the training set, with the expectation of utilizing that knowledge learned explicitly at the testing. The implementation instance of a FSL model consists of a support set and a query set [6]. The Support set consists of N class with K samples each. If K < 10, the learning algorithm is called Few-shot learning, while K = 1 is referred to as One-shot learning. The query set, i.e., the prediction set, is a small subset of independent samples representing all N classes. The FSL model learns from the support set and evaluates the performance of the query set, which the model has never seen before. Overall FSL models aim to learn the similarity and differences between objects during the training and utilize this knowledge to classify the test images in the query sample, which is from an unknown class.

2.2 Convolutional Neural Network (CNN)

The modern CNNs, drive many automated applications in the domain of artificial intelligence by mimicking the human learning process. CNN has become one of the most sort out algorithms in deep learning, especially in computer vision, to solve technically challenging problems such as face detection and automated driving [7]. They can deliver high performance in analytical tasks without human intervention. These multi-layer models automatically extract features from input images using convolution layers and learn patterns for the classifier using neurons with hidden layers. A deep CNN consists of millions of trainable parameters, requiring a significantly larger dataset for promising performance. In practice, extra-large datasets such as ImageNet, which contains 14 million images, are ideal for training deep CNN.

Moreover, although deep CNNs can perform classification tasks with high accuracy, they are computationally demanding and require significant processing power during training. However, deep CNNs are prone to easy convergence to local optimum, gradient vanishing, overfitting, and gradient exploding. Moreover, shallow CNNs such as LeNet [8] exhibited promising performance with sufficient accuracy at image classification. With this view, we have designed a shallow CNN with only 1.4 million trainable parameters for the experiment presented in this chapter.

3   Methodology

The aim of this research is to present an automated self-learning system, where remotely captured UAV based aerial images are analyzed to detect stagnant water. The main constraint of this research is the limitation in the total size of the dataset i.e. the ability of the learning algorithm to work with a smaller dataset. Further, the trained model should be able to carry out testing, independent of the geographical locations and the model of the drone, yet they will depend on the minimum quality of the captured image, especially the resolution. With this scope, in this research, we first capture drone images of different geographical locations and later compare and contrast the performance of the CNN model and FSL model for detecting water areas in these aerial images under the constraint of a smaller dataset.

3.1 Dataset

We custom-created a dataset using aerial images captured with drone cameras of different models and resolutions. The dataset consists of 347 images representing the "with water" and "without water" classes. Although the input images in the dataset have different resolutions, we resize the input images to 256×256 during the model training. For CNN, we split the dataset into training, validation, and testing sets with a 60:20:20 ratio, with the dataset, shuffled before splitting. For FSL, we used 20 samples per class, i.e., K $= 20$, as the support set, whereas the quarry set consisted of 5 images per single iteration.

3.2 CNN Model Architecture

The sequential CNN model constructed for this research is given in Fig. 1. The model consists of 7 trainable layers, with the breakdown of 5 convolutional layers and two fully connected layers. Each convolutional layer is separated by a max-pooling layer, while the input to the model is a 256x256 RGB image. The model adopts sigmoid activation, Adam optimizer with adaptive learning rate, and categorical cross-entropy loss function with 'accuracy' as the performance metric. The dropout layer, with 0.3, helps to prevent overfitting. The model was trained over 100 epochs with batch sizes of 16. Finally, the trained model was tested using never before seen test data.

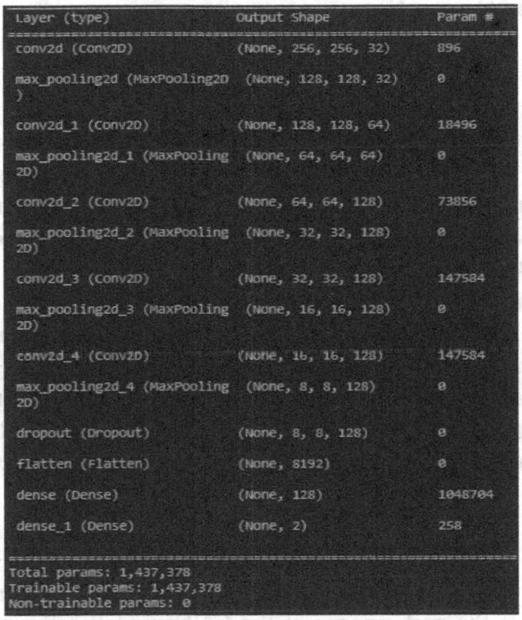

```
Layer (type)                    Output Shape           Param #
=================================================================
conv2d (Conv2D)                 (None, 256, 256, 32)   896

max_pooling2d (MaxPooling2D     (None, 128, 128, 32)   0
)

conv2d_1 (Conv2D)               (None, 128, 128, 64)   18496

max_pooling2d_1 (MaxPooling     (None, 64, 64, 64)     0
2D)

conv2d_2 (Conv2D)               (None, 64, 64, 128)    73856

max_pooling2d_2 (MaxPooling     (None, 32, 32, 128)    0
2D)

conv2d_3 (Conv2D)               (None, 32, 32, 128)    147584

max_pooling2d_3 (MaxPooling     (None, 16, 16, 128)    0
2D)

conv2d_4 (Conv2D)               (None, 16, 16, 128)    147584

max_pooling2d_4 (MaxPooling     (None, 8, 8, 128)      0
2D)

dropout (Dropout)               (None, 8, 8, 128)      0

flatten (Flatten)               (None, 8192)           0

dense (Dense)                   (None, 128)            1048704

dense_1 (Dense)                 (None, 2)              258
=================================================================
Total params: 1,437,378
Trainable params: 1,437,378
Non-trainable params: 0
```

Fig. 1. CNN Architecture

3.3 FSL Model Architecture

For the implementation of FSL, we specifically use FSL modeling with Reptile Algorithm [9], which can learn quickly and classify new tasks with minimal training. This implementation is nearly identical to the original implementation by ADMoreau [10], with few modifications to meet the task-specific requirements. Table 1 tabulates the fine tuned parameters used at the final implementation.

Table 1. The parameters used in our FSL model.

Parameter	Value	Parameter	Value
Learning rate	0.001	Eval iteration	5
Meta step size	0.05	Inner iteration	4
Inner batch size	16	Eval interval	1
Eval batch size	16	Train shots	20
Meta iteration	2000	Shots	5

4   Results and Discussion

In this experiment, we compared the learning performance of the CNN and FSL methods for detecting the presence and absence of pools of water using aerial images. To improve the reliability of the evaluation, the results were averaged at least over five iterations for each model. Also, the dataset split was randomized for each training loop.

For the CNN model-based classification of the presence and absence of water in the input aerial images, the trained model showed a training accuracy of 97.47% and a testing accuracy of 84.29%. Figure 2 shows the training and test accuracy and the loss for over 100 training epochs (x-axis). The model output for selected test images is shown in Fig. 3, along with the predicted class probability value for each image. The predicted class probability value will be a float between 0 and 1, representing the probability of the sample belonging to the negative class. The detection probability value 0 indicates that the input image belongs to the "presence of water," while 1 indicates the "absence of water" in the aerial image. As of Fig. 3, the performance of the trained shallow CNN is significantly accurate, and it can detect the presence and absence of water from UAV images.

Furthermore, performance metrics such as Accuracy, Precision, Recall, and F1 score can be easily calculated using the predicted and actual label values as presented in [11]. Table 2 summarizes the performance of the CNN model in the middle column (2nd Column). The following notations are used in Table 2. True positive (TP): Both actual and predicted labels are positive, False positive (FP): actual negative label, predicted as positive, True negative (TN): Both actual and predicted labels are negative, False negative (FN): Actual positive label is predicted as negative.

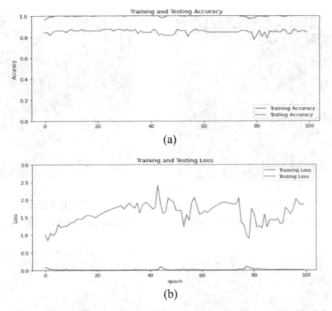

Fig. 2. (a) Training and Testing Accuracy Variation with Epochs; and (b) Training and Testing Loss Variation with Epochs for CNN

Table 2. Performance Comparison for CNN and FSL

Parameter	CNN/(%)	FSL/(%)
Accuracy = (TP + TN)/(TP + FP + TN + FN)	84.29	80.00
Precision = TP/(TP + FP)	83.33	80.00
Recall = TP/(TP + FN)	89.74	80.00
F1 score = 2 * (Precision × Recall)/(Precision + Recall)	86.42	80.00

The observations and results for the FSL-based water area detection model are illustrated in Fig. 4 and 5 and Table 2 (3rd column). According to Fig. 4, the FSL shows improved performance in terms of classification accuracy with epochs. The accuracy of the train and test instances closely follow each other at high epochs. Nevertheless, the maximum accuracy observed at the training and testing, at 2000 epochs, is, on average, around 80%. Close inspection of the model output, i.e., the classification results of the test images, as shown in Fig. 5, indicates that FSL fails to a certain degree to accurately classify images as with and without water pooling areas.

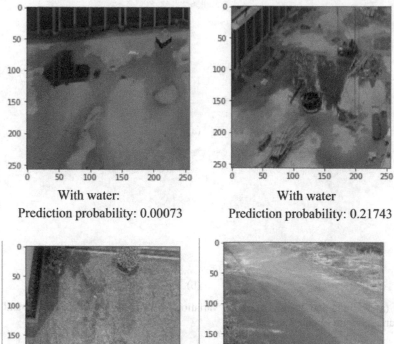

With water:
Prediction probability: 0.00073

With water
Prediction probability: 0.21743

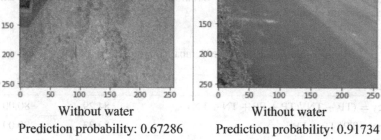

Without water
Prediction probability: 0.67286

Without water
Prediction probability: 0.91734

Fig. 3. Water Detection on Test Images with CNN

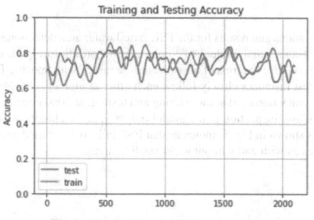

Fig. 4. FSL Training and Testing Accuracy

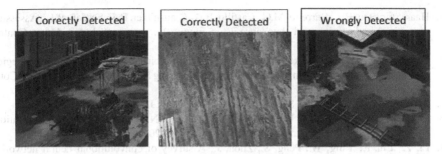

Fig. 5. Water Detection on Test Images with FSL

5 Conclusion

The primary goal of this research is to develop an automated system for detecting water areas in inaccessible locations for PHIs. To address the inaccessibility constraint, we propose to use UAV based aerial images. Another bottleneck associated with this research is the inability to develop a larger dataset with millions of images for the model training. Hence, for the purpose of automating the water region detection, we propose shallow CNN and FSL as potential learning methods as they have shown promising performance with smaller datasets. Also, we have created a new aerial image dataset for the task at hand. The shallow CNN model resulted in a training accuracy of 97.47% and a testing accuracy of 84.29% whereas the FSL resulted in a training accuracy of 82% and a testing accuracy of 79%. Overall, the shallow CNN model outperformed the FSL model for the given task with an accuracy improvement of 5.29% at testing.

The performance improvement at detection was evident during the visual observations. With these observations, we conclude that shallow CNN is better than FSL for the task of detecting the presence of water in aerial images. Hence, we conclude that it is possible to utilize deep learning algorithms with aerial images, to detect potential mosquito breeding areas. Also, we conclude that it is possible to train a deep learning model from scratch using a smaller dataset to meet task specific applications. Eventually, the research findings and the proposed methodology would help PHIs to detect water retaining places inaccessible to humans. Then by eradicating these misquote breeding sites, many lives will be saved from dengue.

Acknowledgment. The authors wish to acknowledge the University Research Grant, URG/2021/14/E, of University of Peradeniya for providing the support to purchase the camera drone.

References

1. Carrasco-Escobar, G., Moreno, M., Fornace, K., et al.: The use of drones for mosquito surveillance and control. Parasit. Vectors **15**, 473 (2022)
2. Hardy, A., Makame, M., Cross, D., Majambere, S., Msellem, M.: Using low-cost drones to map malaria vector habitats. Parasit. Vectors **10**, 29 (2017)

3. Haas-Stapleton, E.J., Barretto, M.C., Castillo, E.B., Clausnitzer, R.J., Ferdan, R.L.: Assessing mosquito breeding sites and abundance using an unmanned aircraft. J. Am. Mosq. Control Assoc. **35**, 228–232 (2019)
4. Mylvaganam, P., Dissanayake, M.B.: Deep learning for arbitrary-shaped water pooling region detection on aerial images. In: 2022 Moratuwa Engineering Research Conference (MERCon), Colombo Sri Lanka, 27 July 2022 (2022)
5. Zhao, K., Jin, X., Wang, Y.: Survey on few-shot learning. J. Softw. **32**, 349–369 (2021)
6. Liu, Y., Zhang, H., Zhang, W., Lu, G., Tian, Q., Ling, N.: Few-shot image classification: current status and research trends. Electronics **11**, 1752 (2022)
7. Li, Z., Liu, F., Yang, W., Peng, S., Zhou, J.: A survey of convolutional neural networks: analysis, applications, and prospects. IEEE Trans. Neural Netw. Learn. Syst. **33**(12), 6999–7019 (2022)
8. Lecun, Y., Bottou, L., Bengio, Y., Haffner, P.: Gradient-based learning applied to document recognition. Proc. IEEE **86**(11), 2278–2324 (1998)
9. Nichol, A., Achiam, J., Schulman, J.: On first-order meta-learning algorithms. arXiv abs/1803.02999 (2018)
10. https://keras.io/examples/vision/reptile/
11. Kanesamoorthy, K., Dissanayake, M.B.: Prediction of treatment failure of tuberculosis using support vector machine with genetic algorithm. Int. J. Mycobact. **10**, 279–284 (2021)

Spatial-Temporal Graph Neural Network Based Anomaly Detection

Ruoxi Wang[1]([✉]), Jun Zhan[1], and Yun Sun[2]

[1] Wuhan Fiberhome Technical Services Co., Ltd., Wuhan, China
xwang@fiberhome.com
[2] Hubei University of Technology, Wuhan, China

Abstract. Multivariate time series anomaly detection is an important task in the monitoring system. In practical applications, an efficient and accurate anomaly detection method is particularly important. Recently, the method of anomaly detection based on prediction has made significant progress, but there are still limitations. This paper proposes a paradigm for multivariate time series anomaly identification based on pre-training. The strategy of pre-training is to use Transformer's encoder to learn the dense vector representation of multiple time series through autoregressive task, so as to enhance the predictability of time series. In the prediction module, we learn the feature dependence of time series through graph attention network, and design an interactive tree structure that takes full advantage of the unique characteristics of time series to capture its time dependence. In addition, our method is well interpretable and allows users to infer the root cause of exceptions. We have proved the effectiveness of our model through extensive experiments. It is significantly superior to the most advanced model in three real data sets.

Keywords: Time series · Anomaly detection · Pre-training · Transformer encoder · Graph neural network

1 Introduction

Anomaly detection is an active research topic in data mining filed, which has been used in database logs, images, network attacks, and time series and so on. Several-variable time sequence made up of multiple univariable time sequence of the same entity, in which each univariable time sequence delegate the monitoring worth of a transducer in the system. In this paper, we mainly study the anomaly detection of multivariate time series, which is widely used to monitor the status of various sensors in industrial, manufacturing and information technology systems. Efficient and accurate identification of outliers from temporal data helps provide alarm to possible events in time and constantly watch transducer systems.

According to the different number of features of time series data, it can be divided into univariate and multivariate time series anomaly detection. Preliminary study centered on the detection of univariate time sequence anomalies, which are considered as

© The Author(s), under exclusive license to Springer Nature Switzerland AG 2023
Z. Hu et al. (Eds.): ICCSEEA 2023, LNDECT 181, pp. 459–471, 2023.
https://doi.org/10.1007/978-3-031-36118-0_42

abnormal when the data deviates from the population distribution at a particular point. Traditional machine learning methods are mostly used for single-variable time series anomaly detection [1–3]. By monitoring each sensor separately, single-variable time series identification techniques can identify irregularities in multivariate time sequence, but because sensors' inherently complicated relationships, this approach has not received sustained attention. With the comprehensive digitization of society, networked sensors are becoming more prevalent in the real-world system [4], as well as the eventual replacement of univariate time series anomaly detection by multivariate time series anomaly detection.

The development and neural network usage has effectively advanced the study of both data mining and pattern recognition. [5, 6]. By taking into account various sensor characteristics at the same time, the unsupervised anomaly detection method based on deep learning has exceeded the univariate time series detection method [7]. Reference [8] uses LSTM to learn its complex several-variable time sequence with dependent on time, and uses prediction deviation as abnormal fraction, and obtains good performance. Reference [9] proposed a multi-scale convolutional cycle encoder, constructed a multi-scale feature matrix to characterize multiple levels of system states with different time steps, used the convolutional encoder to encode the correlation between sensors, and developed a convolution network of long-term and short-term memory based on attention to capture the time pattern. Reference [10] proposed a mutation autoencoder based on LSTM for multi-mode anomaly detection. Decoding entails calculating the anticipated distribution based on the potential space's representation, whereas encoding entails projecting the observed value and dependent on time of each time step onto the possible area. Reference [11] proposes an unsupervised anomaly detection framework based on adversarial auto encoders. The key idea of this framework is to use two discriminators to conduct adversarial training on auto encoders, learn the normal mode of multivariate time series, and then use reconstruction errors to detect anomalies.

In recent years, graphic neural networks have achieved great success in processing graphic data due to their excellent permutation, local connectivity and composability. Numerous real-world issues, such as the multi-media network, chemical structure, and transportation network, may be abstracted as non-Euclidean data with graph structure. It is also possible to think of multivariate time series as graph-structured data, with sensors acting as nodes with hidden relationships. Reference [12] proposes a general graphical neural network framework specifically for multivariate time series data. This framework automatically extracts the orientation relationship between variables mainly through the graphics learning module. This module uses the mixed jump propagation layer and the extended initial cover with capture the multi-sensor dependency and time-dependency relationship in the time series. The entire framework learns together in an end-to-end framework. GRELEN: Multivariate Time Series Anomaly Detection from the Perspective of Graph Relational Learning, multivariate time series anomaly detection from the perspective of graph relational learning, By combining variational auto encoder and graph neural network, the feature extraction and correlation between sensors are respectively used to obtain the composite anomaly measure. Transformer Learning for Multivariate Time Series Anomaly Detection in the Internet of Things For the purpose of detecting multivariate time series anomalies in IoT, learn graph architectures using transformer. The framework automatically learns diagram structure, diagram convolution, and modeling time dependencies based on transformer's architecture. A novel graph convolution is introduced to describe the abnormal information flow between network nodes and a multi-branch attention mechanism to overcome the quadratic complexity barrier.

In order to deal with the problems still existing time sequence with several variables anomaly detection, In this paper, we suggest an unsupervised Pre-Training based Interactive Spatial-Temporal Graph Neural Network (PT-ISTGNN). Specifically, the network consists of pre-training module and prediction module. Among them, the pre-training module uses unlabeled data to carry out unsupervised representation learning of multivariate time series based on improved Transformer [13] encoder, with the purpose of weakening the influence of abnormal fluctuations in training data. The prediction module attempts to explicitly model the correlations between different sensors while modeling the time dependencies within each time series. We treat each univariate time series as a separate feature, construct a fully connected adjacency matrix in all sensors, and then learn the correlation among all sensors through GATv2. In order to make better use of the unique characteristics of time series, we construct an interactive tree structure based on time convolution, which not only realizes a larger receptive field, but also avoids the limitation of TCN's ability to extract time dynamics only from the features of the upper layer. The natural advantage of tree structure can gradually accumulate information from the upper layer and spread it downward, discover the underlying link between long-term and short-term reliance, and promote the ability to predict time series.

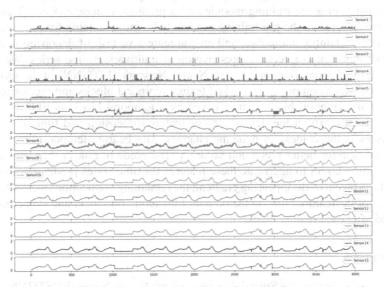

Fig. 1. A fragment of a multivariate time series. The horizontal axis represents the length of the time series, and the vertical axis represents the monitoring value of the sensor.

2 Formalization and Preparatory Knowledge

2.1 Problem Formulation

Multivariate time series anomaly detection aims at detecting anomalies at the entity level
[9]. In this paper, the input sequence of multivariate time series anomaly detection is
composed of sensor data that changes with time. Input sequence $X = \{x_1, x_2, \cdots, x_n\} \in R^{n \times m}$, where n is the length of the time series and m is the number of sensors that generate
the multivariate time series. Figure 1 displays a collection of multivariable time sequence
data with a length of 4000 and consists of 15 sensor monitoring readings.

Because there are so much time sequence data, we just grab the data
$\{x_{t-w}, x_{t-w+1}, \cdots, x_{t-1}\} \in R^{w \times m}$ of w time stamps before time et to predict the value at
time t in each iteration, and the prediction result is defined as $\hat{x}_t$. Our goal is to detect the
abnormal data in the test set. The deviation between the real value x_t and the predicted
value $\hat{x}_t$ is regarded as the abnormal score. The larger the abnormal score is, the greater
the possibility of abnormal occurrence at time T. If the abnormal score exceeds a specific
threshold, the abnormal occurrence at time t is determined.

2.2 Overall Architecture

Due to the flaws in the current prediction algorithms: 1. Abnormal data in the training
set will increase the unpredictability of the model; 2. The unique characteristics of
multivariate time series are ignored. As shown in Fig. 2, PT-ISTGNN proposed in this
paper consists of a pre-training module and a prediction module. Here, Transformer
encoder is adopted in the pre-training module to be more suitable for processing time
series data containing anomalies, and an autoregressive task of denoising is designed
to extract the intensive vector representation of multivariate time series. We propose
an Interactive Spatial-Temporal Graph Neural Network (ISTGNN) in the module for
prediction, which comprised in stacked feature-temporal blocks and prediction layers.
Each feature-temporal block is interleaved by a temporal block as well as a feature
block to jointly extract the dependency of feature dimension and time dimension of
time series in the context of dynamic changes. The Feature-temporal blocks can be
further stacked for better prediction accuracy. The prediction layer then uses two 1 ×
1 convolution layers to aggregate these features for time series prediction. In order to
avoid the problem of disappearing gradient, the remaining connections were added from
the input of the feature block to the output of the temporal block, and skip connections
w added after each feature block.

2.3 ISTGNN

We designed an Interactive Spatial-Temporal Graph Neural Network (ISTGNN) for multivariate time series prediction, which studied the inter-feature dependency and dependent on timing of multivariate time series respectively from the feature dimension and
time dimension.

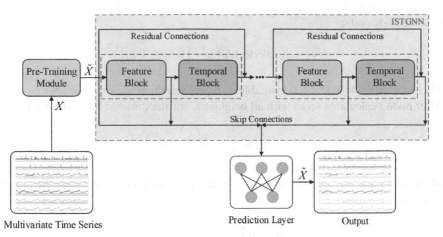

Fig. 2. Proposed model structure, consisting of the pre-training module and ISTGNN, which consisted of the stacked Feature-Temporal block and the prediction layer. The prediction layer projects the hidden features onto the desired dimensions to obtain the prediction results. Residual joins and skip joins are added to the model to avoid the problem of disappearing gradients.

2.3.1 Feature Block

From a graph-based perspective, each sensor is seen as a single node. In most cases, the multivariate time series does not provide explicit prior information about the structure of the graph. We introduce a randomly initialized sensor embedding vector $V = \{v_1, \cdots, v_m\} \in R^{m \times d_v}$, where $v_i \in R^{d_v}$ represents the characteristics of sensor i, it is trained alongside the model's other components. In a real scenario, if two sensors behave similarly, the two sensors behave similarly. Ideally, the learned embedding vector can well represent the behavior characteristics of sensors, so we use these embedding to calculate the similarity of behaviors between sensors:

$$e_{j,i} = \frac{v_i^\top v_j}{\|v_i\| \cdot \|v_j\|} \tag{1}$$

$j \in \{1, 2, \cdots, m\} \setminus \{i\}$ demonstrates that all sensors, excluding sensor i are potential candidates for sensor i. For the first k candidate sensors, we create a directed edge between sensor I and those, and then learn the inherent, nonlinear dependence between nodes through an improved Graph Attention Network (GAT).

GAT is one of the most advanced architectures for graphical learning. In GAT, each node can model the relationship between nodes in any intent by processing its neighbors. However, GAT calculations use a restricted static attention, where the ranking of attention scores is not conditional on the query node. In controlled problems, static attention can even prevent GAT from fitting training data []. On this basis, GATv2 model proposes a more expressive variation of dynamic graph attention than GAT.

Generally, given a graph with n nodes, the input is a set of node representations $\{h_i \in R^d | i \in \{1, \ldots, n\}\}$. In some GNN architectures before GAT, the weights of all neighbors are regarded as the same. To solve this problem, GAT uses score function e to score node pairs:

$$e_{ij} = \text{LeakyReLU}\left(a^{\mathrm{T}} \cdot [Wh_i \oplus Wh_j]\right) \tag{2}$$

where $a \in R^{2d'}, W \in R^{d \times d'}$ is learnable, $\oplus$ represents the connection between two nodes. After node i calculates score with all neighbors, softmax is used for normalization:

$$\alpha_{ij} = \frac{\exp(e_{ij})}{\sum_{l=1}^{L} \exp(e_{il})} \tag{3}$$

where L denotes the number of adjacent nodes for node i. GAT calculates the output of each node as follows:

$$h_i' = \sigma\left(\sum_{j \in L} \alpha_{ij} \cdot Wh_j\right) \tag{4}$$

GATv2 points out that the problem with standard GAT lies in scoring function, where α and W are used in order, and they are likely to collapse into a linear layer. To solve this problem, GATv2 modifies the computational order of linear transformations in attention to get more expressive attention. Specific changes are as follows:

$$e_{ij} = a^{\mathrm{T}}\text{LeakyReLU}\left(W[h_i \oplus h_j]\right) \tag{5}$$

Considering that the dependency between sensors is complex and changeable in reality, the features contain more complex interactions. In this article, we use GATv2's improved dynamic attention because we observe that dynamic attention performs more satisfactorily on all the data sets used in this work.

2.3.2 Temporal Block

The possible causality between several multivariate time series' sensors from the sensor dimension is captured in the preceding section. The modeling of the dependency relationship of multivariate time series along the time dimension is the focus of this section.

Time sequence is a special kind of serial data. To be able to make better use of its characteristics, we arrange multiple T-nodes into a full binary tree structure with depth of h, as shown in Figure [], there are 2h − 1 basic building nod at the h-th level. By gradually down sampling at different levels, deeper level features will contain finer scale temporal information from the shallower level. In this way, we can capture long - and short-term time dependencies at different time scales, thereby reducing the impact of abnormal data and enhancing the predictability of the original multivariate time series. It is theoretically possible to extract features at a coarser level by increasing the height of the tree, but we do not need to do this because it does not help the task of time series prediction, but rather increases the complexity of the model.

T-Node is mainly composed of split and interactive learning. First, the original sequence is down-sampled into two subsequences based on the parity of the subscript, which have a larger time scale than the original sequence, but retain most of the information. In order to extract isomorphic and heterogeneous information from each subsequence, we designed a simple module composed of two one-dimensional convolution layers, and then compensated the information loss in the down sampling process through the interactive learning strategy.

Specifically, the output vector $\vec{X}$ of the feature block is first divided into two subsequences: $\vec{X}_{even}$ and $\vec{X}_{odd}$ according to the parity of the subscript. The information exchange between two subsequences is realized by mutual learning of affine transformation parameters. The interactive learning process consists of two steps.

In the first step, The two subsequences are scaled, and the scaling factor is discovered. First, $\vec{X}_{even}$ and $\vec{X}_{odd}$ are transformed to the hidden state using two different one-dimensional convolution modules respectively, then normalized by the softmax function, and finally element products are performed with $\vec{X}_{odd}$ and $\vec{X}_{even}$ respectively.

$$\vec{X}_{odd}^{s} = \text{softmax}\left(\theta_a\left(\vec{X}_{even}\right)\right) \odot \vec{X}_{odd}$$

$$\vec{X}_{even}^{s} = \text{softmax}\left(\theta_b\left(\vec{X}_{odd}\right)\right) \odot \vec{X}_{even} \tag{6}$$

where, θ represents one-dimensional convolution module, distinguished by subscript. $\odot$ is the Hadamard product.

In the second step, update the scaling features $\vec{X}_{odd}^{s}$ and $\vec{X}_{even}^{s}$ to get the final output of the interactive learning module. Firstly, the two scaling features are further transformed to other two hidden states through two one-dimensional convolution modules, and then the addition and subtraction operation is carried out with itself.

$$X_{odd}' = \vec{X}_{odd}^{s} - \theta_c\left(\vec{X}_{even}^{s}\right)$$

$$X_{even}' = \vec{X}_{even}^{s} + \theta_d\left(\vec{X}_{odd}^{s}\right) \tag{7}$$

In particular, compared with the extended convolution commonly used for time series modeling, the above architecture achieves a larger receptive field by extracting the basic information from the two down sequences. After completing all of the above steps, we reassemble all of the leaf nodes in the tree in parity order, concatenate them into a new sequence representation, and finally add them to the original time series via residual concatenation.

2.4 Anomaly Scoring

Similar to the predicted loss target, the anomaly fraction is calculated as the squared difference between the anticipated value $\hat{x}_t$ and the actual value x_t of the timestamp t, indicating the degree of deviation between the model's actual value and its anticipated value. The timestamp with a higher score has a higher possibility of abnormal occurrence. Specifically, the outlier score St of timestamp t is:

$$s_t = \sum_{i=1}^{m}\left(\hat{x}_t^i - x_t^i\right)^2 \tag{8}$$

If S_t exceeds a fixed threshold, we consider the timestamp t to be an exception.

3 Experiments

In this part, we will compare the proposed model against state-of-the-art techniques in order to evaluate the model's performance using three actual exception detection data sets. NASA has gathered and made available the satellite data sets MSL and SMAP [14]. There are training and test subsets for each data set, and for both test subsets, there are label files with a list of all noted anomalies. The largest public data set presently being used to test time sequence with multiple variables anomaly detection is an application server data set (SMD [15]). It contains 28 different servers. Each server has 38 characteristics, representing different indicators of the server (CPU load, network usage, etc.). The first half of the data is used for training, and the other half is used as a test set (Table 1).

Table 1. Statistics of the data set.

Dataset	#entities	#features	Train	Test	Anomaly (%)
MSL	55	25	58317	73729	10.72
SMAP	27	55	135183	427617	13.13
SMD	28	38	708405	708420	4.16

To be able to verify the effectiveness of the time sequence with multiple variables anomaly detection algorithm proposed in this paper, As a means of comparison, the following typical detection methods are chosen:

LSTM-NDT [8]: An unsupervised threshold setting method is suggested for multivariate time series anomaly detection, and a pruning strategy is used to improve the model performance.

LSTM-VAE [10]: LSTM is combined with VAE, and LSTM is used to replace VAE feedforward network.

OmniAnomaly [15]: By learning the robust representation of multivariate time series, the normal mode is captured, and the input data is reconstructed by using random variable join and plane normalization flow.

MAD-GAN [16]: The fundamental model of GAN learning is LSTM-RNN, which captures the temporal dependency, and GAN discriminators and generators are used to detect anomalies, and the scores of anomalies are calculated by combining the discrimination results of test samples and the reconstruction deviations.

GDN [17]: The unique characteristics of each sensor are captured, the relationship between sensor pairs is learned, and the future behavior of sensors is predicted based on the attention function of adjacent sensors in the figure.

MTAD-GAT [18]: Time stamps' interaction with different time series is dynamically learned using two parallel GAT layers, and the prediction-based model and the reconstruction-based model are simultaneously optimized.

3.1 Training Settings

We implement the algorithms and all its variants using PyTorch (v1.9.0 with Python 3.8.10). All experiments are run on a Windows machine with NVIDIA GeForce RTX 3080 GPU. We have taken GATv2, which is provided in the PyTorch Geometric library (v2.0.1), which is published code by the original author. For multivariate time series prediction, sliding Windows with sizes of 50 and 80 were used in MSL and SMAP data sets, while historical Windows with sizes of 100 were used in SMD data sets. In the pre-training module, the model dimension was set to 128 and the number of multi-head attention heads was set to 8. We used the RAdam optimizer to train the model with 500 epochals at an initial learning rate of 0.001. For the prediction module, the height of the tree in the Temporal block was 3. We trained 50 batches of the model, the batch size was set to 64, and the initial learning rate was 0.0002. In addition, we use early stop strategy and dropout strategy to prevent overfitting. The dropout rates of the two modules are 0.1 and 0.4, respectively.

Table 2. Performance comparison of the 3 datasets using Precision (P), Recall (R), and F1-Score (F1). The top results are highlighted in bold and the secondary results are underlined.

Methods	MSL			SMAP			SMD		
	P	R	F1	P	R	F1	P	R	F1
LSTM-NDT	0.5934	0.5374	0.5640	0.8965	0.8846	0.8905	0.5684	0.6438	0.6037
LSTM-VAE	0.5257	0.9546	0.6780	0.8551	0.6366	0.7298	0.7922	0.7075	0.7842
OmniAnomaly	0.8867	0.9117	0.8989	0.7416	0.9776	0.8434	0.8334	0.9449	0.8857
MAD-GAN	0.8517	0.8991	0.8747	0.8049	0.8214	0.8131	0.9045	0.8541	0.8786
GDN	0.7729	0.9456	0.8506	0.8142	0.8623	0.8376	0.8995	0.9356	0.9172
MTAD-GAT	0.8754	0.9440	0.9084	0.8906	0.9123	0.9013	0.9396	0.9283	0.9339
PT-ISTGNN	0.9521	0.9543	**0.9532**	0.9456	0.9132	**0.9291**	0.9170	0.9673	**0.9415**

The precision and recall of all data sets surpass 0.935, as shown in Table 2, and our model regularly achieves the best F1 performance, which is incomparable with other baselines. This model has excellent generalization capacity. We can infer the following conclusions from Table 2:

1) In SMAP data sets, LSTM-NDT performs the poorest, while LSTM-NDT performs the poorest in MSL and SMD data sets. This is understandable given that LSTM-NDT ignores the dependence between sensors and only takes into account time series with only one variable's temporal pattern, which prevents it from making reliable predictions when the data set contains a lot of sensors. Lstm-vae simply combines LSTM with VAE and only models the time-dependent relationship. However, LSTM-NDT is obviously superior to LSTM-NDT in both data sets, probably because LstM-NDT is a reconstruction-based model. Compared with the prediction-based model, it has a better ability to capture the short-term data change rule of time series.

2) On the contrary, Omni Anomaly uses random method to model the dependence relationship between sensors, while MAD-GAN uses confrontation training to learn the dependence relationship between sensors, both of which achieve good performance. Unfortunately, none of them perform well in time-dependent modeling because they overlook the low-dimensional representation of the time dimension. These two techniques both rely on rebuilt models. In the process of training and learning, they will try to reassemble the data in the window as best as possible. In fact, the training data contains abnormal data.

3) Different from above, both GDN and MTAD-GAT are anomaly detection methods based on graph learning. Additionally, GDN ignores time series' time dependency, which has a significant impact on the performance of its model. Graph attention networks are used by MTAD-GAT to combine prediction-based and reconstruction-based models by learning associations on the temporal and feature dimensions, respectively. The assumption that every sensor in the data set is dependent on every other sensor ignores the complicated partial directional dependency across sensors and is not appropriate for many real-world scenarios.

Our model extracts the intensive vector representation of multivariate time series through pre-training to reduce the negative impact of abnormal data in training data and enhance the predictability of abnormal time series. Then, we consider both feature dependence and time dependence of time series, and use an interactive tree structure to take advantage of the uniqueness of time series ignored by other models. All these factors make our model superior to other models.

We decided to show some of the data in order to more clearly illustrate the impact of the suggested approach. As seen in Fig. 3, the black dotted line indicates the threshold value adaptively produced by the model in accordance with the distribution law, and the red line shows the square error between the predicted value and the actual value, that is, the abnormal score. The blue line segment depicts the actual anomaly distribution of the data, whereas the orange line segment describes the anomaly prediction distribution on the threshold. It can be seen that our model can well catch exceptions in the sequence. This paper deals with the abnormal data in the training data through the method of pre-training. Figure 4 illustrates how the prediction-based model learns as much as it can about the normal distribution of time series data, but it is not very successful at forecasting the original data, thus widening the gap between the predicted value and the true value in the abnormal interval, and then we can use the appropriate threshold to make abnormal judgment.

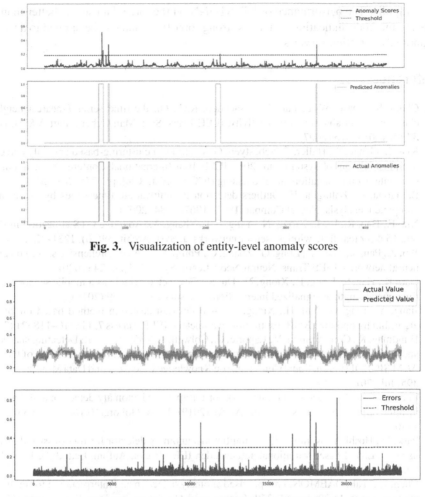

Fig. 3. Visualization of entity-level anomaly scores

Fig. 4. Data analysis based on sensor dimension

4 Conclusion

Multivariate time series anomaly detection can greatly help people find and eliminate the abnormal behavior of equipment in time. In this work, we propose a novel unsupervised multivariate time series anomaly detection architecture, namely pre-trained Interactive Spatiotemporal graph neural network (PT-ISTGNN), which firstly weakens the adverse effects of the anomalies in the training data through the pre-training strategy, so that the model can better learn the normal pattern of time series. Secondly, starting from the features of time series, GATv2 and an interactive tree structure are used to learn the dependence between features and time dependence respectively. Additionally, PT-ISTGNN has demonstrated strong anomaly interpretation capabilities in an experiment based on a generic model, which enables users to identify anomalies' underlying causes more rapidly. The performance of PT-ISTGNN on the three data sets is better than the existing methods, indicating that it has strong robustness and can be applied to a variety of anomaly detection services.

References

1. Chaovalitwongse, W.A., Fan, Y.J., Sachdeo, R.C.: On the time series k-nearest neighbor classification of abnormal brain activity. IEEE Trans. Syst. Man Cybern. Part A Syst. Hum. **37**(6), 1005–1016 (2007)
2. Kiss, I., Genge, B., Haller, P., Sebestyen, G., et al.: Data clustering-based anomaly detection in industrial control systems. In: 2014 IEEE 10th International Conference on Intelligent Computer Communication and Processing (ICCP). IEEE, Cluj, pp. 275–281 (2014)
3. Baragona, R., Battaglia, F.: Outliers detection in multivariate time series by independent component analysis. Neural Comput. **19**(7), 1962–1984 (2007)
4. Xia, F., Hao, R., Li, J., Xiong, N., Yang, L.T., Zhang, Y.: Adaptive GTS allocation in IEEE 802.15.4 for real-time wireless sensor networks. J. Syst. Archit. **59**(10), 1231–1242 (2013)
5. Wu, Z., Pan, S., Chen, F., Long, G., Zhang, C., Philip, S.Y.: A comprehensive survey on graph neural networks. IEEE Trans. Neural Netw. Learn. Syst. **32**(1), 4–24 (2020)
6. Gao, K., Han, F., Dong, P., Xiong, N., Du, R.: Connected vehicle as a mobile sensor for real time queue length at signalized intersections. Sensors **19**(9), 2059 (2019)
7. Jiang, Y., Tong, G., Yin, H., Xiong, N.: A pedestrian detection method based on genetic algorithm for optimize XGBoost training parameters. IEEE Access **7**, 118310–118321 (2019)
8. Hundman, K., Constantinou, V., Laporte, C., Colwell, I., Soderstrom, T.: Detecting spacecraft anomalies using lstms and nonparametric dynamic thresholding. In: Proceedings of the 24th ACM SIGKDD International Conference on Knowledge Discovery and Data Mining, pp. 387–395, July 2018
9. Zhang, C., et al.: A deep neural network for unsupervised anomaly detection and diagnosis in multivariate time series data. In: AAAI (2019). https://doi.org/10.1609/aaai.v33i01.330 11409
10. Park, D., Hoshi, Y., Kemp, C.C.: A multimodal anomaly detector for robot-assisted feeding using an LSTM-based variational autoencoder. IEEE Robot. Autom. Lett. **3**(3), 1544–1551 (2018)
11. Chen, X., et al.: DAEMON: unsupervised anomaly detection and interpretation for multivariate time series. In 2021 IEEE 37th International Conference on Data Engineering (ICDE), pp. 2225–2230. IEEE, April 2021
12. Wu, Z., Pan, S., Long, G., Jiang, J., Chang, X., Zhang, C.: Connecting the dots: multivariate time series forecasting with graph neural networks. In: Proceedings of the 26th ACM SIGKDD International Conference on Knowledge Discovery and Data Mining, pp. 753–763, August 2020
13. Vaswani, A., et al.: Attention is all you need. In: Advances in Neural Information Processing Systems, 30 (2017)
14. Entekhabi, D., et al.: The soil moisture active passive (SMAP) mission. Proc. IEEE **98**(5), 704–716 (2010)
15. Su, Y., Zhao, Y., Niu, C., Liu, R., Sun, W., Pei, D.: Robust anomaly detection for multivariate time series through stochastic recurrent neural network. In: Proceedings of the 25th ACM SIGKDD International Conference on Knowledge Discovery and Data Mining pp. 2828–2837, July 2019
16. Li, D., Chen, D., Jin, B., Shi, L., Goh, J., Ng, S.K.: MAD-GAN: multivariate anomaly detection for time series data with generative adversarial networks. In: International Conference on Artificial Neural Networks, pp. 703–716, September 2019

17. Deng, A., Hooi, B.: Graph neural network-based anomaly detection in multivariate time series. In: Proceedings of the AAAI Conference on Artificial Intelligence, vol. 35, no. 5, pp. 4027–4035, February 2021

18. Zhao, H., et al.: Multivariate time-series anomaly detection via graph attention network. In: 2020 IEEE International Conference on Data Mining (ICDM), pp. 841–850, November 2020

Methods of Topological Organization Synthesis Based on Tree and Dragonfly Combinations

Volodymyr Rusinov, Oleksandr Honcharenko, Artem Volokyta$^{(\boxtimes)}$, Heorhii Loutskii, Oleksandr Pustovit, and Artemii Kyrianov

Igor Sikorsky Kyiv Polytechnic Institute, Kyiv 03056, Ukraine
artem.volokita@kpi.ua

Abstract. The paper is devoted to consideration of several ways to synthesis new topological organizations based on different types of trees and Dragonfly sequences. Various combinations of combining standard topologies with Dragonfly topology, which allows obtaining new fault-tolerant topologies with specified properties, are proposed. The characteristics of the synthesized topologies are investigated, their advantages and disadvantages are shown. The presented variants of topologies can be the basis for the design of fault-tolerant survivable scalable systems. Developing a system using this approach will increase fault tolerance and topological characteristics that directly influence the overall performance of distributed applications.

Keywords: Dragonfly topology · topology · fault tolerance

1 Introduction

Problem Statement. When creating a high-performance computing system, an important part of the work is the development of topological organization. This step allows timely prediction of system characteristics without running a full system model, which allows for proper equipment selection and routing algorithms development. It is known, that most clusters use central processors together with several graphics accelerators to solve computing problems. Usually, a large part of the PCI express lines is already in use, and therefore it is important to predict in advance the number of connections for each node, which will affect the message queue that will be formed. Existing solutions allow for some degree of fault-tolerance, usually, using some variation of standard topological organizations such as tree or mesh which inherits a greater cost. An approach is proposed, based on combining two well performing topological organizations, that will yield a new hybrid topology with characteristics inheriting from Dragonfly and Tree topologies.

Actuality of the Topics Covered. Today, trees and fat trees are widely used as a method of network organization in data centers. Topological organizations such as 3D-mesh [1], DragonFly [2], 6D-tor [3], LCQ [4] are also used in modern supercomputers. All these topologies have fault-tolerant properties due to the presence of several alternative routing paths. Therefore, the task of developing new topologies that will have many balanced routes (one cost patch) is urgent [5, 6]. One option for research on tree topology improvement is a combination with Dragonfly using different scaling schemas. The use of different scaling schemas allows to vary the degree of nodes, diameter and number of alternative paths for routing. The article proposes to use a new method that allows to obtain specified characteristics when combining tree-based topology with a Dragonfly topology. Using Dragonfly and Tree topologies, we expect to generate a low-cost, fault-tolerant topology with low degree and low average diameter, that translates into lower latency in real-world applications.

Analysis of the Rest of the Research and Publications. Currently, there are a number of works devoted to topologies based on the Dragonfly sequence [7–10]. Several studies have investigated the performance of the Dragonfly topology. For example, the performance of the Dragonfly topology with that of other commonly used network topologies in HPC systems, such as fat-tree and torus were compared. It was found that the Dragonfly topology outperformed the other topologies in terms of network diameter, bisection bandwidth, and power consumption [11]. Various optimization techniques have been proposed to improve the performance of the Dragonfly topology. For example, a routing algorithm was proposed that takes into account both the available bandwidth and the congestion level of the network [12]. They found that their algorithm improved the network throughput by up to 30% compared to traditional routing algorithms. In another study, a dynamic load-balancing technique that balances the network traffic across different paths in the network was proposed, resulting in improved network performance [13].

Seeing the Previously Unresolved Parts of the Critical Problem. One of the widespread modifications of the tree is the fat tree, whose indicators allow scaling of fault-tolerant topologies. However, the possibility of using hybrid topologies based on Dragonfly sequences and different types of trees has not been considered before.

Aim of Airticle. The purpose of the study is to develop a method of synthesizing new topologies based on trees and Dragonfly, and to analyze the characteristics of the synthesized topological organizations for different levels of fault tolerance.

Methodology. The article uses an approach based on analysis of initial topologies binary tree and dragonfly, thereby outlining characteristics that can be improved. The study employed an algorithm, based on shortest path finding algorithm, to establish parameters, such as degree or topological traffic, as well as introducing a betweenness parameter, that demonstrates how system based on proposed topology will react to faults. In conjunction, these two approaches show how such system will work in real-life applications, such as datacenter applications or scientific computing. To further illustrate the capabilities of proposed topology, the results are presented using tables and charts. This study provides insights into developing hybrid topologies and shows paths to numerically demonstrate fault-tolerant properties of topological organizations. This study warrants further research into demonstrated topologies, which may be prone to additional overheads on higher scaling steps.

Theory. Most of the topologies used in supercomputers actually have their roots in simple topological organizations such as meshes. So, you can make the transition from simple to complex topologies in several ways:

Hierarchical Approach. A hierarchical topology is formed on the basis of a cluster, which contains a small number of interconnected nodes, in which the connections between the clusters also correspond to one given rule.

Cartesian Product of Topologies. Allows obtaining new topologies with a combination of properties of both initial topologies.

Combining topologies. With the help of additional connections, it is possible to preserve the advantages of some topology and expand its functionality due to the connections inherent in another topology.

The main parameters of different rank (R) topologies include the number (N) of topology nodes, degree (S) of topology, diameter (D) of topology, average diameter (mD) of topology, topological traffic (T), cost (C).

As the initial basic researched topological organizations De Bruyne trees and graphs were used. Consider an ordinary binary tree. Scaling of such a topology is carried out as follows: there is a root vertex, in the next step two new vertices, called tiers, are added to it. At each step of scaling up to the last tier, two vertices are added to each of the vertices, forming a new tier.

The first several ranks parameters of this topology are described in the Table 2 (Tables 1, 3 and 4).

Table 1. Binary tree parameters on r step

Parameter	Apex amount	Degree	Diameter	Cost
Growth equation	2^r-1	3	$2*(r-1)$	$3*(2^r-1)* *2*(r-1)$

Table 2. Tree parameters for the first 9 ranks

N	S	D	C	mD	T	Rank
9	5	3	135	1,75	0,7	2
381	5	13	24765	8,975425	3,59017	7
1533	5	17	130305	12,76907	5,107627	9

However, the tree still has problems that need to be solved. For example, the rapid growth of diameter and topological traffic. One of the options for solving this problem is to stop scaling the tree while adding new links, which is implemented in the thick tree topology. Another option, which is considered in this work, is the synthesis of a hybrid topology based on Dragonfly connections.

However, the tree still has problems that need to be solved. For example, the rapid growth of diameter and topological traffic. One of the options for solving this problem is to stop scaling the tree while adding new links, which is implemented in the thick tree topology. Another option considered in this paper is the synthesis of a hybrid topology based on the tree and the Dragonfly topology.

Let's select several groups in the initial cluster. First, the method of synthesis by permuting numbers was considered, where the cluster dimension N is required to connect N vertices.

However, it can be optimized by "getting rid" of hanging vertices (Fig. 1).

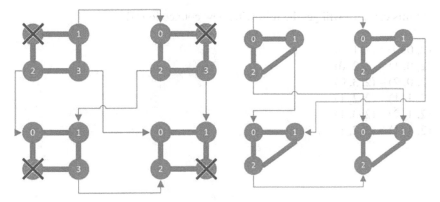

Fig. 1. Optimized synthesis method

In this way, it is possible to get a smaller size of the dragonfly subgroup. Since the levels in the tree have a dimension of 2K, the ideal group sizes are N = 2K–1. (N-d: 1, 3, 7...). Of course, the cluster will not always decompose perfectly. There will be situations when there will not be as many clusters at the tree level as needed for linking or vice versa.

Consider the method of building a tree with horizontal dragonfly connections. For the first step, let's take a simple cluster. Then we add another level. At the second level, we have 2 new clusters. One-to-one communication between levels. The minimum group in a cluster is 1 node. Therefore, the connection between clusters at this level is also "one-to-one". These connections are depicted by the green line in Fig. 2.

When adding the next level, we will get 4 more clusters. To determine further connections, let's denote clusters as (A, B), where A is the number of the level, B is the number of the cluster in the level. Then the full number of each vertex will be written as (A, B, C). We will add new connections for the vertex (A, B, C) to the vertex (A*, B*, C*) according to the following rules:

$$A^* = A$$

$$B^* = C \text{ when } C < B < \text{ or } C + 1$$

$$C^* = B \text{ when } B < B < \text{ or } B - 1$$

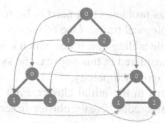

Fig. 2. Connections between clusters

In this case, we will get the following new connections (Fig. 3):

- (2, 0, 0) – (2, 1, 0)
- (2, 0, 1) – (2, 2, 0)
- (2, 0, 2) – (2, 3, 0)
- (2, 1, 1) – (2, 2, 1)
- (2, 1, 2) – (2, 3, 1)
- (2, 1, 1) – (2, 2, 1)

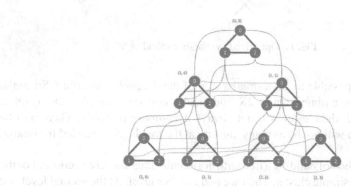

Fig. 3. Additional connections of the third level.

For simplicity, let's introduce such a concept as a topology profile. The vertices of the profile are clusters, and the connections between clusters reflect their type.

Depending on how the further construction of the topology will take place, namely steps 3, 4 and the following, the topology parameters will change. This article discusses four options for further scaling.

Table 3. Notations of the topology profile

Label	Description	Expanded view (for a triangle cluster)
▲ 0	Cluster	
▲1 — ▲2	One-to-one communication	
▲1 ▲2 ▲3 ▲4	Dragonfly-full connectivity	

2 Synthesis of Topologies

The first option was a similar continuation of linking nearby clusters. Nearby-grouping consists of grouping and connecting the closest clusters at each level. Figure 4 shows how the profile of such a topology will look.

A second proposed scaling method is combo-grouped scaling. In fact, it can be considered as a combination of the previous two. The idea is to use near and far scaling through the level. Accordingly, the topology profile will look like Fig. 5.

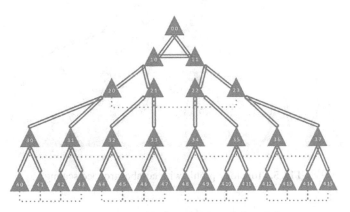

Fig. 4. Topology profile using Nearby-grouping

With this grouping of clusters, we will get the following characteristics of the topology.

Table 4. Nearby-grouped tree parameters

N	S	D	C	mD	T	Rank
9	5	3	135	1,75	0,7	2
21	6	4	504	2,185714	0,728571	3
45	6	5	1350	2,891919	0,963973	4
93	6	7	3906	3,94086	1,31362	5
189	6	9	10206	5,290724	1,763575	6
381	6	11	25146	6,869388	2,289796	7
765	6	13	59670	8,60734	2,869113	8
1533	6	15	137970	10,44953	3,483176	9

One of the key characteristics of the system is fault tolerance, one of the important aspects of which is the bypass of faulty nodes during routing. Naturally, this parameter is difficult to analyze in its pure form, but there is such an indicator as the mediation coefficient - the number of shortest routes passing through a particular vertex.

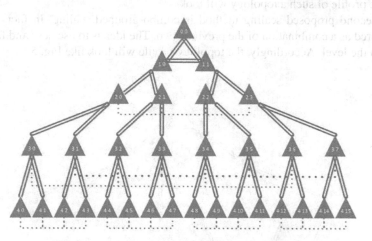

Fig. 5. Topology profile with combo-grouped scaling.

Topology parameters are listed in Table 5 (Table 6).

Results. Figures 6 and 7 show diagrams of topological characteristics in fallows versus the number of vertices. Topologists who look through the information find on the graphs in the same format:

1. Pure tree with cluster
2. Nearby-grouped tree
3. Far-grouped tree

Table 5. Combo-grouped tree parameters

N	S	D	C	mD	T	Rank
9	5	3	135	1,75	0,7	2
21	6	4	504	2,185714	0,728571	3
45	6	5	1350	2,883838	0,961279	4
93	6	7	3906	3,679523	1,226508	5
189	6	7	7938	4,309411	1,43647	6
381	6	9	20574	5,44817	1,816057	7
765	6	11	50490	6,814718	2,271573	8
1533	6	13	119574	8,446934	2,815645	9

4. Combo-grouped tree

Accordingly, the more evenly the values are distributed over the network, the more equal alternative routes exist and the more evenly the traffic is distributed, while the peaks indicate a potential overload of specific nodes.

Since it was previously established that the combined tree shows the best results in terms of topological characteristics, in the future we will analyze it, comparing it with a pure solution without horizontal links. However, before proceeding to a direct comparison of the data, it makes sense to consider in more detail the essence and methods of calculating the intermediation coefficient.

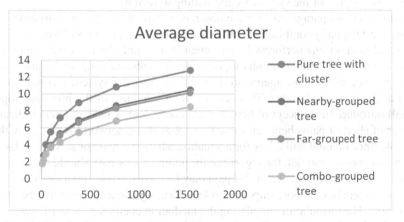

Fig. 6. Graph of the average diameter changes depending on the number of nodes

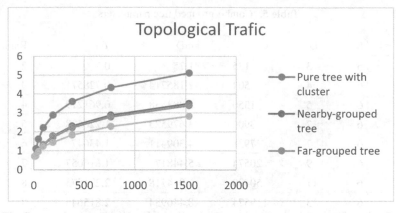

Fig. 7. Graph of the Topological traffic changes depending on the number of nodes

3 Fault Tolerance Analysis

Fault-tolerant design is an innate feature of both Tree topology and Dragonfly topology. There are numerous examples [14] showing how Trees, including simple binary ones, are used as a basis for fault-tolerant organizations, as they both allow for simple routing algorithms, aware of any potential faulty nodes on their path, and allow for creation of complex topologies. Dragonfly topology performs exceptionally well in many modern cloud systems [15, 16] allowing multiple endpoints to be interconnected using a switch, thus increasing fault tolerance at the lowest level. Going higher up, using different sets of parameters (p, a, h) we can achieve different levels of fault tolerance, depending on the cost and degree of the system we are willing to forfeit.

In the previous paragraph we have observed behavior of different variations of proposed Tree-Dragonfly combinations, and the design that employs both "long" and "short" horizontal connections performs better, albeit at a slightly higher cost. Therefore, we propose analysis of Tree-Dragonfly topology with combined horizontal connections for fault-tolerance using some metric to demonstrate how the system reacts to potential faults at the most used nodes. To assess the fault tolerance of the proposed topology, we will introduce the concept of betweenness coefficient. This coefficient indicates the number of shortest paths between some two nodes in the network. A design with solid fault tolerance in mind will allow for a situation where if a node or several nodes with a high betweenness ratio fail, the algorithm for finding a new path should calculate a new shortest path, and the distance of this path may be higher. Nodes with the highest mediation coefficient have a more important role in forming connections with other nodes in the system. The formula for calculating the mediation coefficient is as follows:

$$b_m = \sum_{i \neq j} \frac{B(i, m, j)}{B(i, j)}$$

where B(i,j) is the total number of shortest paths between nodes i and j and B(i,m,j) is the number of shortest paths between i, j passing through node m. Considering that the shortest routes may be unknown and a search algorithm is used in the network, then the mediation of a node can be expressed by the probability of its finding by this algorithm. For this purpose, the concept of the coefficient of dominance of the largest intermediary is introduced, which is calculated according to the formula:

$$CPD = \frac{1}{n-1} \sum_i (B_{max} - B_i)$$

where B_{max} is the largest value of the coefficient of dominance of the largest intermediary.

To search for a path, the modified Floyd-Warshall algorithm was chosen, including capability for memorizing the path between nodes. The Floyd-Warshall algorithm compares all possible paths in the graph between each pair of nodes. For some graph G(V, E), we construct a distance matrix, set the distance of all vertices to themselves as 0. We search for the distance from one vertex to another, and in a separate matrix we mark the node involved in the search for the shortest path. Thus, the output will be two matrices: a matrix of distances between vertices and a mediation matrix. For the proposed topology at the second scaling step, intermediary vertices are marked in green [17].

A Tree-Dragonfly with combined horizontal connections with 21 nodes has several intermediary nodes with high betweenness values. Table below demonstrates how this topology's characteristics change with increasing number of faults. Among other metrics, an analysis of betweenness coefficient for this topology with increasing number of faulted nodes is performed.

Table 6. Amount of shortest paths that go through given nodes

Disabled nodes\ Node number	0	1	2	3	4	5	6	7	8	9	10	11	12	13	14	15	16	17	18	19	20
All nodes work	39	37	59	15	27	41	11	15	13	11	23	21	13	21	19	23	25	7	7	9	5
Node 2 is disabled	29	27		31	43	29	25	23	9	11	23	21	13	21	19	23	25	7	7	9	5
Node 4 is disabled	43	9		33		15	31	25	9	17	17	29	17	17	23	23	25	7	7	9	5
Node Ois disabled		1		5		15	23	55	9	25	17	33	19	17	25	27	25	7	7	9	5
Node 7 is disabled		33		5		15	13		5	25	17	33	17	17	23	35	13	17	7	9	5
Node 15 is disabled		31		5		15	1		25	23	3	25	25	17	29		11	25	7	9	5
Node 1 is disabled				5		15	1		25	23	3	25	25	17	29		11	25	7	9	5

As we can see from the table, using a combined topology allows to utilize nodes on the lower levels of Tree topology more efficiently. Although, in the default Tree-Dragonfly topology topmost nodes remain the most active, as soon as one of them gets deactivated, nodes on lower level get more throughput, assuming the responsibility of the higher-level nodes. Most importantly, we can say that under high stress of roughly 25% of nodes disabled, the overall system remains operational, albeit with lower topological metrics.

4 Results

To analyze the proposed solution, it is suggested to analyze the impact of the modification on the topological characteristics, namely the SD product and the topological traffic. Table 7 presents a comparison of the SD characteristics for the considered topologies. The combined topology is presented in 2 versions depending on the type of level from which the alternation starts: f (far) and n (near). The data is presented in absolute values and in relative values (in parentheses after the absolute values).

Table 7. SD comparison

N	Pure cartesian production cluster x tree	Nearby grouped	Far grouped	Combo grouped (F-type)	Combo grouped (N-type)
9	15	15 (–0)	15 (–0)	15 (–0)	15 (–0)
21	25	24 (–1)	24 (–1)	24 (–1)	24 (–1)
45	35	30 (–5)	30 (–5)	30 (–5)	30 (–5)
93	45	42 (–3)	42 (–3)	42 (–3)	36 (–9)
189	55	54 (–1)	54 (–1)	42 (–13)	48 (–7)
381	65	66 (+1)	66 (+1)	54 (–11)	60 (–5)
765	75	78 (+3)	78 (+3)	66 (–9)	72 (–3)
1533	85	90 (+5)	90 (+5)	78 (–7)	84 (–1)
Summary relative:	0	–1	–1	–49	–31

In the Table 8 the topological traffic of the considered network is described. The evaluation is proposed to be carried out by calculating the average topological traffic for each of the proposed solutions, followed by the calculation of the optimization rate, which will show how many times the average traffic of the obtained topology is better than the traffic of the usual one.

Table 8. Topological traffic comparison

N	Pure cartesian production cluster x tree	Nearby grouped	Far grouped	Combo grouped (F-type)	Combo grouped (N-type)
9	0,7	0,7	0,7	0,7	0,7
21	1,102857	0,728571	0,728571	0,728571	0,728571
45	1,610909	0,96397	0,961279	0,961279	0,963973
93	2,208415	1,31362	1,287751	1,226508	1,172121
189	2,874772	1,763575	1,711115	1,43647	1,518762
381	3,59017	2,289796	2,215403	1,816057	1,896795
765	4,338417	2,869113	2,77907	2,271573	2,387342
1533	5,107627	3,483176	3,382849	2,815645	2,931435
Middle traffic:	2,6916	1,7640	1,7208	1,4945	1,5374
Optimization rate:	1	1,5259	1,5642	1,8010	1,7508

5 Discussion

Analyzing the obtained results, several key points can be underlined. First of all, although the near and far topologies do not differ in terms of SD, there is a slight difference in topological traffic between them. It is natural for the traffic of the far-grouped version to be slightly better, because the connections on the tree are balanced by the long-distance connections connecting its different branches. Another aspect is that layering matters. Thus, starting rotation from the far-grouped layer gives a reduction in SD of 49 points, while the N option provides only a 31 point gain. Similarly, the traffic of the F variant is better on average (1.4945 vs. 1.5374). Presumably, the reason for this gain is that the first dragonfly-connected layer on the 3rd level of the tree is the far one, followed by the near one, but at this stage of scaling, the difference between them is minimal, and the tree is not branched enough for the sequence to give a niticable gain. However, the further transition to switching leads to a sharp increase in long-distance connections, which are critically needed at this stage, and therefore to a decrease in diameter and traffic. This makes this variant of the topology the most suitable for practical use in the future.

6 Conclusion

This article proposes a method for synthesizing a tree topology with horizontal connections, and analyzes various construction options. It was found that the application of the method allows to significantly reduce the diameter (up to 4) at the cost of increasing the degree of topology by 1. In addition, for the best of the presented options, an analysis of the mediation coefficient was carried out in order to determine the stability of the topology to failures. It was revealed that the introduction of additional connections can help in achieving up to -2 diameter at rank 9 for both Nearby-grouped and Far-grouped trees and up to -4 diameter for combo-grouped trees. It is also applied for average diameter. The results show a small improvement in cost for combo-grouped trees but it is not very high. The topological traffic has also improved. However, it comes at a cost of $+1$ degree in the topology. While doing fault-tolerance analysis it was possible to remove about 30% of nodes before the topology was broken.

These results are quite promising. First of all, because a similar method can be applied to other tree topologies - such as FatTree, often used in data centers and supercomputers along with Dragonfly. At the same time, the use of Dragonfly groups as independent computing metaclusters allows you to organize a flexible system for redistributing tasks and resources, because each of them has many connections with other groups above and below the level, which makes interlevel connections extremely dense. The prospects for the application of this property are yet to be explored. The third potential area of application of the described method is the optimization of topologies that are naturally mapped to or decomposed into trees. The organization of the Dragonfly groups themselves remains a separate issue, since in this study the method was applied to an elementary cluster of 3 vertices. There are many prerequisites for the fact that the development of this method in application to a real cluster of at least 8–9 vertices will make it possible to further condense the links, thereby increasing the reduction in diameter and fault tolerance.

References

1. Alverson, R., Roweth, D., Kaplan, L.: The gemini system interconnect. In: 2010 18th IEEE Symposium on High Performance Interconnects, pp. 83–87 (2010). https://doi.org/10.1109/HOTI.2010.23
2. Kim, J., Dally, W.J., Scott, S., Abts, D.: Technology-driven, highly-scalable dragonfly topology. ACM SIGARCH Comput. Archit. News 36(3), 77–88 (2008). https://doi.org/10.1145/1394608.1382129
3. Ajima, Y., Sumimoto, S., Shimizu, T.: Tofu: a 6D Mesh/Torus interconnect for exascale computers. Computer (Long. Beach. Calif). 42(11), 36–40 (2009). https://doi.org/10.1109/MC.2009.370
4. Khan, Z.A., Siddiqui, J., Samad, A.: Linear crossed cube (LCQ): a new interconnection network topology for massively parallel system. IJCNIS 7(3), 18–25 (2015). https://doi.org/10.5815/ijcnis.2015.03.03
5. Guan, K.C., Chan, V.W.S.: Cost-efficient fiber connection topology design for metropolitan area WDM networks. J. Opt. Commun. Netw. 1(1), 158 (2009). https://doi.org/10.1364/JOCN.1.000158
6. Hadeed, W., Abdullah, D.B.: Load balancing mechanism for edge-cloud-based priorities containers. Int. J. Wirel. Microwave Technol. (IJWMT) 12(5), 1–9 (2022). https://doi.org/10.5815/ijwmt.2022.05.01
7. Kim, J., Dally, W., Scott, S., Abts, D.: Cost-efficient dragonfly topology for large-scale systems. IEEE Micro 29(1), 33–40 (2009). https://doi.org/10.1109/MM.2009.5
8. Camarero, C., Vallejo, E., Beivide, R.: Topological characterization of hamming and dragonfly networks and its implications on routing. ACM Trans. Archit. Code Optim. 11(4), 1–25 (2015). https://doi.org/10.1145/2677038
9. Shpiner, A., Haramaty, Z., Eliad, S., Zdornov, V., Gafni, B., Zahavi, E.: Dragonfly+: low cost topology for scaling datacenters. In: 2017 IEEE 3rd International Workshop on High-Performance Interconnection Networks in the Exascale and Big-Data Era (HiPINEB), pp. 1–8 (2017). https://doi.org/10.1109/HiPINEB.2017.11

10. Volokyta, A., et al.: Extended DragonDeBrujin topology synthesis method. Int. J. Comput. Netw. Inform. Secur. (IJCNIS) **14**(6), 23–36 (2022). https://doi.org/10.5815/ijcnis.2022.06.03

11. Hoefler, T., Schneider, T., Lumsdaine, A.: Characterizing the influence of system noise on large-scale applications by simulation. In: Proceedings 2010 ACM/IEEE International Conference High Performance Computing Networking, Storage Analysis, pp. 1–12 (2010)

12. Wang, H., Yuan, X., Kandemir, M.: Congestion-aware routing algorithms for Dragonfly networks. IEEE Trans. Comput. 2831–2845 (2014)

13. Bland, W., Brightwell, R., Grant, R., Miller, P.: Dynamic load-balancing in a Dragonfly network. In: Proceedings 2011 International Conference High Performance Computing Networking, Storage Analysis. IEEE, pp. 1–12 (2011)

14. Loutskii, H., et al.: Topology synthesis method based on excess De Bruijn and dragonfly. In: Hu, Z., Petoukhov, S., Dychka, I., He, M. (eds.) ICCSEEA 2021. LNDECT, vol. 83, pp. 315–325. Springer, Cham (2021). https://doi.org/10.1007/978-3-030-80472-5_27

15. Neelima, P., Reddy, A.R.M.: An efficient load balancing system using adaptive dragonfly algorithm in cloud computing. Clust. Comput. **23**(4), 2891–2899 (2020). https://doi.org/10.1007/s10586-020-03054-w

16. Polepally, V., Shahu Chatrapati, K.: Dragonfly optimization and constraint measure-based load balancing in cloud computing. Clust. Comput. **22**(1), 1099–1111 (2017). https://doi.org/10.1007/s10586-017-1056-4

17. Hougardy, S.: The Floyd-Warshall algorithm on graphs with negative cycles. Inf. Process. Lett. **110**(8–9), 279–281 (2010). https://doi.org/10.1016/j.ipl.2010.02.001

Hardware Modified Additive Fibonacci Generators Using Prime Numbers

Volodymyr Maksymovych[1], Krzysztof Przystupa[2], Oleh Harasymchuk[1], Mariia Shabatura[1(✉)], Roman Stakhiv[1], and Viktor Kuts[1]

[1] Lviv Polytechnic National University, Bandery 12, Lviv 79013, Ukraine
mariia.m.mandrona@lpnu.ua

[2] Department of Automation, Lublin University of Technology, Nadbystrzycka 36, 20-618 Lublin, Poland
k.przystupa@pollub.pl

Abstract. The article presents hardware modified additive Fibonacci generators, which use modular addition, with the base of the prime number. Generators differ from the classical presence in their composition of the logic circuit, which is the base of arithmetic addition of an additional component, thus enhancing the chaotic nature of the formation of random numbers. The analysis of statistical characteristics of these generators for large values of arguments is carried out. The implementation of the proposed generator structures allows, in comparison with known devices significant increasing the repetition period of the generated pseudo-random sequence for the vast majority of initial settings of structural elements of the circuit while providing satisfactory statistical characteristics of the output sequence. This will contribute to the improvement of the characteristics of existing systems that include generators of pseudorandom sequences and to the expansion of their scope of use.

Keywords: pseudorandom bit sequence generator · statistical tests · two-level frequency synthesizers

1 Instruction

Currently, the pseudorandom bit sequence generators (PRBSG) and pseudorandom number generators (PRNG) are gaining considerable popularity. This is primarily because they are used in systems of information technical protection, providing cybersecurity for computer networks, forecasting, computer games etc. As a result of such popularity, new and more refined demands for the quality of the generated sequences are constantly being made to such generators. Such sequences should, by their characteristics, be as close as possible to truly random sequences, satisfy the passage of the most famous sets of statistical tests, have the maximum possible repetition period, and be unpredictable and resistant to known attacks. The requirements for the generators themselves are significantly increasing; in particular, they require increasing their speed and simplicity of software and hardware implementation. PRNG and PRBSG having satisfactory statistical characteristics [1–5] are often involved in solving the problems of cyber security, such us cryptographic key generation, password hashing, streaming encryption, authentication, network protocol and key generation. Therefore, in this direction are active work to modify existing and find new methods and ways to build such generators in order to improve their performance and expand the scope.

The choice to improve the basic structure of AGF was made based on the results of our previous investigations [5–13], which confirm the possibility of improving their characteristics, in particular, through the introduction of an additional logic scheme. Unlike many known generators, such as generators based on linear feedback shift registers (LFSR), today, AGF with the additional logic scheme, is unpredictable.

The goal of this particular paper is to develop and investigate the characteristics of the new basic structure of AGF, which differs from the classical AGF in the ability to operate with an arbitrary module of the recurrent equation, e.g. with a module whose value is a prime number. In addition, the structure of AGF remains an additional structural element - a logic scheme, which adds to the arithmetic process an additional component, thus enhancing the chaotic nature of the formation of pseudorandom numbers. Exactly this innovation which is proved in this work, allows significantly improved characteristics of pseudorandom sequences generators of this type.

In this paper the authors consider the method of investigating the characteristics of generators based on creating two models – hardware and abstract, proving the identity of the results obtained on their basis.

The abstract model is used to research the characteristics of generators with a large number of discharges of their structural elements, which significantly reduces the research time. The hardware model allows going to the hardware implementation of generators, for example, when they are implemented in the base crystal of a programmable logic integrated circuits, which confirms the direction of this development in the hardware implementation of devices.

1.1 Relates Works

Many researchers and scientific institutions have been developing and researching pseudorandom number and bit sequence generators for a long time [1, 2, 14–27]. The range of their applications depends on the quality of designed and implemented generators. Moreover, the areas of their use are highly diverse.

Random number sequences are essential in computing. Password salts, ASLR offsets TLS nonces, DNS source port numbers and TCP/IP sequence numbers all rely on random numbers. Moreover, in cryptography random sequences, it found in keys generation, encryption systems and cryptosystems under attack. If there were no randomness, all crypto systems would be insecure due to their predictability.

Pseudorandom number generators and bit sequences and approaches to their construction are significant and critical for solving information protection problems [18].

Random numbers are essential input materials for the Internet of Things (IoT) functions [21–23]. The random number generator (RNG) is capable for being used for ensuring cryptographic security in low-power branches of Smart Grid [23].

A large number of PRBSG and PRNG are known, differing in their characteristics, construction methods, and areas of potential application. Fibonacci generators play an important role in the vast majority of PRNG. The principles of additive Fibonacci generators (AFG) construction and the area of their use are discussed in many papers [1–5]. Those aimed at hardware implementation use modular addition, with a base equal to the power of two [23–26]. However, with this approach, their statistical characteristics, and therefore crypto stability, are unsatisfactory, as proven by various studies. It is considered that such generators may only be part of the pseudorandom sequence generators (PRSG) or information security systems as components or additional units. Therefore, some modifications of the primary known methods are often used to obtain better results. In particular, [27] describes a high-speed crypto-graphically secure pseudorandom sequence generator built on a combination of three connected Fibonacci generators. Its repetition period is much more extended than of a conventional Fibonacci generator with a delay. His output is unpredictable. The generated sequence successfully passes the most severe random test kits.

Summarizing the analysis of existing pseudorandom sequences generators based on AGF, their main drawback is limited functionality due to the impossibility of ensuring their functioning based on a recurrent equation with an arbitrary modulus. This work is aimed at overcoming this shortcoming under the condition of ensuring satisfactory statistical characteristics of the output sequence.

The analysis of AFG structures indicates their similarity to the structures of modified two-level frequency synthesizers proposed by us in [28, 29]. This allows creating a new type of modified additive Fibonacci generators (MAFG) that uses modular addition with a prime number.

The purpose of the work is the hardware implementation of MAFG, which uses modular addition, with the prime number basis.

1.2 Structural Circuits and Researching of MAFG

In this paper, we will consider only one of the possible options for building an AFG generator, which based on the equation

$$x_i = (x_{i-2} + x_{i-1}) mod \ m \tag{1}$$

Structural circuits of two MAFG versions are shown in Fig. 1 and Fig. 2.

The MAFG version 1 contains: two adders SM1, SM2; three registers Rg1–Rg3; one multiplexer MS and one logical element OR. The MAFG version 2 has the additional logic scheme LS.

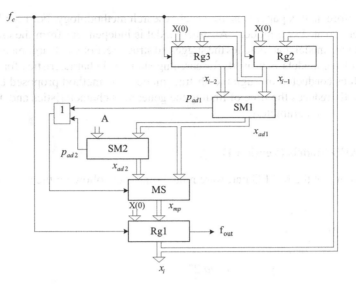

Fig. 1. Structural circuit of MAFG (version 1)

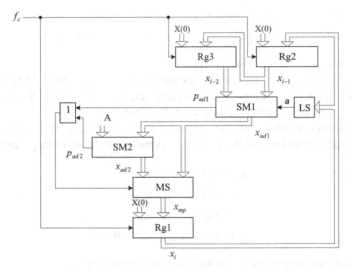

Fig. 2. Structural circuit of MAFG (version 2)

2 Methodology for the Researching of MAFG Statistical Characteristics

AFG, like any cryptographic pseudorandom sequence generators, used arguments which values are very large. Herewith, there are certain problems in the procedure of determining their statistical characteristics.

Therefore, in this paper was used such research methodology. For both MAFG versions were created two models. Abstract model is independent from the circuit design and hardware model that matches the suggested structure. For small arguments values the identity of the models is proved. Researching statistical characteristics for arguments' large values conducted through the abstract model. The method proposed by us allows significantly reduces the time of finding the generators characteristics and, thus, allows to optimize their parameters.

2.1 MAFG Models (Version 1)

The version 1 of the MAFG hardware model functions following the algorithm:

$$x_{i-2} = x_{i-1},$$
$$x_{i-1} = x_i,$$
$$x_i = x_{mp},$$
$$x_{ad1} = (x_{i-2} + x_{i-1}) mod\ 2^n$$
$$if\ (x_{i-2} + x_{i-1}) < 2^n\ then\ p_{ad1} = 0\ else\ p_{ad1} = 1$$
$$x_{ad2} = (x_{ad1} + A) mod\ 2^n$$
$$if\ (x_{ad1} + A) < 2^n\ then\ p_{ad2} = 0\ else\ p_{ad2} = 1$$
$$if\ (p_{ad1} = 0)\ and\ (p_{ad2} = 0)\ then\ x_{mp} = x_{ad1}\ else\ x_{mp} = x_{ad2}$$

where: x_1, x_{i-1}, x_{i-2} – numbers in registers,

Rg1-Rg3, x_{ad1}, x_{ad2} – numbers on the sum outputs of the adders SM1 and SM2,

p_{ad1} Pad1, Pad2 x_{ad2} – numbers on the carry outputs of the adders SM1 and SM2,

x_{mp} x_{mp} – number on the multiplexer MS output,

n n – number of binary bits of the generator structural components (Fig. 1).

The abstract model of MAFG version 1 operating on the following expressions:

$$x_{i-2} = x_{i-1},$$
$$x_{i-1} = x_i,$$
$$x_i = x_{mp},$$
$$x_{ad1} = (x_{i-2} + x_{i-1}) mod\ m \tag{3}$$

where m is an arbitrary number, in this case a prime number.

Figure 3 presents the dependencies of the pseudorandom numbers X X, formed by the MAFG versus the iteration p number i, for both the hardware and abstract models at the same initial value $X(0)$.

Thus X X and $X(0)$ are determined from the formulae:

$$X = m^2 . x_{i-1} + x_{i-2} \tag{4}$$

$$X(0) = m^2 . x_i(0) + x_{i-1}(0) + x_{i-2}(0) \tag{5}$$

where $x_i(0), x_{i-1}(0), x_{i-2}(0)$ are the initial values of the numbers x_i, x_{i-1}, x_{i-2}

Figure 4 represents the dependences of the repetition periods of the MAFG pseudorandom number sequence versus the initial values X(0) $X(0)$.

The obtained results (Figs. 3, 4) show completely identical picture for both the hardware and abstract models of pseudorandom number sequence formation. Similar results were obtained for the other prime numbers m m.

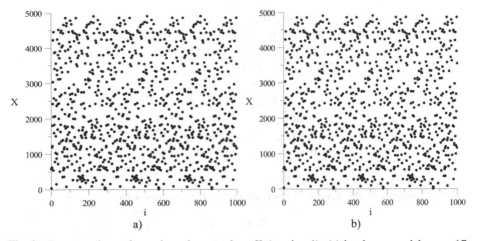

Fig. 3. Current values of pseudorandom numbers X (version 1): (a) hardware model: $m = 17$, $n = 5, A = 2^n - m = 15, X(0) = 0 \div m^3 - 1$; (b) abstract model: $m = 17, X(0) = 0 \div m^3 - 1$

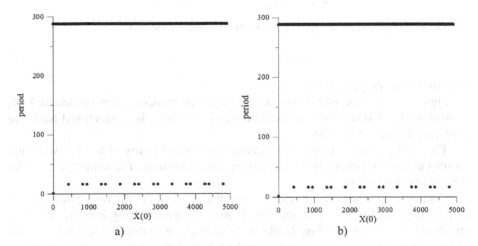

Fig. 4. Repetition periods versus $X(0)$ (version 1): (a) hardware model: $m = 17, n = 5, A = 2^n - m = 15, X(0) = 0 \div m^3 - 1$; (b) abstract model: $m = 17, X(0) = 0 \div m^3 - 1$

2.2 MAFG Models (Version 2)

The hardware model of MAFG version 2 functions by the following algorithm:

$$x_{i-2} = x_{i-1},$$
$$x_{i-1} = x_i,$$
$$x_i = x_{mp},$$
$$a = a_0 \oplus a_1 \oplus \ldots \oplus a_3$$
$$x_{ad1} = (x_{i-2} + x_{i-1} + a) \bmod 2^n,$$
$$\text{if } (x_{i-2} + x_{i-1} + a) < 2^n$$
$$\text{then } p_{ad1} = 0 \text{ else } p_{ad1} = 1 \tag{6}$$
$$x_{ad2} = (x_{ad1} + A) \bmod 2^n$$
$$\text{if } (x_{ad1} + A) < 2^n$$
$$\text{then } p_{ad2} = 0 \text{ else } p_{ad2} = 1$$
$$\text{if } (p_{ad1} = 0) \text{ and } p_{ad2} = 0)$$
$$\text{then } x_{mp} = x_{ad1} \text{ else } x_{mp} = x_{ad2}$$

where a_i is the value of x_i binary bits.

The abstract model version 2 operates on the next expressions:

$$x_{i-2} = x_{i-1},$$
$$x_{i-1} = x_i,$$
$$x_i = x_{mp}, \tag{7}$$
$$a = a_0 \oplus a_1 \oplus \ldots \oplus a_3$$
$$x_{ad1} = (x_{i-2} + x_{i-1} + a) \bmod m.$$

The quantity of number x_i bits, taken into account in the expression

$$a = a_0 \oplus a_1 \oplus \ldots \oplus a_3 \tag{8}$$

may vary in the range of $1 \div n$.

Figure 5 shows the plot of the current values of pseudorandom sequences X X, formed by the MAFG versus the iteration step i i, for both the abstract and hardware models at the same X(0) $X(0)$.

The results shown on Figs. 5, 6 confirm the whole identity of the hardware and abstract models of pseudorandom number sequence formation. The same results for the other prime numbers m.

Comparison of the research results of the characteristics of the two options for the construction of MAGF (Fig. 1 and Fig. 2) shows that the adding into the generator structure logic scheme LS (Fig. 2) allows significantly increasing the repetition period of the output pseudorandom sequence for the vast majority of initial settings of circuit structural components.

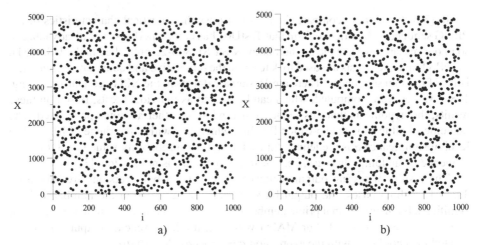

Fig. 5. Current values of pseudorandom numbers XX (version 2): (a) hardware model: $m = 17$, $n = 5, A = 2^n - m = 15, a = a_0 \oplus a_1 \oplus a_2 \oplus a_3, X(0) = 2$; (b) abstract model: $m = 17$ $m = 17$, $a = a_0 \oplus a_1 \oplus a_2 \oplus a_3, X(0) = 2$

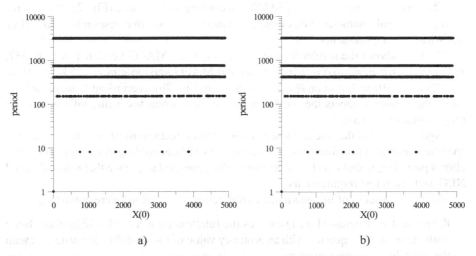

Fig. 6. Repetition period versus $X(0)$ $X(0)$ (version 2): (a) hardware model: $m = 17, n = 5, A = 2^n - m = 15, a = a_0 \oplus a_1 \oplus a_2 \oplus a_3, X(0) = 0 \div m^3 - 1$; (b) abstract model: $m = 17, a = a_0 \oplus a_1 \oplus a_2 \oplus a_3, X(0) = 0 \div m^3 - 1$.

3 Methodology of Estimation Pseudorandom Sequence

The quality of PRBSG and PRNG can be evaluated using a variety of tests that can be graphical or statistical. Statistical test packages are currently being developed, based on different verification algorithms and giving a complete picture of the generated sequence quality. Most popular tests: NIST, Diehard, Dieharder, FIPS 140, TestU01, ENT, AIS 20, AIS 31, Crypt-X, SPRNG.

In [30], the authors compare the tests to assess the statistical characteristics of the PRNG sequences. In article inform that TestU01 has certain performance disadvantages, so in our research we use the methodology NIST SP 800-22 published in the USA [31].

NIST statistical tests are used to determine the qualitative and quantitative characteristics of random sequences. There are 15 statistical tests in the NIST set. During testing, 188 P-values were computed, which can be treated as individual tests. Based on these results, the hypothesis that this sequence is random is either accepted or rejected. The outcome of all tests is the so-called P-value, which lies within the range of [0, 1]. The level of significance α is predefined for each test. If the probability value $P \geq \alpha$, then the sequence is random, otherwise, i.e. $P < \alpha$, the sequence is not random [31].

The results of testing output sequence of MAFG of both versions are shown on Table 1. In both cases, the output bit sequence was taken from the number x_i least significant bit, using Fermat prime number $m = 65537$ and the same initial settings $x_i(0) = x_{i-1}(0) = = x_{i-2}(0) = 1$. For MAFG version 2 the logic scheme output signal was created in accordance with the expression $a = a_0 \oplus a_1 \oplus ... \oplus a_{16}$.

Table 1 shows that MAFG version 1 did not pass eight tests (shown "–"). Comparing MAFG version 1 and version 2 shows that the second passed all NIST tests. This shows that the MAFG version 2 has passed all NIST tests.

Therefore, the construction of MAFG, according to version 2 (Fig. 2), allows providing statistical characteristics of the generated sequence that responds to all NIST statistical test requirements.

Figure 7 shows the results for the output sequence MAFG version 1 $m = 65537$, $x_i(0) = x_{i-1}(0) = x_{i-2}(0) = 1$ in Fig. 7a) and MAFG version 2 $m = 65537$, $x_i(0) = x_{i-1}(0) = x_{i-2}(0) = 1$, $a = a_0 \oplus a_1 \oplus ... \oplus a_{16}$ in Fig. 7b). According to methodology NIST the sequence meets the condition randomness when test value falls within the range when (0.98–1.0).

Figure 7a shows that the sequence is not random because most tests lie outside the specified interval. Therefore, the construction of MAFG, according to version 2 (Fig. 7b), allows providing statistical characteristics of the generated sequence that responds to all NIST statistical test requirements.

So, in the process of modulation, two theoretical innovations were confirmed:

- the presented structure (Fig. 1) ensures the functioning of the MAGF in accordance with the recurrent equation with an arbitrary value of the modulus, in particular, with the modulus – a prime number;
- the introduction of an additional logic circuit (Fig. 2) into the structure of the MAGF ensures a significant improvement of its main characteristics.

Table 1. NIST results testing MAFG

№	Statistical Test	MAFG version 1		MAFG version 2	
		p-value	status	p-value	status
1	Frequency	0.943242	+	0.292291	+
2	Block Frequency	0.044220	+	0.622168	+
3	Cumulative Sums	0.571161	+	0.911247	+
4	Runs	0.000000	-	0.377007	+
5	Longest Run	0.000000	-	0.350485	+
6	Rank	0.971006	+	0.757790	+
7	FFT	0.128132	+	0.192724	+
8	Non-Overlapping Template	0.000000	-	0.626709	+
9	Overlapping Template	0.000000	-	0.368587	+
10	Universal	0.000000	-	0.289667	+
11	Approximate Entropy	0.000000	-	0.096000	+
12	Random Excursions	0.000000	-	0.958867	+
13	Random Excursions Variant	0.253205	+	0.938062	+
14	Serial	0.000000	-	0.177628	+
15	Linear Complexity	0.653773	+	0.566688	+

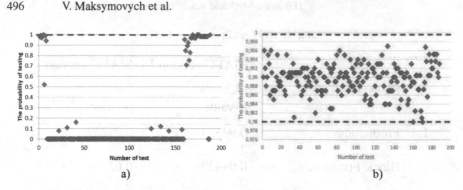

Fig. 7. Statistical portrait of PRNG: (a) MAFG version 1, (b) MAFG version 2

4 Summary and Conclusion

The proposed method of researching the characteristics of generators is based on creating two models – hardware and abstract. It allows proving the operation of MAGF according to the given algorithm, based on a recurrent equation with an arbitrary modulus. It simply moves to hardware implementation of generators, for example, their implementation in programmable logic integrated circuits.

The proposed approach for hardware implementation of modified additive Fibonacci generators allows determining their statistical characteristics at very large values of arguments. It is proved that adding to the structure of generator the logic scheme allows significantly increasing the repetition period of output pseudorandom sequence for most initial settings of the circuit basic components. This extends the possible range to use sequences generated by such generators. The obtained results of this article may be helpful for other researchers who are working in this area.

Further research in this direction will be the generalization of obtained results on the AGF structures with an arbitrary number of registers and the implementation of these structures on programmable logic integrated circuit.

Acknowledgment. The studies carried out in within the project Research & Development of Ukraine «Generator» №0122U000954.

References

1. Schneier, B.: Applied Cryptography: Protocols, Algorithms, and Source Code in C, p. 578. Wiley, New Jersey, USA (2007)
2. Gorbenko I.D., Gorbenko Y.I.: Applied Cryptology: Theory. Practice. Application, p. 754. Fort Publishing House, Kharkiv, Ukraine (2012)
3. Maksymovych, V., Kostiv, Y., Mandrona, M., Harasymchuk, O.: Generator of pseudorandom bit sequence with increased cryptographic immunity. Metall. Min. Ind. **6**, 24–28 (2014)
4. Maksymovych, V., et al.: Simulation of authentication in information-processing electronic devices based on poisson pulse sequence generators. Electronics **11**(13), 2039 (2022)
5. Karpinski, M., Maksymovych, V., Harasymchuk, O., Sawicki, D., Shabatura, M., Jancarczyk, P.: Development of additive Fibonacci generators with improved characteristics for cybersecurity needs. Appl. Sci. **12**(3), 1519 (2022)

6. Maksymovych, V., Karpinski, M., Harasymchuk, O., Kajstura, K., Shabatura, M., Jancarczyk, D.: A new approach to the development of additive fibonacci generators based on prime numbers. Electronics **10**(23), 2912 (2022)
7. Maksymovych, V.N., Mandrona, M.N.: Comparative analysis of pseudorandom bit sequence generators. J. Autom. Inf. Sci. **49**(3), 78–86 (2017)
8. Shabatura, M.M., Maksymovych, V.M., Harasymchuk, O.I.: Dosimetric detector hardware simulation model based on modified additive fibonacci generator. Adv. Intell. Syst. **938**, 162–171 (2019)
9. Maksymovych, V.N., Mandrona, M.N., Harasymchuk, O.I.: Designing generators of poisson pulse sequences based on the additive fibonacci generators. J. Autom. Inf. Sci. **49**(12), 1–12 (2017)
10. Maksymovych, V.N., Harasymchuk, O.I., Kostiv, Y.M., Mandrona, M.N.: Investigating the statistical characteristics of poisson pulse sequences generators constructed in different ways. J. Autom. Inf. Sci. **49**(10), 11–19 (2017)
11. Maksymovych, V., Kostiv, Y., Garasimchuk, O., Mandrona, M.: A study of the characteristics of the fibonacci modified additive generator with a delay. J. Autom. Inf. Sci. **48**, 76–82 (2016)
12. Maksymovych, V., Harasymchuk, O., Opirskyy, I.: The designing and research of generators of poisson pulse sequences on base of fibonacci modified additive generator. In: Zhengbing, H., Petoukhov, S., Dychka, I., He, M. (eds.) ICCSEEA 2018. AISC, vol. 754, pp. 43–53. Springer, Cham (2019). https://doi.org/10.1007/978-3-319-91008-6_5
13. Maksymovych, V., Shevchuk, R., Shabatura, M., Sawicki, P., Harasymchuk, O., Zajac, T.: Combined pseudo-random sequence generator for cybersecurity. Sensors **22**(24), 9700 (2022)
14. Milovanovic, E.I., Milovanovic, I.Z., Stojcev, M.K., Stamenkovic, Z., Nikolic, T.R.: Concurrent generation of pseudo random numbers with lfsr of fibonacci and galois type. Comput. Inform. **34**(4), 941–958 (2015)
15. Chen, J., Su, J., Kochan, O., Levkiv, M.: Metrological software test for simulating the method of determining the thermocouple error in situ during operation. Measure. Sci. Rev. **18**(2), 52–58 (2018)
16. Fang, M.T., Chen, Z.J., Przystupa, K., Li, T., Majka, M., Kochan, O.: Examination of abnormal behavior detection based on improved YOLOv3. Electronics **10**(2), 197 (2021)
17. Away, Y., Noor, R.S.: FPGA-based design system for a two-segment fibonacci lfsr random number generator. Int. J. Electric. Comput. Eng. **7**(4), 1882–1891 (2017)
18. Dichtl, M., Golić, J.D.: High-speed true random number generation with logic gates only. In: Paillier, P., Verbauwhede, I. (eds.) CHES 2007. LNCS, vol. 4727, pp. 45–62. Springer, Heidelberg (2007). https://doi.org/10.1007/978-3-540-74735-2_4
19. Barker, E.B., Kelsey, J.M.: Recommendation for random bit generator (RBG) constructions. Gaithersburg, USA (2016)
20. Baldanzi, L., et al.: Cryptographically secure pseudo-random number generator IP-core based on SHA2 algorithm. Sensors **20**(7), 1869 (2020)
21. Kietzmann, P., Schmidt, T.C., Wählisch, M.: A guideline on pseudorandom number generation (PRNG) in the IoT. CSUR **54**(6), 1–38 (2021)
22. Souaki G., Halim K.: Random number generation based on MCU sources for IoT application. ATSIP, Morocco, pp. 1–6 (2017)
23. Fujdiak, R., Mlynek, P., Misurec, J., Masek, P. Design of low-power random number generator using signal quantization error in smart grid. TSP, pp. 7–10 (2016)
24. Park, S., Kim, K., Nam, C.: Dynamical pseudo-random number generator using reinforcement learning. Appl. Sci. **12**(7), 3377 (2022)
25. Hu, Z., Dychka, I., Sulema, Y., Radchenko, Y.: Graphical data steganographic protection method based on bits correspondence scheme. IJISA **9**(8), 34–40 (2017)
26. Hu, Z., Dychka, I., Onai, M., Zhykin, Y.: Blind payment protocol for payment channel networks. IJCNIS **11**(6), 22–28 (2019). https://doi.org/10.5815/ijcnis.2019.06.03

27. Hu Z., Gnatyuk S., Okhrimenko T., Tynymbayev S., Iavich M.: High-speed and secure PRNG for cryptographic applications. IJCNIS **12**(3), 1–10 (2020). https://doi.org/10.5815/ijcnis.2020.03.01

28. Knut, D.: The Art of Computer Programming, p. 386. Fundamental algorithms. Massachusets, USA (1998)

29. Maksymovych, V., Stakhiv, R., Stakhiv, M.: Modified structure of two-level digital frequency synthesizer for dosimetry. Measur. Equip. Metrol. **80**(1), 17–20 (2019)

30. Hurley Smith D., Hernandez Castro J.: Great expectations: a critique of current approaches to random number generation testing & certification. In: Proceedings 4th International Conference, SSR Darmstadt, Germany, pp. 143–163 (2018)

31. NIST SP 800-22: A Statistical Test Suite for Random and Pseudorandom Number Generators for Cryptographic Applications (2010). https://nvlpubs.nist.gov/nistpubs/Legacy/SP/nistspecialpublication800-22r1a.pdf

Data Acquisition System for Monitoring Soil Parameters

Tetiana Fedyshyn[1]([✉]), Krzysztof Przystupa[2], Tetiana Bubela[1], and Iryna Petrovska[1]

[1] Information Measuring Technologies Department, Lviv Polytechnic National University, 79013 Lviv, Ukraine
tetiana.i.fedyshyn@lpnu.ua
[2] Department of Automation, Lublin University of Technology, 20–618 Lublin, Poland

Abstract. Monitoring of soil parameters is important for farmers to make the right management decisions. This is especially useful for small farms, where the farmer needs to have up-to-date information about the condition of his land. Laboratory soil testing is a time-consuming and labor-intensive process. Therefore, the development of a mobile application for Android and iOS as a cyber-physical system for monitoring the main soil parameters (humidity, temperature, acidity, density) is important and useful. The data acquisition system proposed by the authors for a cyber-physical system that receives information from sensors ensures control over the processes of growing crops. User data and location information are stored in real time in the program's cloud database. In order to predict the planned yield with the most economical use of available resources, it is important to choose the correct dates for sowing seeds and harvesting. Therefore, the authors proposed to implement a neural network in a cyber-physical system. A neural network model for predicting grain yields as part of a cyber-physical system is proposed. The authors conducted an analytical study of the influential factors on soil condition and the consequences that variations in these factors may lead.

Keywords: Monitoring · Soil · Cyber-Physical System · Neural Network

1 Introduction

There are no generally accepted recommendations for the formation of the structure of soil parameters and methods of their research to meet the needs of information monitoring systems. Classical physicochemical methods are usually implemented in laboratories and are not suitable for field conditions [1]. Precision analytical spectroscopic methods are being actively developed to control the physical and chemical parameters of soils [2, 3], but they are expensive and can be realized only in laboratory conditions. It should be noted that methods using bioindicators are rapidly developing today [4, 5]. However, they are considered labor-intensive and not suitable for operational monitoring. The control of such a parameter as soil moisture is so important that an international soil moisture network has been created to provide users with the necessary information. It is also especially important to control the impact of soil moisture dynamics on modeling hydrological processes associated with heterogeneity of environmental factors (e.g.,

© The Author(s), under exclusive license to Springer Nature Switzerland AG 2023
Z. Hu et al. (Eds.): ICCSEEA 2023, LNDECT 181, pp. 499–513, 2023.
https://doi.org/10.1007/978-3-031-36118-0_45

topographic features, soil properties, land use types, and previous precipitation) [6]. To correct for the effects of cross-sensitivity, impedance spectroscopy is used to measure soil moisture with increased accuracy based on a multivariate analysis of the results of experimental studies of various natural soils. However, the proposed correction procedure is time-consuming and unsuitable for implementation in a portable monitoring tool. Real-time continuous monitoring possesses a great potential to revolutionize field measurements by providing first-hand information for continuously tracking variations of heterogeneous soil parameters and diverse pollutants in a timely manner and thus enable constant updates essential for system control and decision-making [7]. Therefore, the creation of inexpensive and effective monitoring systems based on cyber-physical systems (CPS) is particular relevant today. Such solutions are already being proposed and implemented on the basis of PCs [8], but access to PCs is not always possible and does not provide mobility. A reliable data acquisition system for CPS should be user-friendly and mobile. Mobile applications are being developed, for example, for fire monitoring and plant identification [9, 10]. Therefore, the implementation of the data acquisition system for CPS in the form of a mobile application proposed by the authors is a good solution to the problem of soil monitoring.

1.1 Research Methods

To achieve the goal set in the article, namely the development of a data acquisition system for CPS, we have identified certain tasks and methods for solving them. First of all, it is the justification of the choice of parameters for soil monitoring. For this purpose, the methods of interdisciplinary theoretical analysis and system analysis were used. The methods of system engineering were used to create the CPS. For the CPS data collection and transmission system, the technology of selecting a set of object parameters from the relevant measurement arrays and databases was used. The data collection from sensors is proposed to be realized using Wi-Fi technology, which provides a data transfer rate of more than 100 Mbit/s, and users can move between access points within the coverage area of the Wi-Fi network using devices equipped with Wi-Fi client transceivers and access the Internet. The reliability of the presented results is ensured by conducting a large volume of measurements and experimental studies. The accuracy of the presented measurement results was evaluated using probability theory and mathematical statistics. Predictive modeling is ensured by the use of neural networks, namely the architecture of neural recurrent networks of the Long Short Term Memory type [11].

2   Analytical Investigation of Influencing Factors on Soil Condition

In order to provide objective information and make the right decisions, soil monitoring should provide initial, current and periodic data on the main characteristics of the soil cover [12]. Mandatory evaluation criteria are granulometric and mineralogical composition, data on the humus state of the soil, evaluation data on the formation of soil regimes (Table 1).

Humans use soils very efficiently, not only without destroying them but even increasing their fertility or turning naturally useless lands into fertile land. At the same time, in

Table 1. Indicators, evaluation criteria and processes monitored during operational soil monitoring

Indicators, evaluation criteria	Controlled processes	Term
Nutritional mode: −content of mobile forms of macro- and microelements; −used reserves of macronutrients	Nutrient supply of soil	Every year

Water mode:
      −soil moisture;
      −reserves of productive moisture;
      −WGV regime on reclaimed
          lands
	Features of accumulation and consumption of moisture in the soil	Once a month

Temperature mode: − soil temperature; − heating index of the sum of active and effective temperatures	Heat exchange in soils, heat supply of soil	Every year

Assessment of agroecological condition:
  − balance of nutrients;
  − density of radioactive contamination;
  − content of mobile forms of heavy metals;
  − pesticide content
	Potential of soil fertility in terms of nutrient content and production of ecologically clean agricultural products	Every 5 years
		Every year
		Every 5 years
		Every 5 years

Estimation of intensity of manifestations of erosion: − development of water linear and planar erosion; − deflation and occurrence of dust storms; − irrigation erosion	Development and intensity of erosion processes	Every 15-20 years
Assessment of physical and chemical condition of soils: − pH salt and water; − hydrolytic acidity; − cation exchange capacity; − the degree of saturation of the bases	Change of physicochemical properties of soil, manifestations of physicochemical degradation	Every 5 years

the history of human civilization, more productive soils have been irrevocably destroyed and lost than are currently being plowed up around the world.

The analysis shows that the following main influencing factors on soil characteristics can be identified:

− fertilizer input;
− solar radiation;
− soil cultivation.

In turn, the main parameters of operational soil monitoring are:

- pH;
- temperature;
- humidity;
- hardness.

2.1 Evaluation of Soil Acidity Changes

Changes in soil acidity significantly affect the availability of nutrients to plants. Excessively high (more than 9.0) and excessively low (less than 4.0) soil pH have a toxic effect on plant roots. Within these pH indicators, the behavior of individual nutrients is determined, namely their deposition or conversion into available or inaccessible to plants forms. Thus, in very acidic soils (pH 4.0–5.5) such elements as iron, aluminum and manganese pass into easily accessible for assimilation by plants forms, in addition, their concentration reaches a toxic level [13]. At high acidity of the soil, its filtration capacity, capillarity and permeability deteriorate. As a result of starvation, under certain conditions, cultivated plants can die even without good reason.

On very alkaline soils (pH 7.5–8.5), on the contrary, there is a significant decrease in the availability of plants such as iron, manganese, phosphorus, copper, zinc, boron and most trace elements. Optimal for plants, the reaction of soils with a pH of 6.5 allows most nutrients to remain in a form accessible to plants in the soil solution, which usually prevents their lack. A sharp change in soil acidity can be provoked by: acid rain, insufficient fertilizer application or active use of chemical plant protection products.

2.2 Characterization of the Thermal Regime of the Soil

The question influence of soil temperature on plant growth in the context of global climate change is even more relevant, which stimulates the growing interest scientists in studying the mechanisms and influence of soil temperature on plant ecosystems. Virtually all processes that occur in the soil, from the weathering of primary minerals to plant nutrition and the accumulation of organic carbon, are highly dependent on soil temperature. The main factor influencing the temperature of the soil is solar insolation, which depends on the geographical location, exposure of the area, season and the presence of clouds [14]. The amount of heat absorbed by the soil surface depends on the color of the soil, the presence of plant residues or vegetative plants, organic matter content and evaporation. The ability of the soil to supply nutrients to plants depends on the total amount of the element in the solid phase of the soil (its stock) and a much smaller fraction that is in equilibrium between the soil solution and exchange sites on clay particles and organic matter (exchange fixed soil absorption complex) (Fig. 1).

Therefore, soil temperature is one of the determining factors affecting plant nutrition, which should not be underestimated when creating a monitoring system.

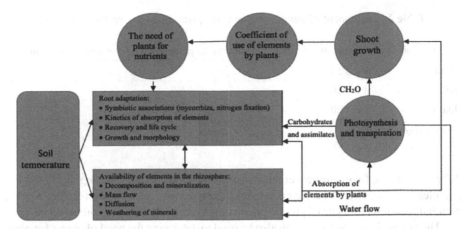

Fig. 1. Conceptual diagram of the interaction between plants and soil factors that control the availability and absorption of nutrients

2.3 Moisture in the Soil as One of the Most Important Parameters of the Data Acquisition System for Soil Monitoring

As the practice of recent years shows, the risks arid conditions are increasing. Therefore, modern technologies require both scientists and farmers to significantly improve them, and in particular in finding a set of measures aimed at the rational use natural resources of moisture.

The evaporation surface significantly affects the moisture content in the soil.The smoother the surface, the less moisture evaporates. The water content in the soil also depends on the exposure of the land [15]. Significantly affects the moisture content in the soil relief. In high places evaporation is more intense than in low places. One of the factors influencing the water regime of the soil in different areas is the composition of the local flora. It should be noted the positive impact of the forest near the fields. Soil moisture affects the solubility, movement and effectiveness of organic and mineral fertilizers, the degree of soil contamination with pesticides and other products of man-made origin, the extent to which agricultural plants absorb chemicals harmful to human health. The amount of water that a plant can consume is determined by a combination of factors:

- The amount of water available to the plant
- Depth of roots in the soil layer
- Plexus roots in the soil

The amount of water available to the plant is determined by the pore diameter (Table 2). It is important to avoid soil compaction, which compresses the pores and reduces access to water. The loam contains approximately 20 mm of plant-accessible water per 10 cm of soil, but the amount that can be used depends on the depth of the roots and their plexus.

Knowing the moisture reserve and the forecast of average precipitation, you can predict yields and use fertilizer effectively. The plant's need for water is different in

Table 2. The amount of water available to the plant depending on the type of soil

Type of soil	The amount of water available to the plant per 10 cm of soil profile, mm
sand	about 10
silt	about 20–25
loam	about 20
brown clay	about 10–15

different periods, for example, cereals consume the most moisture during the ejection of panicles [16, 17]. Therefore, for cereals it is necessary to decide on the place of sowing, which depends on the type of precursors and moisture content in the soil before sowing.

The soil moisture approach can also be used to determine the need of plants for phosphorus and potassium. This will also determine the potential yield, limited by moisture conditions, and, accordingly, the expected removal of mineral components from the soil with the crop.

2.4 Influence of Fertilizers on the Fertility of Irrigated Soil

One of the important indicators of potential soil fertility is the presence of organic matter in it. In compliance with the requirements for the quality of irrigation and agricultural techniques, the accumulation of organic matter is accelerated. Experience has shown that under the action of fertilizers, the amount of roots of crops - one of the sources of humus - increases most significantly in the arable layer and depends not only on the amount of fertilizer, but also on the method of application. Alfalfa has the greatest role in enriching the soil with organic matter under irrigation and fertilizer application. The cultivation of intermediate crops is also effective in this respect. In the field, where three harvests were obtained in a year, 40 t / ha and more of organic mass remain in the soil only due to roots and post-harvest residues.

By increasing the activity of biological processes and chemical transformations in the soil, irrigation accelerates and expands the cycle of substances, which is one of the essential features of irrigated agriculture.

One of the most important factors in enhancing biological processes in the soil during irrigation is the acceleration of mineralization of organic matter and broken down by ammonifying bacteria, resulting in the formation of ammonia, which is then nitrified.

3 The Prognostication Process as a Component of a Cyber-Physical Monitoring System in the Agricultural Sector

Modern achievements, the disclosure of the essence of many biological phenomena, and the development of control methods make it possible to adjust the processes of crop formation and product quality. In order to monitor the condition of soils, a data acquisition system is proposed, which is part of a cyber-physical system.

Scientific management methods involve forecasting, planning and organizing production. This makes it possible to transfer the production process of a certain type of crop production to a scientific, strictly controlled quality basis and thereby implement elements of one of the most promising areas of scientific and technological progress in crop production - harvest programming [18]. Programming involves the development of a program, i.e., the optimal quantitative ratio of regulated factors, taking into account low-regulated and unregulated weather conditions, which in the technological process system ensure the planned yield with the most economical use of available resources. Prognostication as an integral part of yield programming involves the development of a forecast, i.e. a probable idea of the theoretically possible yield, which is provided by the resources of climatic factors, soil fertility, fertilizers, crop protection products, etc.

Time series prediction is considered to be one of the most widely used methods in data science, used in various industries such as finance, supply chain management, manufacturing, and inventory planning. Predictive models built on machine learning have been widely used in time series projects required by various businesses to facilitate the predictive allocation of time and resources. It's worth mentioning that before you start training a neural network, you need to do one more important thing - divide the data sample into two parts, the test and the training. Test and training partitioning involves a model validation procedure that allows you to simulate how the model will perform on new, unknown data. Here's how the procedure works (Fig. 2):

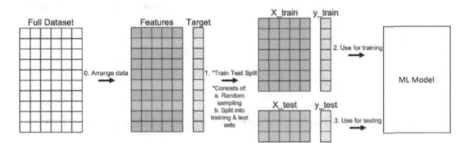

Fig. 2. Splitting the data sample into training and test data

If you do not use a test section and instead train and test the model on the same data, it will not be able to model behavior on new data, i.e. the network will not be able to make predictions. Neural networks have been attracting a lot of attention for a long time. They are a computational system of interconnected nodes and act similarly to the way the brain functions. This system combines large raw data sets, and by finding patterns, they can solve very complex problems and classify input data or even make complex predictions.

At the current stage, the most advanced modeling concepts are used to build predictive models: methods of fractal and intelligent analysis of system evolution, fuzzy logic tools, and artificial neural networks [19]. In other words, the forecast establishes the potential of grain crops in specific soil and climatic conditions, and the program realizes the potential of the forecast through management.

The task of soil monitoring is to control the dynamics of the main physical, chemical, biological and other soil processes under natural conditions and anthropogenic loads [20]. For objective information support and correct decision-making, soil monitoring should provide initial, current and periodic data on the main characteristics of the soil cover. Therefore, the issue of the impact of soil temperature on plant growth under global climate change is becoming even more relevant, which stimulates an increase in the interest of scientists in studying the mechanisms and impact of soil temperature on plant ecosystems [21].

The conducted research and monitoring have shown the high efficiency of using the artificial neural network model for long-term forecasting of non-stationary random processes on the example of winter wheat yield. The optimal version of the predicted model is shown in Fig. 3.

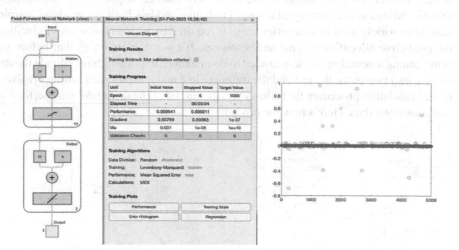

Fig. 3. A neural network model for forecasting grain yields as part of the CPS

It has been established that assessing the condition of soils based on operational monitoring of their main parameters is necessary for building predictive mathematical models of soil quality [22]. This will make it possible to formulate correct management decisions to maintain the parameters of land at the proper level and to take timely corrective measures. CPSs combine information from intelligent sensors distributed in the physical environment to better understand the environment and perform more accurate actions [23]. In a physical context, actuators make changes to the user's environment based on the data received. In a virtual context, CPSs are used to collect data on users' virtual activities, such as the use of social networks, blogs, and e-commerce sites. Cyber-physical systems then respond to such data in a certain way, predicting the actions or needs of users in general [24]. Therefore, a subsystem was created that would allow for rapid testing in the open field and instant response to changes in its important parameters, and in combination with neural network-based forecasting, help the user make the right management decisions to stabilize and improve the quality of the crop.

Based on the importance of monitoring soil conditions, a subsystem for collecting information for a cyber-physical system for monitoring agricultural production was developed, which involves measuring such basic parameters as soil moisture, temperature, acidity, and density. The developed subsystem provides visualization of measurement information in a form understandable to the user. The subsystem combines the implementation of software, hardware solutions, and forecasting based on a neural network.

4 Implementation of a Software Solution for a Data Acquisition System for Monitoring Soil Conditions for Agricultural Production Management

In order to adapt the generalized structure of the CPS to the task of managing grain production, it is divided into sub-tasks, which are preparation of agricultural land for sowing; production (cultivation) process; certification of products of such production,

Fig. 4. Interrelation of CPS elements with the stages of grain cultures production

etc. For each of the levels of the SPS, the structural elements will undergo certain modifications, but the unification of requirements for them will be ensured both at the level of the object of study (soil, water, air, etc.) and in relation to the finished product (grain). In general, to build the CPS of grain production, the main stages of this process are structured (Fig. 4), at each of which the CPS will make decisions necessary to perform certain management actions to optimize this process. In order to solve this problem, a system has been created that will allow for prompt testing in the open field and instant response to changes in its important parameters. Thus, for the CPS structure being developed (Fig. 4), the user entities of such a system are identified, which should be agricultural enterprises.

At the same level, it is necessary to clarify the set of classes of tasks $Z = \{z1, z2,..., zi,..., zn\}$, which the designed system is focused on (research, technological, managerial, etc.), and physical objects described by the real parameters and characteristics to which these tasks belong (in this case, the main objects of the CPS are the parameters of soil, air, crops, and harvest). A software product has been developed that allows working with wireless networks, flexibly adjusting the parameters of data acquisition, receiving information about the process in a wireless network in real time, providing the ability to save finished files for their further use, and normalizing input parameters. It is proposed to use Wi-Fi technology, which provides a data transfer rate of more than 100 Mbps, and users can move between access points within the coverage area of the Wi-Fi network using devices equipped with Wi-Fi client transceivers and access the Internet. The system for collecting and transmitting CPS data is characterized by the selection of a set of object parameters from the relevant measurement arrays and databases. To implement the proposed technology, we have created software adapted to the task of agricultural production of grain crops from the standpoint of creating a cyber-physical system. At the first stage of the cultivation technology, namely, determining the place of grain in the crop rotation, the user must specify the predecessor for the crop to be sown. The next step is to obtain measurement information about soil moisture and temperature. After processing the measurement information, the program will give a result as to whether this crop can be sown. The next steps involve making decisions about the amount of fertilizer applied, the readiness of the crop for sowing, and the calculation of the mass rate of sowing. At the final stage, the cyber-physical system will give instructions on how to harvest the crop correctly, depending on its degree of readiness, by processing the measured data.

Reliability and precision of the results are ensured by the use of verified measuring instruments, multiple repeated measurements of the main soil parameters. In addition, the uncertainty budget [25] for the measurement of moisture (Tables 3), temperature (Tables 4), acidity, and density was calculated for the developed measurement data collection system. The sources of uncertainty in the measurement of the main soil characteristics were identified.

The relative total standard uncertainty of the acidity measurement was also calculated and is 3.12%, and for soil density - 2.24%. Mobile devices have long been an excellent, reliable addition to existing measurement tools, and in some places have replaced PCs [26–28]. If we take the example of expertise, mobile devices allow you to avoid being tied to your workplace [29]. That is why it was expedient to develop a mobile application for Android and iOS, which will become indispensable in solving the problems of agricultural production control (Fig. 5).

When installing the mobile version of the application on a smartphone, tablet, or device running Android or iOS, users will be able to control all production processes. View them separately and receive recommendations at each stage of production.

Table 3. Uncertainty budget of soil moisture measurement

№	Source of uncertainty	Characteristics of accuracy	Distribution type	Standard uncertainty	Sensitivity factor	Contribution
1	soil temperature	1	normal	2.11	1.00	2.11
2	time deviation	0.5	normal	1.54	1.00	1.54
3	sensor immersion depth	0.1	normal	0.88	1.00	0.88
4	measurement point	0.1	normal	0.03	1.00	0.03
5	soil acidity (pH)	0.12	normal	1.20	1.00	1.20
6	soil density	0.04	normal	0.80	1.00	0.80
7	oxygen in the soil	0.17	normal	0.58	1.00	0.58
Relative standard uncertainty of type A				0.77%		
Relative standard uncertainty of type B				3.14%		
Relative total standard uncertainty				3.23%		

4.1 Discussion of Research Results

The result of this investigation is the development of a data acquisition system for CPS monitoring of soil parameters in the form of a mobile application running on Android and iOS. In order to identify the most important monitoring parameters, an analytical study of the influencing factors on soil condition was conducted and a list of them was established, namely, the main ones include humidity, temperature, acidity, and density. The sources of uncertainty in measuring the main soil characteristics were identified and common sources were identified, namely the depth of the sensor, the measurement location, and the oxygen content in the soil. The lowest uncertainty value was obtained for soil density –2.24%, and the highest value was obtained for moisture content –3.23%. The contribution of uncertainty sources to the total standard uncertainty of measurement of each monitoring parameter was determined.

Table 4. Uncertainty budget of soil temperature measurement

№	Source of uncertainty	Characteristics of accuracy	Distribution type	Standard uncertainty	Sensitivity factor	Contribution
1	soil moisture	0.2	normal	0.16	1.00	0.16
2	soil temperature	0.5	normal	1.54	1.00	1.54
3	time deviation	0.1	normal	0.88	1.00	0.88
4	sensor immersion depth	0.1	normal	0.03	1.00	0.03
5	measurement point	0.12	normal	1.20	1.00	1.20
6	soil acidity (pH)	0.04	normal	0.80	1.00	0.80
7	soil density	0.17	normal	0.58	1.00	0.58
Relative standard uncertainty of type A				1.09%		
Relative standard uncertainty of type B				2.34%		
Relative total standard uncertainty				2.58%		

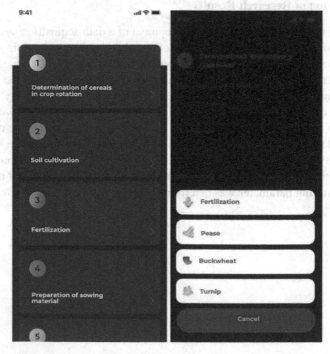

Fig. 5. Interface of a software program for a mobile platform

5 Conclusions

The authors substantiated the relevance of the creation of CPS for monitoring the condition of soils during the implementation of agrotechnological processes. The disadvantages of the known methods used to control the cultivation of crops were analyzed, based on which the research objectives were formulated. On the example of grain cultivation, the interrelationships of CPS elements with the stages of grain production were established. This structure can become the basis for other types of agricultural products. The authors have created a data acquisition system for CPS, which contains channels for measuring such important soil parameters as moisture, temperature, acidity, and density. The software for implementing the proposed data acquisition system was created with the ability to predict using a neural network. It is implemented in the form of a mobile application on a smartphone with the Android and iOS operating systems. It has been established that assessing the condition of soils based on the operational monitoring of their main parameters is necessary for building predictive mathematical models of soil quality. This will make it possible to formulate correct management decisions to maintain the parameters of land at the proper level and to implement corrective measures in a timely manner. For the developed information gathering system, the uncertainty budget of the measurement of the main soil parameters was calculated, which is of value in assessing the metrological characteristics of the proposed system. The contribution of uncertainty sources to the total standard measurement uncertainty of each of the monitoring parameters was determined. Temperature, time, and acidity give the largest contribution to the total standard uncertainty of humidity measurement. Time and acidity contribute the most to the total standard uncertainty of soil temperature and density measurements. Temperature and time contribute the most to the total standard uncertainty of the acidity measurement.

Acknowledgment. The authors express their gratitude to the staff of the Department of Information Measuring Technologies of Lviv Polytechnic National University, Ukraine, for their help and full assistance in preparing these materials.

References

1. Haluschak, P.: Laboratory methods of soil analysis. Canada–Manitoba Soil Surv. **132** (2006)
2. Dong, T., et al.: Rapid and quantitative determination of soil water-soluble nitrogen based on surface-enhanced raman spectroscopy analysis. Appl. Sci. **8**(5), 701 (2018)
3. Xing, Z., Du, C., Tian, K., Ma, F., Shen, Y., Zhou, J.: Application of FTIR-PAS and raman spectroscopies for the determination of organic matter in farmland soils. Talanta **158**, 262–269 (2016)
4. Philip, C., Juan, C., Aciego, P., Wu, Y., Xu, J.: Microbial indicators of soil quality in upland soils. chapter: Mol. Environ. Soil Sci. Part Ser. Progr. Soil Sci. 413–428 (2012)
5. Zornoza, R., Acosta, J., Bastida, F., Domínguez, S., Toledo, D., Faz, A.: Identification of sensitive indicators to assess the interrelationship between soil quality, management practices and human health. Soil **1**, 173–185 (2015)
6. Doolittle, J.A., Brevik, E.C.: The use of electromagnetic induction techniques in soils studies. Geoderma, 223–225 (2014)

7. Fan, Y., Wang, X., Funk, T., Rashid, I.: A critical review for real-time continuous soil monitoring: advantages, challenges, and perspectives. Environ. Sci. Technol. **56**(19), 13546–13564 (2022)
8. Mashud, M.A., Hossain, M.S., Islam, M.N., Islam, M.S.: Design and development of PC based data acquisition system for radiation measurement. Int. J. Image Graph. Sig. Process. (IJIGSP) **5**(7), 34 (2013)
9. Andrea, R., Wahyuni, A.I., Safitri, N.F.: Supryani.: android based forest fire monitoring system. Int. J. Inform. Eng. Electr. Bus. (IJIEEB) **14**(3) (2022)
10. Süzen, A.A., Gürfidan, R., Kayaalp, K., Şimşek, M.A.: Real-time tree counting android application and central monitoring system. Int. J. Inform. Technol. Comput. Sci. (IJITCS) **12**(2) (2020)
11. Rao, K.S., Devi, G.L., Ramesh, N.: air quality prediction in visakhapatnam with LSTM based recurrent neural networks. Int. J. Intell. Syst. Appl. (IJISA) **11**(2), 18–24 (2019)
12. Hellinger, A., Seege, H.: German Agenda Cyber Physical Systems. Berlin, pp. 23–26 (2011)
13. Yatsuk, V., Bubela, T., Pokhodylo, Y., Yatsuk, Y., Kochan, R.: Improvement of data acquisition system of objects physic-chemical properties. In: Proceedings of the 9th IEEE International Conference on Intelligent Data Acquisition and Advanced Computing Systems: Technology and Applications (IDAACS '2017). Bucharest, Romania, pp. 41–46 (2017)
14. ISO/IEC 7498–1:2004 Information Technology. The interconnection of open systems. Basic reference model. Part 1. The reference model
15. Hu, Z., et al.: Statistical techniques for detecting cyberattacks on computer networks based on an analysis of abnormal traffic behavior. Int. J. Comput. Netw. Inf. Secur. (IJCNIS) **12**(6), 1–13 (2020)
16. SO/IEC 27033–1:2009 Information technology. Security techniques. Network security. Part 1: Overview and concepts: ISO/IEC JTC 1/SC 27 Information security, cybersecurity and privacy protection
17. Lee, J., Bagheri, B., Kao, H.: A cyber-physical systems architecture for industry 4.0-based manufacturing a systems, Manufacturing Lett. **3**, 18–23 (2015)
18. Hu, Z., Dychka, I., Sulema, Y., Radchenko, Y.: Graphical data steganographic protection method based on bits correspondence scheme. Int. J. Intell. Syst. Appl. (IJISA) **9**(8), 34–40 (2017)
19. Hoang, D.D., Paik, H.Y., Kim, C.K.: Service-oriented middleware architectures for cyber-physical system. Int. J. Comput. Sci. Netw. Secur. **12**(1), 79–87 (2012)
20. Melnyk, A.: Cyber-physical systems: problems of creation and direction of development. J. Lviv Polytech. Natl. Univ. Comput. Syst. Netw. **806**, 154–161 (2014)
21. Kochan, R., Kochan, O., Chyrka, M., Jun, S., Bykovyy P.: Approaches of voltage divider development for metrology verification of ADC. In: IEEE 7th International Conference on Intelligent Data Acquisition and Advanced Computing Systems (IDAACS), vol. 1, pp. 70–75 (2013)
22. Kvaterniuk, S., Petruk, V., Kochan, O., Frolov, V.: Multispectral ecological control of parameters of water environments using a quadrocopter. Sustain. Product. Nov. Trends Energy Environ. Mater. Syst. 75–89 (2020)
23. Chen, J., Su, J., Kochan, O., Levkiv, M.: Metrological software test for simulating the method of determining the thermocouple error in situ during operation. Measure. Sci. Rev. **18**(2), 52–58 (2018)
24. Twardowski, P., Legutko, S., Krolczyk, G.M., Hloch, S.: Investigation of wear and tool life of coated carbide and cubic boron nitride cutting tools in high speed milling. Adv. Mech. Eng. **7**(6), 1687814015590216 (2015)
25. Thao, N.X.: Evaluating water reuse applications under uncertainty: a novel picture fuzzy multi criteria decision making method. evaluating water reuse applications under uncertainty:

a novel picture fuzzy multi criteria decision making medthod. Int. J. Inf. Eng. Electr. Bus. (IJIEEB) **10**(6), 32−39 (2018)

26. Khanna, N., Pusavec, F., Agrawal, C., Krolczyk, G.M.: Measurement and evaluation of hole attributes for drilling CFRP composites using an indigenously developed cryogenic machining facility. Measurement **154**, 107504 (2020)

27. Michałowska, J., et al.: Monitoring the risk of the electric component imposed on a pilot during light aircraft operations in a high-frequency electromagnetic field. Sensors **19**(24), 5537 (2019)

28. Przystupa, K.: Reliability assessment method of device under incomplete observation of failure. In: 18th International Conference on Mechatronics-Mechatronika (ME), pp. 1–6 (2018)

29. Przystupa, K.: Selected methods for improving power reliability. Przegląd Elektrotechniczny **94**(12), 270–273 (2018)

Investigation of Drawbacks of the Software Development Artifacts Reuse Approaches based on Semantic Analysis

Olena Chebanyuk[1,2]([✉])

[1] Department of Informatics, New Bulgarian University, Sofia, Bulgaria
olena.chebaniuk@npp.nau.edu.ua
[2] Software Engineering Department, National Aviation University, Kyiv, Ukraine

Abstract. Development of reliable software systems requires effective activities for implementing of software development lifecycle processes. One of the activities, which helps to raise an effectiveness of software development lifecycle processes (requirement analysis, software designing, development and testing), is reusing of software development artifacts. Organization of quick and successful reuse of software development artifacts for different problem domains requires: (i) considering peculiarities of functional and non-functional requirements for software; (ii) using of quality (full and non-contradictable domain models), obtained after domain analysis; (iii) involving effective approaches for estimation of reuse risks. Many existing reuse approaches and practices do not guarantee effective reusing of software development artifacts because there are no well-working tools supporting full reuse process including a semantic analysis of requirement specification and processing requests to repository of software components.

Paper presents the analysis of implementing software assets reusing approach based on domain engineering. Key feature of reuse is a semantic analysis of software development artifacts [2, 12]. Analysis is organized in the next way: (i) case studies for software artifacts reuse approach based on domain engineering are considered; (ii) steps of reuse approach are described and corresponding peculiarities and drawbacks are outlined.

Conclusions contain the results of analysis and recommendations for implementation of approaches for development of software systems in information defense area.

Keywords: Domain Engineering · Domain Analysis · Information Defense · Object Constraint Language · Semantic Analysis

1 Instruction

Many software systems require components that provide elements of functionality in area of information defense. For example, pay systems use channels for secured data transitions, messengers require protected voice communications, photo albums need authorized access for user data, medical systems require additional defense of personal dates, etc. [3, 6].

As a result, software development artifacts that require functionality of information defense are used often. An effective reuse of such artifacts may reduce the development time, complexity, and cost of the project. That why, it is important involving of fundamentals software development reuse approaches in the process of development software systems when different software development artifacts from information defense area are used.

One of such fundamentals are domain engineering approaches that are based on two main stages domain analysis for gathering of information about problem domain, and application engineering processes, aimed to provide effective reuse procedures.

Centrally, theoretical fundamentals of reuse approaches need additional précising to consider peculiarities and specifics of chosen problem domain.

2 Literature Review

Semantic analysis of information usually is used when user obtains answers according to his questions or requests. For example – asking from ontologies, communication with chat-bots. Theoretical background for tools for supporting requests is a predicate logic of the first or second order. Using this foundation, such standards OCL, OWL, SQL, and techniques that are based on them, for example LINQ and other query-based technologies have been developed [11]. Such technologies may work quickly and provide clear results of searching. Well-known drawback of these technologies – that they need high-quality specialist for writing correct queries.

From the other hand, technologies for natural language processing, allow to obtain information about the semantic of the text [12]. They may help to user to perform more difficult text-processing operations for example find answers or extract necessary facts from text [12] or recognize text from images[13], for example using Tesseract API.

3 Research Task

To estimate drawbacks of approaches allowing to obtain an information about semantic of software development artifacts. In software engineering practices, such an operation usually performed by adding to software models OCL constrains.

One approach to estimate software development artifacts reuse effectiveness is proposed in [2] for the problem domain "information system defense" <Case study is related to sub-domain "solving of information conflicts in local network" of.

Estimate effectiveness of every step of the chosen reuse approach from the point of view of convenience for development of software system in information defense area.

4 Experimental Part

The idea of the proposed experiment is to consider sub-domain from problem domain "development software systems in information defense area" and investigate the process of estimating effectiveness of software development artifacts reuse according to the proposed approach for software development artifacts reuse represented in [2].

Chosen problem domain is "Security system of access control for disk recourse". This problem domain is investigated when complex application of implementation of new secure protocol was created.

4.1 Domain Analysis

Design a problem domain tree. Problem domain tree for sub-processes "Security system of access control for disk recourse" is represented on the Fig. 1.

Defined problems in designing of domain tree for considering problem domain: names of the processes are not defined in standards for security systems designing. It may be cause for different understanding of contexts of processes for different specialists. For example, using some regional standards [10], different specialist may consider different terms as key terms, and as a result different problem domain trees they may be designed (see Fig. 1a) and 1b)).

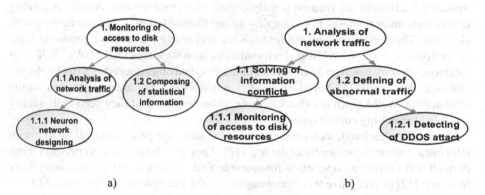

a)   b)

Fig. 1. Different variants of domain tree for problem domain "Security system of access control for disk recourse"

4.2 Designing of Repository Table

In order to investigate considering domain large repository of software assets was designed. Markets of software components contain many software modules to realize cryptography protocols. Science libraries propose description of algorithms for implementation of different aspects of security and recommendations how to archive the best security features in local networks. Reports, security standards, practical level in enterprise environments, IBM Red Books, and other types of artifacts were sources to compose a repository of software assets in this area [5]. Table 1 represents small fragment of repository table used for considering project.

Defined problems. Different experts may pay attention to different key points of processes when OCL expressions are designed. As a result, OCL expressions reflecting different aspects of requirements may be designed. From other side, some other important facts may skipped. In order to find skipped conditions it is recommended to involve

Table 1. Repository table of software assets for problem domain "Security system of access control for disk recourse

Processes	Keywords	Description of software asset	OCL code and expressions [7]
1.1 Analysis of network traffic	access matrix, neuron network, intrusion detection system	Software modules of network audit	OCL1.1.1 Context comp_access (User u):boolean Pre: u.verifyred = true Result: return true;
1.1.1 Neuron network designing	neuron network architecture, initial studying, structure verification	Software module of newton network studying	OCL are given for concrete software components
		Software module for treat prevention	OCL are given for concrete software components
1.2 Processing of statistical information about network traffic	log journal, statistical characteristics of traffic, virus signature, attack signature	Software modules of outgoing traffic calculating	OCL 1.2.1 Context Router Set{time1,time2,...,timen} -> iterate(i: Integer, sum: float = 0 I sum + time(i))
		Authentication of users by software modules of firewall	OCL 1.2.2 Context firewall inv: Comp implies Comp.Mac-> forAll(v I v.oclIsKindOf(File))

effective text processing approaches, for example [4], into analysis of requirement specification.

In addition, proposed approach [2] supports four types of OCL expressions. The same ideas may be expressed using different types of OCL expressions. For example, OCL 1.1.1 from the Table 1, may be written from the context of class "User".

Context User (computers c):boolean

Pre: c.policy = true

Result: return true

This OCL expression describes the same condition as the OCL 1.1.1, namely from which computers user can get access to some resource. From the point of view of network audit procedures semantic of this operation is the same.

It is noticed, that procedures of monitoring or controlling access to resources require procedure of selecting objects from some collections. It is recommended, to add new types of OCL for processing collection of objects for model of analytical representation of OCL expressions. Proposed model is flexible and allow doing it.

In addition, the recommendation for domain expert is to express the same condition using all possible types of OCL expressions. This recommendation expects that domain expert has high experience of OCL expressions designing and increases the time of domain analysis.

It is recommended to add to practices of company using plug-in or environment for automated verification of designed OCL expressions as well as using of methods of extracting facts from text [9]. Selected plug-ins must be convenient for usage, and be as close as possible to family of OCL standards.

4.3 Software Specification Analysis

4.3.1 Requirement Analysis. Designing of Software Specification for Considering Problem Domain with Semantic Attributes

Small part of designed requirement specification, prepared in authors' investigation is represented in the Table 2.

Designing of requirement specification touches to the same problem as designing of repository table. When requirement engineer designs OCL expressions different types of OCL expressions may be used to express the same semantic. Recommendation for requirement engineer – is to read existing OCL expressions and try to use the same type of OCL that were used by domain experts for verifying the same condition expressed in natural language. Additional requirement for requirement engineer also high level of knowing of OCL and experience of work with the corresponding plug-in or environments.

4.4 Comparison of Semantic Attributes

4.4.1 Defining Common Keywords in Repository Tree and Requirement Specification

The list of the common keywords: "access matrix, log journal, and statistical characteristics of traffic" (Table 1 and Table 2). This point fully available for the chosen application domain.

4.4.2 OCL Expressions Comparison

Results of OCL expressions comparison for problem domain repository tree and requirement specification are represented in the Table 3.

Comparison operation is performed realizing ideas explained in model of approximate OCL comparison. Words that are differ in project and repository (reference) OCL expressions are marked by yellow color (Table 3).

OCL 1.2.2 and OCL F2.1, OCL 1.2.1 and OCL F2.1.1, OCL 1.2.2 and OCL F2.1.1 and OCL 1.2.2 and OCL F3.1 have different types – they are not equal.

Estimation of comparison model – for the chosen problem domain it needs precision of conditions for comparison of collections used in OCL expressions. Other words – it needs to design comparative model for the new type of OCL expressions working with collections of objects. Analysis of the Table 3 shows that coefficients for OCL comparison in paper [2] were chosen correctly because those OCL expressions that looks like the same from human point of view are appeared equal after calculations.

Table 2. Requirement specification of security model for monitoring access to disk in local network

RQ code	RQ description and limitation in natural language	Keywords and OCL for comparison	OCL code and expressions [7]
F1	Plug-in for monitoring access to disk resources	1.1 access matrix	OCL F1.1.1 Context disk_access (User u):boolean
F1.1	Access for disk memory must be given only to verified user	OCL 1.1.1	Pre: u.checked = true Result: return true;
F 1.2	If assess to specific folder for domain user and your security group is allowed in domain computer then user can modify files in this folder	1.1 access matrix OCL 1.1.1	OCL F1.1.2 Context GroupPolicy inv: Comp implies User.ID-> forAll(v I v.oclIsKindOf(User))
F2	Calculate a network traffic statistics	1.2 log journal OCL 1.2.1 OCL 1.2.2	OCL F2.1 Set{time1,time2,…,timen} -> iterate(i: Integer, sum: Integer = 0 I sum + time(i))
F2.1	If computer has more than 10000 open connections than access to it is blocked	1.2 statistical Characte-ristics of traffic OCL 1.2.1 OCL 1.2.2	OCL F2.1.1 Context NetworkTraficc::block(stat acess, User u): EBoolean pre: stat.con> 10000 post: result = true
F3	Every 10 s save collection of users that get access to database(their logins, and access matrixes)	1.2 log journal OCL 1.2.1 OCL 1.2.2	OCL F3.1 Context buffer Set{ Tuple{login:'Jordi Cabot', 11111}, Tuple{login:'Marco Brambilla',00000,Tuple(login:'Massimo Tisi',11001)

Table 3. Comparison of project and reference OCL expressions.

Reference OCL expressions (from repository)	Project OCL expressions	Comparison results, explanations, and recommendations for reuse
Context comp_access (User u):boolean Pre: u.checked = true Result: return true;	Context disk_access (User u):boolean Pre: u.checked = true Result: return true;	OCL 1.1.1 and OCL F1.1.1 are considered approximately equal *Solution 1* - Software modules of network audit can be reviewed for reuse to realize RQ F1.1
Context comp_access (User u):boolean Pre: u.checked = true Result: return true;	Context GroupPolicy inv: Comp implies User.ID-> forAll(v | v.oclIsKindOf(User))	OCL 1.1.1 and OCL F1.1.2 have different types and considered not equal There is no software asset for reuse
Context Router Set{time1,time2,...,timen} -> iterate(i: Integer, sum: float = 0 | sum + time(i))	Set{time1,time2,...,timen} -> iterate(i: Integer, sum: Integer = 0 | sum + time(i))	OCL 1.2.1 and OCL F2.1 are considered approximately equal *Solution 1* is used
Context Router Set{time1,time2,...,timen}-> iterate(i: Integer, sum: float = 0 | sum + time(i))	Context buffer Set{Tuple{login:'Jordi Cabot',11111},Tuple{login:'Marco Brambilla',00000}}	OCL 1.2.1 and OCL F3.1 are not considered approximately equal

5   Comparative Analysis of the Proposed and Other Approaches

This chapter represents comparative analysis of different approaches allowing analyzing semantic of software development artifacts according to defined.

Table 4 summarizes advantages and disadvantages of different software engineering approaches. The first row represents drawbacks defined during the analysis represented in the previous point. The seconds' row – drawbacks of the proposed approach, the third and the fourth rows describe drawbacks of other approaches.

Paper about MoC tools represents EmS formal design tools, i.e. the formal description of the semantics of models (including the execution semantics). Different scenarios of semantic comparison of specification and design documents are described [14].

Paper [15] proposes an ontology-based method to reason about the correctness of SysML models that include OCL constraints. In order to do this Transformation language from is proposed SysML models with OCL constraints to OWL id proposed. Some operation need to be implemented in mind of stakeholder, this factor influences to different representation of the same semantic.

Analysis of the table shows that defined drawbacks that were recognized in authors' approach are typical for development of approaches that process semantic of software development artifacts.

Table 4. Defined drawbacks of approaches using semantic comparison of software development artifacts,

	Proposed approach	MoC Tools [14]	Sys ML OWL [15]
Representation different results of systematization information about domain (p 3.1)	+	+	+
Different varians to consider the same semantics aspects when domain model is designed (p3.2)	+	+	+
Different varians to represent the same semantics when domain model is used (p3.3)	+	+ −	+
Not correct semantic model for comparison	−	+ −	-

6 Conclusions and Recommendations

6.1 Domain Analysis Area

In order to design domain model reusable in application engineering stage the next problems need to be solved:

1) Different variants of domain models may represent the same semantic. (Examples for domain trees: different domain experts may assign different names of the processes or different structure of domain trees may represent the same semantic (Fig. 1), In texts or OCL the same semantics may be described by different expressions.

   The solution is the next: follow naming rules - for naming of classes, attributes, and defining keywords using definitions from international (regional or company) standards. In addition, it is necessary to think about possible synonyms of used terms. Alternative way to solve this problem is to involve approaches for natural languages text processing [12]. They may be realized in different ways – reusing of cloud services, taking software complexes for processing of texts, etc. From the other hand such factors as grammar errors or different styles for text representation, different encodings, quality of texts in images etc.
   Solution is the next: it is recommended to use definitions from enterprise standards where it possible; (ii) keep a style of the text allowing formal extraction of answers to questions (avoid dialects, non-formal style of facts representing, grammar errors).

2) Follow strict rules for representing of semantic for software development artifacts. In order to simplify the process of OCL expressions recognition in application engineering stage it is recommended for domain expert to write OCL expressions using all possible types of them.

6.2 Evaluation of Drawbacks of the Proposed Approach

1) It is recommended to add the new type of OCL expression allowing to process collections. Represented approach proposes the flexible model for analytical representation of OCL expression (based on graph representation) with possibility to extend supported types of OCL expressions. In addition, the comparison model allows setting up different weights for comparison coefficients.

2) Estimation of comparison model. Chosen coefficients for comparison of OCL expressions for considering problem domain are set correctly. It is proved by the fact that the from human point of view and results of calculation, preformed using proposed model. Are matched.

6.3 Application Engineering Area

It is recommended for requirement engineer to obtain high qualification with using of plug-ins or environments for verification of OCL expressions in automated mode. In addition, composition of such type a specifications needs more time and qualification in comparison with approaches allowing composing simple requirement specification.

Advantage of the proposed approach – pre and post conditions allow providing a good background for designing of UML sequence (or activity) diagrams.

OCL expressions may be used as initial information for automated generation of metamodel (class diagrams + OCL expressions) [8]. It allows to save time of performing processes of software designing and testing.

Additional activity recommended for domain expert and for stakeholders is to involve approaches and software systems for extracting important facts from requirement specification and performing other operations of text processing [1], for example documentation about software modules.

References

1. Andonov, F., Slavova, V., Petrov, G.: On the open text summarizer. Inform. Content Process. **3**, 278–287 (2016). http://www.foibg.com/ijicp/vol03/ijicp03-03-p05.pdf
2. Chebanyuk, O.: An approach to software assets reusing. In: Zlateva, T., Goleva, R. (eds.) Computer Science and Education in Computer Science. CSECS 2022. Lecture Notes of the Institute for Computer Sciences, Social Informatics and Telecommunications Engineering, vol. 450. Springer, Cham (2022). https://doi.org/10.1007/978-3-031-17292-2_6
3. Gnatyuk, S.: Critical Aviation Information Systems Cybersecurity, Meeting Security Challenges Through Data Analytics and Decision Support, NATO Science for Peace and Security Series, D: Information and Communication Security. IOS Press Ebooks, vol. 47, no. 3, pp. 308–316 (2016)
4. Ivanova, K.: NLA-Bit: a basic structure for storing big data with complexity O(1). Big Data Cogn. Comput. **5**(1), 8 (2021). https://doi.org/10.3390/bdcc5010008
5. Lisboa, L., Garcia, V.D., Lucrédio, D., et al.: A systematic review of domain analysis tools. Inform. Softw. Technol. **52**(1), 1–13 (2016)
6. Odarchenko, R.R, Abakumova, A., Polihenko, O., Gnatyuk, S.: Traffic offload improved method for 4G/5G mobile network operator. In: Proceedings of 14th International Conference on Advanced Trends in Radioelectronics, Telecommunications and Computer Engineering (TCSET-2018), pp. 1051–1054 (2018)

7. OMG standard Object Constraint Language 2.3.1 (2020). https://www.omg.org/spec/OCL/2.3.1/About-OCL/
8. Pérez, B., Porres, I.: Reasoning about UML/OCL class diagrams using constraint logic programming and formula. Inf. Syst. **81**, 152–177 (2018)
9. Slavova, V.: Emotional valence coded in the phonemic content–statistical evidence based on corpus analysis. Cybern. Inform. Technol. **2**, 3–21 (2020)
10. TZI. https://tzi.ua/ua/nd_tz_1.1-003-99.html
11. Akuma, S., Anendah, P.: A new query expansion approach for improving web search ranking. Inform. Technol. Comput. Sci. 2, 1–16 (2022). https://doi.org/10.5815/ijitcs.2023.01.05
12. Almutiri, T.F.: Nadeem markov models applications in natural language processing: a survey I.J. Inform. Technol. Comput. Sci. **2,** 1–16 (2022). https://doi.org/10.5815/ijitcs.2022.02.01
13. Goel, V., Kumar, V., Jaggi, A.S., Nagrath, P.: Text extraction from natural scene images using OpenCV and CNN I.J. Inform. Technol. Comput. Sci. **9**, 48–54 (2019). https://doi.org/10.5815/ijitcs.2019.09.06
14. Diallo, P.I.: A framework for the definition of a system model MoC-based semantics in the context of tool integration. Embedded Systems. Université de Bretagne occidentale - Brest (2014). English. ffNNT : 2014BRES0067ff. fftel-03258224f
15. Lu, S., Tazin, A., Chen, Y., et al.: Detection of Inconsistencies in SysML/OCL models using owl reasoning. SN Comput. Sci. **4**, 175 (2023). https://doi.org/10.1007/s42979-022-01577-0